ENERGY AT THE SURFACE OF THE EARTH

An Introduction to the Energetics of Ecosystems

Student Edition

This is Volume 27 in

INTERNATIONAL GEOPHYSICS SERIES

A series of monographs and textbooks

Edited by J. VAN MIEGHEM (July 1959–1976) and ANTON L. HALES

A complete list of the books in this series appears at the end of this volume.

ENERGY AT THE SURFACE OF THE EARTH

An Introduction to the Energetics of Ecosystems

Student Edition

DAVID H. MILLER

Department of Geological and Geophysical Sciences
The University of Wisconsin—Milwaukee
Milwaukee, Wisconsin

1981

ACADEMIC PRESS
A Subsidiary of Harcourt Brace Jovanovich, Publishers
New York   London   Toronto   Sydney   San Francisco

ACADEMIC PRESS, INC.
111 Fifth Avenue, New York, New York 10003

United Kingdom Edition published by
ACADEMIC PRESS, INC. (LONDON) LTD.
24/28 Oval Road, London NW1 7DX

LIBRARY OF CONGRESS CATALOG CARD NUMBER: 80–989

ISBN 0–12–497152–0

PRINTED IN THE UNITED STATES OF AMERICA

81 82 83 84   9 8 7 6 5 4 3 2 1

To the memory of F. A. Brooks, radiation and engineering meteorologist, innovative micrometeorologist at the University of California at Davis, and to C. F. Brooks, meteorologist and climatologist at Harvard University, founder of the American Meteorological Society

CONTENTS

Chapter IV
Diffuse Solar Radiation

Chapter V
Total Incoming Solar Radiation

Chapter VI
Incoming Longwave Radiation

Chapter VII
Radiant Energy Absorbed by Ecosystems

Chapter VIII
Surface Temperature of Ecosystems

Chapter IX
Longwave Radiation Emitted by Ecosystems

Chapter X
Resultants of the Upward and Downward Radiation Fluxes

Chapter XI
Fixing of Carbon by Ecosystems

Chapter XII
The Release of Carbon Fixed in Ecosystems

Chapter XIII
Broad-Scale Transformations of Fossil Energy

Chapter XIV
Phase Changes of Water in Ecosystems: I. Freezing and Thawing

Chapter XV
Phase Changes of Water in Ecosystems: II. Vaporization

Chapter XVI
The Flux of Sensible Heat from Ecosystems

Chapter XVII
Substrate Heat Flux in Terrestrial Ecosystems

Chapter XVIII
Substrate Energy Storage in Aquatic Ecosystems and Its Place in Their Energy Budgets

Chapter XIX
Potential and Kinetic Energy in Ecosystems

Chapter XX
Energy Budgets at Different Depths in Ecosystems

Chapter XXI
Ecosystem Contrasts

Chapter XXII
Energy Conversions at Nodes

Chapter XXIII
Integrating the Energy Fluxes

PREFACE

This book presents one way of looking at the manner in which the biological, physical, and cultural systems that mantle the landmasses of our planet receive, transform, and give off energy, which is an essential condition of existence that takes many forms. Energy conversions establish the climate in which these systems operate.

The principal forms of energy that are converted at the ecosystem scale include radiant, latent, mechanical, chemical and fossil, and thermal. We begin with radiant energy absorbed by ecosystems, a phenomenon that is independent of their surface temperature and that can be looked on as a burden or a gift, depending on circumstances. An increase in such absorption raises surface temperature, as described in the fulcrum chapter of the book, Chapter VIII. This increase in turn sets into action outflows of energy that by the first law of thermodynamics are equal in energy units, although not necessarily equal in quality to the inflows. While the second law comments that quality is likely to suffer in such a transaction, our principal tool of analysis is the first-law equivalence, which can be stated as a simple accounting in watts per square meter of ecosystem area. These temperature-dependent fluxes of energy are discussed in the chapters following Chapter VIII; the final chapters deal with vertical stratification and areal contrasts in energy budgets, the augmented energy budget of the city, and the responses that serve to keep the budget balanced.

Anyone who looks at the landscape perceptively, whether in wildlands, cultivated areas, or the city, can see energy in movement or transformation everywhere; awareness of the environment as a functioning entity is enhanced, I believe, by recognizing the manifestations of energy in it. I first encountered this way of analyzing nature when studying snowmelt floods that had occurred and that conceivably might occur in the Sevier River in Utah; subsequently I applied energetics to another system—the clothing that protects a soldier from a hostile thermal environment. In both cases, seemingly so far apart, much the same energy fluxes are important, and it appeared that an organizing concept that could make sense out of such different situations had much to commend it as a means of studying a set of objects in which I had long had interest, the ecosystems that comprise the variegated surface of the earth. It is true that energy budgets are routinely cast for thermodynamic systems in such artifacts as engines and houses, others for atmospheric motion systems, others within leaf cells, others encompassing an entire planet or star; but here we select for energetic analysis only those systems that are located at the outer active surface of the lands, confronting sun and atmosphere and functioning at the scale of ecosystems. Smaller systems (e.g., leaves) are considered in passing, and the scale is enlarged only slightly to take in cities, those fascinating interminglings of human and natural systems, still *terrae incognitae* to science and yet whose working and very survival are basically expressed in their energetics. The uncertain future of fossil energy poses questions for wildland and agricultural ecosystems too, but nowhere more than for the vulnerable modern city.

My aim is not to tell everything about any one energy flux or any ecosystem, but to try to develop a proportioned and numerically illustrated treatment that will help the reader see each flux in its true setting and observe ecosystems coupled into their environments. Because real measurements carry more conviction, I have preferred observed data, many of them from sites I have visited or worked in, over modeled or assumed or asserted quantities or mere symbols divorced from numerical content. Similarly, I have not repeated oversimple formulas for the fluxes, but prefer that the reader who finds it necessary to estimate should refer to the original articles, where qualifications and cautions were set out by the field worker.

Study of the energetics of systems at the surface of the earth draws upon the content of several disciplines, each of which has its own objects of study and its own way of viewing the rest of creation. No single discipline—not meteorology, not hydrology, not ecology, not geography—and no practitioner field—forestry, agronomy, architecture, city planning, or engineering—can encompass the subject essayed in this book. Although these fields make use of energetics as a mode of investigation, they do so only in an auxiliary role. Accordingly, I make no attempt to summarize the principles or content of any discipline, but rather select what it can contribute toward interface energetics. Contributions to the resulting synthesis, if such it is, have come from many

disciplines of science and practice and I hope the synthesis will give
back to each field as much as it takes; for example, the climate of the
soil and the structure of the atmosphere are governed by energy fluxes
at the interface between these adjoining media.

Spatial contrasts in energy reflect contrasts in geology, soil conditions,
hydrology, topography, solar input, and atmospheric coupling, as well
as in cultural practices, and at the same time heighten contrasts in the
landscape. Energetics analysis helps us to perceive the true variety of
the world, and beyond these hoped-for contributions to individual disciplines I hope that ecosystem energetics may contribute in a small way
to the riddles of food production and overpeopling of the earth, energy
and water resources, urbanization, and the imperiled environments of
life.

Different forms of energy have different kinds of utility and are the
provinces of different disciplines, but there exists behind this diversity
of appearance a unity of essence. The energy problems besetting the
world are not likely to yield to single disciplines, each regarding its own
form of energy as if it had little to do with other forms and generating
its own specialized data. In the service of a broader integration I have
tried to use data from these disciplines to make sense of it all. How well
the resulting picture captures nature is for the reader to say.

ACKNOWLEDGMENTS

While working in research organizations, I received valuable advice from such co-workers as the late Paul Siple and W. L. D. Bottorf; Frank Snyder, Walter Wilson, Robert Gerdel, K. R. Knoerr, R. A. Muller, Arnold Court, S. E. Rantz, and H. W. Anderson; and in Australia, Alec Costin and Frank Dunin. In academic associations I have learned from the late Carl Sauer; Joseph Kittredge, Morris Neiburger, Dan Luten, J. R. Mather, Forest Stearns, H. H. Lettau, Douglas Carter, and the late B. M. Varney, Fritz Prohaska, and Katherine Wurster in the United States; and Alan Tweedie and Ann Marshall in Australian universities, as well as many others, at Newcastle in particular. With especial gratitude I recall visits that C. C. Wallén and Marcel de Quervain made to research installations and other encouragement from scientists of a community that is truly worldwide in scope—a fact most truly felt when traveling abroad. These include Rudolf Geiger, Wolfgang Weischet, Hermann Flohn, V. Conrad, M. M. Yoshino, M. I. Budyko, B. A. Aizenshtat, Iu. L. Rauner, and the late B. L. Dzerdzeevskii and S. P. Khromov. Invaluable presences over the years have been John Leighly, Helmut Landsberg, and the late Warren Thornthwaite.

Fulbright periods in Australia made writing possible; my appreciation for awards in 1966, 1971, and 1979 goes to the Australian-American Educational Foundation and Newcastle University, as well as to Alec Costin, James Auchmuty, E. A. Linacre, students at Macquarie and Newcastle, and, again over the years, to Alan Tweedie, Vice-Principal at Newcastle.

For painstaking technical review of the manuscript I am indebted to Alan Tweedie and for helpful comments on early drafts of several chapters to an anonymous reviewer. For critical analyses of logic and sense and meticulous editorial advice I am indebted more than I can say to my wife, Enid, my companion through these years.

Chapter 1

INTRODUCTION

THE ENERGY BUDGET

Energy, a basic quantity in the universe, is present at the surface of the earth in many forms, so diverse that recognition of its continuity was a major intellectual feat of 19th century scientists. Familiar changes in its form are the absorption of solar energy by plant leaves, emission of longwave radiation by all surfaces, formation of chemical energy from radiant, its release in decomposition of organic material, and energy going into latent form when water evaporates or snow melts.

Transformations imply inputs and outputs of energy. Inputs include the delivery of energy as sunshine to an ecosystem, the sensible heat it extracts from warmer air, and the return of heat in winter from warmer layers of an aquatic system to its cold surface. Outputs from ecosystems include energy that is radiated away or carried off in the atmosphere or carried off in the harvest.

Budget accounting of inputs and outputs shows that when all forms of energy flow are considered, the inputs balance the outputs. This statement of continuity is the first law of thermodynamics, that energy is not created or destroyed. It says nothing about the quality of energy of different forms; that is a matter of the second law, which indicates the ability of a particular form of energy to do useful work. Solar and chemical energy have higher quality than thermal energy at ecosystem temperatures or longwave radiation (Moore and Moore, 1976, p. 73; Lönnroth *et al.*, 1980). These considerations suggest direction of transformation, but our concern here is primarily with first-law accounting. Determining an energy budget for a surface is a matter of striking an account of all the inputs and outputs, and the necessity for them to be in balance provides a quantitative check on the measurements. The concept is straightforward. Problems arise in applying the budget over time and space.

Variability over Time and Space

The budget by definition is never out of balance; if any input increases, one or several outputs increase to the identical amount. While the balance remains, the mix of its constituents changes. Some variations are regular, like summer and winter, and seasonal budgets show how an ecosystem responds to a change in energy loading. Sudden changes, like the arrival of a cold wave, show in a different way how the system responds. Variability is generated both by extraterrestrial changes, which are relatively regular, and by systems in the earth's atmosphere since the clouds and wind of a passing storm cause fluctuations in energy budgets at the underlying ecosystems.

Spatial differences in the energy budget occur at many scales, among which we will be primarily concerned with those of ecosystems and secondarily with differences within ecosystems and contrasts among them. Energy fluxes differ from place to place and thereby depict in a physical way the variety we see in the world around us.

Components

Let us list the basic energy fluxes that are found in ecosystems:

(1) Solar (or shortwave) radiation absorbed by an ecosystem;

(2) Longwave radiation from the atmosphere absorbed by an ecosystem;

(3) Photosynthetic conversion of solar energy;

(4) Energy released by decomposition and from fossil sources;

(5) Emitted longwave (thermal) radiation from leaves;

(6) Heat converted in evaporation and in melting snow;

(7) Heat taken into the ground at certain times and released at others;

(8) Sensible-heat flux from the surface into the air;

(9) Conversions of kinetic and potential energy, such as wind energy.

Sum: All of the above total to zero if we assign a positive value to inputs and a negative value to outgoes.

Our emphasis is on the rates of transformations and flows—the dynamics of energy. The watt appears more often than the joule, and the watt per square meter still more often (see Table I).

TABLE I

Approximate Conversions to Older Metric and Traditional Units

---

1 joule (J) = 0.24 gram-calorie (cal) = 0.28×10^{-3} watt-hours (W h) = 0.95×10^{-3} Btu

1 MJ m^{-2} = 24 gram-cal cm^{-2} = 24 langley (ly) = 88 Btu ft^{-2}

1 watt (W) = 1 J sec^{-1} = 0.24 cal sec^{-1} = 3.4 Btu hr^{-1}

1 W m^{-2} = $\dfrac{1}{698}$ ly min^{-1} = 0.086 ly hr^{-1} = 2.06 ly day^{-1} = 0.32 Btu ft^{-2} hr^{-1}

THE LOCUS OF ENERGY TRANSFORMATION

Energy transactions are studied in many kinds of systems, initially in steam engines, later in natural systems. In any thermodynamics investigation it is first necessary to delimit a system as an object of study. We can then distinguish its internal processes from the energy inputs and outflows that express its relations with the rest of the world.

In studying energy transactions at the surface of the earth, the most convenient system to define is the ecosystem. This is a biological concept that expresses the structure of the earth's green mantle and can be expanded to include aquatic systems, desert surfaces where plants are sparse, winter snow cover, and urban systems. Ecosystems are as convenient for energy studies as for water studies because they are reasonably homogeneous pieces of the earth's surface. Many of them are stratified in the vertical, but are effectively uniform in the horizontal and delimited by sharp edges (Fig. 1). A person flying over Wisconsin sees a mosaic of contrasting ecosystems—woodlots, corn, alfalfa, and oat fields; each unit is uniform horizontally. In wildlands too ecosystems of reasonable homogeneity make up the whole landscape. Rocky ridges, meadows, pine stands, and brush fields: each is internally uniform.

For reasons that are not entirely clear, the spatial scales of ecosystems in human landscapes are of about the same size, i.e., a horizontal extent of a hundred to a few hundred meters (Miller, 1978). In agricultural lowlands sizes are related to cultivation practices and the size of the total farm or unit of land management. In uplands ecosystem sizes are related to dissection of the surface, which produces slope facets of differing exposure to sun and wind; the different inputs of water and energy support different kinds of ecosystems. These entities can be compared on a unit-area basis by use of data expressed as energy flux density (watts per square meter).

Fig. I.1.   Mosaic of riparian and small crop-ecosystems of unmechanized agriculture on the alluvial floor of Waipio Canyon of the island of Hawaii. Further contrast is provided by slope ecosystems.

The stratification of most ecosystems in the vertical presents a choice in defining the locus of energy studies. We can adopt a volume approach and measure all flows entering or leaving a three-dimensional space that, for example, encompasses the root zone, trunk space, and canopy of a forest stand. Or we can adopt a surface or interface approach, by definition a two-dimensional system without capacity to store energy that experiences only inflows and outflows, including the diurnal movement of heat from the surface into the soil body by day and out of it by night.

Either approach is satisfactory as long as system boundaries are consistently adhered to. Generally speaking, unless special attention is given to the changes in vertical fluxes through an ecosystem in order to separate ecosystem functions at different levels, the interface approach is more common. It allocates energy storages to the adjacent media of soil body and local air and internalizes other energy phenomena. Primary emphasis is on the energy fluxes that couple the ecosystem as an entity to its environment, the sun and atmosphere.

ENERGY IN ECOSYSTEMS

The basic features of the energy budget of an ecosystem are those that define (a) its coupling with its environment, (b) the level of energy inputs (hence of outputs), (c) the kinds of energy transformations, especially those biologically controlled, and (d) the mix of energy outputs, which can be regarded as yields. These features are related; a high level of inputs makes available ample energy to be transformed, and the kinds of transformation determine the mix of yields. Variations in a particular energy flux often affect the density of other fluxes, and this leads us to consider interaction or competition among fluxes. We also must consider the associations of an energy flux with such mass fluxes as those of water, nitrogen, or carbon dioxide, the utilization of which can be examined from the standpoint of production or energy conversion in the ecosystem canopy (Miller, 1979).

Transformations and flow rates of energy vary over space, and this book speaks particularly to the contrasts among the energy budgets of ecosystems in their infinite variety over the diverse land surfaces of the planet. Spatial patterns at the ecosystem scale are seen in their associations with geologic, hydrologic, climatic, and cultural characteristics of their environment and thus express other forms of coupling of ecosystem and environment.

The energy budget becomes concrete when we look at ecosystems as segments of the surface of the earth, and still more concrete when we examine specific ecosystems at which real measurements of the energy fluxes have been made for specific practical purposes. The illustrations that follow represent ecosystems exhibiting various couplings with their environments and their levels and kinds of energy transformations and outputs.

June in Denmark

A grassland system represents the efficient agriculture of Denmark, which turns rain and solar energy into milk for cheese and meat for bacon to the benefit of the Danish balance of trade and standard of living. The grass in a university research site near Copenhagen absorbs, through a long June day, a 24-hr average of 330 W m^{-2} of longwave radiation and 193 W m^{-2} of solar energy, of which 90 are photosynthetically active (left-hand bar graph of Fig. 2). The total absorbed energy, $+523$ W m^{-2}, defines the energy level and generates a mean grass temperature of 20°C.

This quantity of energy, 523 W m^{-2} over the daily cycle, goes in three nonbiological directions (the grass radiates away 390 W m^{-2}, gives off

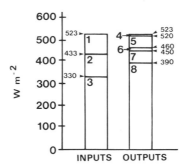

Fig. I.2. Energy inputs and outputs of a grass ecosystem at Højbakkegaard Experimental Farm in Denmark (56°N latitude) in June (mean of years 1955–1964). Inputs: (1) absorbed photosynthetically active radiation; (2) absorbed photosynthetically inactive solar radiation; (3) absorbed longwave radiation; total absorbed radiation = 523 W m^{-2}. Outputs: (4) net photosynthetic energy conversion by the grass; (5) evapotranspiration or latent-heat conversion by grass; (6) net 24-hr intake of heat from grass into the soil body; (7) net flux of sensible heat from the grass into the air; (8) longwave radiation emitted by the grass; total outputs = −523 W m^{-2}. Measurements by the Hydrotechnical Laboratory, The Royal Veterinary and Agricultural University, Copenhagen (see discussion in Chapter XXIII). Numbers on bars are cumulated sums.

60 to the cooler air, and puts 10 into subsoil storage; see Fig. 2) and in two correlated biological directions. One of these is net photosynthesis, 3 W m^{-2}, with which is associated a flow of 60 W m^{-2} of latent heat that powers the daily evaporation of 2.1 mm of water. The investigators feel that the photosynthetic term might perhaps be larger if June rainfall did not tend to fall short of the energy potential. Even drawing on water stored in the soil, early summer usually experiences a moisture stress (Marshall and Holmes, 1979, p. 302) that reduces biological productivity by 15 to 20%. The energy budget of this system thus aids research on drought effects on the yield of milk and meat from Danish farms.

June and December

Many differences between the winter and summer environments of this ecosystem are portrayed by comparing its December and June energy budgets. In December, radiant energy absorbed by the grass amounts to 305 W m^{-2}, of which only 5 W m^{-2} is photosynthetically active (Fig. 3). In June, as we saw earlier, 523 W m^{-2} is absorbed, of which 90 W m^{-2} is photosynthetically active. The difference in biological value is manyfold, in lower-quality energy half again as much as in winter.

To remain in energy balance in December at a temperature of +1°C (indicating that it must emit 325 W m^{-2} as longwave radiation), the

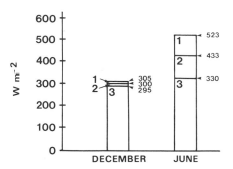

Fig. I.3. Levels of radiant energy absorbed by a grass ecosystem at Højbakkegaard, Denmark, in December and June (mean of years 1955–1964). (1) absorbed photosynthetically active radiation; (2) absorbed photosynthetically inactive solar radiation; (3) absorbed longwave radiation. The seasonal contrast is most marked in the forms of radiant energy that are active biologically.

grass extracts 10 W m^{-2} of sensible heat from relatively warm air off the North Sea and raises 10 W m^{-2} from the relatively warmer underlying soil. Its situation vis-a-vis the atmosphere and soil is quite different from summer when it was richly endowed with radiant energy and, even after radiating at a rate appropriate to its surface temperature, still had energy to warm the soil and to warm and moisten the air. The budget is balanced in both seasons, but the differences in its level and in the mix of inputs and outputs express seasonal differences in functioning and yields. (See Chapter XXIII)

Sierra Ecosystems

Mountain ecosystems produce amenity, forage, and fiber, but the chemical energy in their biological products is surpassed by the potential energy in their water output. The rate at which the snow cover of spring will yield up this water is important to hydrologic engineers who manage hydropower production, flood control, and irrigation, the objectives of a 10-year investigation in the basin of Castle Creek, California (Chapter XIV).

The meadow snow absorbed an average of +406 W m^{-2} radiation over a sunny 24-hr period (16–17 April 1946) and received another 44 W m^{-2} from warm air (Fig. 4). It radiated away 310 W m^{-2} at its relatively low temperature and converted the rest, 140 W m^{-2}, into the energy of liquid water molecules. This energy conversion changed the physical state of 36 kg of snow per square meter of the meadow, which [area = 50 hectares (ha)] amounted to 18,000 tonnes of water, most of which went down Castle Creek in a wave that peaked at 1900 hours.

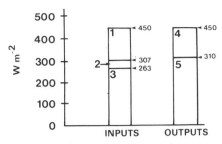

Fig. I.4. Energy budget and production of meltwater from snow cover at Central Sierra Snow Laboratory, 24-hr means on 16–17 April 1947. Inputs: (1) absorbed solar radiation; (2) latent and sensible atmospheric heat delivered to the snow surface by turbulent exchange; (3) atmospheric energy delivered to the snow surface by longwave radiation; total input of energy = 450 W m^{-2}. Outputs: (4) Net melting of snow (gross daytime melting less refreezing at night); (5) radiation emitted by the snow in long wavelengths; total = -450 W m^{-2}. Conversion of energy in melting snow = 140 W m^{-2}; meltwater equivalent = 140 J sec^{-1} m^{-2} × 86,400 sec/335 kJ kg^{-1} = 36 kg m^{-2} day^{-1} = 36 mm day^{-1}.

The potential energy of 18,000 tons at an altitude of 2 km above sea level was 360×10^3 MJ, or in electric-utility language 10^5 kW h, say \$1000 in value.

Suppose the land manager wishes to slow down the melting of his frozen asset, the snow cover, to take advantage of its greater value later in the year for irrigation. He asks how much more slowly snow cover in a lodgepole pine ecosystem melts than it does on the meadow on an average spring day. Considering only daytime hours, he finds the absorption of solar and longwave radiation to be 440 W m^{-2} in the pines and 550 in the meadow. The snow in the forest receives less solar energy but more longwave radiation; the intake is 0.2 less than on the meadow.

Comparisons

Energy inputs from the sun and atmosphere to these ecosystems vary a great deal. Expressed in terms of radiant energy absorbed, the Sierra meadow at 406 W m^{-2} is intermediate between the dark Danish winter at 305 W m^{-2} and the long days of summer at 523 W m^{-2}. Transformations and outputs of energy also differ. At this season, the snow-covered meadow puts out only longwave radiation and meltwater. The clover-grass ecosystem in June produces a mix of yields: biomass, water vapor, warmed soil and warmed air, as well as longwave radiation.

Casting ecosystem energy budgets gives us a useful set of indicators of their coupling with their environment, their yields, and their roles in the larger landscape (Drozdov, 1978). Entropy may indeed act as the "manager" in processes in nature (Sommerfeld, 1956, p. 41), but energy, acting as the bookkeeper, tells us how the credits and debits pile up in each transaction with the environment.

A Note on Energy Units

We need only a few units to depict the storages, transformations, and fluxes of energy in ecosystem environments. The unit of energy in the Système Internationale (S.I.) is the joule (J). Its dimensions of (mass × length squared)/(time squared), or ML^2T^{-2}, derive from Newton's force (mass × acceleration or $M \cdot LT^{-2}$) acting over distance (L), and the joule represents the work of accelerating 1 kg at a rate of 1 m sec^{-2} over 1 m, or 1 kg m^2 sec^{-2}.

In describing heat storage in a unit area of an ecosystem, the S.I. uses joules per square meter (J m^{-2}). For large storages it is convenient to use megajoules per square meter (MJ m^{-2}); for example, the quantity of heat taken into substrate storage during summer in Lake Ontario is of the order of 2000 MJ m^{-2} and in nearby cornfields only 125 MJ m^{-2}.

Rates of transformations or flows of energy require the time dimension: 1 joule per second = 1 watt. In describing energy flows in an ecosystem we use watts per square meter of ecosystem area (W m^{-2}). The input of solar energy to the Danish ecosystem was 193 W m^{-2} averaged over a 24-hr day in summer and 10 W m^{-2} in winter. The solar constant is approximately 1375 W m^{-2}; few flux densities exceed this rate. A basic threshold in ecosystems, the freezing point of water at 273°K, is associated with the emission of approximately 315 W m^{-2} of radiant energy, and the limiting upper temperature of an extensive evaporating surface, 33°C, is associated with emission of radiation at a rate of approximately 500 W m^{-2}.

REFERENCES

Drozdov, A. V. (1978). Landschaftskundliche Aspekte bei Bilanzuntersuchungen. *Petermanns Geogr. Mitt.* **122**, 13–16.

Lönnroth, M., Johansson, J. B., and Steen, P. (1980). Sweden beyond oil: Nuclear commitments and solar options. *Science* **208**, 557–563.

Marshall, T. J., and Holmes, J. W. (1979). "Soil Physics." Cambridge Univ. Press, London and New York.

Miller, D. H. (1978). The factor of scale: Ecosystem, landscape mosaic, and region. *In* "Sourcebook on the Environment" (K. A. Hammond, G. Macinko, and W. B. Fairchild, eds.), pp. 63–88. Univ. of Chicago Press, Chicago, Illinois.

Miller, P. C. (1979). Quantitative plant ecology. *In* "Analysis of Ecological Systems," (D. J. Horn, G. R. Stairs, and R. D. Mitchell, eds.), pp. 179–231, Ohio State Univ. Press, Columbus, Ohio.

Moore, J. W., and Moore, E. A. (1976). "Environmental Chemistry." Academic Press, New York.

Sommerfeld, A. J. W. (1956). "Thermodynamics and Statistical Mechanics" (F. Bopp and J. Meixner, eds.; J. Kestin, transl.), Lectures on Theoretical Physics, Vol. 5. Academic Press, New York.

Chapter II

SURFACE CHARACTERISTICS OF ECOSYSTEMS

The reception of incoming energy fluxes, their transformations, and the mix of outgoing fluxes all depend on physical characteristics of the ecosystem, which couple the ecosystem with its environments of sun, air, and substrate. A dark, absorbent system is closely coupled with the sun, as is shown in its warmth and in the quick response of its energy budget to increases or decreases in solar radiation. An open pine forest on a ridge is more closely coupled with the air blowing through it than is a low, sheltered meadow. A lake surface is more closely coupled with large storages of heat and water in the lake basin than the surface of a mantle of snow or the elevated canopy of a forest are coupled with their respective substrates.

Among the characteristics of an ecosystem that affect its exchanges of energy and water with the substrate, atmosphere, and space, we first look at the size and structure of the intricate three-dimensional zone that it presents to the sky and atmosphere. As in architecture and biology, we cannot completely separate structure and function. In order to understand how an ecosystem functions, scientists had to learn how to describe its geometry and structure. In the millennia in which man has lived in and among ecosystems, many of their important physical characteristics have gone unperceived, unseen, or incorrectly estimated.* Many of the basic parameters of our environment, other than area (as the unit based on a day's plowing), have never been measured.

---

* For example, one of the first comprehensive attempts to outline the distribution of heat over the earth (by Zenker and Voeikov (1889) not quite a century ago) used a value for albedo of snow cover (0.16) that is only a quarter of its true value.

ECOSYSTEM SIZE AND STRUCTURE

Origins and Area

Leaves live side by side in the millions, shading, sheltering, and warming one another, forming a new entity that we see by looking straight down into it as the sun and atmosphere do. Grassland, with its many openings to the sky, suggests high absorptivity. The temperature of leaf surfaces varies from the cool deeps in the shade to warm surfaces in the sun but out of the wind.

Besides looking down on an ecosystem from above, however, we shall get down in it and see how its structure influences its reception of incoming matter or energy. What we know about the exchanges of a single leaf is not easily extrapolated into information about the exchanges of a community of leaves, "because the [single] leaf cannot readily be oriented with respect to incident energy and air movement in a way that is typical of an array of leaves on a whole plant" (Slatyer, 1967, p. 270). The structure of the community of leaves affects two vector processes—solar radiation and wind—that control the energy budget of individual leaves.

"The geometric and elastic properties of the community have a profound influence on the internal climate of the community through their influence on the ventilation properties of air movement" (Lemon, 1967), which is measured by the coefficient of turbulent diffusivity in square meters per second.

We cannot discuss here all the internal workings within a community of organisms. Physiology is outside our scope, yet we should not forget that an ecosystem is both environment and environed. It is the environment of the organisms that comprise it, and it is environed by the outside world, which we here simplify to sun, air, and ground.

The ecosystem (which might be a forest glade, a cornfield, or a wood lot) is above the size scale of a leaf, an insect, an animal, and a human being as a locus of energy transactions. Horizontal uniformity is the principal criterion. Areal extent influences development of the unique atmospheric boundary layer of an ecosystem, hence its coupling with the atmosphere. Size tends to be small in urban landscapes and larger in areas of mechanized flatland agriculture. As size increases, the interaction with neighboring ecosystems diminishes.

Edges themselves have ecological significance; they are discontinuities in the horizontally uniform properties inside ecosystems and also form special habitats. Some are artificial, others natural, and are generated by many factors, not all of which are obvious. They may nevertheless be persistent over time, as are certain boundaries between coastal sage and chaparral in southern California (Bradbury, 1978). Distinct edges help define ecosystems as objects of study.

Many disciplines study such fundamental elements as pedons or soil bodies, plant communities, forest sites, geomorphological slope facets, lake basins, glaciers, crop fields, yield-generating units or sediment or pollution sources in hydrology, and local boundary layers in micrometeorology. The fact that many of these pieces of the surface of the earth are more or less human-sized might enhance their appeal; their size is within our power to grasp, and we need not resort to an abstraction like a map. To some they resemble organisms, especially if the investigator can regard their limits as containing the processes he wants to study. Some investigators speak of "units of environment." Such demarcations might account for the attraction we feel toward islands, or forest glades, small lakes, or marshes. Edges themselves hold special interest as habitats that give easy access to two worlds for their small inhabitants, and experience unique energy transfers.

While each discipline defines these sectors of the earth within its own framework of self-assigned tasks, the individual approaches tend to converge—in the study of energy–mass transfers, for example. These approaches form powerful explanatory models in physiology and ecology, forestry and agronomy, hydrology, meteorology, geomorphology, and geography.

Depth

Land form, especially in dissected topography, affects the income of radiant energy and the power of the atmosphere to ventilate an ecosystem, illustrated in the contrasts of temperature in concave and convex topography. It also affects the substrate. The most obvious case is that of a basin in the land surface that fills with water. The capacity of an aquatic ecosystem to store energy and matter is related to the depth and shape of the basin (Hutchinson, 1957). The stratification that often develops in concave topography is at its maximum in lake basins with seasonal isolation of the lower layer of water. Different depths around the edges of a lake basin may be occupied by different plant communities, and this zonation gives a diversity "attractive to different forms of wildlife" (Weller, 1978).

Energy and mass transformations take place at different depths. A snow cover is penetrated only a few centimeters by solar radiation, while in a forest or lake some radiation penetrates to meters. The earth–atmosphere interface is often a thick zone throughout which energy interactions occur, a biota-filled volume, to use Reifsnyder's (1967, p. 132) term, or the soil–plant–air continuum, to use Philip's (1966).

Depth of an ecosystem is often associated with more complete absorption of radiant energy, opportunity for recycling water, CO_2, or nutrients, and capacity to store energy or forms of matter that might have future value to ecosystem functioning. The high respiration of trees as compared with low plants has been ascribed to "the maintenance costs of a trunk-and-branches system.... Probably over half their budget [of tree stands] is spent on defense expenditures related solely to competition" (Wilson, 1967).

Thickening of the interface is brought about by the master converters of energy, the plants. They reach downward for water and warmth and upward to filter the air for CO_2 molecules for an energy-fixing process and to optimize the trapping of photons of radiant energy coming from the empyrean. The energy-fixing assimilation of CO_2, absorption of radiant energy, leakage of water vapor, and dissipation of kinetic energy are all facilitated by depth. Early efforts to understand the functioning of cropland without considering distributions of processes with depth reached a limit that is analogous to the limits reached by efforts to understand the atmosphere before we could measure it in depth.

Structure

Exchanges and conversions in ecosystems are most intense near the outer boundary, but not necessarily immediately at it. Layers of uniform intensity of interaction come into existence parallel to this outer surface; a stratified structure results. Geiger contrasts the dark, cold forest floor with the zone of light, warmth, and activity in the upper tree crowns; the cold, dark, oxygen-depleted water of the hypolimnion of an aquatic ecosystem contrasts with the upper layer that is light-pervaded, warm, oxygenated, and seething with life.

Differences between ecosystem layers govern the fluxes and transformations of water (Miller, 1977, Chapter VI), which moves generally downward under gravitational forces; energy is freer to move in any direction, upward and out as well as downward, and layers communicate more easily with respect to energy.

The parameters of structure, or "arrangement in space of the components of vegetation" (Fosberg, in Peterkin, 1967) that involve no preconceptions about environmental adaptations, include:

Height of plants, or depth of the community below its outer surface;
Branching habit;
Size of stems, especially in trees;

Size and density of crowns, the members that face the atmosphere and sun;

Thickness and density of canopy;

Layering or stratification of canopy, which is a characteristic appropriate for one-dimensional modeling, if horizontal uniformity permits;

Depth, density, spacing, and stratification of root systems.*

"What is needed is a map of the interfaces between the three domains of the SPAC, i.e., of the soil–plant, the soil–atmosphere, and the plant–atmosphere interfaces" (Philip, 1966). The soil–plant and the plant–atmosphere interfaces are highly complicated in form and difficult to measure and describe. Many attempts to develop predictive methods have been "vitiated by the failure of workers to pay as much attention to the accuracy of the geometry of their models as they do to the mechanisms of the transfer processes" (Philip, 1966); architectural analyses of plants (Hallé *et al.,* 1978; Ross, 1975a) will prove helpful in this question.

The soil of an ecosystem provides essential storage space for energy, water, and other substances. As this storage is alternately drawn down and replenished, it evens out the responses of the ecosystem to fluctuations in radiation and other inputs.

Competition among the members of a plant community influences its structure and use of energy. Evans (1963) cites a case in which six pasture grasses growing singly exhibited best growth when the water table was 0.30–0.35 m below the ground surface, but when growing together responded quite differently; optimal depths of different species then varied from 0.06 to 1.1 m. For total energy conversion by the whole community, the optimal depth of water table was 0.65m—different from that for any pure stand of individual grass.

Leaf Area and Geometry

Leaf width is important in turbulent exchanges, which are greatest for narrow leaves; conifer needles, however, that tend to clump have a somewhat greater critical dimension. Total surface area of leaves in an ecosystem is several times greater than the ground area. The ratio between these two areas, or leaf-area index (LAI), is a convenient

---

* Other parameters have also been used. Stem density, the summation of stem diameters per hectare of forest site, has been a useful interim measure of foliage area and density and seems to be better related to the penetration of shortwave radiation than indexes that give disproportionate weight to large trees (Miller, 1959). Such a dimension is an empirical expedient, pending the measurement of more physically relevant characteristics. Tree density and height can be combined into a ventilation index (Szeicz *et al.,* 1979).

number to denote the density of photosynthesizing surfaces in terms of the unit horizontal areas (square meters) in which incoming radiation and other energy fluxes are expressed. Leaf-area index is relevant to both nonradiative and radiative fluxes of energy and, in fact, can be determined by inversion techniques from below-canopy measurements of the solar beam, as well as by such indexes of tree vigor as its quantity of sapwood (Waring *et al.*, 1980).

Leaf area is a basic factor in gross photosynthesis, respiration, and net production and is used as an indicator of cultural practices needed. Some communities seem to produce best at an index around 3, and this value is maintained during as much of the growing season as possible. A low index in the beginning of the season shows how much photosynthetically active radiation is being wasted on bare soil. Evergreen forest ecosystems, on the other hand, may display indexes up to 40–50 in thermally moderate sites with adequate soil moisture (Waring *et al.*, 1978).

A profile of leaf area from the top to the bottom of the ecosystem indicates the depths where photosynthesizing surfaces are concentrated. It is expressed in surface area per unit layer of the canopy and is sometimes called a density because of the inverse cubic meter term. It can be visualized by thinking of a canopy as a set of shelves so spaced that a certain shelf area is located in each cubic volume or as a frequency curve (Gary, 1978). It is generally employed in studies of ventilation of plant communities (Ross, 1975b) and can be determined by camera techniques (Aber, 1979).

Figure 1 (Lemon, 1967) shows leaf-area profiles in corn in New York and clover in New Zealand. Corn foliage was most abundant in a layer from 0.3 to 0.6 of the depth of the community (2.8 m) where leaf-area

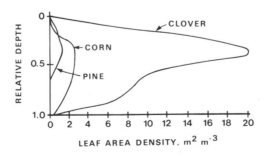

Fig. II.1.   Leaf-area profiles, normalized to relative depth, in clover, corn, and pine [data from Lemon (1967), Rauner (1968)]. Total leaf areas of the three ecosystems were 4, 4.5, and 6.2 m^2 m^{-2}, respectively; their depths were 0.5, 2.8, and 6.5 m.

density was about 2.5 m² m⁻³. (The leaf-area index was 4.5 m² m⁻² of ground area.) Clover foliage was most concentrated (at a high density of 20 m² m⁻³) at a shallow depth in the layer from 0.2 to 0.5 of total depth (which was 0.5 m). These profiles express fundamental differences in the structure of the two ecosystems, which affect such processes as the ways the crop extracts momentum from the air and photosynthetically active radiation suffuses the chloroplasts.

Turbulent and radiative fluxes in the foliage profile bring about characteristic profiles of the concentrations of vapor, CO_2, and temperature that form the immediate environment of leaf surfaces at different depths. "The micrometeorological and plant processes, the community structure and the community climate are all rigidly coupled" (Lemon, 1967). A change in ventilation, for instance, results in compensating changes in energy and mass exchanges and in air temperature and humidity in each layer. Much modeling of ecosystem processes makes use of one-dimensional profiles in which radiation, vapor, and CO_2 are passed from layer to layer.

Profiles in 18-year pine near Kursk, measured in energy-budget investigations (Rauner, 1968), had leaf-area of 6.2 m² m⁻² and skeletal area (branches and stems) of 0.8 m² m⁻². Leaf-area density was a maximum at 1.3 m² m⁻³ at a depth of 0.3. In deciduous trees leaf-area density peaked at 1.4 m² m⁻³ at a depth of 0.2 (Rauner, 1976).

Leaf angle affects the entry of radiation into an ecosystem, and so it has to be measured and modeled also. Vertical leaves present active surfaces at the most effective angles to the direct solar beam when the sun is low in the sky (Anderson and Denmead, 1969); see Table I. Vertical collectors of solar energy have a similar advantage in middle and

TABLE I

Receipts of Direct Solar Radiation on Foliage of Leaves of Different Inclinations[a,b]

Solstice		Horizontal (W m⁻²)	Vertical (W m⁻²)
Winter			
Leaf-area index	1	71	150
	3	107	198
Summer			
Leaf-area index	1	230	215
	3	348	380

[a] Latitude 35°S.
[b] Data from Anderson and Denmead (1969).

high latitudes. Under a high sun vertical leaves receive less radiation than horizontal ones in ecosystems with sparse foliage (LAI = 1).

Optimum leaf angle also depends on the light-saturation properties of the species. Where photosynthetically active radiation is weak, horizontal leaves have the greatest net photosynthesis; where it is intense, a nearly vertical position is best (Chang, 1968, p. 32). The complex geometry of leaf area and inclination angles can be worked out by a joint field-computer approach, like the work in the Estonian Academy of Sciences at Tartu (Ross, 1975a, b; Nil'son, *et al.*, 1977) or the National Institute of Agricultural Sciences in Tokyo (Uchijima, 1974). Leaf width, area, and inclination of the principal species in a wet-meadow tundra ecosystem were employed in modeling its energy budget and photosynthesis (Miller *et al.*, 1976). Widths varied from 2.9 mm in *Carex*, inclined at 72°, to 5.6 mm in *Salix*, inclined at 24°, and each of these species contributed about 0.2 m^2 m^{-2} to total leaf-area index, which attained only 1.0 to 1.2 m^2 m^{-2} in the most favorable of the four years of the study. Even in this thin ecosystem, 0.1–0.2 m in depth, it was advantageous to stratify the canopy in modeling the receipt and transformation of energy.

Ecosystems are not easily measured, particularly in dimensions related to the physics by which they are coupled with their environments. Combined geometrical and statistical approaches, e.g., Ross (1975a) and Nil'son *et al.* (1977), appear to be fruitful in describing canopy architecture. They can be used with data on the optical properties of leaves and tested by field measurements in different ecosystems. In the following pages we outline some characteristics that govern each physical coupling.

FACTORS IN COUPLING OF AN ECOSYSTEM WITH THE SUN

The sun, a major but variable source of energy for all ecosystems, is the only source for the energy-conversion process of photosynthesis, which yields the highest-grade output that ecosystems produce; it also provides lower-grade (longer in wavelength) energy that is important in other aspects of ecosystem energy budgets. In most places the solar flux regularly goes to zero, and physiological processes are accommodated to this unique regime and to the cycle of the year.

Angle of Incidence

The direct solar beam and much of the incoming flux of diffuse radiation are directional. This fact makes important the angles of leaf inclination and also the general ecosystem attitude, as expressed by slope azimuth and steepness. The azimuth and altitude of the sun and the orientation and slope of an ecosystem are connected by definite trigonometric relations, which are discussed in Chapter III. In general, the geometry of individual plants is a part of their strategy for intercepting light. For example, the angles of branching in a particular low-latitude tree is such as to produce "the maximum effective leaf surface" (Honda and Fisher, 1978) presented to the solar beam, rather than simply the maximum area of leaf surface.

Absorption of Solar Radiation by Ecosystem Surfaces

Absorption of photons of the solar flux by inorganic substances differs somewhat from wavelength to wavelength, but not radically. This is not so for leaves, which are closely coupled with solar energy in the photosynthetically active spectral region and only loosely coupled in the solar infrared region (wavelengths $0.7-3.0$ μm). Another aspect of this selective reflection involves energy-mediated transfers of information, like flower patterns visible only to ultraviolet-sensing insects that are potential pollinators.

Absorption is facilitated by transmission of the solar flux deep into an ecosystem, with several contacts with leaves. The quasi-cavity nature of ecosystem canopy is important in attaining optimum absorption. This will be discussed at greater length in Chapter XX, along with the role of transmitted radiation at different depths within ecosystems, including ground vegetation and hydrologic processes.

Absorption varies throughout the year as foliage covers more of the ground surface and then dies back and as snow comes and goes. It may be as small as 0.25 in winter and spring, indicating a weak coupling with the sun, but 0.80–0.85 for photosynthetically active radiation during the main growing season. Reflection and absorption of solar radiation will be discussed in Chapter VII.

An absorptive ecosystem in the direct solar beam responds quickly when a cloud interrupts the beam. Its surface temperature drops, photosynthesis decreases, and its energy outflows abruptly diminish. Similarly, but more slowly, a change in absorptivity changes the coupling of a system with the sun.

Absorption is spread over a considerable depth of many ecosystems, depending on the vertical distribution of leaf area. This fact has benefits in energy retention within the system that can be seen in comparing a forest with a desert surface, where concentrated absorption in a thin layer leads to a rapid rise in surface temperature, outgoing longwave radiation, and sensible-heat flux. A solar pond stores heat by localizing absorption away from the outer surface of the pond. At a small scale a film of water on leaves traps photons by internal reflection and increases absorptivity.

Absorptivity changes during the year as foliage surface changes; new stems push up, new leaves unfold and grow, finally mature, and then grow old. This sequence is less obvious in evergreens, but there is still measurably more absorption of energy in fall than spring. A typical variation in absorption is shown in measurements made on a field of grain from February through October by radiation meteorologists in Wien (Fig. 2) (Dirmhirn, 1964, p. 131). The effect of snow cover is seen in February and of bare soil in March and after the grain is cut in September and October (stubble). Midsummer values of 0.82–0.85 are typical of many crops during active growth.

Also shown is the absorption of energy in the visible range, i.e., of photosynthetically active radiation (curve 2). In the months of active photosynthesis this absorptivity is greater than for total shortwave, which includes the relatively low absorption in solar infrared wavelengths. After the grain ripens in August, it absorbs less photosynthetically active radiation.

RADIATIVE COUPLING WITH THE ATMOSPHERE

A landscape element at the surface of the moon derives its energy through its couplings with the sun and substrate. In contrast, a landscape element at the surface of the earth lies under a deep ocean of air, which contains molecules that emit radiation.

This radiation is strongly absorbed in all plant communities. Absorptivities are of the order of 0.96–0.98 (Lorenz, 1966), partly because the

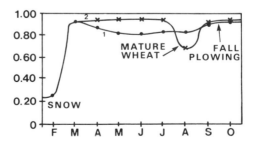

Fig. II.2. Absorptivity of solar radiation by wheat field in Austria through a crop season (Dirmhirn, 1964, p. 131). Curve (1) absorptivity over total solar spectrum; (2) absorptivity of radiation in the visible range (i.e., photosynthetically active radiation).

spaces among leaves form a set of blackbody cavities. Conversely, little incoming longwave radiation is reflected, perhaps 0.10 at desert surfaces and a third of this in most ecosystems (Chapter VII). Changes in the incoming radiative flux are therefore strongly registered in most ecosystems; a decrease of 50 W m^{-2} when a hole in a nocturnal cloud deck moves across the sky produces an immediate effect in surface temperature that is often visible as dew forms.

COUPLING OF AN ECOSYSTEM WITH THE ATMOSPHERE BY TURBULENCE

The face presented by an ecosystem to the restlessly moving atmosphere and the shelter or exposure of its topographic site determine the importance of the turbulent heat fluxes in its overall energy budget. Can it extract warmth from the air at night? Can it easily get rid of excess heat when it is baking under the midday sun? Is it exposed to episodic bursts of kinetic energy when the wind rises?

Exposure and Ventilation

Exposure of an ecosystem to the wind depends on its topographic location. The streamlines of motion in storms tend to sweep over dissected terrain, bunching over the ridges where the air speeds up and spreading out and slowing in the valleys where different ecosystems live (Aulitzky, 1967).

Local valley circulations in periods between storms are weaker but more persistent than storm winds and produce a constant spatial pattern of mechanical energy in sheltered and exposed areas. Drainage and pooling of cold air at night produce other patterns that affect energy inputs and outputs of ecosystems. The spatial variability in the wind field also determines the delivery pattern of rain and especially snow, with consequent effects in ecosystem energy budgets.

The contact between the ecosystem and the lower layer of air is affected by its being sheltered by taller vegetation or topography, by its permeability and roughness, its size, and the fetch of the wind over it. The contact between ecosystem and atmosphere also depends on such atmospheric qualities as stability, the vertical profile of wind speed as a result of upwind coupling of air and surface, and the thickness of the boundary layer.

This thickness depends on the fetch and thermodynamic stability of the boundary layer which results from the inflow of sensible heat from the underlying surface. Inversion layers in the lowest 1 or 2 km of the atmosphere isolate the planetary boundary layer from the free air and tend to force it to find an equilibrium with surface heat fluxes. Coupling becomes a mutual process in which vertical mixing is sustained by conversion of horizontal kinetic energy, as expressed in the familiar Richardson number. If this energy is reduced by topographic barriers coupling is reduced.

Roughness

Foliage characteristically is open to ventilation because the inflow of atmospheric carbon dioxide is as necessary to the plant as the receipt of chemically effective radiation from the sun. Openness itself generates turbulence in the air stream, particularly where some members of an ecosystem project into the air.

The roughness length z_0, a parameter defined from the logarithmic profile of wind speed above a surface, is an index to atmospheric coupling of an ecosystem. The value of z_0 is a few millimeters at smooth vegetation, 140 mm at the canopy of root crops (Sutton, 1955, p. 107), and several meters at forest stands (Rauner, 1961) and over residential areas of a city (Morgan et al.., 1977).

The roughness coefficient depends on the number of obstacles in the airstream and their height. The ratio of height to roughness length z_0, determined from ingenious field experiments, "varies from about 5 for tall trees, to 10 for grasses, and to 30 for sand grains" (Lettau, 1967) for average densities of obstacles. This relation can also be applied to determine roughness of urban interfaces (Nicholas and Lewis, 1980). A more detailed study of the relation of roughness to height (Garratt, 1977) includes a variety of measures of spacing between obstacles, related to their silhouette areas, shape, and slenderness. Values of z_0 from a variety of field experiments show its wide range (Garratt, 1977):

sparse grass
  0.0012 m
grassland
  0.001 m
wheat, increasing both in height and in the obstacle parameter
  $0.005 \rightarrow 0.015 \rightarrow 0.05 \rightarrow 0.07$ m
corn
  0.06m
vineyard
  0.02 m (wind parallel to rows)
  0.12 m (wind at right angles to rows)

TABLE II

System-Scale External Resistance[a] at Wind Speed of 2 m sec^{-1}

Roughness length z_0 (m)	Resistance (sec m^{-1})
0.001	170
0.01	90
0.1	30

[a] Data from Slatyer (1967, p. 50).

scattered trees and shrubs 10–20 m apart
  0.45 m
scattered trees and shrubs 5–10 m apart
  0.8 m
savanna, semiarid, low-tree
  0.3 m
forest, wet schlerophyll
  5.0 m
forest, subtropical rain forest
  1.3 m

Roughness varies with wind speed. At certain critical speeds the flutter of leaves may tend to increase it, so that the elastic properties of plant communities are important in atmospheric coupling. Variation of roughness with wind speed also is caused by the bending of plant stalks and waves that travel across the canopy.

The roughness coefficient is related to the external resistance of a crop to the outward diffusion of vapor molecules (see Table II). Ecosystems with rougher surfaces display less external resistance to diffusion of vapor or to inward diffusion of CO_2.

Roughness changes with ecosystem leaf area and height through the

TABLE III[a]

Season	z_0 (m)	Season	z_0 (m)
Winter	0.015	Summer	0.36
Spring	0.09	Autumn	0.22

[a] Data from Kung (1963).

year. Kung (1963) calculated z_0 from data on agricultural land use in southern Wisconsin (Table III). His calculations of mean roughness at latitude 45°N across North America give 0.023 m in winter and 0.41 m in summer. "Vegetation cover, rather than topography, represents the most efficient roughness structure" (Kung, 1963, p. 49).

Roughness (z_0) of an ecosystem for momentum is determined by measuring wind speeds at two or more heights (z) above an extensive uniform surface and graphically relating them to the term $\ln(z/z_0)$ to derive z_0. Well-matched anemometers are necessary, but the most difficult condition is to confine profile measurements within the boundary layer of air that is in equilibrium with the particular surface, so that its turbulence pattern does not carry the effects of upstream obstacles or ecosystems of different roughness (Thom, 1975).

The penetration of eddies into the foliage and the network of branches is indexed by diffusivity. Lemon (1967) measured a mean daytime diffusivity coefficient of 0.1 m^2 sec^{-1} in a cornfield at 0.2 depth from the outer surface, which decreased with greater depth into the community. This coefficient was derived from simultaneous measurements of the gradients of vapor, CO_2, temperature, and momentum, and the flux divergence of radiation. Calculations in forest from profile measurements in young stands of pine and spruce gave a value of 1 m^2 sec^{-1} (Reifsnyder, 1967), reflecting the rougher top.

Coupling by Conversion of Kinetic Energy

This coupling is important in tall vegetation, which raises its energy-dissipating surfaces high into the airstream. Important dimensions of ecosystems are the mechanical strength and flexibility (or elasticity) of stems and branches. Lodging due to rain load and wind is a common problem in farming, reducing production and rendering harvest operations more costly. Windfall of trees is an infrequent but important factor in forest succession, an example being hurricane felling of the forests of New England.

Human settlement began early to create openings in forested lands and conversely to plant windbreaks around farmsteads and shelterbelts between fields on the prairie. Besides altering the field of kinetic energy so that the wind deposits snow in the places desired, shelterbelts also absorb kinetic energy; if properly designed, they filter out large eddies in the airstream. Ecosystems of taller vegetation act as energy dissipaters, like riffles in a stream.

The roughness of a mosaic of ecosystems is illustrated by the hedge landscapes bordering the North Sea. In winter the hedges, though leafless, still produce a roughness value of 0.9 m at Quickborn, near Hamburg (van Eimern *et al.*, 1964, p. 17). Leafing out in summer raises the roughness of this flat landscape to 1.45 m. Shelter of the individual fields between the hedges changes the coupling of both plants and animals with the atmosphere, and it may also be that the roughened landscape extracts more heat from the marine air in winter (Fig. 3).

COUPLING WITH THE SUBSTRATE

An ecosystem's energy budget is affected by access to the heat storage capacity of biomass, soil, rock, and water. Storage in biomass is usually small, except in forest stands,* so that we are considering substrate capacity primarily. Capacity that is turned over in the diurnal cycle usually is at shallow depths within the root zone; capacity turned over in the

Fig. II.3.   Heckenlandschaft of the coastal plain of the North Sea, near Hamburg, showing the small size of ecosystems in this long-settled region. Many field borders have hedges that provide shelter from ocean winds.

* In which it may support evaporation at night when it has ceased in nearby pasture ecosystems (Hicks *et al.*, 1975).

annual cycle extends deeper. In any case, heat stored in the substrate is more or less captive since lateral movement, except in aquatic ecosystems like streams, is usually minor.

Thermal Admittance

The thermal properties of a substrate include thermal conductivity (λ) and volumetric heat capacity (ρc) and can be combined into one expression, the thermal admittance, which indexes the daily (or annual) heat pulse that the substrate can take in and later pay out.

Lettau (1967) has shown that experiments with micrometeorological processes can be carried on outside the laboratory by altering one coupling between a landscape element and its environment and observing the response. One method is to suddenly remove the shade from a surface and see how that surface responds to the pulse of solar energy. The response depends, if the atmospheric coupling does not change radically, on the coupling with the substrate shown by the coefficient of thermal admittance $\sqrt{\lambda \rho c}$.

An immobile substrate is at a disadvantage in competing with the atmosphere for heat made available at the interface. If it is fluid, however, it can compete well with the air because the efficient process of turbulent conductivity can operate beneath, as well as above, the surface. Thermal admittance is large, partly because specific heat c is larger for water than for rock, but principally because turbulent conduction operates faster than molecular conduction.

Measurement of the three terms of thermal admittance is not difficult in the laboratory, though the bulk density ρ has to be determined without compacting the soil taken from the field. The specific heat c of the bulk soil can be determined from the proportions of mineral and organic particles, air-filled pores, and water. Thermal conductivity is determined by measuring heat flow at a known temperature gradient. Alternatively, thermal diffusivity, determined by observing the weakening and lag of a heat pulse moving into the soil, can be converted to admittance by taking its square root and multiplying by volumetric heat capacity. These parameters, however, are more difficult to measure under field conditions of stratification and spatial inhomogeneities typical of many soils. Information on the thermal admittance of natural surface covers, such as sod, mulch, snow, and aerated soil, is poor, especially in view of the effects of varying moisture content at different levels.

Interaction of Heat and Moisture in the Soil

Interesting interactions between fluxes of mass and energy are exemplified in the influence of moisture content on the thermal properties of the soil. A moist soil has greater density and greater specific heat per unit mass because water (ρ = 1000 kg m^{-3}, c = 4167 J $°K^{-1}$ kg^{-1}) has replaced air (ρ = 1.3 kg m^{-3}, c = 1000 J $°K^{-1}$ kg^{-1}) in the pores, so that heat capacity per unit volume or in a layer of given depth is much greater. A moist soil can store at least twice as much heat as a dry soil. Moist soil displays greater thermal conductivity, to which are added fluxes of heat by nonmolecular processes. The essential quantity of thermal admittance is increased in all three factors when a soil gains moisture.

A moist soil competes more strongly with the atmosphere for the radiation absorbed at the interface between soil and air than a dry soil. It can take heat to considerable depths. Less heat, on the other hand, lingers in the shallow layers during the critical weeks of spring when the farmer is anxious to start his field work and finds the top soil still too wet and cold.

Many land-management practices alter the coupling of an ecosystem with its substrate, even though their primary purpose might be to remove weed competition, assure better seed germination, improve water intake and control, or increase soil aeration. Water management in cropland, for instance, makes available water films up to 30–40 molecules thick on soil particles, filling small pores with water while leaving big pores full of air. One object of drainage of fields in the Midwest is to accelerate spring warming by concentrating the substrate heat flux in the upper layers.

The covering of bare soil by growing foliage results in a drop in the daytime heat intake to a fraction of its earlier 3–4 MJ m^{-2}. This is a familiar progression in annual crops and occurs in forest soil when the trees leaf out.

ECOSYSTEMS AT THE ACTIVE SURFACE OF THE EARTH

"The interface between the atmosphere and the solid earth is one of the most interesting boundaries in nature..." (Landsberg and Blanc, 1958). Yet in many important respects it is unknown to us. Well back in the 19th century Voeikov drew scientific attention to what he named "the outer active surface" of the earth as a zone of the most intense energy and mass exchanges and consequent concentration of physical, chemical, and biological phenomena. Since that time the importance of this interface has at times been recognized in principle, but practice has not followed. Except when we are inside a building or a forest, this active surface is within our daily view; a teacher can ask students to look out of the window at the lawn, the tree tops, or the pavement of the parking lot to see different kinds of active surface. Still, we have only sketchy knowledge of the basic characteristics of this surface that govern its coupling with the sun, soil, and air.

The increased interest in climatic processes (U.S. National Academy of Sciences, 1980, pp. 21, 35), and especially in ecosystems as units for study promises to redress some of these areas of neglect. Whether they are considered as energy converters or as processors of water, carbon, nutrients, or urban wastes, it is necessary to understand their energy metabolism and to see how this metabolism is supported by energy inputs.

Transient properties of ecosystems, such as their deformation in high winds, external wetting by rain, or increased internal moisture when soil moisture rises after rain, affect their coupling with the environment. Ecosystem characteristics play an important role in determining their own energy scenarios.*

Basic factors that influence the ventilation, radiation budget, and heat and water budgets of an ecosystem through its coupling with the sun and air include (a) its area; (b) its geologic and geomorphologic site, which affects its ventilation and the capacity of its substrate to store nutrients, heat, and water; (c) its geographic location and altitude, which affect its inputs of energy; (d) species interactions in biological competition, regeneration, and succession; and (e) the mixed layer of the atmosphere. The influence of these factors depends on the closeness of couplings between the ecosystem and its environment. In this chapter we have identified such couplings as absorptivity of solar and longwave radiation, roughness, ventilation, and thermal admittance as being important in energy modeling of an ecosystem. This identification forms a prelude to considering the specific energy fluxes themselves and gives us a means of evaluating the ways an ecosystem responds to the forcing functions from the outside world.

REFERENCES

Aber, J. D. (1979). Foliage-height profiles and succession in northern hardwood forests. *Ecology* **60**, 18–23.
Anderson, M. C., and Denmead, O. T. (1969). Shortwave radiation on inclined surfaces in model plant communities. *Agron. J.* **61**, 867–872.
Aulitzky, H. (1967). Significance of small climatic differences for the proper afforestation

---

* A still more variable property of an ecosystem, its surface temperature, is so much a result of energy inputs and a cause of energy outputs that it will be treated by itself in Chapter VIII.

of highlands in Austria. *In* "Forest Hydrology" (W. E. Sopper and H. W. Lull, eds.), pp. 639–653. Pergamon, New York.

Bradbury, D. E. (1978). The evolution and persistence of a local sage/chamise community pattern in Southern California. *Assoc. Pac. Coast Geogr., Yearb.* **40,** 39–56.

Chang, J.-H. (1968). "Climate and Agriculture," Aldine, Chicago, Illinois.

Dirmhirn, I. (1964). "Das Strahlungsfeld im Lebensraum." Akad. Verlagsges., Frankfurt a.M.

Evans, L. T. (1963). Extrapolation from controlled environments to the field. *In* "Environmental Control of Plant Growth" (L. T. Evans, ed.), pp. 421–427. Academic Press, New York.

Garratt, J. R. (1977). "Aerodynamic Roughness and Mean Monthly Surface Stress Over Australia," *Aust. C.S.I.R.0. Div. Atmos. Phys. Tech. Pap.* No. 29.

Gary, H. L. (1978). The vertical distribution of needles and branchwood in thinned and unthinned 80-year-old lodgepole pine. *Northwest Sci.* **52,** 303–309.

Hallé, F., Oldeman, R. A. A., and Tomlinson, P. B. (1978). "Tropical Trees and Forests. An Architectural Analysis." Springer Publ., New York.

Hicks, B. B., Hyson, P., and Moore, C. J. (1975). A study of eddy fluxes over a forest. *J. Appl. Meteorol.* **14,** 58–66.

Honda, H., and Fisher, J. B. (1978). Tree branch angle; Maximizing effective leaf area. *Science* **199,** 888–890.

Hutchinson, G. E. (1957). "A Treatise on Limnology. Vol. I: Geography, Physics, and Chemistry." Wiley, New York.

Kung, E. C.-T. (1963). Climatology of aerodynamic roughness parameter and energy dissipation in the planetary boundary layer over the Northern Hemisphere. *In* "Studies of the Effects of Variations in Boundary Conditions on the Atmospheric Boundary Layer" (H. H. Lettau, ed.). *Univ. Wis. Dep. Meteorol., Annu. Rep., 1963* pp. 37–96.

Landsberg, H. E., and Blanc, M. L. (1958). Interaction of soil and weather. *Soil Sci. Soc. Am., Proc.* **22,** 491–495.

Lemon, E. (1967). Aerodynamic studies of CO_2 exchange between the atmosphere and the plant. *In* "Harvesting the Sun" (A. San Pietro, F. A. Greer, and T. J. Army, eds.), pp. 263–290. Academic Press, New York.

Lettau, H. H. (1967). Problems of micrometeorological measurements (on degree of control in out-of-doors experiments). *In* "The Collection and Processing of Field Data" (E. F. Bradley and O. T. Denmead, eds.), pp. 3–40. Wiley (Interscience), New York.

Lorenz, D. (1966). The effect of the long-wave reflectivity of natural surfaces on surface temperature measurements using radiometers. *J. Appl. Meteorol.* **5,** 421–430.

Miller, D. H. (1959). Transmission of insolation through pine forest canopy, as it affects the melting of snow. *Mitt. Schweiz. Anst. Forstl. Versuchsw.* **35,** 57–79.

Miller, D. H. (1977). "Water at the Surface of the Earth." Academic Press, New York.

Miller, P. C., Stoner, W. A., and Tieszen, L. L. (1976). A model of stand photosynthesis for the wet meadow tundra at Barrow, Alaska. *Ecology* **57,** 411–430.

Morgan, D., Myrup, L., Rogers, D., and Baskett, R. (1977). Microclimates within an urban area. *Ann. Assoc. Am. Geogr.* **67,** 55–65.

Nicholas, F. W., and Lewis, J. E., Jr. (1980). Relationships between Aerodynamic Roughness and Land Cover in Baltimore, Maryland. U.S. Geol. Surv. Prof. Paper 1099-C.

Nil'son, T., Ross, V., and Ross, Iu. (1977). Nekotorye voprosy arkhitektoniki rastenii i rastitel'nogo pokrova. *In* "Propuskanie Solnechnoi Radiatsii Rastitel'nym Pokrovom," (L. Riives, ed.), pp. 71–144. Akad. Nauk. Eston. SSR, Inst. Astron. Fiz. Atmos., Tartu.

Peterkin, G. F. (comp. 1967). "Guide to the CheckSheet for IBP Areas, including a Classification of Vegetation for General Purposes," by F. R. Fosberg, IBP Handbook No. 4, Section IBP/CT (Conservation of Terrestrial Biological Communities). Blackwell, Oxford.

Philip, J. R. (1966). Plant water relations: some physical aspects. *Annu. Rev. Plant Physiol.* **17,** 245–258.

Rauner, Iu. L. (1961). O teplovom balansa listvennogo lesa v zimnii period. *Izv. Akad. Nauk SSSR, Ser. Geogr.* **4,** 83–90.

Rauner, Iu. L. (1968). Biometricheskie pokazateli lesnoi rastitel'nosti v sviazi s izucheniem ee radiatsonnogo rezhima. *In* "Aktinometriia i Optika Atmosfery" (V. K. Pyldmaa, ed.), pp. 335–342. Izd. Valgus, Tallinn.

Rauner, Iu. L. (1976). Deciduous forest. *In* "Vegetation and the Atmosphere. Vol. 2: Case Studies" (J. L. Monteith, ed.), pp. 241–262. Academic Press, New York.

Reifsnyder, W. E. (1967). Forest meteorology: the forest energy balance. *Int. Rev. For. Res.* **2,** 127–179.

Ross, Iu. K. (1975a). Radiatsionnyi Rezhim i Arkhetektonika Rastitel'nogo Pokrova. Gidrometeorol. Izd., Leningrad.

Ross, J. (1975b). Radiative transfer in plant communities. *In* "Vegetation and the Atmosphere. Vol. 1: Principles" (J. L. Monteith, ed.), pp. 13–55. Academic Press, New York.

Slatyer, R. O. (1967). "Plant-Water Relationships." Academic Press, New York.

Sutton, O. G. (1955). "Atmospheric Turbulence," 2d ed. Wiley, New York.

Szeicz, G., Petzold, D. E., and Wilson, R. G. (1979). Wind in the subarctic forest. *J. Appl. Meteorol.* **18,** 1268–1274.

Thom, A. S. (1975). Momentum, mass and heat exchange of plant communities. *In* "Vegetation and the Atmosphere. Vol. 1: Principles" (J. L. Monteith, ed.), pp. 57–109. Academic Press, New York.

Uchijima, Z. (1974). Micrometeorology of cultivated fields. *In* "Agricultural Meteorology in Japan" (Y. Mihara, ed.), pp. 41–79. Univ. Press of Hawaii, Honolulu.

U.S. National Academy of Sciences (1980). "A Strategy for the National Climate Program," National Academy of Sciences, Washington, D. C.

van Eimern, J., Karschon, R., Razumova, L. A., and Robertson, G. W. (1964). Windbreaks and shelterbelts. *WMO Tech. Note* No. 59.

Voeikov, A. I. (1889). Der Einfluss einer Schneedecke auf Boden, Klima und Wetter. *Geogr. Abhandl.* **3**(3), 317–435.

Waring, R. H., Thies, W. G., and Muscato, D. (1980). Stem growth per unit of leaf area: a measure of tree vigor. *Forest Sci.* **26,** 112–117.

Waring, R. H., Emmingham, W. H., Gholz, H. L., and Grier, C. C. (1978). Variation in maximum leaf area of coniferous forests in Oregon and its ecological significance. *Forest Sci.* **24,** 131–140.

Weller, M. W. (1978). Management of freshwater marshes for wildlife. *In* "Freshwater Wetlands: Ecological Processes and Management Potential" (R. E. Good, D. F. Whigham, R. L. Simpson, and C. C. Jackson, Jr., eds.), pp. 267–284. Academic Press, New York.

Wilson, J. W. (1967). Ecological data on dry-matter production by plants. *In* "The Collection and Processing of Field Data" (E. F. Bradley and O. T. Denmead, eds.), pp. 77–127. Wiley (Interscience), New York.

Chapter III

DIRECT SOLAR RADIATION

Give me the splendid silent sun, with all his beams full-dazzling.
Walt Whitman
"Leaves of Grass," 1855

The direct beam of solar radiation, when present, is the most prom-
inent, exploitable, and directional* of all energy inputs into ecosystems
at the surface of the earth. It is also the most variable, and its variations
are a principal generator of variations in other energy fluxes.

Two characteristics of the beam are important to ecosystems: (a) its
high thermodynamic quality, which means that its energy is easily convertible to other useful forms, and (b) its changing pattern of intensity.

NATURE AND SIGNIFICANCE OF THE SOLAR BEAM

Because the mass of hydrogen that constitutes the bulk of the sun
exceeds the critical mass, thermonuclear reactions proceed deep within
it and produce energy that works its way by radiative and convective
processes to the surface, from whence it is radiated outward into space.
The energy flux density of this outward radiation is 70×10^6 W m^{-2},
and diminishes outward with the square of the distance from the sun.

* The Latin word "radius," from which "radiare" (to emit beams) is derived, means the
spoke of a wheel, and the idea of emitting rays in all directions has continued into present-
day usage. Of all the beams the sun radiates, an ecosystem receives a particular beam,
and the directional nature and orienting properties of this beam give it unique value to
the system.

31

At the mean distance of the earth it is about* 1375 W m^{-2}. This is the energy flux received on the solar cells that a satellite of the earth stretches out to power itself.

The sun emits photons at frequencies ranging from very high to moderate, which correspond to wavelengths less than 0.3 μm to several thousand μm. The energy at wavelengths greater than 4 μm is small. The energy e of a photon is equal to Planck's constant h (6.64 × 10^{-34} J sec) times the photon's frequency v [(speed of light c)/(wavelength λ)] and therefore to the reciprocal of its wavelength

$$e = hv = hc\lambda^{-1}.$$

The solar spectrum can be divided on the basis of these effects of photons in ecosystems.

1. Wavelength greater than 1.0 μm.	Heating effects only; 0.29 of total flux. At λ = 1.2 μm, photon energy is 1.6 × 10^{-19} J
2. 1.0–0.70 μm.	Has elongating effect on plants; 0.22 of total flux (bands 1 and 2 make up the solar infrared)
3. 0.70–0.61 μm.	Strongly absorbed by chlorophyll (the red region) and manifests greatest quantum efficiency in photosynthesis (Milthorpe and Moorby, 1979, p. 98); also affects photoperiodic activity of plants; 0.11 of total
4. 0.61–0.51 μm.	Low photosynthetic activity; peak sensitivity of the human eye (green, yellow); 0.18 of total; at λ = 0.6 μm the energy of a photon is 3.3 × 10^{-19} J
5. 0.51–0.41 μm.	Strongly absorbed by chlorophyll (blue); 0.11 of total
6. Ultraviolet (UV) segment A, 0.40–0.315 μm.	Fluorescence in plants; 0.07 of total, but less studied than UV-B (Silberglied, 1979)

* The exact figure is not yet certain. A recent estimate gives the range 1368–1377 W m^{-2} (Feygel'son, 1977) and a 6-month set of measurements averaged 1376 W m^{-2} (Hickey et al., 1980).

| 7. UV-B, 0.315–0.280 μm. | Antirachitic radiation with production of vitamin D; peak effect on sunburn of skin occurs at 0.297 μm; germicidal activity; energy content about 6×10^{-19} J per photon; 0.015 of total flux |
| 8. UV-C, Shorter than 0.28 μm. | Germicidal; erythemal; wavelengths shorter than 0.26 μm are lethal to some organisms; 0.005 of total flux from the sun* |

From the ecological standpoint it makes sense to divide the solar spectrum into infrared, visible, and ultraviolet regions, bands 1–2, 3–5, and 6–8, respectively. The incongruity of applying a biological criterion, the sensitivity of the human eye, to divide the range of a physical flux by distinguishing the "visible" range (bands 3–5) is only apparent because the intensity of photons in each of the three ranges has a distinct role in ecosystem development. Each range conveys a different kind of coupling of sun to ecosystem and a different value to human systems; windows in a house can selectively admit visible radiation and exclude solar infrared. One device can split the beam caught by a tracking mirror into a component for interior illumination and another used for heating or photovoltaic conversions (Metz, 1976).

Measurement of the flux density of solar radiation at many wavelengths produces a pattern that characterizes all radiating surfaces. Intensity increases rapidly with wavelength up to a peak and then decreases more slowly as wavelength continues to increase. Intensity plotted against wavelength forms a curve skewed to the right (Fig. 1). Although the quantity of energy per photon decreases over this spectrum as wavelength lengthens, the numbers of photons change in such a way as to produce the characteristic flux density increase, peaking, then decrease in total energy flux. Energy flux is a nonlinear function of wavelength. This relation was discovered at the beginning of the 20th century by Max Planck (1901) in one of the great turning points in science. It combines wavelength, the temperature of the radiating surface, and three constants:

the speed of light c, 3.00×10^8 m sec^{-1},
Boltzmann's constant k, 1.38×10^{-23} J deg^{-1}, and
Planck's constant h, 6.64×10^{-34} J sec.

* This radiation can be produced artificially, but little in the natural beam at wavelengths shorter than 0.29 reaches the earth's surface under the atmosphere as presently constituted.

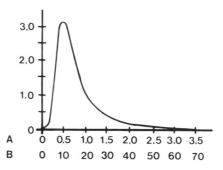

Fig. III.1. Planck's relation, for two temperatures important in environmental studies, of the intensity of radiation emitted (flux density over a spectral band of 1 μm) at different wavelengths. Scale A shows wavelengths (μm) for emission from a surface at 6000°K (solar), scale B at 300°K (terrestrial). Ordinate is flux density per μm wavelength band (units are langleys $\sec^{-1} K^{-5} \times 10^{-16}$ or $4.19 \ W \ m^{-2} K^{-5} \times 10^{-12}$). (Kondratyev, 1969, Fig. 1.4).

For the specific case of the sun at 6000° K, the relation discovered by Planck has the expression (Grum and Becherer, 1979, p. 106)

$$E_\lambda = 2\pi c^2 h \lambda^{-5}/\exp[(hc/k\lambda 6000) - 1]$$

in which E_λ represents flux density of radiation at wavelength λ. Wavelength in both the exponent (λ^{-1}) and the numerator (λ^{-5}) accounts for the nonlinear shape of the curve of radiation intensity as a function of wavelength at a temperature of 6000° K. Two standard operations can be performed on it: It can be differentiated to locate the wavelength at which emission is the strongest, and it can be integrated to determine the total energy flux from the surface at all wavelengths.

(a) Differentiation shows that maximum intensity of radiation from a 6000° K surface lies at 0.55 μm wavelength. This is Wien's displacement law, which was in fact identified before the more basic Planck's law. If the differentiation is carried out for different temperatures, Wien's law is expressed as

$$\lambda_{max}T = 2897 \times 10^{-6} \quad m \ deg \qquad (or \ 2897 \quad μm \ deg).$$

This relation expresses the fact that hot surfaces display peak radiation intensities at short wavelengths. Solar radiation is easily distinguished from radiation from a surface at a terrestrial temperature like 300° K.

(b) Integrating Planck's equation gives the total flux of radiant energy over all wavelengths. The form of this relation was also found prior to the discovery of the exact expression of Planck's law. It is

$$E = \epsilon\sigma T^4.$$

Here E is the total flux of energy of all wavelengths in watts per square meter; σ is the Stefan–Boltzmann constant, which takes care of the numerical and dimensional conversions from temperature to energy flux, and has the value 5.673×10^{-8} W m^{-2} deg $^{-4}$; and ϵ is emissivity, representing the degree to which real surfaces depart from being ideal or blackbody radiators. As seen earlier, its value approaches unity for many ecosystems.

This equation shows why the radiation flux density emitted at the surface of the sun is so much larger than that emitted by an ecosystem. The ratio of 20:1 between the solar temperature of 6000° K and the ecosystem temperature of 300° K is taken to the fourth power. The ecosystem radiates about 450 W m^{-2}, the sun 70×10^6 W m^{-2}. At the great distance of a terrestrial ecosystem from the sun, the second figure is reduced to a value at most only three times the first, but varies with changing direction of the solar beam, and is zero at night.

GEOMETRY OF THE SOLAR BEAM

The solar beam as a vector quantity serves to orient organisms, and its continual but regular changes through the day and year provide information but also pose a problem for ecosystems utilizing its energy. Two coordinates describe the angular position of the sun: azimuth (angle from north) and altitude (angle from the horizon). These are represented as the two polar coordinates on Fig. 2, which shows the apparent path of the sun through the sky at eight dates (five values of solar declination) over an ecosystem at 45° north latitude.

Azimuth

The solar azimuth varies through the cycle of the day, and also the year, except that it is 180° at local noon every day. At 45° N latitude the azimuth of the sun at sunrise varies from 56° (northeast horizon) at the summer solstice to 124° (southeast) at the winter solstice (Fig. 2). The rising and setting of the summer sun far poleward of east and west brings light in north windows and onto the north edges of forest bodies. The rate of change in azimuth varies slightly during the day, but not

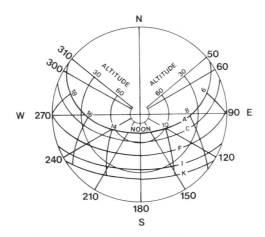

Fig. III.2. Paths of the sun across the sky at different seasons at 45°N latitude. The reader is invited to take a mental position at the center of the diagram and orient it. Then at the summer solstice the sun rises to your right, in the northeast at azimuth 55°, and travels along path A. At 0800 it is located at azimuth 92°, nearly due east, and at an altitude of 38° (shown by concentric circles with the horizon at the outer edge of the diagram and the zenith at the center). At solar noon, the solar azimuth is 180° and the altitude 68° (Brown, 1973). Path A, 22 June; C, 12 August and 1 May; F, equinoxes; I, 3 November and 9 February; K, winter solstice.

so much as to prevent honeybees from extrapolating it during cloudy periods (Gould, 1980).

Azimuth and Altitude

At the winter solstice the sun (Fig. 2) rises in the southeast (azimuth 125°, altitude 0°); by 1000 hours it has traveled across the sky to azimuth 152° and altitude 16°, changing its azimuth more than its altitude above the horizon. By noon (azimuth 180°) it has climbed no higher than 21° above the horizon, and ridges still hide it from many mountain ecosystems. Changes after noon are symmetrical. Azimuth is important for row-crop ecosystems; east–west rows have different solar-energy regimes than north–south rows.

The task of a sun-tracking device, whether in a plant or a solar furnace, is to follow the sun's azimuth and altitude. Some flowers of the high Arctic manage this problem at hours of highest sun (*Dryas*) or through much of the day (*Papaver*) (Keran, 1975) with corollas that are spherical or parabolic collectors, depending on the size of the gynoceium or other part to be heated. Pollinating insects that bask in the corolla reach temperatures 6–15°C above air temperature.

The motion of the sun produces a complex movement of shadows, which turn and change shape at the same time. The shadow line in forest openings can be determined graphically, or computed, to deter-mine the number of hours of sun and shade experienced by vegetation at different places in the opening (Satterlund, 1977). Graphical solutions portray the complex pattern of shade in a deep alpine valley, where winter sun is at a premium (Garnett, 1935) and where human settle-ments are located only where solar energy can reach.

Altitude

For horizontally extended ecosystems remote from ridges that might shade them, solar azimuth is less important than sun height or altitude (h) above the horizon because the vertical component (S') of the solar beam on their top surfaces is a sine function of altitude

$$S' = (I_0/R^2) \sin h.$$

(The flux density of the solar beam is represented by I_0/R^2, where R is the seasonally changing distance from the sun in proportion to its mean distance and I_0 is the mean annual flux density of the beam, taken here as 1375 W m^{-2}.)

As Fig. 2 shows, the winter sun at the 45th parallel is never higher than 21°. Since sin 21° is 0.36, only a third of the energy flux in a square meter of the direct beam is received on a square meter of ecosystem canopy. The ratio at noon (sin 68° = 0.93) in summer is 0.93. At high angles the sine function changes little, and the vertical component of the beam remains about the same for several hours of midday.

The change in altitude during the course of the day provides, like the change in azimuth, a sampling of beam angles for leaves of different orientations and vertical angles in an ecosystem. Low sun altitude at the beginning and end of the day is important to vertical leaves, for instance.

The two astronomical factors that govern sun altitude h are combined in a neat expression:

$$\sin h = \sin \phi \sin \delta + \cos \phi \cos \delta \cos t.$$

The three independent variables represent the areal variation in the angle of the solar beam due to latitude ϕ, the seasonal variation or de-lineation δ, and time of day t, the hour angle from solar noon. The change during the morning hours in most of the earth is from 0 to 40 or 50° (or more), a rate of 6–8° hr^{-1}. The mean flux increase during four morning hours at the month when it is fastest is shown in Table I. The sun in equatorial regions frequently encounters a sky full of moisture

TABLE I

Increase in Energy Flux Density of Vertical Component of Direct Beam during Morning[a]

Station	Latitude	Change in flux density (mW m^{-2} sec^{-1})[b]	Remarks
Kinshasa	4°S	19	Equatorial
Keetmanshoop	27°S	41	Desert site
Tashkent	41°N	25	Interior
Pavlovsk	60°N	13	Near Leningrad
Resolute Bay	75°N	5	High Arctic

[a] From data of isopleth diagrams in Berliand (1965).

[b] The rate of change (sec^{-1}) of the vertical component of the direct beam (mW m^{-2}) can also be written mJ m^{-2} sec^{-2}; the analogy to an acceleration (the sec^{-2} term) emphasizes its role as stimulus or forcing function in an ecosystem energy budget.

and haze, and the direct beam increase is small. In high latitudes the solar input on ecosystems is more gradual—and prolonged; the small diurnal variation in solar altitude reduces the frequency of inversions and ground frosts (Corbet, 1969).

A convenient reference point in the day is solar noon, when $t = 0$ and $\cos t = 1$, and the equation becomes $\sin h = \sin \phi \sin \delta + \cos \phi \cos \delta$.

$$90 - h = \phi - \delta, \qquad h = 90 - \phi + \delta.$$

At noon the sun altitude is equal to 90° less the latitude plus the declination. At Milwaukee's latitude ($+43°$), $h = 90° - 43° + \delta = 47° + \delta$. At Christmas, when $\delta = -23.5°$ (the sun being in the southern hemisphere), $h = 47° - 23.5° = 23.5°$, quite low in the sky. At midsummer's day, when $\delta = +23.5°$, $h = 47° + 23.5° = 70.5°$, not fully vertical but still very high.

Most solar collectors are fixed flat plates that cannot be changed to remain perpendicular to the noonday solar beam through its seasonal changes; otherwise the tilt would be 66° from the horizontal in December and 20° from it in June. If we split the difference and tilt the collector south at an angle equal to the latitude, it will collect 12.2 MJ m^{-2}* on an average summer day (Duffie and Beckman, 1976). This compromise tilt improves on a horizontal collector in winter (6.5 MJ m^{-2}) and receives less (18.0 MJ m^{-2}) in summer when the space-heating demand is less.

* 1 kW h = $10^3 \times 60 \times 60$ J = 3.6 MJ.

Length of Day

Figure 2 also shows the number of hours the sun path lies above the horizon: 9 hr in winter, 15 hr in summer. The length of the light and dark periods with their respective roles in physiological processes and the changing ratio of dark to light are major direct and indirect factors in the energy budget of an ecosystem. At Christmas (northern hemisphere) the days have ceased getting shorter. By Easter they are longer than the nights, and light has definitely overcome darkness (Table II). Day length provides an informational energy pulse that helps the internal cycles of organisms stay in phase with each other and anticipate cyclic changes in the environment that lag after changes in the solar beam. This is the circadian rhythm.

Horizontal planes on the earth poleward of the tropics never experience the vertical sun, but daily exposure to the slanting beam in summer lengthens poleward as the noon altitude diminishes. The net result is that solar radiation totaled through a summer day is nearly the same at all latitudes in the summer hemisphere. In winter, on the other hand, days shorten, and sun altitude diminishes with increasing latitude; the two trends reinforce one another, and the daily pulse of solar energy is much less at a high latitude than a low. The latitude of an ecosystem is more important in winter.

INCIDENCE OF THE SOLAR BEAM ON ECOSYSTEMS UNDER THE CHANGEABLE ATMOSPHERE

The altitude h of the sun determines not only how direct is the impact of the energy flux in the solar beam on a leaf or ecosystem, but also how long a path the beam must travel through the often turbid atmosphere before it reaches the ecosystem. In contrast to the precise trigonometric

TABLE II[a]

	Sunrise[b]	Sunset[b]
15 December	0716	1618
15 March	0604	1758
15 June	0412	1932 (15.3 hr length)
15 September	0531	1802
15 December	0716	1618

[a] Milwaukee (43°N latitude).
[b] Central standard time.

relation governing solar altitude, there is little precision in the attenuation of the beam.

The solar beam is depleted exponentially along its path; the longer
the path the greater the depletion, but the amounts are highly dependent on photon energies, i.e., wavelengths, and the composition of the
air. The atmosphere is most transparent in the wavelength range to
which our eyes are sensitized, and it takes an effort of the imagination
to visualize it as opaque or gray, yet for many other wavelengths it is.
Many photons of the highest intensity are absorbed in photochemical
reactions in the high atmosphere. Others are absorbed in heating molecules of oxygen and water vapor without chemical change, and others
are refracted by air molecules or cloud droplets. Dust takes a further
toll.

Path Length through the Overlying Atmosphere

The basic expression of attenuation of a beam with initial intensity
$E_{\lambda,\infty}$ at a given wavelength is Beer's law:

$$E_\lambda/E_{\lambda,\infty} = \exp\left(-\csc h \int_0^\infty k_\lambda \rho \, dz\right),$$

in which E_λ is the fraction reaching the surface at $z = 0$ (where z is the
geometric height), k_λ the coefficient of absorption and scattering, ρ air
density, and $\csc h$ expresses the approximate path length as a function
of solar altitude.

Many forms of depletion, expressed by k_λ, are amenable to mathematical formulation but in expressions difficult to work out numerically,
and others are yet little understood. The interactions of photons of different energy contents with the many kinds of gas molecules, particles,
droplets, and clouds of particles and droplets in the earth's often murky
atmosphere are highly complicated, but all deplete the beam. Absorbed
photons go no farther, but some that are scattered out of the beam reach
underlying ecosystems from area sources, the sky and cloud bases, and
will be discussed in the next chapter.

Water droplets scatter the solar beam, but clouds of droplets have a
more complicated effect that includes absorption. They weaken the
beam and frequently block it, so that the disk of the sun is not visible
and beam radiation vanishes from the environment of an ecosystem.

Path length through the atmosphere depends on sun altitude, and
Beer's law can be generalized by a transmission coefficient τ taken to
the exponent m representing path length. The path of the ray from ze-

nithal sun to an ecosystem near sea level is taken as $m = 1$,* i.e., a unit or minimal depth of atmosphere. (A value less than 1 is possible at mountain sites.) As the sun altitude h increases, m increases with csc h, down to low altitudes. At sun altitudes previously mentioned, m is as follows for ecosystems near sea level:

90°	1.00	45°	1.4
70°	1.06	20°	2.9.

Over longer paths more opportunity exists for scattering and absorption to select certain wavelengths, in general reddening the beam. Photographers like the "long interesting shadows" and the "warmer golden-orange light" of low sun (Sealfon, 1977).

Beam flux density is equal to an exponential diminution of the extra-terrestrial flux density (taken here as 1375 W m^{-2}):

$$\text{incident flux density} = 1375\ \tau^m,$$

in which τ is a generalized transmission coefficient for all solar wavelengths and is about 0.8 in clean air and 0.7 in the average. Under an atmosphere of average transmission characteristics, the flux density incident on an ecosystem growing on a slope perpendicular to the beam is $(1375)(0.7)^{1.06} = 940$ W m^{-2} on a summer noon at the 45th parallel when sun altitude is 70° and $m = 1.06$. When the sun is low and the path is long, a change in atmospheric transparency has a large effect on the flux incident on an ecosystem.

Scattering and Absorption in Clean Air

Although a "clean" atmosphere is harder and harder to find in view of the broadening scale of atmospheric pollution, it can serve as a starting point. Rayleigh scattering, most marked in the blue and ultraviolet wavelengths, causes the beam to be depleted in a clean dry atmosphere as shown in Table III.

The next step is to consider an atmosphere with absorptivity by ozone, oxygen, and water vapor. If the vapor content is 10 kg in an atmospheric column of 1-m^2 cross section, absorption further depletes the solar beam as shown in Table IV. The remnant beam when sun altitude is 90° is therefore 1.0 less 0.09 scattered and 0.09 absorbed, for an input of 0.82 of the extraterrestrial flux. Ecosystems at sea level receive $(0.82)(1375)$ W m^{-2}, or 1128 W m^{-2} through clean, relatively dry air.

Absorption by atmospheric gases other than ozone averages, over the

---

* "m" is sometimes called "air mass," but this usage is confusing because the term has had an entirely different meaning in synoptic meteorology.

TABLE III

Depletion of the Direct Beam by Scattering in a Clean Dry Atmosphere[a]

Sun altitude	m	Depletion of beam
90°	1.0	0.09
70°	1.06	0.10
45°	1.4	0.12
20°	2.9	0.22

[a] Data from Schulze (1970, pp. 53, 55).

year about 2 W m^{-2} by CO_2, 4 W m^{-2} by oxygen (distributed through the depth of the atmosphere), and 80 W m^{-2} by water vapor (mostly in the lower layers of the atmosphere) (Robinson, 1966).

Flux Density under Atmospheric Aerosols

The term "aerosols" takes in a wide range of kinds, shapes, sizes, and concentrations of droplets and particles in the air, and these attributes control their interactions with photons in the solar beam. "The distribution and properties of aerosols are highly variable and not well known" (Kondratyev, 1972, p. xii). Empirical indexes of attenuation are obtained by measuring beam intensity at the surface and comparing it with what it would be under clean dry air, taking into account the path length m. In the early days of air-mass analysis in synoptic meteorology, when observations aloft were few, it was felt that an index of turbidity, representing the effects of water vapor and aerosols, would identify airstream sources, and a contemporary book on synoptic meteorology (Chromow, 1940) presents the data of Table V.

Aerosol and gas absorption, added to the depletion caused by scattering, removes fractions at different indexes of turbidity as shown in

TABLE IV

Depletion of the Direct Beam by Absorption in a Clean Atmosphere

Sun altitude	m	Absorption by O_2 and O_3	Absorption by water vapor
90°	1.0	0.02	0.07
70°	1.06	0.02	0.07
45°	1.41	0.02	0.08
20°	2.9	0.03	0.10

TABLE V

Linke Turbidity Index under Different Midlatitude Air Streams[a]

	Moscow	Karadag (Crimea)	Washington, D. C.	Madison, Wisconsin
Arctic and maritime polar air	2.5	2.6	2.6	2.2
Continental polar air	3.1	2.7	2.8	2.6
Continental tropical air	3.5	3.1	3.7	3.5

[a] Data from Chromow (1940, p. 225).

Table VI. At low sun altitudes, more than half the initial flux density is lost in transit.

In general, depletion of the whole solar beam is not excessive, otherwise we could not see the planets or the sun itself, but turbidity is a hindrance that meteorologists could overcome only in part by working on high mountains. Langley (1884), in reporting results of an arduous set of solar measurements at Mt. Whitney, stated that in attempting to measure extraterrestrial solar energy "we are as though at the bottom of a turbid and agitated sea, and trying thence to obtain an idea of what goes on in an upper region of light and calm." Radiation modeling found that a fivefold increase in aerosols "blocks out beam radiation completely" (Lettau and Lettau, 1969), and the sun would become invisible in a cloudless sky! Visible radiation as a fraction of total solar is a function of turbidity (McCartney and Unsworth, 1978). Depletion on heavily polluted days in Milwaukee was greater in morning haze, but on days of moderate pollution greater in the afternoon when solar heating mixed particles through a deep layer (Bridgman, 1980).

Summer aerosol turbidity over the U.S. is double that of winter in airstreams from the same source regions. High turbidity is associated

TABLE VI

Depletion of the Direct Beam in Turbid Atmospheres[a]

Sun altitude	Linke turbidity index		
	1.90	2.75	3.75
90°	0.17	0.23	0.31
70°	0.18	0.24	0.32
45°	0.22	0.29	0.39
20°	0.37	0.49	0.60

[a] Data from Schulze (1970, p. 55).

with anticyclonic stagnation (frequent in the southern Appalachians) and maritime tropical air in summer (which in addition carries high vapor concentrations). Low values occur after front passages, more because of arrival of fresh polar air than because of any supposed cleansing effect of rain.

Interruption of the Solar Beam by Clouds

A thick cloud scatters (and also absorbs) the solar beam to the vanishing point (Fig. 3). The optical thickness depends on the quantity of liquid water in the cloud droplets and how finely divided the water is, i.e., the surface/volume ratio. For example, the incident beam is reduced to 0.1 by a 90-m path through stratus and altocumulus clouds (Coulson, 1975, p. 53). It is reduced to 0.01 of its incident strength after only 70 m in fairweather cumulus and 35 m in *Cumulus congestus*. In clouds of the planetary boundary layer, liquid-water content seems to be the principal factor (Stephens *et al.*, 1978). Terrain-bound clouds have a systematic influence on the beam, as do clouds that form in a diurnal regime.

The duration of beam radiation at a site is not hard to measure, the

Fig. III.3. Even small clouds can interrupt the direct beam. Ecosystems in the Australian interior are shaded in this picture.

most ingenious device being a burning-glass sphere that concentrates the beam to burn a spot on a card; as the sun moves, a burned line is traced across the card, the length of which measures the hours of sunshine, though not the energy flux density.

Duration of sunshine in the conterminous United States ranges from a mean of 0.41 in December in the Northeast to a mean of 0.87 in August in the Southwest. In any one region the difference between extensive sheets of stratiform clouds in winter storms and smaller (but deeper) convective clouds in summer is about 0.2 (Angell and Korshover, 1975). Long-term variability is due chiefly to variability in cloudiness (Angell and Korshover, 1978). Duration reached maxima in the Great Plains and in the analogous zone in the Soviet Union in the mid-1930s (Berliand, 1976). The variance of annual direct-beam mean flux densities reflects these differences in duration and also reflects variations in atmospheric transparency.

Periods of low input occurred around 1915 and 1965 at Leningrad and since about 1950 at Iakutsk (Berliand *et al.*, 1972). Low years run 0.05 below the average.* From the standpoint of ecosystem energy conversions, these are years when solar radiation is less in quantity, has a more isotropic, less directional geometric character, and is somewhat richer in the high-frequency end of the spectrum.

INCIDENCE OF THE SOLAR BEAM ON ECOSYSTEMS OF DIFFERENT ATTITUDES

The receiving surfaces of ecosystems lie at many angles of orientation relative to the direction of the solar beam. It is convenient to class these as (1) perpendicular (or "normal") to the direction of the beam, which presupposes either a transient state or a heliotropic plant that tracks the ever moving sun; (2) horizontal surfaces on which the beam flux density is multiplied by the sine of the solar altitude; (3) vertical walls; and (4) sloping surfaces. On these surfaces the incident flux density may be more or less than that on a horizontal surface, depending on the angle at which the slope faces the sun.

Normal-Incidence Solar Radiation

Although attenuated over a long path, the low sun in the early morning delivers a strong energy flux to walls, animals, and vertical leaves

---

* Part of this loss is recovered as diffuse radiation; the net decline in total solar radiation is 0.14 of the beam variation in low latitudes and 0.20–0.24 of it in high (Budyko, 1974, p. 298).

that receive its full impact. At 0600 the solar beam on an average day in June delivers 320 W m^{-2} to a normal surface at Karadag (45°N), four times the flux on a horizontal surface. The curve rises sharply to more than 500 W m^{-2}, where it holds for several hours, then drops as sharply to zero at sunset. A sun-tracking surface adds the most energy when the sun is low in the sky at the beginning and end of the day. The difference at 45°N is substantial over the whole year: 172 W m^{-2} average flux against 100. It is still larger at a higher latitude: 114 against 55 W m^{-2} at Pavlovsk (latitude 60°N) (Kondratyev, 1965, p. 310).

Solar-energy installations that concentrate radiation in order to obtain high temperatures must use the direct beam. These installations include solar furnaces, in which the incoming beam is concentrated several hundredfold,* and devices for raising steam or obtaining high-temperature heat (100–400°C), which operate at smaller concentrations, often with single-axis collectors. Pointing errors are important but can be reduced to 1 min of arc (Hughes, 1980).

Basking. The direct beam can provide valuable heat to organisms, even in winter, if they present the right angle to it (Fig. 4). Sun scald of tree trunks results from solar heating of the sun-facing side. Peccaries in Arizona reverse the nocturnal heat loss of 155 W m^{-2} and gain 23 W m^{-2} during the day (Zervanos and Hadley, 1973). Lizards that bask on sunny days can remain active all winter (Congdon *et al.*, 1979). Insects in the summer Arctic and at high altitudes in the midlatitudes make use of the direct beam, especially as focused by the Arctic poppy mentioned earlier. This symbiosis is an ingenious conversion of solar energy to warm an insect enough for its wing muscles to function so it can fly with the pollen it has picked up. A mutual dependence of plant and insect that has evolved over the ages is powered by the direct beam.

Incidence of the Direct Beam on Horizontal Surfaces

Ecosystems that present an extensive horizontal surface to the sky and sun, perhaps a wheat field in Kansas, experience the vertical component (sin h) of the solar beam. The dominating influence of solar altitude on both the sine effect and path length is displayed in empirical relations with the flux density of the vertical component of the solar beam. Approximate values of the clear-weather energy flux at several sun altitudes at Russian stations are shown in Table VII (Berliand, 1965).

---

* A solar furnace of moderate size at the U.S. Natick Laboratories in Massachusetts produces an energy flux density of 4×10^6 W m^{-2} (Davies and Cotton, 1957), which comes within a factor of 15 of the initial flux density at the sun.

Fig. III.4. Birds in cold climates locate the spots where convection is weak and solar energy absorbed in their dark plumage is retained. Air temperature $-10°C$ on a sunny January day in Milwaukee. At right angles to the solar beam dark plumage is most effective (Lustick *et al.*, 1980).

TABLE VII

Relation of Direct Solar Radiation on a Horizontal Surface to the Height of the Sun[a]

Sun altitude (deg)	Direct-beam solar radiation on a horizontal surface (W m^{-2})	
	Under an atmosphere of average transparency	Extraterrestrial
0	0	0
10	90	246
20	250	465
30	400	780
40	520	875
50	660	1042
60	760	1180
70	830	1265

[a] Data from Berliand (1965).

TABLE VIII

Boundary-Layer Stability at
Different Sun Altitudes[a]

Sun altitude (deg)	Pasquill's stability class
night	-2
0–15	0
15–35	1
35–60	2
>60	3

[a] From Luna and Church (1972).

Associations of solar altitude with direct-beam input indicate absorption of radiant energy and approximate the generation of sensible heat and the destabilization of the lower air. This chain of relations hangs together well enough to provide a rough approximation of the stability classes used in calculating the dispersion of pollutants. Large values of Pasquill's index (see Table VIII) denote instability that grows as the day progresses. Estimation can be further simplified by approximating sun altitude by shadow length (Lavdas, 1976). Formation of clouds in a diurnal regime that follows the diurnal regime of evapotranspiration (Weischet, 1980) and the upward mixing of particles (Bridgman, 1980) tend to modify the geometrical simplicity of the march of solar altitude and reduce afternoon flux.

Daily Totals.   The total flux received during a day, represented by the area under a daily curve, has two dimensions—the midday peak and the day length. A winter day is short, and its peak flux is weak compared with that of a summer day. The total energy input varies with the prod-

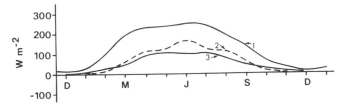

Fig. III.5.   Flux density ($W\ m^{-2}$) of solar radiation through the year at Iakutsk (62°N latitude) (means 1954–1959). (1) Direct solar beam on a surface perpendicular to it; (2) vertical component of the direct beam; (3) diffuse solar radiation, also on a horizontal surface. (Data from Barashkova *et al.*, 1961, p. 447).

uct of these two ratios: $\frac{1}{3}$ the day length at latitude 60° times $\frac{1}{10}$ the peak intensity, or about $\frac{1}{30}$.

The factors of solar altitude, day length, and atmospheric clarity join together to shape the annual regime of daily totals of the vertical component. Figure 5 shows three fluxes through the year at Iakutsk, Siberia (62°N), a high-latitude station with a large seasonal variation: direct beam at normal incidence (1) and its vertical component (2). It also shows diffuse solar radiation (3) scattered downward out of the solar beam (to be discussed in Chapter IV). The range in the normal-incidence beam (curve 1) from December to June is from 13 W m^{-2} to 250, twenty times as much. The vertical component of the direct beam is a little greater than the input of diffuse radiation (curve 3).

Direct-Beam Flux on Sloping Ecosystems

The orientation (or aspect) of a sloping ecosystem, measured by its azimuth angle, and the changing azimuth of the sun as it journeys across the sky, along with the pair of elevation angles (the altitude of the sun and the steepness of the slope) are members of an equation for solar flux on the ecosystem (Robinson, 1966, p. 41). The simplified equation sets A as the difference between the azimuth angles of the sun and the slope, and steepness as E.

$$\cos \beta = \cos E \sin h + \sin E \cos h \cos A,$$

in which $\cos \beta$ is the function converting the flux in the solar beam to the flux normal to the slope in question. (β is the angle between the direction of the sun and a line normal to the slope; for a horizontal surface it corresponds to the zenith angle of the sun.)

This formula is laborious to use, and many workers have compiled graphs and tables of its solution for many slopes, locations, and times of year. An algorithm to calculate daily totals (Swift, 1976) covers more cases and requires as inputs only the Julian date, latitude, and slope steepness and orientation.

Shading from the direct beam by neighboring slopes also must be taken into consideration in deeply dissected country. East–west valley floors in the Alps below 20° slopes receive 0.46 of the radiation incident on a freely exposed flat surface, but north–south valley floors receive 0.89 (Dirmhirn, 1964, p. 193). In valleys that lie east–west the sun-facing slope (called *adret* in the Alps) has a suite of ecosystems distinct from the opposite slope (or *ubac*). In north–south valleys the early morning heating of the east–facing slopes gives biological and geophysical processes an earlier timing than that on the west-facing slopes, and this timing often affects the circulation of the mountain-valley breeze system. Lee and Baumgartner (1966) mapped the solar climate at 22,000 grid points over 350 km² of mountains in Germany and found that over about half the area solar inputs ranged from 0.95 to 1.05 of that to an unshaded horizontal ecosystem; 2% of the area received more than 1.20 of the input to a horizontal surface. South-facing ecosystems in a mountain basin received 0.05 more beam radiation than a meadow during the snow-melting season in the Sierra Nevada (latitude 39°N, elevation 2.0 km), and those facing north received up to 0.10 less (Fig. 6). The resulting difference of three weeks in their melting regimes prolongs the period of meltwater yield from the whole basin. Similarly, east-facing ecosystems begin and end their daily pulses of meltwater generation 2–3 hr earlier than west-facing ecosystems; their juxtaposition prolongs the diurnal meltwater wave in the stream that drains both slopes.

SPATIAL PATTERNS AND ECOLOGICAL CONSEQUENCES

Microscales, Local, and Mesoscale Patterns

At microscales the interplay of light and shadow in the foliage is an environmental condition that the photosynthesis apparatus readily deals with; gardeners know the value of dappled light for certain species. Leaves in more continuous shade deep in the foliage tend to have structures that accommodate the special energy input. The moving sun flecks

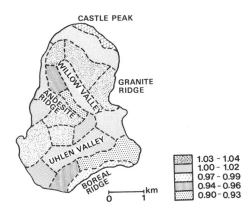

Fig. III.6. Relative direct-beam flux on topographic facets of the basin of the Central Sierra Snow Laboratory as of 1 May (flux on horizontal surface = 1.00). Statistically significant differences were found between facets for ablation of the spring snow cover.

beneath forest canopy bring bursts of direct-beam flux to forest-floor vegetation.

At a larger scale, a man in the mountains under a 1000 W m^{-2} solar beam* and a pedestrian on a hot day in the city under a 700 W m^{-2} beam soon learn the rudiments of the local pattern of the direct flux. Handel's wish that ". . . where'er you walk/ Trees shall crowd into a shade" refers to strong summer sunshine; in contrast, almost without thinking, we seek the sunny side of the street in winter.

Mesoscale patterns of beam energy, discussed earlier as slope effects, have ecological consequences in stands of Western white pine on 20% slopes in Northern Idaho (noon sun altitude in summer 60–65°):

(1) The *site* index is greatest on northeast slopes, being 64 at azimuth 46° compared to 61 at azimuth 226°. (The mean daily values of the component of the solar beam perpendicular to these two slopes differ by about 40 W m^{-2} in summer.)

(2) The basal-area growth of *individual* pines, in contrast, is optimum on south slopes (azimuth 201°) (Stage, 1976), an unexpected fact confirmed by analyzing slope aspect, steepness, and habitat effects in other common tree species.

Urban dust domes reduce solar flux density by 10–20%, and more in the ultraviolet, and were one of the first biological costs of the Industrial Revolution. Comparison of visible-range beam radiation in central Milwaukee with that at the field station of the University of Wisconsin-Milwaukee 50 km north showed 0.24 lower flux density in the city. Greatest depletion was in the blue, where it was 0.35 (Bridgman, 1978, 1980).

Larger-Scale Patterns

Persistent clouds cut off the solar beam in many coastal lands and mountains. Advected coastal fogs and stratus reduce solar impact for 20 km or farther inland in central California; in northern Japan they handicap rice growing. A rapid decrease with distance of a spatial correlation of sunshine duration is found at stations in the southeastern and southwestern United States and is ascribed to different solar regimes at coastal and inland places (Angell and Korshover, 1975).

Mountain ranges often support different cloudiness regimes than do lowlands and produce spatial differences in the direct-beam radiation

---

* Strong sun in the Andes is uncomfortable because of the radiation gradient between the sun-exposed and the shaded parts of the body (Prohaska, 1970). A similar problem in balancing the temperature in an office building is caused by the directional solar beam.

TABLE IX

Mean Annual Values of Vertical Component of Direct-Beam Solar Radiation at Stations near the 60th Parallel North[a]

Station	Longitude	Mean annual vertical component of solar beam	Remarks
Lerwick	1°W	40	Atlantic storms
Bergen	5°E	42	West coast location; a hearth of wave cyclone investigation
Helsinki	25°E	58	
Leningrad	30°E	45	Urban site
Voeikovo	30°E	50	Rural site east of Leningrad
Ivdel'	60°E	54	
Sytomino	62°E	56	
Vanavara	102°E	55	
Iakutsk	128°E	68	Continental interior
Okhotsk	143°E	55	Foggy coast of the Pacific

[a] Data from Berliand (1964). Data given in watts per square meter.

received by ecosystems. Convective clouds occur from time to time in the summer in the Sierra Nevada but almost never over the Central Valley. Conversely, frequent anticyclonic circulation aloft in winter favors radiation fog several hundred meters thick that keeps valley ecosystems in deep gloom for days or even weeks at a time; meanwhile, the ridges bask in intense sunshine (Miller, 1977, p. 356). Terrain-bound clouds and fog show up plainly in satellite imagery, e.g., in the Rhinegraben and Münster embayment (Wiessner and Fezer, 1979).

The 60th parallel of north latitude (which coincidentally was the locale

TABLE X

Duration of Photosynthetically Active Beam Flux at Flux Densities Exceeding 300 W m[-2] at Different Latitudes[a]

Latitude	21 December	21 June
0°	9.0	9.0
20°N	7.0	9.8
40°N	3.0	10.5
60°N	0	11.0

[a] Data from Shul'gin (1967, p. 96). Data given in hours.

of much of the pioneering research on solar radiation) traverses several cloudiness regions, which cause substantial differences in the vertical component of the solar beam (Table IX).

Considering only the photosynthetically active fraction of the direct beam at the winter and summer solstices, different latitudes can be compared in terms of hours of flux densities exceeding 300 W m^{-2} (Table X). The poleward decrease in winter is larger than the poleward increase in summer when the whole hemisphere is well supplied with solar energy.

However, this season of high solar-beam input in the north lasts only a short time; as the sun turns back southward, days begin to shorten. High-latitude ecosystems are geared to a season of strong solar energy that lasts only a brief time. The latitude of an ecosystem expresses itself not in the June peak flow of energy, but rather in the shortness of the growing season.

We have spoken principally of the changes in direct-beam flux through the day and year and from place to place. These are a central fact of life in the environment of ecosystems and often the most variable quantity in their energy budgets.

REFERENCES

Angell, J. K., and Korshover, J. (1975). Variation in sunshine duration over the contiguous United States between 1950 and 1972. *J. Appl. Meteorol.* **14**, 1174–1181.

Angell, J. K., and Korshover, J. (1978). A recent increase in sunshine duration within the contiguous United States. *J. Appl. Meteorol.* **17**, 819–824.

Barashkova, E. P., Gaevskii, V. L., D'iachenko, L. N., Lugina, K. M., and Pivovarova, Z. I. (1961). "Radiatsionnyi Rezhim Territorii SSSR," Gidrometeorol. Izd., Leningrad.

Berliand, T. G., ed. (1964). "Aktinometricheskii Spravochnik Zarubezhnye Strany Ezhegodnye Dannye." Gidrometeorol. Izd., Leningrad.

Berliand, T. G. (1965). Sutochnyi khod solnechnoi radiatsii v osnovnykh klimaticheskikh zonakh zemnogo shara. *Tr. Gl. Geofiz. Obs.* **179**, 3–27.

Berliand, T. G. (1976). O vekovom khode prodolzhitel'nosti solnechnogo na territorii Velikikh Ravnin S SH A. *Tr. Gl. Geofiz. Obs.* **365**, 29–35.

Berliand, T. G., Rusin, N. P., Efimova, N. A., Zubenok, L. I., Mukhenberg, V. V., Ogneva, T. A., Pivovarova, S. I., and Strokina, L. A. (1972). Issledovanie radiatsionnogo rezhima i teplovogo balansa zemnogo shara. *Tr. Vses. Meteorol. S"ezda, V* **3**, 57–77.

Bridgman, H. A. (1978). Direct visible spectra and aerosol optical depths at urban and rural locations during the summer of 1975 at Milwaukee. *Sol. Energy* **21**, 139–148.

Bridgman, H. A. (1980). Diurnal variations in the spectrum of direct beam visible radiation at urban and rural locations at Milwaukee. *Arch. Meteorol., Geophys. Bioklimatol., Ser. B* **28**, 101–113.

Brown, J. M. (1973). Tables and conversions for micrometeorology. U.S. Forest Serv., Gen. Tech. Rep. **NC-8**.

Budyko, M. I. (1974). "Climate and Life" (Engl. transl. ed. by D. H. Miller). Academic Press, New York.

Chromow [Khromov], S. P. (1940). "Einführung in die synoptische Wetteranalyse." Springer-Verlag, Vienna.

Congdon, J. D., Ballinger, R. E., and Nagy, K. A. (1979). Energetics, temperature and water relations in winter aggregated *Sceloporus jarrovi* (Sauria: Iguanidae). *Ecology* **60**, 30–35.

Corbet, P. S. (1969). Terrestrial microclimate: Amelioration at high latitudes. *Science* **166**, 865–866.

Coulson, K. L. (1975). "Solar and Terrestrial Radiation. Methods and Measurements." Academic Press, New York.

Davies, J. M., and Cotton, E. S. (1957). Design of the Quartermaster solar furnace. *Sol. Energy* **1**(23), 16–22.

Dirmhirn, I. (1964). "Das Strahlungsfeld im Lebensraum." Akad. Verlagsges., Frankfurt a.M.

Duffie, J. A., and Beckman, W. A. (1976). Solar heating and cooling. *Science* **191**, 143–149.

Feygel'son, Ye. M. (1977). Chronicle, International symposium on radiation (Garmisch–Partenkirchen, August 19–28, 1976). *Izv. Atmos. Ocean. Phys.* **13**, 239–241.

Garnett, A. (1935). Insolation, topography and settlement in the Alps. *Geogr. Rev.* **25**, 601–617.

Gould, J. L. (1980). Sun compensation by bees. *Science* **207**, 545–547.

Grum, F., and Bechner, R. J. (1979). "Optical Radiation Measurements, Vol. 1, Radiometry." Academic Press, New York.

Hickey, J. R., Stowe, L. L., Jacobowitz, H., Pellegrino, P., Maschhoff, R. H., House, F., and VonderHaar, T. H. (1980). Initial solar irradiance determinations from Nimbus 7 cavity radiometer measurements. *Science* **208**, 281–283.

Hughes, R. O. (1980). Effects of tracking errors on the performance of point focusing solar collectors. *Sol. Energy* **24**, 83–92.

Keran, P. G. (1975). Sun-tracking solar furnaces in High Arctic flowers: significance for pollination and insects. *Science* **189**, 723–726.

Kondratyev, K. Ya. (1965). "Aktinometriia." Gidrometeorol. Izd., Leningrad.

Kondratyev, K. Ya. (1969). "Radiation in the Atmosphere." Academic Press, New York.

Kondratyev, K. Ya. (1972). Radiation in the atmosphere. *W.M.O.* WMO No. 309. Geneva

Langley, S. P. (1884). Researches on solar heat and its absorption by the earth's atmosphere. A report of the Mount Whitney expedition. *U.S. War Dep., Signal Serv., Prof. Pap.* No. 15.

Lavdas, L. G. (1976). A groundhog's approach to estimating insolation. *J. Air Poll. Cont. Assoc.* **26**, 794.

Lee, R., and Baumgartner, A. (1966). The topography and insolation climate of a mountainous forest area. *For. Sci.* **12**, 258–267.

Lettau, H., and Lettau, K. (1969). Shortwave radiation climatonomy. *Tellus* **21**, 208–222.

Luna, R. E., and Church, H. W. (1972). A comparison of turbulence intensity and stability ratio measurements to Pasquill stability classes. *J. Appl. Meteorol.* **11**, 663–669.

Lustick, S., Adam, M., and Hinko, A. (1980). Interaction between posture, color, and the radiative heat load in birds. *Science* **208**, 1052–1053.

McCartney, H. A., and Unsworth, M. H. (1978). Spectral distribution of solar radiation. I: Direct radiation. *Q. J. R. Meteorol. Soc.* **104**, 699–718.

Metz, W. D. (1976). An illuminating new use for solar energy. *Science* **194**, 1404.

Miller, D. H. (1977). "Water at the Surface of the Earth." Academic Press, New York.

Milthorpe, F. L., and Moorby, J. (1979). "An Introduction to Crop Physiology." 2nd ed., Cambridge Univ. Press, London and New York.

Planck, M. (1901). Ueber das Gesetz der Energieverteilung im Normalspectrum. *Ann. Phys., Ser.* 4 **4**, 553–563.

Prohaska, F. (1970). Distinctive bioclimatic parameters of the subtropical–tropical Andes. *Int. J. Biometeorol.* **14**, 1–12.

Robinson, N., ed. (1966). "Solar Radiation." Am. Elsevier, New York.

Satterlund, D. R. (1977). Shadow patterns located with a programmable calculator. *J. For.* **70**, 262–263.

Schulze, R. (1970). "Strahlenklima der Erde." Steinkopff, Darmstadt.

Sealfon, P. (1977). Capturing the essence of natural sunlight. *New York Times,* July 10, Sect. D, p. 22.

Shul'gin, I. A. (1967). "Solnechnaia Radiatsiia i Rastenie." Gidrometeorol. Izd., Leningrad.

Silberglied, R. E. (1979). Communication in the ultraviolet. *Ann. Rev. Ecol. Syst.* **10**, 373–398.

Stage, A. R. (1976). An expression for the effect of aspect, slope, and habitat type on tree growth. *For. Sci.* **22**, 457–460.

Stephens, G. L., Paltridge, G. W., and Platt, G. M. R. (1978). Radiation profiles in extended water clouds. III: Observations. *J. Atmos. Sci.* **35**, 2133–2141.

Swift, L. W., Jr. (1976). Algorithm for solar radiation on mountain slopes. *Water Resour. Res.* **12**, 108–112.

Weischet, W. (1980). Klimatologische Interpretation von METEOSAT-Aufnahmen. *Geogr. Rdschau.* **32**, 80–84.

Wiesner, K. and Fezer, F. (1979). Terrestrisch beeinflusste Wolkenformen auf Satellitenbildern Mitteleuropas. *Erdkunde* **33**, 316–328.

Zervanos, S. M., and Hadley, N. F. (1973). Adaptational biology and energy relationships of the collared peccary (*Tayassu tajacu*). *Ecology* **54**, 759–774.

Chapter IV

DIFFUSE SOLAR RADIATION

A large fraction of the solar energy scattered out of the direct beam as it passes through the atmosphere nevertheless reaches ecosystems at the earth's surface. The flux they receive tends to be stronger in the short wavelengths than is beam radiation, and they receive it from the whole dome of the sky. It causes no glare and casts no shadows and penetrates deep into ecosystem canopies. It remains for many hours daily at an intensity well-suited to photosynthetic energy conversion and reduces the abrupt change in radiant-energy loading that otherwise would occur when a moving line of shade crosses the photosynthetic apparatus in a leaf.

Scattering takes several forms. Radiation of short wavelengths is effectively scattered by air molecules. Particles in the atmosphere scatter differently than air molecules, and small particles scatter differently than large ones. Further differences result from the chemical nature of aerosols, which affects their refractive index. Droplets in the atmosphere have different sizes and refractive indexes and occur in concentrations up to the dense crowding of water droplets in clouds, which are effective scattering and reflecting agents. Most complex are the interactions between photons and the many sizes and shapes of aerosols in different layers of the atmosphere. Ecosystems under such a changeable atmosphere receive a variable energy flux, difficult to calculate but easy to measure by a shaded pyranometer (Fig. 1).

THE DIFFUSE FLUX GENERATED BY RAYLEIGH SCATTERING

Rayleigh scattering by air molecules is the dominant process in clean, dry air. At a sun height of 45° this mode of scattering delivers about 55 $W \, m^{-2}$ to the earth's surface, considerably smaller than the flux densities

Fig. IV.1. Low angle of the shade ring of a pyranometer on a midwinter day at latitude 56°N is shown by Dr. Christensen at the Højbakkegaard Experiment Station near Copenhagen, where diffuse solar radiation keeps photosynthesis going almost all winter.

under turbid or cloudy air. The Rayleigh formula for the volume coefficient of scattering is

$$\sigma_\lambda = 32\pi^3(n_\lambda - 1)^2/3N\lambda^4,$$

in which n_λ is the refractive index, N the number of particles per unit volume, and λ the wavelength (Kondratyev, 1969, p. 177). This coefficient is used in Beer's law to determine scattered flux density.

The coefficient of scattering of beam radiation in a unit volume of air is very small for wavelengths in the solar infrared portion of the spectrum—a tenth of the coefficient of scattering of red light and a hundredth of that of blue light. A secondary factor is the number of molecules per unit volume; the downward flux is much less at high altitudes—at 3 km it is about one-third less than near sea level (Flach, 1965).

The scattering of photons around a molecule is principally forward and backward, and about half of the flux scattered out of the solar beam's vertical component reaches the ground. Two parts of the sky generate large fluxes: the lower elevations near the horizon (where more air

molecules are present) and an area around the sun. An area opposite
the sun generates substantially less flux than other parts of the sky.
[Details of the instantaneous patterns may be found in Robinson (1966)
and Kondratyev (1969).] This lack of isotropy complicates the calculation of the diffuse flux incident on sloping ecosystems (Kondratyev,
1977, p. 18).

Brooks (1959, p. 47) suggested that about 0.45 of the flux received at
the surface could be considered as "localized in a large cone angle
around the sun" and treated geometrically like the solar beam. Its movement across the sky is like the sun's path discussed in Chapter III. This
basic pattern can still be discerned when haze process is added, but it
is obliterated by cloud scattering. The balance (0.55) of the diffuse flux
at the ground can be considered as derived at approximately uniform
intensity over the hemisphere. It illuminates north slopes and walls,
enters north windows, and bathes shade plants in an alternative flux
of energy.

At low sun angles the predominant flux of diffuse solar radiation,
even on a horizontal surface, comes from the lower bands of the sky.
As the sun moves higher, the upper bands of the sky begin to contribute
more, and when the sun is at 45°, the relative contribution from each
band of the sky is about the same as from a uniformly overcast sky
(Uchijima et al., 1976). This distribution pattern is employed to determine
the diffuse flux received by leaves at different inclinations in an ecosystem. The total flux, however, does not change much with sun height
after it gets higher than 10°; from a height of 20 to 70° the flux increases
only from 50 to 70 W m^{-2} (Sivkov, 1968, p. 138).

The diffuse flux is increased by reflection from the earth's surface. At
the usual surface absorptivity of 0.75 the diffuse flux at a solar altitude
of 37° is 63 W m^{-2}; with a new snow cover (0.30 absorptivity) it is
increased to 95 W m^{-2} (Sivkov, 1968, p. 81; Robinson, 1966).

Scattering by air molecules, depending largely on the inverse fourth
power of the wavelength of the incident radiation, is strongest in the
ultraviolet and visible parts of the solar spectrum. In fact, if radiation
of wavelength 0.2 μm were not absorbed by ozone aloft, it would be
almost completely removed from the solar beam by molecular scattering.
Strong attenuation affects radiation up to 0.29 μm in high mountains
and up to 0.31 μm in lowlands at 50° sun height.

Diffuse ultraviolet radiation can contribute up to three-quarters of the
total UV, explaining why it is easy to sunburn while in the shade. The
small heating effect of these wavelengths gives little warning of their
photochemical strength. The UV fraction also serves as a rule in the
communication convention of bees (Brines and Gould, 1979). Rayleigh

TABLE I

Flux Density of Diffuse Solar Radiation under Different Atmospheres[a]

Atmospheric condition	Sun Height		
	20°	45°	70°
Rayleigh ($T = 1$)	30	56	62
Rural atmosphere ($T = 2.75$)	70	95	100
$T = 4$ (summer)	90	120	130
"low transmission"	105	150	170

[a] Data from Dirmhirn (1964), Sivkov (1968), and Schulze (1970). (T is the Linke turbidity index.) Data given in watts per square meter.

scattering is important in the visible spectrum. Solar infrared wavelengths go largely unscattered and reach ecosystems with a mainly directional character.

DIFFUSE RADIATION UNDER A HAZY ATMOSPHERE

The relatively straightforward geometry of attenuation in Rayleigh scattering does not express what happens in a hazy atmosphere, in which photons of different energy content meet particles and droplets of all sizes, shapes, and refractive indices. The size of particles that are larger than air molecules reduces selectivity in the spectral patterns of the scattered light. For instance, an increase in salt particles turns the sky white, especially near the horizon. Furthermore, as aerosol concentration increases, shown perhaps by diminishing visual range, Rayleigh scattering contributes relatively less and less to the total result and is very weak at ranges as low as 5 km (Bullrich, 1964, p. 150).

Haze makes a unit volume of air a more effective scattering agent and substantially increases the flux density of diffuse radiation at the earth's surface, its influence being greater on the diffuse than on the beam flux (McCartney, 1978). Table I shows that flux density under haze in a cloudless sky increases to 100 W m^{-2} or more, compared with about 50 W m^{-2} under clean air. Kalitin (1943; also cited in Kondratyev, 1969, p. 379) found that at an altitude of 20° the diffuse flux increased from 90 W m^{-2} to 145 W m^{-2} as the vertical component of the beam decreased, due to aerosol attenuation, from 240 W m^{-2} to 165 W m^{-2}, a recovery of 55 of the 75 W m^{-2} lost. From a standpoint of photosynthesis, most of the loss in the direct beam in a turbid atmosphere is "recovered" within the diffuse radiation (McCree and Keener, 1974); changes in turbidity within the present range are not likely to reduce photosynthetic rates of crops more than 5%.

A gradual increase in atmospheric dust may be evident in increases in the mean cloudless-day diffuse flux at 16 cities in the USSR (at sun altitudes 40–45°), from 70 W m^{-2} in 1961–1962 to 90 in 1969–1970 (Berliand, 1975, p. 404) and has been discussed as a factor in climatic change.

Difference in flux density distinguishes the eastern from the western United States. Western deserts and mountains lie under clean, dry low-density air as long as man-made pollutants are kept under control. Sight distances are long, shadows are black, and the atmospheric environment has a directional dimension that is lacking in the more isotropic radiation field of the eastern humid lowlands. In conditions of very high turbidity urban atmospheres produce a flux that increases from 100 W m^{-2} at a sun height of 20° to 210 W m^{-2} at 70° (Meinel and Meinel, 1976).

DIFFUSE RADIATION UNDER A CLOUDY ATMOSPHERE

Clouds of small ice crystals or water droplets scatter beam radiation and, even when thin, become powerful sources of diffuse solar radiation. A wisp of cloud floating in a clear sky generates a strong diffuse flux, obvious in its whiteness and intensity. Single clouds on the side of a pyranometer station away from the sun can reflect so strongly as to drive the pen off the chart. Crest clouds, more distant but longer lasting, reflect on the instrument over several hours and produce a "padding" of the trace (Ives, 1946). Moving clouds near the sun transmit a large part of the solar beam if optically thin, scatter it if thicker, and increase hemispheric radiation by 0.05–0.15, or up to 15 W m^{-2} (Robinson, 1977). Cloud shape affects their scattering power, and the sometimes intricate topography of their top surfaces seems especially important (McKee and Klehr, 1978).

As cloud cover spreads over the sky, say from 0.3 to 0.7, the diffuse flux increases by about half, from 130 W m^{-2} to 200 (Barashkova et al., 1961, p. 61). The increase in cirrus deriving from contrails may have increased diffuse radiation in parts of the United States, but the lack of observations does not permit a clear answer.

When clouds thicken, the brightness of their bases diminishes. The bright edges of a small cumulus surround a gray center that marks a thicker part of the cloud. Little light emerges from the bottom of a deep cumulonimbus—sometimes less than 0.1 of the flux from a cloudless sky. Multiple scattering turns much of the radiation upward, and absorption also occurs and reduces the diffuse flux emerging from the cloud base. Some clouds might absorb 20–40%, but the "component responsible for this large absorption, and its variability, is mainly unknown" (American Meteorological Society, 1972).

The simple exteriors of clouds belie their real complexity, for clouds are populations of many sizes of droplets that interact variously with radiation, depending on relative wavelength and droplet size, which are in the same range of magnitude. Moreover, this population of droplets is continuously changing; some grow and others vanish. These changes go fast in the brief life of a convective cloud and spatial differences are large. These dynamic processes in cloud microphysics change the diffuse flux that reaches the ecosystem below. One relevant cloud dimension is the ratio h/L, in which h is the geometric depth and L a free-path length directly proportional to scattering radius of droplets and inversely proportional to the liquid-water mass per unit volume; both factors being liable to rapid fluctuation. An increase of h/L from 1 in thin clouds to 10 in thicker ones can result from a change in droplet population or water content, or from overall growth in depth, or from all three factors in combination. This change in h/L from 1 to 10 when the sun altitude is 45° reduces the relative transmission of energy to the ground from 0.85 to 0.30 (Robinson, 1966, p. 62).

These diverse effects of clouds on the diffuse flux at the underlying surface are usually integrated by cloud type, a classification that implicitly contains approximate information on water content, droplet size distribution, and depth. Table II shows such mean data for two sun altitudes and for half- and full-sky cover by three cloud types. A nonlinear effect is present; i.e., doubling the cloud cover does not double the flux density under it. A study at Davos, Switzerland, found maxi-

TABLE II

Diffuse Radiation from Different Types of Clouds[a]

	Solar altitude 20°		Solar altitude 45°	
	0.5 cover	Overcast	0.5 cover	Overcast
Cirrus	55	90	85	160
Altocumulus	75	100	130	220
Cumulonimbus	65	55[b]	120	140[b]
Fog	—	60	—	125

[a] Data from Kondratyev (1969, p. 383); Fog data from Haurwitz (1948, cited in Brooks, 1959, p. 54). Data given in watts per square meter.
[b] Even less under deep cumulonimbus.

mum flux when the fraction of sky covered was between 0.7 and 0.9, depending on type (Bener, 1963). The flux under an overcast sky is almost always less than under broken clouds; overcast conditions often indicate great cloud depth.

An increase in surface albedo increases the downward diffuse flux even more under clouds than under a cloudless sky because it increases the opportunities for multiple reflection. This effect is important in polar regions and, along with extensive cloud cover and low sun path, explains why the diffuse flux dominates the shortwave radiation input to high-latitude ecosystems.

Clouds, like haze, increase the effect of sun altitude on the diffuse flux. The flux increase per degree of increased sun altitude, instead of about 1 W m^{-2} under cloudless skies and some turbidity, becomes 3 W m^{-2} per degree of sun altitude under cirrus and cumulonimbus and 4 W m^{-2} or more under altocumulus [from data in Kondratyev (1969, p. 383)]. Generalized observational data from the extensive radiation network in the USSR (Table III) show that the diffusing effect of middle-level clouds is particularly strong at sun altitudes greater than about 40°.

SPECTRAL COMPOSITION OF THE DIFFUSE FLUX

Selective Scattering

Attenuation of the solar beam in the atmosphere is sensitive to wavelength, particularly those wavelengths that involve air molecules in Rayleigh scattering and small particles in Mie scattering. Large particle and

TABLE III

Diffuse Solar Radiation from Different Sky Conditions[a]

Sun altitude	Cloudless-Sky average	Through overcast sky		
		Cirrus	Middle	Low clouds
5°	35	35	25	12
10°	45	70	60	45
20°	70	125	115	90
30°	90	170	160	140
40°	100	210	230	180
50°	115	240	290	220
60°	125	250	335	250

[a] Data from Barashkova *et al.* (1961). Data given in watts per square meter.

water droplet scattering acts more uniformly on photons of all wavelengths. The spectral composition of the diffuse flux reaching an eco-system therefore depends on atmospheric composition, but usually involves a shift toward the blue end of the spectrum.

The wavelength of peak radiation intensity is shifted from 0.55 μm in the undepleted beam to approximately 0.45 in the diffuse flux, so that the diffuse flux under a cloudless sky is enriched in photosynthetically active radiation. At a wavelength of 0.7 μm the intensity of the diffuse flux is less than half its maximum intensity, and at 1.0 μm it is only 0.15 of the maximum (Schulze, 1970, p. 119).

The diffuse flux under clouds is depleted in the short wavelengths and relatively strong in wavelengths longer than 0.55 μm (Dirmhirn, 1964, p. 91). However, snow cover, with high reflectivity in wavelengths shorter than 1.0 μm, augments diffuse flux from clouds above it in the shorter visible wavelengths (Dirmhirn, 1964, p. 100, 130).

In general, the spectral composition of diffuse solar radiation is at least as valuable for photosynthesis in ecosystems as is the direct beam; in clear weather it is more valuable although inadequately observed (Grum and Becherer, 1979, p. 118). Forest meteorologists feel that the photo-synthetically active fraction of the diffuse flux is important enough to warrant separate measurement.

Illumination

The frequencies important in illumination and photosynthesis come close to coinciding, both lying in the 0.4–0.7 μm region. The energy flux in this region is not a uniform fraction of the total flux of shortwave radiative energy. However, Kondratyev points out that the general re-lationships of radiant-energy flux to sun height and cloud interception, the two principal influences, hold in general also for the 0.4–0.7 μm region.*

People sometimes like the amenity of direct-beam sunlight in a room, but too much of it brings shifting shadows, glare, and overheating, and it "cannot be relied on as a major source of illumination in buildings" (Hopkinson et al., 1966, p. 24). Skylight or diffuse solar radiation is more useful and can be maximized by building orientation, screens, and win-dow design.

---

* Illumination is measured in lux. 1 lux (lx) = 1 lumen (lm) m^{-2}; if 680 lm are taken as 1 W for typical relation of visible flux to total solar flux, then 680 lm m^{-2} = 680 lx = 1 W m^{-2} (Gates, 1965). For comparison, illumination from a full moon at 60° altitude is 0.7 lx, or 1 mW m^{-2}. This illumination, beautiful on a night ride in the country and useful in the Northeastern power blackout of 1965 (missing in 1977), is only one-twentieth of the flux from the sun 6° below the horizon.

VARIATIONS OVER TIME

Diffuse radiation reaches ecosystems even when the sun altitude is negative, that is, in the dawn and dusk periods. This diffuse flux runs about 500 lx at sunset or sunrise and 15 lx at a sun altitude of $-6°$, which is defined as the end of civil twilight, and signals time for many animals. A summer trip from Berkeley (38°N) to Whitehorse, Yukon (61°N), found that the morning song of robins began from 55 to 117 min before and ended from 13 to 32 min before actual sunrise (Miller, 1958). These lengths of time before sunrise increased steadily as the observer traveled poleward because the flatter angle of the sun's path at high latitudes brings it near the horizon for a longer time before it actually emerges; twilight is longer at high latitudes than at low.

Several nocturnal primates keep their internal clocks in time by noting a level of about 2 lx (or approximately 3 mW m^{-2}), which in Los Angeles occurs 12–24 min after sunset (Kavanau and Peters, 1976). Clouds, rain, and occasional extreme temperatures affected this relation less than other radiation fluxes; twilight is more stable from day to day than is illumination before sunset.

Diurnal Regime

The flat response of the diffuse flux density to changes in sun altitude beyond 10° or so, especially in clear weather, indicates a relatively steady flow of energy to ecosystems through the principal part of the day. The mean hourly fluxes at two stations in June (Table IV and Fig. 2) show this flat maximum. Even in December, the flux at Tbilisi remains within 0.15 of its peak intensity for a third of the daylight period. Figure 2 also shows diurnal marches at a station (Kinshasa) under equatorial cloudiness.

The daily pulse of energy in diffuse radiation is a function of day length, and the hourly coefficient where the midday sun altitude is 20° and snow lies on the ground is approximately 0.25 MJ m^{-2} (Barashkova

TABLE IV

Mean Hourly Values of Flux Density of Diffuse Shortwave Radiation in June[a]

	00	02	04	06	08	10	12	14	16	18	20	22	24	Mean
Tbilisi 42°N	0	0	0	65	155	210	215	205	145	60	0	0	0	88
Iakutsk 62°N	0	0	20	90	165	185	220	200	165	90	35	0	0	100

[a] Data for Iakutsk from Barashkova *et al.* (1961); Tbilisi data from Tsutskiridze (1967, p. 53). Data given in watts per square meter.

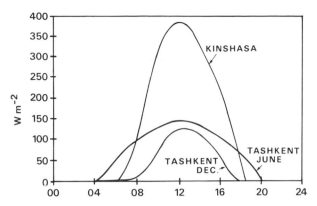

Fig. IV.2. Diurnal marches of diffuse solar radiation at Kinshasa (4°S latitude) in December and at Tashkent (41°N latitude) in December and June (W m^{-2}). [Data from Berliand (1965)].

et al., 1961, p. 65). At a place without snow (Milwaukee at the equinox) where cloudiness exceeds 0.6 and the midday sun height is 45°, the ratio is 0.45 MJ m^{-2}, and the energy pulse in a 14-hr day totals 6.3 MJ m^{-2}. Under the clouds of the intertropical convergence zone at Kinshasa the pulse is 10.3 MJ m^{-2}. Daily pulses of diffuse radiation exceed 20 MJ m^{-2} at Arctic stations (Berliand, 1965). Aperiodic changes in this energy pulse occur when cloud systems of varying density move over a place and bring a shift in spectral composition of the radiant-energy input.

Annual Regimes

Flux densities of diffuse solar radiation in summer average 80–100 W m^{-2} over a 24-hr day, except that in mountains and deserts they are smaller. Those in winter are quite small, as received on horizontal surfaces (Table V). Solar collectors that are tilted toward the equator intercept at a more favorable angle the part of the diffuse flux that comes from the region of the sky near the sun.

Changes in diffuse radiation from winter to summer and from year to year are illustrated at the Blue Hill Observatory of Harvard University (latitude 42°N, altitude 195 m) (Fig. 3). The June average is 93 W m^{-2}, with a standard deviation of 10 W m^{-2}; the December average is 26 W m^{-2}, and the standard deviation is again about 0.1 of the mean. Annual values during the IGY period at USSR stations varied only slightly from the long-term mean (Pleshkova, 1965). At Voeikovo, near Leningrad, 1957 ran 4% above, 1958 5% below, and 1959 right at the long-term mean.

TABLE V

Mean Monthly Values of the Flux of Diffuse Shortwave Radiation[a]

Location	Latitude (deg.)	Month													Reference
		D	J	F	M	A	M	J	J	A	S	O	N	D	
Iakutsk	62°	6	11	28	49	82	101	98	90	68	47	29	15	6	Barashkova *et al.* (1961)
Eskdalemuir	55°	12	16	34	55	91	119	125	124	101	70	37	20	12	Berliand (1964)
Wien	48°	18	28	44	63	87	102	106	95	82	64	46	29	18	Sauberer and Dirmhirn (1958)
Tbilisi	42°	31	41	53	76	88	91	85	96	80	60	50	38	31	Tsutskiridze (1967)
Blue Hill	42°	26	29	42	55	70	83	94	95	87	59	41	30	26	Berliand (1964)
Tashkent	41°	30	36	46	58	67	69	60	47	48	47	43	33	30	Barashkova *et al.* (1961)

[a] Data given in watts per square meter.

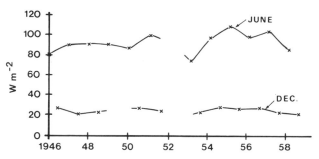

Fig. IV.3.   Mean diffuse solar radiation at Blue Hill Meteorological Observatory (42°N latitude) in June and December for several years (W m^{-2}). [Data from Berliand (1965)].

SPATIAL VARIATION

Small Scale

Diffuse solar radiation impinges all day on all terrestrial ecosystems and cannot be shaded out as can the direct beam. Microscale variations in intensity are produced when a part of the sky is blocked out by a steep ridge or neighboring taller vegetation if the block is not snow covered.

Mesoscale variations associated with slopes are generally small. Diffuse radiation tends to be more uniform than the direct beam on different slopes and provides illumination of slopes and walls in shadow. Its energy input to a north wall is not too different from that on a south or west wall (Dirmhirn, 1964, p. 178). Its intensity on south and east–west slopes relative to the flux on a horizontal surface is shown in Table VI. At low sun the south slope receives less than a horizontal surface because part of its sky is blocked out, an effect that can be computed (Robinson, 1966).

Larger Scale

Variations in the diffuse flux occur in mountain ranges and coastal lands that generate special cloud regimes. Diffuse solar radiation contributes less energy at high altitudes than low because fewer scattering molecules are interposed between the sun and the earth's surface. Diffuse insolation was 31 W m^{-2} in an average of several clear days at an elevation of 4 km in the Pamir mountains, and 54 W m^{-2} on the desert to the west. In cloudless summer weather it averages 55 W m^{-2} at Kaz-

TABLE VI

Relative Diffuse Radiation on Slopes in the Northern Hemisphere[a,b]

		Slope steepness	
Orientation	Sun altitude	20°	45°
South slope	10°	0.98	1.06
	30°	0.95	1.16
East or west slope	10°	1.19	1.04
	30°	1.55	1.22

[a] Data from Mukhenberg (1965).
[b] Figures include radiation reflected from opposite slope.

begi Observatory (3.7 km) in the Caucasus and 90 W m^{-2} at Tbilisi (0.4 km) (Tsutskiridze, 1967, p. 59). The gradient is 10 W m^{-2} km^{-1}. Estimates of the diffuse solar flux to different altitude zones in the eastern Alps (Dirmhirn, 1964, p. 94) show a decrease of about 8 W m^{-2} km^{-1} in winter and 15 W m^{-2} km^{-1} in summer. Diminution of the smoothing effect of the diffuse flux sharpens the contrasts in energy budgets of ecosystems on different slopes at high altitudes.

The western United States gets less energy from the flux of diffuse radiation than do the eastern states. The deserts of central Asia receive about 0.1 less than do comparable regions of the Soviet Union. Otherwise, yearly mean fluxes show only a weak spatial pattern. A poleward decrease occurs but is reversed at the Arctic coast. As known since the earliest polar meteorological expeditions, Arctic stations receive much diffuse radiation due to the prevalence of snow cover and thin layers of stratus, a cloud type that seems to be caused by the small diurnal variations of radiation in the Arctic summer (Herman and Goody, 1976).

We tend to think of diffuse solar radiation as the sky at dawn or the north light in an artist's studio, but it actually quietly carries during the day a substantial quantity of energy from the overarching vault of the sky. Its midday flux density ranges up to 200–300 W m^{-2} and is a function of sun height, aerosol nature and concentrations, and the microphysical characteristics of clouds; these factors also affect the degree to which it is enriched in high-energy photons and the photosynthetically active fraction of the solar flux. It is continuous during the day and in addition begins before sunrise and lingers after sunset. In its largely nondirectional aspect it resembles energy fluxes we shall discuss in later chapters.

REFERENCES

American Meteorological Society, Atmospheric Radiation Working Group (1972). Major problems in atmospheric radiation: An evaluation and recommendations for future efforts. *Bull. Am. Meteorol. Soc.* **53,** 950–956.

Barashkova, E. P., Gaevskii, V. L., D'iachenko, L. N., Lugina, K. M., and Pivovarova, Z. I. (1961). "Radiatsionnyi Rezhim Territorii SSSR." Gidrometeorol. Izd., Leningrad.

Bener, P. (1963). Der Einfluss der Bewölkung auf die Himmelstrahlung. *Arch. Meteorol., Geophys. Bioklimatol., Ser. B* **12,** 442–457.   •

Berliand, M. E. (1975). "Sovremennye Problemy Atmosfernoi Diffuzi i Zagriazneniia Atmosfery." Gidrometeoizd., Leningrad.

Berliand, T. G., ed. (1964). "Aktinometricheskii Spravochnik Zarubezhnye Strany Ezhegodnye Dannye." Gidrometeorol. Izd., Leningrad.

Berliand, T. G. (1965). Izmenchivost' solnechnoi radiatsii, postupaiushchei k poverkhnosti zemli. *Tr. Gl. Geofiz. Obs.* **179,** 28–44.

Brines, M. L., and Gould, J. L. (1979). Bees have rules. *Science* **206,** 571–572.

Brooks, F. A. (1959). "An Introduction to Physical Microclimatology." Associated Student Store, Davis, California.

Bullrich, K. (1964). Scattered radiation in the atmosphere. *Adv. Geophys.* **10,** 99–260.

Dirmhirn, I. (1964). "Das Strahlungsfeld im Lebensraum." Akad. Verlagsges., Frankfurt a.M.

Flach, E. (1965). Klimatologische Untersuchung über die geographische Verteilung der Globalstrahlung und der diffusen Himmelsstrahlung. *Arch. Meteorol., Geophys. Bioklimatol., Ser. B* **14,** 161–183.

Gates, D. M. (1965). Energy, plants, and ecology. *Ecology* **46,** 1–13.

Grum, F., and Becherer, R. J. (1979). "Optical Radiation Measurements, Vol. 1, Radiometry." Academic Press, New York.

Herman, G., and Goody, R. (1976). Formation and persistence of summertime Arctic stratus clouds. *J. Atmos. Sci.* **33,** 1537–1553.

Hopkinson, R. G., Petheridge, P., and Longmore, J. (1966). "Daylighting." Heinemann, London.

Ives, R. L. (1946). Cloud reflection effects in pyrheliometer records. *Bull. Am. Meteorol. Soc.* **27,** 155–159.

Kalitin, N. N. (1943). Rasseiannaia radiatsiia bezoblachnogo neba. *Dokl. Akad. Nauk SSSR* **39,** No. 8.

Kavanau, J. L., and Peters, C. R. (1976). Activity of nocturnal primates: Influences of twilight zeitgebers and weather. *Science* **191,** 83–86.

Kondratyev, K. Ya. (1969). "Radiation in the Atmosphere." Academic Press, New York.

Kondratyev, K. Ya. (1977). Radiation Regime of Inclined Surfaces. *WMO Tech. Note* No. 152 (WMO No. 467), Geneva.

McCartney, H. A. (1978). Spectral distribution of solar radiation. II: global and diffuse. *Q. J. R. Meteorol. Soc.* **104,** 911–926.

McCree, K. J., and Keener, M. E. (1974). Effect of atmospheric turbidity on the photosynthetic rates of leaves. *Agric. Meteorol.* **13,** 349–357.

McKee, T. B., and Klehr, J. T. (1978). Effects of cloud shapes on scattered solar radiation. *Mon. Weather Rev.* **106,** 399–404.

Meinel, A. B., and Meinel, M. P. (1976). "Applied Solar Energy: An Introduction." Addison-Wesley, Reading, Massachusetts.

Miller, R. C. (1958). Morning and evening song of robins in different latitudes. *Condor* **60,** 105–107.

Mukhenberg, V. V. (1965). Nekotorye osobennosti prikhoda solnechnoi radiatsii na na-
klonnye poverkhnosti. *Tr. Gl. Geofiz. Obs.,* **179,** 108–117.
Pleshkova, T. T. (1965). Rasseiannaia radiatsiia i ee anomali v period MGG i MGS na
territorii SSSR. *Tr. Gl. Geofiz. Obs.* **179,** 88–97.
Robinson, N., ed. (1966). "Solar Radiation." Am. Elsevier, New York.
Robinson, P. J. (1977). Measurements of downward scattered solar radiation from isolated
clouds. *J. Appl. Meteorol.* **16,** 620–625.
Sauberer, F., and Dirmhirn, I. (1965). Das Strahlungsklima. *In* "Klimatographie von
Österreich" (F. Steinhauser, O. Eckel, and F. Lauscher, eds.), Vol. 3, No. 1, pp. 13–102.
Österr. Akad. Wiss., Denkschr. Gesamtsakad., Wien.
Schulze, R. (1970). "Strahlenklima der Erde." Steinkopff, Darmstadt.
Sivkov, S. I. (1968). "Metody Rascheta Kharakteristik Solnechnoi Radiatsii." Gidrometeo-
rol. Izd., Leningrad.
Tsutskiridze, Ia. A. (1967). Radiatsionnyi i Termicheskii Rezhimy Territorii Gruzii. *Tr.
Zakavkaz. Nauchno-Issled. Gidrometeorol. Inst.* **23** (29).
Uchijima, Z., Inoue, K., and Inayama, M. (1976). The climate in growth chamber (7).
Angular distribution of diffuse radiation in vinyl houses. *J. Agric. Meteorol.* **32,** 127–136.

Chapter V

TOTAL INCOMING SOLAR RADIATION

It is reasonable to combine the radiant-energy fluxes described in the two preceding chapters because they have similar spectral compositions and diurnal regimes and compensating aperiodic variations. Clouds and aerosols that reduce or cut off the direct solar beam usually increase diffuse solar radiation, so the sum* of the two fluxes varies less than either one does individually.

However, special qualities of each flux for ecosystems make it important to know the relative contribution that each one makes to the total solar input.

(1)   The direct beam can be focused and transformed into thermodynamically efficient, high-temperature operating cycles (solar furnaces, flower corollas).

(2)   The direct beam is more sensitive than diffuse radiation to slope of the receiving surface and produces greater differentiation among ecosystems.

(3)   The two fluxes penetrate ecosystems differently and must be separately entered in the subroutines of photosynthesis and energy-budget models.

(4)   The fluxes display different spectral distributions.

Diffuse radiation carries more photosynthetically active radiation (0.65), the direct beam less (0.42); the solar infrared that makes up much of the direct beam has little photosynthetic value.†

---

* See note on terminology at end of chapter.

† This fact is worth noting because animals like us seem to appreciate the direct beam, with its heating and orienting values, much more than we do diffuse radiation. We tend to think of weather dominated by diffuse radiation as featureless or gloomy.

71

THE RELATIVE CONTRIBUTIONS OF THE COMPONENT FLUXES

The diverse geometries and spectral compositions of these two fluxes make it desirable to know their relative magnitude. In clear weather this relation depends on sun angle and the concentration of scattering aerosols. In cloudy weather it depends on the weakening or interruption of the direct beam. Reduction of the direct beam generally means a strengthening of the diffuse flux—up to a point; seldom is the gain in diffuse as large as the loss in the beam.

Clear skies let most of the direct beam through and scatter relatively little. At sea-level sites the diffuse fraction is typically 0.15–0.20, varying with wavelength (Table I).

Considering the total solar spectrum, the beam contributed 0.96 in clear midday weather at 1.6-km altitude (Table II). A morning hour at a coastal site on successive days in summer found the direct beam contributing 395 W m^{-2} on one day and zero on another while diffuse radiation was about the same on both mornings (Table II), and the direct beam contributed 0.58 and 0, respectively, of the total solar input.

Turning to daily totals (Table III), the direct beam and diffuse radiation in mostly clear weather in Moscow averaged 126 W m^{-2} and 144 W m^{-2}, respectively, or 0.47 and 0.53 of the total. The direct beam contributed less than half of the total because of its low angle much of the day. The same effect is seen in the sunny day (2 July) at Hamburg (Table III). The direct flux is small or absent on cloudy days; whether the diffuse flux is increased or not depends on cloud thickness.

In general, an increase of an hour in daily duration of sunshine in the average day brings an increase of approximately 0.03 in the relative share of the direct beam in the total (Flach, 1965, Fig. 11). The direct/diffuse

TABLE I

Contribution of Clear-Sky Diffuse Solar Radiation to Total Solar Radiation at Different Wavelengths[a]

Wavelength (μm)	Fraction arriving as diffuse	Fraction arriving in direct beam	Wavelength (μm)	Fraction arriving as diffuse	Fraction arriving in direct beam
0.3	0.72	0.28	0.7	0.09	0.91
0.4	0.33	0.67	1.0	0.05	0.95
0.5	0.19	0.81	2.0	0.02	0.98
0.6	0.13	0.87			

[a] Data from Schulze (1970, p. 119). Sun height = 60°.

TABLE II

Midday Solar Fluxes in Summer

Location	Sky condition	Hour ending	Direct W m^{-2}	Direct Fraction	Diffuse W m^{-2}	Diffuse Fraction	Total W m^{-2}
Blue Canyon,[a] 39°N, 13 July 1963[b]	Clear	1150	1020	0.96	45	0.04	1065
Hamburg,[c] 54°N							
2 July 1974	Sunny	1000	395	0.58	286	0.42	681
1 July 1974	Cloudy	1000	0	0	256	1.0	256

[a] Blue Canyon measurements by R. A. Muller and the author (unpublished).
[b] Sun height 70°, path 0.89, altitude 1.6 km. Extraterrestrial flux at this hour was 1275 W m^{-2}.
[c] Hamburg measurements by Deutscher Wetterdienst (Germany, 1974).

ratio is also "a rather sensitive measure of the overhead degree of turbidity" and might be used to monitor aerosols (Deirmendjian, 1980).

Integration over a month incorporates the regularly changing height of the sun, kinds and concentrations of aerosols, the irregularly changing cloud cover and depth, and concentration and size distribution of cloud droplets. These factors govern the direct and diffuse fluxes at a mid-latitude station subject to cyclonic cloudiness in winter (June) (Table IV). The range in diffuse radiation flux density over the months of the year is only a third of the range in the direct beam, which responds much more to the summer increase in sun altitude. The direct-beam fraction varies from about 0.45 in winter to 0.65 in summer. The cloud shields

TABLE III

Whole-Day Averages of Solar Fluxes in Summer[a]

Location	Sun altitude	Direct W m^{-2}	Direct Fraction	Diffuse W m^{-2}	Diffuse Fraction	Total W m^{-2}
Moscow University						
6 June 1976, mostly sunny	55° at noon	126	0.47	144	0.53	270
3 June 1976, rainy	55° at noon	0	0	48	1.00	48
Hamburg Observatory						
2 July 1974, 0.76[b]	58° at noon	134	0.50	136	0.50	270
1 July 1974, 0.09[b]	58° at noon	10	0.10	93	0.90	103

[a] Data from Monthly observational summaries by MGU (Moscow, 1976) and Hamburg Observatory (Germany, 1974).
[b] Sunshine duration as fraction of that astronomically possible.

TABLE IV

Monthly Mean Values of Direct, Diffuse, and Total Solar Radiation near Melbourne[a,b]

	December		March		June		September		December		Range in all 12 months
	W m^{-2}	Fraction	W m^{-2}	Fraction	W m^{-2}	Fraction	W m^{-2}	Fraction	W m^{-2}	Fraction	
Direct	204	0.66	129	0.63	38	0.46	104	0.61	204	0.66	200
Diffuse	108	0.34	75	0.37	42	0.54	67	0.39	108	0.34	71
Total	312		204		80		171		312		

[a] Data from Funk (1963).
[b] Aspendale, 38°S Latitude.

of winter cyclones produce the same effect in other regions, e.g., in central Asia, where the direct beam contribution at Tashkent varies from 0.48 in winter to 0.75 in summer (Barashkova et al., 1961, p. 69).

In equatorial regions themselves massive convective clouds block the direct beam more than the cloud sheets of well-spaced midlatitude cyclones do. A traveler, coming to an airport like Singapore's, descends through a deep troposphere crowded at every level with enormous clouds of every shape, to reach a surface at which only approximately 0.4 of yearly total solar radiation arrives direct from the sun, high though it is every day. Sun height and cloud types in high latitudes are radically different from those of equatorial lands, but the direct beam contributes no more—hardly half the yearly total.

The diffuse fraction is large on days when only a small fraction of the extraterrestrial flux Q gets through a cloud-filled atmosphere; for daily data at Canadian stations the overall relation is

$$D/Q_0 = 1.2 - 1.2Q_0/Q,$$

in which D is diffuse and Q_0 total solar radiation at the surface (Tuller, 1976) (Q_0/Q is the Angot number). When the surface receives half the extraterrestrial flux density (the range of this ratio is from 0.2 to 0.8 by days, 0.3 to 0.7 in monthly means), the formula becomes

$$D/Q_0 = 1.2 - (1.2)(0.5) = 1.2 - 0.6 = 0.6,$$

and diffuse radiation forms 0.6 and beam radiation 0.4 of the total. Chang (1980) suggests that the sum of the two ratios is more stable than either one alone.

Albedo is secondary to atmospheric factors in explaining the D/Q_0 ratio (Tuller, 1976); turbidity is important, at least in the visible range (McCartney, 1978), as, of course, is cloud cover, the annual ratio varying from 0.28 at 19 desert sites to 0.63 at 22 polar sites (Chang, 1980). The ratio varies with the season, and at high altitudes (3 km) is not consistent (Neuwirth, 1979).

Ecosystems at high elevations lie under a relatively thin blanket of diffusing molecules, and m (the optical path for molecular scattering) is reduced by $p/101.3$, where p is the local pressure in kilopascals and 101.3 kPa is mean sea-level pressure. This formula does not apply to depletion by water vapor and particles, which are concentrated in the lower layers of the atmosphere.

As sun height increases beyond 15–20°, the scattering path is shortened, and diffuse solar radiation increases more slowly than the vertical component of the direct beam, which follows the sine function. The relative fraction of diffuse in the total, which is 1.00 before dawn, decreases to 0.2 or less at midday.

FLUX DENSITIES OF TOTAL SOLAR RADIATION INCIDENT ON ECOSYSTEMS

Clear-Day Radiation

Sun-earth geometry (Chapter III) is the only factor in flux density of the vertical component of incoming solar radiation outside the atmosphere. A horizontal surface at 43°N latitude (Milwaukee) along the earth's orbit around the sun experiences variation from about 140 W m^{-2} in December to 490 in June, a range of $\times 3.5$; the range would be greater except that the earth is closer to the sun in the northern hemisphere winter than in summer. These means over the 24 hours incorporate the effects of longer day length in June as well as the more direct incidence.

At this latitude each factor experiences an approximate 1.85 times increase from December to June; 1.85×1.85 results in the 3.5 increase in the mean flux. At higher latitudes both factors are larger and so is the summer increase over winter.

Downward scattering retrieves about half the energy attenuated by scattering but none of that absorbed by atmospheric gases, and cloudless-sky solar radiation at the earth's surface, incorporating both beam and diffuse components, amounts to around 0.8 of the extraterrestrial flux density. The ratio is smaller in winter than summer because the longer paths of the direct beam in winter offer more opportunity for absorption and outward scattering.

In a way it is a source of wonder that the atmosphere, as deep and massive as it is, transmits any energy to the underlying surface. It is, in fact, opaque in some spectral bands, but in the wavelengths where solar emission is strong these opaque or partially absorbing bands are not numerous or wide. The molecules composing the bulk of our atmosphere just do not vibrate to most of the solar photons, which pass through with minimum absorption.

Clear-day radiation at the nearest station (e.g., Madison, 102 W m^{-2} in December and 370 in June, Baker and Klink, 1975, p. 18) forms the base from which solar radiation on days of average cloudiness at a study site is calculated. It represents a stable upper limit of solar energy input for the season and location (incorporating latitude and the transparency of the regional air).

Ecosystems under Clouds

Thick clouds appear to absorb 5–10% of the cloudless-sky radiation, but measurements vary, and the mechanism is not entirely clear (American Meteorological Society, 1972), an uncertainty that reduces the accuracy of estimates made from satellite imagery (Tarpley, 1979). In addition, reflection by clouds increases from an approximate equality with transmission in clouds of 40-m depth (Kondratyev, 1969, p. 303) to 0.85 or more in clouds of 400-m depth, which transmit relatively little. Strong reflection identifies deep clouds to the satellite sensor, which sees them as brighter than snow; weak transmission identifies them to the observer in the gloom at the ground, who may receive 0.05 or less of the cloudless-sky intensity.

Total radiation from the sun at altitude 30° is shown in Table V for mean aerosol depletion and various sky conditions or cloud densities, although scatter in these relations is large because we usually lack information about the cloud parameters (height of base, thickness, droplet-size distribution, vapor density, and so on) that are radiatively important (Liou, 1976).

Observations of the solar disk (Table V) are routine in Europe but rare in the United States. Like other visual observations, they have some systematic observer errors (Gal'perin, 1974) but when corrected can be combined with observations of cloud type and sky cover at a given sun

TABLE V

Flux Densities of Solar Radiation through Different Conditions of Cloudiness at Moscow[a,b,c]

Condition of sky or cloud density	Flux density (W m^{-2})	Fraction
Cloudless	418	1.00
Scattered clouds, not covering sun	397	0.95
Sun's disk visible through		
cirrus	363	0.87
stratus	286	0.68
Sun not visible, because of		
high clouds	307	0.73
low clouds	203	0.49
Thick overcast	98	0.24

[a] Data from Nebolsin, cited in Kondratyev (1969, p. 458).
[b] Latitude 56°N, summer; path = 2.
[c] Extraterrestrial solar flux density approximately 450 W m^{-2}.

height to estimate total solar radiation. The difference between two steps of the cloud density scale (0–4) is small for scattered clouds (0.1–0.3 cover) but amounts to 60 W m^{-2} at 0.4–0.7 cover, 100 W m^{-2} at 0.8–0.9, and 160 W m^{-2} under overcast, at all hours in the year when the path length is 3 (Haurwitz, 1945) at Blue Hill Observatory.

The ratio of the fluxes under cirrus to cloudless-sky fluxes in Table V declines slowly from 0.97 at sky cover 0.3 to 0.92 at sky cover 0.85 and then drops off sharply to 0.77 with overcast. Under alto-cumulus a similar slow decline accelerates at sky cover of 0.85. This break in the relation of flux density to cloud cover at 0.85 appears to be associated with a maximum diffuse flux at 0.8 to 0.9 cloud cover [Bener (1963); see also Chapter IV of this volume]. Overcast is most likely to occur toward the center of a cloud system, where clouds are deepest.

Deep cloudiness is often associated with convergence in cyclonic flow aloft. Subsidence in anticyclonic flow prevents cloud development in the middle troposphere, but low clouds and fog may be present over lowlands. The boundary layer under anticyclonic subsidence over such lowlands as northwestern Europe tends to fill with fog or stratus, contrasting with the adjacent mountains, which rise into the levels of subsiding air and clear skies. The contrast of the North European Plain is beautifully described in early pages of "The Magic Mountain" by Thomas Mann, as young Hans Castorp leaves his ancestral Lübeck and travels up out of the clouds to a sanitarium in the therapeutic sunshine of the Alps. The Sierra at 2.1-km altitude experiences in winter a frequency of 0.81 sunny days (those exceeding 50 W m^{-2} in December), while Davis in the Sacramento Valley experiences only 0.61 such days.

Cyclonic and anticyclonic curvature on the 700-mbar surface above the Sierra in spring shows cyclonic flow associated with weak solar radiation and slow snow melting, especially in the tightly curved flow of

TABLE VI

Flux Density of Mean Solar Radiation on a Horizontal Surface Exceeded on 0.1, 0.5, and 0.9 of the Days at the Solstices and Equinoxes, Argonne Laboratory [a,b]

	20–26 Dec.	22–28 March	21–27 June	20–26 Sept.	20–26 Dec.
Clearest days, 0.1	108	269	351	232	108
Median, 0.5	57	185	291	159	57
Cloudiest days, 0.9	15	49	136	52	15

[a] Data from Baker and Klink (1975, p. 40); latitude 42°N.

[b] Values given in watts per square meter.

a cold low aloft. Anticyclonic flow brings days of strong solar radiation and large meltwater waves moving down the streams every afternoon (Miller, 1955, p. 78).

Statistical treatment of daily radiation data (Table VI) indicates that the clearest 0.1 of days receive amounts similar to the clear-day mean for the latitude as given by Budyko (1974, pp. 46–47). A dark day in summer receives about as much energy as a sunny day in winter, and seasonal variation is sometimes lost in the effects of dense cloud systems. In continental interiors, however, summer is usually less cloudy than winter, and the cloudiness factor reinforces the factors of sun height and day length to produce a several-fold increase in flux density (Fig. 1).

SPECTRAL COMPOSITION OF THE INCIDENT SOLAR FLUX

Ultraviolet. Ultraviolet radiation (0.290–0.385 μm) on a horizontal surface runs 0.06–0.08 of the total at Aspendale (Collins, 1973), the larger fraction being at low sun altitudes. The summer flux density averaged 18 W m^{-2} on all days and 23 W m^{-2} on cloudless days. At least half of the total in cloudless conditions arrived in the diffuse mode.

Photosynthetically active radiation. Radiation in the photosynthetically active wavelengths is a variable fraction of the total and averages about 0.45, depending on exact definition of the range (Ross, 1975, p. 18). This fraction tends to be constant on clear days because scattering, mostly

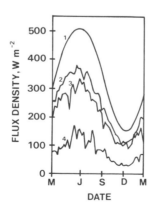

Fig. V.1. Mean annual march of solar energy on a horizontal surface at Lemont, Illinois (Argonne National Laboratory, 42°N latitude). Curve (1) extraterrestrial; (2) at the earth's surface under clear skies; (3) under half-cloudy skies; (4) under overcast (W m^{-2}) (Baker and Klink, 1975, p. 15).

in the short wavelengths, balances absorption, mostly in the long (Williams, 1976).

June means of photosynthetically active radiation run 100–125 W m^{-2} in high latitudes and 125–150 W m^{-2} in southern latitudes of the Soviet Union (Efimova, 1965). Summations for growing seasons variously defined thermally are 1125 MJ m^{-2} at Leningrad over the period when mean air temperature exceeds 5°C and 915 MJ m^{-2} over the period when it exceeds 10°C.

Summer can be defined as the period with sufficient photosynthetically active radiation at the times when leaf cells are warm and moist enough to convert this radiation into other forms of energy. Ecosystems far north of Leningrad receive only 400–500 MJ m^{-2} during the period warmer than 10°C; ecosystems in the Ukraine range from 1350 MJ m^{-2} received in the 10° C season in Polesia, 1580 in the steppe, to 1750 near the Black Sea (Konstantinov et al., 1972). Summer thus defined varies fourfold, from 400 MJ m^{-2} near the Arctic Ocean to 1750 near the Black Sea.

Illumination. Although skylight is the preferred source of illumination in buildings, the direct-beam component sometimes plays a role. Some house plants, for instance, need beam radiation at least part of the day.

Solar infrared radiation. This accompaniment of the more useful flux in the spectral regions of illumination and photosynthetic activity is sometimes unwelcome. Most ecosystems reflect much of it, and some buildings exclude it by special window glass. One approach proposed is to separate it from the illuminating wavelengths at a "cold-mirror" (Duguay, 1977), which reflects visible wavelengths for lighting a building interior and transmits infrared to a heat pipe that produces hot water.

DETERMINING THE COMBINED SOLAR FLUXES ON HORIZONTAL SURFACES

Observations

Measuring total solar radiation presents less difficulty than measuring either beam or diffuse flux separately, for which the sun has to be, respectively, followed or blocked out. Pyranometers absorb incoming radiation from the hemisphere on a horizontal surface under a glass dome that reduces convection and shuts out radiation of wavelengths longer than 3–4 μm. The sensors are alternating black and white concentric circles or wedges, the black segments of which absorb more radiation than the white, or else all-black sensors thermally compared against a heat sink. The temperature difference generates a current of a few to 50 millivolts in thermocouple circuits. Solar photoelectric cells are also used, with provision for their spectral selectivity. For special wavelengths, such as those producing photosynthetic activity, filters are added. The small number of pyranometer stations, even in countries with old meteorological traditions, is unfortunate for many kinds of study because the extrapolation error rises above ± 15% at sites only 50 km from a station (Hay and Suckling, 1979).

Networks of solar radiation sensors exist in many national meteorological organizations and are occasionally organized for special purposes, e.g., studies of ozone pollution, or for establishing spatial differences at a scale untouched by national networks, e.g., on sugar plantations in Hawaii, or the Wisconsin network of Kerr et al. (1968). Individual stations aid research in agriculture, glaciology, and snow hydrology. If radiation data are not used in real-time forecasting, observers may not notice when the sensors have drifted out of calibration (Durrenberger and Brazel, 1976; Fritschen and Gay, 1979, Chap. 5) and some records have experienced this problem.

Estimating Solar Radiation

Geometric and predictable factors in the delivery of total solar radiation are combined in estimating procedures with empirical means of accounting for the effects of turbidity and clouds. The weakness of these methods is that neither aerosols nor clouds are well-observed with respect to their radiative properties. For example, we have neither sufficient data on particle size distributions aloft nor information on cloud density or other parameters.

Many inhomogeneities exist in the real atmosphere, and they may change rapidly, "particularly in regard to cloud content and distribution" (Atwater and Ball, 1978). A crude attenuation model (Miller, 1950) obtained good reconstitutions of hourly solar radiation in clear weather but found that available observations of cloud characteristics* were so primitive as to make difficult any degree of precision in cloudy weather. One solution is to develop an attenuation figure from a nearby radiation station. This method is improved by considering three solar components, the direct beam, circumsolar diffuse, and all-sky diffuse (Revfeim, 1978) and can give hourly radiation on slopes as well as on horizontal

---

* For example, cloud density, even in the visual scale used at Blue Hill Meteorological Observatory (Haurwitz, 1945) is not routine in many countries.

ecosystems, since it considers the nonisotropic nature of the diffuse flux (Kondratyev, 1977, p. 29).

Some estimating methods begin with the daily extraterrestrial flux, as presented by Leighly (Smithsonian Institution, 1966, p. 419), and work downward. Others begin with cloudless-sky solar radiation observations made in the region. Both beginnings were employed in an energy-budget approach to evaporation from Lake Huron (Bolsenga, 1975), in which solar radiation had high importance. Simple formulas, like the Savinov–Ångström equation:

$$Q = Q_c[1 - (a + bn)n],$$

in which a and b are coefficients, n is the fractional cloud cover as a monthly mean, Q_c the cloudless-day radiation, and Q the actual monthly radiation (Budyko, 1974, p. 45), give adequate results for mapping energy budgets (Flocas, 1980). The quadratic in n indicates the greater likelihood of thick clouds when the general sky cover is extensive, as shown, for example, by Satterlund and Means (1978) in the Pacific Northwest.

Estimating methods stretch a small quantity of real data over a large period or region, most reliably in anticyclonic weather (Hay and Suckling, 1979). They are not recommended for short periods and certainly not for individual days for which correlations with existing stations yield an estimate with a standard error no better than ±25% (Kerr et al., 1968) unless daily sums are obtained from hourly estimates using cloud imagery from geosynchronous satellites (Tarpley, 1979).

VARIATIONS IN TOTAL SOLAR RADIATION OVER TIME

Ecosystems must deal with several variations in the solar radiation that reaches them:

(1)   the diurnal rise from zero before sunrise to a midday peak and subsequent decline, marked by a large energy impact during the morning hours;

(2)   a continuous change in the direction of the beam component through the day;

(3)   fluctuations in flux density, directional component, and spectral composition when clouds pass over the sun;

(4)   runs of dark or bright days as cloud systems pass;

(5)   seasonality of flux density and spectral composition, and a change in azimuth angle of the beam early and late in the day;

(6) changes from year to year.

Site-specific depletion, caused by aerosols, also produces time variations. The daily and seasonal variations are astronomical in origin, re-
current, and predictable. The third, fourth, and last items listed are
neither periodic nor predictable, but they do impose episodic fluctuations on the daily and annual regimes of solar energy received by
ecosystems.

Diurnal Regimes

Definitions of day length vary, depending on how much of the twilight
period is included. Radiation flux density in twilight periods influences
some organisms and keeps circadian clocks in time; the dark period
serves this photoperiodic function in others, and in the Arctic this role
may be replaced by the low-sun period of the day (Teeri, 1976). Plants
and animals use these signals "to synchronize their activities with the
seasons" (Adkisson, 1966). The internal clocks are not always known,
but they must keep accurate time in detecting small changes in the relative duration of light and dark; in order to estimate seasonal changes
within a week, the clock must be accurate within 1–3%, yet many insects
estimate the time of year with an accuracy of less than a week.

The other coordinate of the diurnal energy pulse is midday flux density, which in summer reaches 800–1000 W m^{-2} in mountains and deserts and ~600 W m^{-2} in humid lowland climates. These flux densities
decline uniformly in the southern hemisphere in June to zero at the
Antarctic Circle. In the northern hemisphere maxima are found at about
latitude 30° and decline poleward (Berliand, 1964). The sharpness of the
daily curve is most marked on a winter day when two midday hours,
1100–1300, account for about 0.40 of the daily total, but only 0.22 of the
total in a 16-hr summer day (Whillier, 1965). In solar-energy coordination
with the load curve of an urban utility, the midday peak could reduce
the utility load substantially (Kalhammer, 1979).

Frequency distributions of hourly flux densities of total solar radiation
in summer (Table VII) show the relative incidence of high rates at a
station where the sun is never high, in the midwest, and at two dry
sites. Flux densities exceeding 800 W m^{-2} differentiate the stations most
plainly.

The measures, day length and midday flux density, that express the
diurnal march of total solar radiation are illustrated in Table VIII. The
seasonal contrast at high latitudes is exemplified at Iakutsk, where the
winter is nearly as dark as, and far colder than, at the North Pole, the

TABLE VII

Frequencies of Hourly Solar Radiation in Summer Exceeding Several Levels of Flux Density[a]

Location	Exceeding			
	400 W m^{-2}	600 W m^{-2}	800 W m^{-2}	1000 W m^{-2}
Keflavik, Iceland, 65°N	0.26	0.14	0.07	0
Columbus, Ohio, 40°N	0.52	0.34	0.20	0.04
Yuma Test Station, 33°N	0.65	0.52	0.38	0.16
Wake Island, 19°N	0.61	0.47	0.33	0.13

[a] From graphs in Bennett (1965, p. 16), based on short records.

sun is low and the days are short. These astronomical factors form a regime that might be much modifed, even in the mean, by a diurnal regime of cloudiness or turbidity.

Aperiodic Variations

Passing systems of clouds bring bright days and dark days. Bright days are sharply limited by solar geometry and irreducible atmospheric losses, but low values can extend down to very low fluxes under deep clouds. The frequency distribution is cut off at its upper end and skewed out at its low end, as shown in the Sierra Nevada in the snow-accumulation and snow-melting seasons (Table IX). The range in spring is large, from less than 100 to more than 400 W m^{-2}, and the negative skewness expresses a high frequency of days near the clear-day maximum and is found in many climates, e.g., Moscow in spring and summer (Kotova and Poltaraus, 1979).

The frequency distribution of daily solar radiation at Blue Hill Meteorological Observatory shows two modes in January, which correspond to cloudy days under winter cyclones or nearly all clear between cyclones, corresponding to the familiar U-shaped frequency curve of cloud cover. The June distribution has several nodes. Smaller storm systems are less likely to cover the station all day, and convective clouds of summer produce intermediate daily energy pulses. Some stations display a tendency for colder days in winter to be sunnier (Baker and Enz, 1979) but this useful association is not always present (Asbury *et al.*, 1979).

If we take a mean flux density of 300 W m^{-2} to represent a bright day (or an energy pulse of 26 MJ m^{-2}), there is a frequency of 0.58 at the Sierra Laboratory in April and 0.80 in June. At Los Angeles 62% of the days in June receive more than this value, as do about half the days in the humid East (Table X). Regions of fewer than 40% of days above 300

TABLE VIII

Flux Densities of Total Solar Radiation on Average Days in June and December at Selected Stations[a]

Location	Time													24-hr mean (W m⁻²)	Total daily pulse (MJ m⁻²)	Data source
	00	02	04	06	08	10	12	14	16	18	20	22	24			
Iakutsk, 62°N																Barashkova et al. (1961)
June	0	0	20	160	350	485	570	510	380	230	80	10	0	237	20	
December	0	0	0	0	0	10	45	25	0	0	0	0	0	7	0.5	
Edmonton, 54°N																Hay (1977)
June	0	0	5	135	370	560	655	605	445	275	45	0	0	255	22	
December	0	0	0	0	0	70	160	130	15	0	0	0	0	30	2.6	
Madison, 43°N																U.S. Weather Bureau (1964)
June	0	0	0	90	370	600	695	660	475	195	10	0	0	260	23	
December	0	0	0	0	10	140	245	210	70	0	0	0	0	56	5	
Poona, 19°N (short record)																Berliand (1970)
June	0	0	0	17	295	595	730	660	420	115	0	0	0	237	20	
December	0	0	0	0	143	500	685	615	330	10	0	0	0	190	16	
Kinshasa, 4°S																Robinson (1966, p. 138)
June	0	0	0	160	375	600	530	250	0	0	0	0	0	160	14	
December	0	0	0	230	520	700	600	350	0	0	0	0	0	200	17	

[a] Data given in watts per square meter.

TABLE IX

Frequency Distribution of Daily Means of Total Solar Radiation, Central Sierra Snow Laboratory (Lat. 39°N), 1946–1951[a]

24-hr average flux density (W m^{-2})	Frequency		Cumulated frequency	
	January	April	January	April
Less than 50	0.19	0	0.19	0
50–99	0.16	0.08	0.35	0.08
100–149	0.45	0.05	0.80	0.13
150–199	0.20	0.10	1.00	0.23
200–249	0	0.09	1.00	0.32
250–299	0	0.10	1.00	0.42
300–349	0	0.26	1.00	0.68
350–399	0	0.31	1.00	0.99
>399	0	0.01	1.00	1.00
Mean	107	274		
Median	113	310		
Interquartile deviation	73	145		

[a] Source: Miller (1955, p. 9), from published data of the Corps of Engineers and Weather Bureau Cooperative Snow Investigations.

W m^{-2} are found in the lower South and New England, as well as the Pacific Coast (Bennett, 1975).

Runs of consecutive dark or sunny days are significant to ecosystems and solar-energy installations in which radiation conversions are associated with some kind of storage. Successive high-radiation days in summer deplete soil storage; as the soil dries, it heats the air more and provides less vapor to be condensed into clouds that would block some of the remorseless sun; drought is incipient. Runs of low-radiation or high-radiation days, deriving from storm spacing and upper-air steering, are as important in the energy climate of an ecosystem as in its hydroclimate (Miller, 1977, pp. 228–242, 304–316). Days when solar radiation averaged less than 125 W m^{-2} at Baghdad had a year-long frequency of 0.14, but in sequences of 3 days or longer they had a frequency of 0.04 (Abbas and Elnesr, 1974). Consecutive bright days in summer were 4 times as frequent in Phoenix as in Miami (Fritz and MacDonald, 1960). Runs of 3 days or more of sunny weather in St. Paul occur 1.8 times each June (Baker and Enz, 1979).

The parameter of interdiurnal variability takes into account this clustering of dark or sunny days (Table XI). Its values in June range from 0.42 of the mean at Utsunomiya to 0.02 at Cairo. Interdiurnal variability in dry regions is generally less than 0.1 of the mean flux density, but in cyclogenetic regions it exceeds 0.4.

TABLE X

Meridional Profiles of Frequencies of June Days Exceeding 300 W $m^{-2a,b}$

West Coast	Latitude (°N)	Frequency	Interior	Latitude (°N)	Frequency	East Coast	Latitude (°N)	Frequency
Fairbanks	65	0.35	Fort Churchill[c]	59	0.55	Knob Lake	55	0.26
Vancouver[c]	49	0.42	Winnipeg	50	0.43	Caribou	47	0.35
Seattle[c]	47	0.40	St. Cloud	46	0.46	Halifax[c]	45	0.32
Medford	43	0.71	Madison	43	0.52	Blue Hill[c]	42	0.44
Sierra Snow Laboratory	39	0.80	Argonne	42	0.48	Seabrook[c]	39	0.48
Davis	38	0.88	Little Rock	35	0.51	Charleston[c]	33	0.51
Los Angeles[c]	33	0.62	Lake Charles[c]	30	0.71	Miami[c]	26	0.44
Honolulu[c]	21	0.57	Brownsville[c]	26	0.71			

[a] Sources: Machine listings of National Weather Records Center, ESSA, Environmental Data Service; Midwestern data from Baker and Klink (1975); Canadian data from Hay (1977, p. 123).

[b] 24-hr mean flux density of 300 W m^{-2} is equal to daily pulse of 25.9 MJ m^{-2}.

[c] Coastal station.

TABLE XI

Interdiurnal Variability of Total Solar Radiation[a,b]

Station	June	December
Resolute Bay, Canada	56	Polar night
Utsunomiya, Japan	79	36
Cairo, Egypt	12	23
Kinshasa, Zaire	47	65

[a] Data from Berliand *et al.* (1972).
[b] Data given in watts per square meter for mean daily flux density.

Seasonal Regimes

The daily pulses of solar radiation to midlatitude ecosystems change from summer to winter as a result of joint operation of day length and midday flux density (Table VIII). The June day is shorter at Edmonton and Madison than at 62°N, but midday radiation is stronger, giving a daily pulse similar to that at Iakutsk.

June–December differences indicate the range of energy inputs to which ecosystems must accommodate themselves—19.5 MJ m^{-2} at Iakutsk and Edmonton and 18 at Madison. A meridional profile through the interior of North America (Table XII) shows the zonal change in this June-December difference. The astronomical component of this figure is augmented in the middle latitudes, where cloudiness is more extensive in winter than in summer. At Tacubaya, however, the rains begin in June, and daily radiation is reduced by about 1 MJ m^{-2}, blunting the annual march.

Year-to-Year Variability

A sophisticated data-quality study of records in the midwestern United States found that clear-day radiation varies little from year to year "as long as pyranometers remain stable" (Baker and Klink, 1975, p. 6). Year-to-year variation in average-day radiation is caused by differences in the vigor and number of cloud systems. In the period of record at Karadag (1938–1952, 1955–1959) (Barashkova *et al.*, 1961), monthly means in June ranged from 147 W m^{-2} to 226 W m^{-2}, and dispersion of the monthly values is expressed in a standard deviation of 22 W m^{-2} (0.12 of the mean).

Annual means vary less, naturally, from year to year. Based on 24-yr records of sunshine duration, the variation among years in total solar radiation at each site averages 0.03 (Hoyt, 1978), corresponding to ap-

TABLE XII

Differences between Daily Pulses of Total Solar Radiation in June and December in Interior North America[a]

Station	Latitude (°N)	Difference in energy pulse[b]
Resolute Bay	75	26
Fort Churchill	59	20
Edmonton	54	19.5
Moosonee	51	18
Madison	43	18
Little Rock	35	15.5
San Antonio	29	15
Tacubaya	19	5

[a] Data from U. S. Weather Bureau (1964) and Berliand (1964, pp. 74–76, 100). Data given in megajoules per square meter.
[b] $10 \text{ MJ m}^{-2} = 116 \text{ W m}^{-2}$ over 24 hr.

proximately 5 W m^{-2}. Year-to-year variation of photosynthetically active radiation received at Berlin gave a standard deviation equal to 0.071 of the mean and 0.079 of the quantity absorbed in a cornfield (Birke and Rogasik, 1976).

SPATIAL DIFFERENCES IN TOTAL SOLAR RADIATION

Small-Scale Contrasts

Microscale differences in total solar radiation on horizontal surfaces are largely due to shading by trees or ridges, as discussed in Chapter III. Mesoscale differences caused by shading by individual clouds have a cooling and stabilizing effect (Ivanov and Khain, 1975), and more lasting contrasts are produced by topographically bound clouds.

Slopes. The geometry of sloping ecosystems was summarized in Chapter III for beam flux, but these surfaces also are contrasted in their receipts of diffuse radiation.* Ridging the fields in the Iakutsk region

---

* Detailed formulas for determining the total flux are available in Robinson (1966), Meinel and Meinel (1976), and Revfeim (1978), among others. Slopes in drainage basins among other slopes are discussed by Hay (1971) and Swift (1976). While difficulties are presented by the nonisotropy of the diffuse component (Kondratyev, 1977, p. 18) and the size of the flux reflected from adjacent slopes, "the most complicated problem" in such calculations arises from the effects of clouds (Kondratyev, 1977, p. 42)

(62°N Lat.) extends the period of useful solar energy input from four months on a horizontal surface to seven months on a 20° south-facing slope. Differences in total radiation among sloping ecosystems frequently exceed 50 W m^{-2} in the daily mean and at particular hours are still greater and support topographically bound wind systems that shift through the day.

By considering diffuse as well as beam radiation to slope facets in a dissected part of Baffin Island, the sites favorable to glacier formation were identified (Williams *et al.*, 1972). An ecological study in virgin steppe of central Russia measured solar radiation of 285 W m^{-2} on the upland, 235 in a valley bottom, and 305–325 on the opposite south-facing upland (Rauner *et al.*, 1974), a range of 80 W m^{-2} or almost a third of the areal average.

Solar-energy collectors. Flat plates collect the diffuse as well as the beam flux, and tilted collectors producing domestic or farm hot water and space heating use solar radiation efficiently without flux concentration. The angle of tilt for winter efficiency is commonly 15° plus the latitude (see Meinel and Meinel, 1976, for performance data).

Energy storage is as much a problem as is determining the collector area needed (50–70 m^2), and depends on the climatological duration of low-radiation periods. In Wisconsin conditions 50–100 kg of water storage of heat are needed for 1 m^2 of collector (Duffie and Beckman, 1976). Dark days (<50 W m^{-2}) at St. Paul occur in December at a frequency of 6.7 and runs of three or more such days at 1.9 (i.e., two per year) (Baker and Enz, 1979, Table 7).

The fossil-energy costs of steel and aluminum used in collectors are not inconsiderable (Whipple, 1980) and draw attention to the possible contributions of passive solar space-heating (perhaps 10^{11} watts more by the year 2000; this conversion is equivalent to 1 W m^{-2} over the urban areas of this country).

Urban Scales

Sunny days, as defined by the 300 W m^{-2} threshold in June, tend to be 0.1 fewer in a city than outside it when judged by a few station pairs in the United States. Adelaide receives less direct-beam radiation than Flinders University, 160 m higher, but more diffuse in about the same amount (Lyons and Forgan, 1975). Boston, on the other hand, receives 0.15–0.20 less than Blue Hill in total solar radiation. The contrast between Moscow University and a rural station grew larger over the period from 1955 to 1964 (Dmitriev and Abakumova, 1969).

Larger Scales

Differing cloud regimes in lowlands and mountains produce different radiation regimes within fairly short distances, even in the maritime climate of Norway (Skaar, 1980). Davis and the Sierra Snow Laboratory differ, as mentioned earlier with regard to winter fog in the valley, which is nearly cloudless in summer while convective clouds form in the mountains. Davis receives more than 300 W m^{-2} on 0.88 of June days, while at the Snow Laboratory, 140 km northeast and 2 km higher, only 0.80 of these days receive more.

The Wisconsin network (Kerr and Rosendal, 1968; Kerr et al., 1968) identified radical shifts in the spatial pattern of solar radiation. In a given month the tracks of a few storms form a distinct radiation pattern that may display little resemblance to a pattern drawn from the coarser national network. The effect of the Great Lakes on cloudiness is evident in some months. Spatial correlation coefficients of solar radiation in individual days in March averaged 0.8 with the ten stations nearest (within 150 km) to two base stations in north central and west central Wisconsin, but only 0.6 with a base station near Lake Michigan (Kerr et al., 1968).

Coastal and mountain effects are combined across southern Honshu, where heavy winter clouds on the Japan Sea side reduce solar radiation 52 W m^{-2} below that on the lee side, a rate of change of 30 W m^{-2} per 100 km (Nishizawa, 1977). A still steeper gradient exists in the trades on Oahu between an annual mean of 150 W m^{-2} on the windward side of the Koolau Range to 250 W m^{-2} over Pearl Harbor (Ramage, 1979), a distance of only 15 km, but a flux density difference highly important to sugar cane production.

Meridional profiles of solar radiation are featureless in summer over a stretch of 60 degrees of latitude. In winter, on the other hand, contrast in solar energy input between low and high latitudes is approximately 5 W m^{-2} per 100 km.

Few equatorial ecosystems receive large yearly quantities of solar radiation. Folklore about the withering intensity of the equatorial sun has been disproven by measurements; at Djakarta (Berlage, 1948), at 6°S latitude midday flux density on a horizontal surface averaged between 550 and 580 W m^{-2} in different months—about the same as summer intensities at Madison, Edmonton, and Iakutsk (Table VIII). The daily mean flux density of 180–195 W m^{-2} is much smaller than the June means in the United States.

In latitudes lower than 40° strong radiation is more a matter of lack of cloudiness than latitude. The spatial pattern is cellular (Budyko, 1974), and in the dry parts of every continent annual mean flux densities of

210 W m^{-2} extend poleward nearly to latitude 40°. Poleward of latitude 40°, the cloudiness factor takes second place behind latitude except near the oceans. The high-latitude lands receiving less than 100 W m^{-2} annually—the dark part of the earth—are the Arctic and Antarctic coasts. In northern Norway tundra ecosystems receive only 80 W m^{-2}.

Nevertheless, even at the high latitude of Denmark (Sørensen, 1975) and Sweden (Lönnroth et al., 1980) solar energy fluxes are large enough to play a major role in national energy planning (80% in the Swedish plan). Solar cells for high-quality electrical energy, managed forest ecosystems for methanol and high-temperature process heat, and direct solar for space-heating would yield economic energy over a wide range of quality.

Although not a single energy flux, total solar radiation makes a convenient combination because its two components have similar time regimes and spectral quality, and both are important to ecosystems. The basic controls are sun height and atmospheric cloudiness, with aerosol concentration having a minor effect, although the pure geometric effect of sun height on flux density is buffered by the area-source character of the diffuse component and by aperiodic fluctuations caused by passing cloud systems. The variable direction, spectral composition, and flux density of solar energy present a major problem that terrestrial ecosystems meet in many ways.

Note on Terminology: The sum of incoming solar radiation has several names: sun and sky radiation, "global radiation," insolation (*incoming solar radiation*), or simply total solar radiation. Note that "sky radiation" does not refer to radiation emitted by atmospheric gases (the subject of Chapter VI) but only to high-energy photons that are scattered out of the solar beam but are unchanged in wavelength, arriving at the earth's surface as diffuse radiation from the hemisphere of the sky. Global radiation is a confusing term because the concept actually refers to radiation from only the *hemisphere* above an ecosystem. Biologists sometimes measure radiation with spherical receivers, but these data are hard to interpret in terms of other radiation fluxes and are not discussed here.

REFERENCES

Abbas, M. A., and Elnesr, M. K. (1974). Incoming solar radiation measurements in Iraq and their presentation. *Pure Appl. Geophys.* **112,** 753–763.
Adkisson, P. L. (1966). Internal clocks and insect diapause. *Science* **154,** 234–241.
American Meteorological Society, Atmospheric Radiation Working Group (1972). Major problems in atmospheric radiation: an evaluation and recommendations for future efforts. *Bull. Am. Meteorol. Soc.* **53,** 950–956.

Asbury, J. G., Maslowski, C., and Mueller, R. O. (1979). Solar availability for winter space heating: An analysis of SOLMET data, 1953 to 1975. *Science* **206**, 679–681.

Atwater, M. A., and Ball, J. T. (1978). Intraregional variation of solar radiation in the eastern United States. *J. Appl. Meteorol.* **17**, 1116–1125.

Baker, D. G., and Enz, J W (1979) Climate of Minnesota. Part XI—The availability and dependability of solar radiation at St. Paul, Minnesota, *Univ. Minn. Agric. Exp. Stn. Tech. Bull.* No. 316.

Baker, D. G., and Klink, J. C. (1975). Solar radiation reception, probabilities, and areal distribution in the North-Central Region. *Univ. Minn. Agric. Exp. Stn., Tech. Bull.* No. 300; also *North-Central Reg. Res. Publ.* No. 225.

Barashkova, E. P., Gaevskii, V. L., D'iachenko, L. N., Lugina, K. M., and Pivovarova, Z. I. (1961). "Radiatsionnyi Rezhim Territorii SSSR." Gidrometeorol. Izd., Leningrad.

Bener, P. (1963). Der Einfluss der Bewölkung auf die Himmelstrahlung. *Arch. Meteorol. Geophys. Bioklimatol. Ser. B* **12**, 442–457.

Bennett, I. (1965). The Yuma Test Station, Arizona. Hourly and daily insolation records 1951–1962. *U.S. Army Natick Lab., Tech. Rep.* **ES-15**.

Bennett, I. (1975). Variation of daily solar radiation in North America during the extreme months. *Arch. Meteorol., Geophys. Bioklimatol. Ser. B* **23**, 31–57.

Berlage, H. P., Jr. (1948). Solar radiation measurements in the Netherlands Indies, Part II. Recording radiation of the sun and sky from 1926–1941. Batavia, Kon. Mag. Meteorol. Obs., *Meteorol. Geophys. Dienst. Verh.* **34**, 15–47.

Berliand, T. G., ed. (1964). "Aktinometricheskii Spravochnik Zarubezhnye Stranii Ezhegodnye Dannye." Gideometeorol. Izd., Leningrad.

Berliand, T. G. (1970). "Solar Radiation and Radiation Balance Data (The World Network)," Hydrometeorol. Publ. House, Leningrad.

Berliand, T. G., Rusin, N. P., Efimova, N. A., Zubenok, L. I., Mukhenberg, V. V., Ogneva, T. A., Pivovarova, S. I., and Strokina, L. A. (1972). Issledovanie radiatsionnogo rezhima i teplovogo balansa zemnogo shara. *Tr. Vses. Meteorol. S"ezda, V* **3**, 57–77.

Birke, J., and Rogasik, J. (1976). Die langjährigen Globalstrahlungsmessungen des Hauptobservatoriums Potsdam und ihre Nutzung für die Pflanzenproduktionsforschung. *Z. Meteorol.* **26**, 87–93.

Bolsenga, S. J. (1975). Estimating energy budget components to determine Lake Huron evaporation. *Water Resour. Res.* **11**, 661–666.

Budyko, M. I. (1974). "Climate and Life" (Engl. transl. ed. by D. H. Miller). Academic Press, New York.

Chang, J.-H. (1980). Diffuse radiation as related to global radiation and the Angot value. *Arch. Meteorol. Geophys. Bioklimatol. Ser. B* **28**, 31–39.

Collins, B. G. (1973). Ultra-violet radiation at Aspendale, Victoria. *Aust. Meteorol. Mag.* **21**, 113–118.

Diermendjian, D. (1980). A survey of light-scattering techniques used in the remote monitoring of atmospheric aerosols. *Rev. Geophys. Space Phys.* **18**, 341–360.

Dmitriev, A. A., and Abakumova, G. M. (1969). Vliianie gorodskikh aerosolei na koeffitsient prozrachnosti atmosfery. *Vestn. Mosk. Univ., Geogr.* No. 1, 52–58.

Duffie, J. A., and Beckman, W. A. (1976). Solar heating and cooling. *Science* **191**, 143–149.

Duguay, M. A. (1977). Solar electricity: The hybrid system approach. *Am. Sci.* **65**, 422–427.

Durrenberger, R. W., and Brazel, A. J. (1976). Need for a better solar radiation data base. *Science* **193**, 1154–1155.

Efimova, N. A. (1965). Raspredelenie fotosinteticheski aktivnoi radiatsii na territorii Sovetskogo Soiuza. *Tr. Gl. Geofiz. Obs.* **179**, 118–130.

Flach, E. (1965). Klimatologische Untersuchung über die geographische Verteilung der

Globalstrahlung und der diffusen Himmelstrahlung. *Arch. Meteorol., Geophys. Biokli-matol., Ser. B* **14**, 161–183.

Flocas, A. A. (1980). Estimation and prediction of global solar radiation over Greece. *Sol. Energy* **24**, 63–70.

Fritschen, L. J., and Gay, L. W. (1979). "Environmental Instrumentation." Springer-Verlag, New York.

Fritz, S., and MacDonald, T. H. (1960). The number of days with solar radiation above or below specific values. *Sol. Energy* **4**(1), 20–22.

Funk, J. P. (1963). Radiation observations at Aspendale, Australia, and their comparison with other data. *Arch. Meteorol., Geophys. Bioklimatol., Ser. B* **13**, 52–70.

Gal'perin, B. M. (1974). Mean total solar radiation under different cloudiness conditions. *In* "Actinometry, Atmospheric Physics, Ozonometry" (G. P. Gushchin, ed.), pp. 49–52. Israel Program Sci. Transl., Jerusalem. (Transl. of *Tr. Gl. Geofiz. Obs.* **279**.)

Germany, Wetterdienst (1974). Beilage zum Medizin-Meteorologischen Bericht. (Hamburg Observatorium *Monthly*.)

Haurwitz, B. (1945). Insolation in relation to cloudiness and cloud density. *J. Meteorol.* **2**, 154–166.

Hay, J. E. (1971). Computation model for radiative fluxes. *J. Hydrol. (N.Z.)* **10**, 36–48.

Hay, J. E. (1977). An analysis of solar radiation data for selected locations in Canada. *Atmos. Environ., Climatol. Stud.* **32**.

Hay, J. E., and Suckling, P. W. (1979). An assessment of the networks for measuring and modelling solar radiation in British Columbia and adjacent areas of western Canada. *Canad. Geogr.* **23**, 222–238.

Hoyt, D. V. (1978). Interannual cloud-cover variations in the contiguous United States. *J. Appl. Meteorol.* **17**, 354–357.

Ivanov, V. N., and Khain, A. P. (1975). Role of screening of the underlying surface by clouds in cellular convection. *Izv. Atmos. Ocean. Phys.* **11**, 666–667. (Transl. of *Izv. Akad. Nauk* by Am. Geophys. Union.)

Kalhammer, F. R. (1979). Energy-storage systems. *Sci. Am.* **241**(6), 56–65.

Kerr, J. P., and Rosendal, H. E. (1968). Distribution of global radiation in Wisconsin Dec. 1966 through June 1967. *Mon. Weather Rev.* **96**, 232–236.

Kerr, J. P., Thurtell, G. A., and Tanner, C. B. (1968). Meso-scale sampling of global radiation; analysis of data from Wisconsin. *Mon. Weather Rev.* **96**, 237–241.

Kondratyev, K. Ya. (1969). "Radiation in the Atmosphere." Academic Press, New York.

Kondratyev, K. Ya. (1977). Radiation Regime of Inclined Surfaces. *WMO Tech. Note* No. 152 (WMO 467).

Konstantinov, A. R., Goisa, N. I., Oleinik, R. N., and Sakali, L. I. (1972). Teplovoi i vodnyi rezhim sel'kokhoziaistvennykh polie. *Tr. Vses. Meteorol., S''ezda, V* **3**, 254–262.

Kotova, O. M., and Poltaraus, B. V. (1979). Sezonnye osobennosti prikhoda i izmeneniia summarnoi solnechnoi radiatsii v Moskve. *Vest. Mosk. Univ., Ser. Geogr.* 1979, no. 2, 59–67.

Liou, K.-M. (1976). On the absorption, reflection and transmission of solar radiation in cloudy atmospheres. *J. Atmos. Sci.* **33**, 798–805.

Lönnroth, M., Johannson, J. B., and Steen, P. (1980). Sweden beyond oil: nuclear commitments and solar options. *Science* **208**, 557–563.

Lyons, T. J., and Forgan, B. W. (1975). Atmospheric attenuation of solar radiation at Adelaide. *Q. J. R. Meteorol. Soc.* **101**, 1013–1017.

McCartney, H. A. (1978). Spectral distribution of solar radiation. II: global and diffuse. *Q. J. R. Meteorol. Soc.* **104**, 911–926.

Meinel, A. B., and Meinel, M. P. (1976). "Applied Solar Energy: An Introduction."Addison-Wesley, Reading, Massachusetts.

Miller, D. H. (1950). Insolation and snow melt in the Sierra Nevada. *Bull. Am. Meteorol. Soc.* **31**, 295–299.

Miller, D. H. (1955). "Snow Cover and Climate in the Sierra Nevada California," Publications in Geography, No. 11. University of California Press, Berkeley.

Miller, D. H. (1977). "Water at the Surface of the Earth." Academic Press, New York.

Moscow. Moskovskii Gosudarstvennyi Universitet im. M. V. Lomonosov, Geogr. Fak. (1976). Nabliudeniia Meteorologicheskoi Observatorii MGU, Moskva, Iiun' 1976.

Neuwirth, F. (1979). Beziehungen zwischen Globalstrahlung, Himmelstrahlung und extraterrestrischer Strahlung in Österreich. *Arch. Meteorol., Geophys. Bioklimatol., Ser. B* **27**, 1–13.

Nishizawa, T. (1977). Heat balance. *In* "The Climate of Japan" (E. Fukui, ed.), pp. 135–166. Kodansha, Tokyo and Elsevier, Amsterdam.

Ramage, C. S. (1979). Prospecting for meteorological energy in Hawaii. *Bul. Am. Meteorol. Soc.* **60**, 430–438.

Rauner, Iu. L., Ananeva, L. M., and Samarina, N. N. (1974). Radiatsionno-teplovoi rezhim lugovo-stepnykh fitotsenozov na razlichnykh elementakh rel'efa. *In* "Issledovaniia Genezisa Klimata," pp. 257–270. Akad. Nauk SSSR, Inst. Geogr., Moscow.

Revfeim, K. J. A. (1978). A simple procedure for estimating global daily radiation on any surface. *J. Appl. Meteorol.* **17**, 1126–1131.

Robinson, N., ed. (1966). "Solar Radiation." Am. Elsevier, New York.

Ross, J. (1975). Radiative transfer in plant communities. *In* "Vegetation and the Atmosphere. Vol. I: Principles" (J. L. Monteith, ed.), pp. 13–55. Academic Press, New York.

Satterlund, D. R., and Means, J. E. (1978). Estimating solar radiation under variable cloud conditions. *Forest Sci.* **24**, 363–373.

Schulze, R. (1970). "Strahlenklima der Erde." Steinkopf, Darmstadt.

Skaar, E. (1980). Application of meteorological data to agroclimatological mapping. *Int. J. Biometeorol.* **24**, 3–12.

Smithsonian Institution (1966). "Smithsonian Meteorological Tables" (R.J. List, ed.), 6th ed., Smithsonian Institution Publ. No. 4014. Washington, D.C.

Sørensen, B. (1975). Energy and resources. *Science* **189**, 255–260.

Swift, L. W., Jr. (1976). Algorithm for solar radiation on mountain slopes. *Water Resour. Res.* **12**, 108–112.

Tarpley, J. D. (1979). Estimating incident solar radiation at the surface from geostationary satellite data. *J. Appl. Meteorol.* **18**, 1172–1181.

Teeri, J. A. (1976). Phytotron analysis of a photoperiodic response in a high Arctic plant species. *Ecology* **57**, 374–379.

Tuller, S. E. (1976). The relationship between diffuse, total, and extraterrestrial solar radiation. *Sol. Energy* **18**, 259–263.

U.S. Weather Bureau (1964). "Mean Daily Solar Radiation, Monthly and Annual." Washington, D.C. (Republished *in* "Climatic Atlas of the United States," plates 69–70. U.S. Dep. Commer., Environ. Sci. Serv. Adm., Environ. Data Serv., Washington, D,C., 1968).

Whillier, A. (1965). Solar radiation graphs. *Sol. Energy* **9**(3), 164–165.

Whipple, C. (1980). The energy impacts of solar heating. *Science* **208**, 262–266.

Williams, J. G. (1976). Small variation in the photosynthetically active fraction of solar radiation on clear days. *Arch. Meteorol., Geophys. Bioklimatol., Ser. B* **24**, 209–217.

Williams, L. D., Barry, R. G., and Andrews, J. T. (1972). Application of computed global radiation for areas of high relief. *J. Appl. Meteorol.* **11**, 526–533.

Chapter VI

INCOMING LONGWAVE RADIATION

An invisible, little-known energy flux that has sometimes gone un-recognized as the largest supplier of energy to ecosystems almost any-where in the world is entirely different from the two fluxes of radiation and their sum that we have described in Chapters III–V. This is the downward flux of longwave radiation* (wavelengths from 4 to 50 μm), i.e., the downward component of a field of radiant energy that fills the atmosphere. It is neglected partly because of its invisibility, partly be-cause it is reasonably steady,† a characteristic highly valuable to eco-systems receiving it.

MAGNITUDE AND SIGNIFICANCE OF THE FLUX

Quanta emitted as a result of rotational and vibrational motions in triatomic molecules in the atmosphere, though emitted in bands rather than continuously over the spectrum, and though emanating from gases that form only a trace of the total atmosphere, reach the earth's surface at flux densities of 200–400 W m^{-2}. This flux is at times augmented by the radiation emitted by droplets and particles.

---

* *Terminology* Longwave radiation is called "atmospheric" because it is emitted by the atmosphere; also "terrestrial" (as contrasted with solar), but then it is likely to be confused with the upward flux from the earth itself; also "nocturnal" because it is more easily measured at night, but it is larger by day; also "thermal" because its effects are primarily heating.

† This flux is sometimes confused with the net resultant of the upward and downward components of the field of longwave radiation, which in reality is not itself an energy flux but merely the resultant of fluxes from distinct sources and not so easily amenable to physical interpretation. As a small quantity (less than 50 W m^{-2}) in the energy balance, it may receive perfunctory attention in analysis; this is discussed in Chapter X.

These streams of radiative energy do not originate in the bulk constituents of the atmosphere, nitrogen and oxygen, for their molecules have no absorption or emission bands in the spectral region 4–50 μm relevant to radiators having a temperature of 250–300°K (Wien's law). In this spectral region, however, a few triatomic molecules intermingled with the bulk constituents emit photons by rotational transitions at wavelengths longer than 15 μm and by vibrational and rotational–vibrational transitions at wavelengths shorter than 15 μm.

The 4–8-μm band, from CO_2 and H_2O, has an energy content of the order of 50–100 W m^{-2}, mostly generated in the lowest 50 m of the atmosphere, for which water-vapor content and temperature can generally be represented by measurements at the usual shelter height, 1.5–2 m. Radiation at wavelengths greater than 12 μm, in overlapping bands from CO_2 and H_2O, mostly comes from the lowest 0.2–0.6 km of the atmosphere. Its energy content, depending on temperature and humidity of these layers, is approximately four times as great as that in the 4–8-μm band. A minor stream in wavelengths from 9.5 to 10 μm originates in ozone molecules in the stratosphere.

The total inflow of radiation dwarfs other inflows of heat in the energy budgets of almost all ecosystems. By its very size and steadiness, it smooths the conditions under which ecosystems survive and function and helps provide a predictable environment. During the polar night "the environment is composed of thermal [i.e., longwave] radiation, modified at times by turbulent winds" (Gates, 1963). Because energy is seldom supplied by conduction from the substrate and turbulent transport from the stable nocturnal atmosphere at flux densities larger than 50–100 W m^{-2}, longwave radiation dominates the energy budget. The size of this gift renders it unnecessary for an ecosystem to extract more sensible heat from air or soil and thereby affects these other heat fluxes.

When it fails, ecosystems are in trouble. Brooks (1952) calculated a downward flux of only 210 W m^{-2} in the freeze of 22 January 1937 in California. To remain at an above-freezing temperature, an ecosystem must be able to disburse approximately 315 W m^{-2}; when the income dropped, so also did orchard temperature—to a level at which disbursement was 280 W m^{-2}, about $-8°C$, which was catastrophic for the trees.

RADIATION UNDER A CLOUDLESS SKY

Variations in concentration and temperature of atmospheric sources of longwave radiation produce variations in the downward component at the surface of the earth.

"Permanent" Gases

Changes in stratospheric ozone are ecologically less important for their effect on longwave radiation than they are for that on solar ultraviolet. The thermal contribution of ozone is small because the emitter is cold; flux density is 15–20 W m^{-2} (Staley and Jurica, 1972). Tropospheric ozone, which is now recognized on a regional scale, may also affect the longwave flux.

Carbon dioxide, even at its present trace concentrations, emits strongly in several spectral regions, totaling to a flux density of about 70–75 W m^{-2}. A diurnal variation is caused by the variation in its temperature and its concentration, which is usually lower by day, substantially so in valleys (Reiter and Kanter, 1980). If its concentration doubles or triples as a result of human dependence on fossil energy, this flux will increase substantially, as shown by many models [see Ramanathan *et al.* (1979); Manabe and Wetherald (1980)].

Water Vapor

Water vapor, the most variable constituent of the atmosphere, radiates in the 5.5–7.5 μm and >13 μm bands, overlapping somewhat with the CO_2 emission bands, to deliver from half to three-quarters of the clear-sky flux density. Its range, from 150 to 300 W m^{-2}, depends on vapor concentration and temperature, which are so well correlated in many regions that an estimating formula can bury vapor concentration in the temperature variable (Swinbank, 1963; Deacon, 1970). In its 5.5–7.5-μm spectral region most of the energy is emitted by vapor molecules near the ground—within 0.1–0.2 km (Kondratyev, 1969, p. 620), and local vapor sources may be important. Emission in longer wavelengths comes from deeper layers of the troposphere.

If "precipitable" water (the vertically integrated mass of vapor in the atmospheric column) is taken as the independent variable, the downward component of longwave radiation at the underlying surface is as shown in Table I. Because much of the flux originates near the ground, shelter observations of vapor pressure are also appropriate predictors (Table II). The range of temperature occurring at each value of vapor pressure introduces a variation of 40–50 W m^{-2} about the mean relation shown in Table II, but the large effect of water vapor is still clear.

Aerosols

The contribution of aerosol components is generally recognized as augmenting the longwave flux, even when specific effects are not yet identified. Substances with absorption–emission bands within the win-

TABLE I

Downward Longwave Radiation at Different Masses of Atmospheric Water Vapor [a]

Mass of atmospheric vapor (kg m^{-2})	Flux density (W m^{-2})
5	230
10	260
20	320

[a] Data from Barashkova (1960), from 4-yr record at Karadag.

dow from 7 to 14 μm, in which natural constituents of the atmosphere (except stratospheric ozone) do not emit, are particularly important, and some of these may be increasing in concentration as a result of industrialization and chemicalized agriculture (Wang et al., 1976; Wang et al., 1980; Donner and Ramanathan, 1980).

Particles, including some of biological origin, also may generate substantial radiation in the window (Dalrymple and Unsworth, 1978b). Calculations of radiation at wavelengths around 6 μm show that ordinary aerosols augment the water-generated flux of zenithal radiation to the ground by 0.17 (Kattawar and Plass, 1971), even though this spectral region was outside the window.

Source Geometry

Molecules at a particular level absorb certain wavelengths of the descending stream from colder molecules above. They emit radiation at nearly the same wavelengths but with greater intensity since they are warmer. The descending flux increases in density as shown by soundings (Kuhn et al., 1959) and is the basis for many models.

The descending stream coming from the zenith carries less energy than do those at lower angles. At a horizon angle of 50° the increase is 3%, at an angle of 30° it is 7%. The sky directly overhead is the coldest in terms of the radiative energy an ecosystem receives from it, and the 8–12 μm window is most marked (Schulze, 1970, p. 128). Monte Carlo modeling shows that the zenith always emitted less flux than did the lower parts of the sky, an order of magnitude less from a dry, cloudless atmosphere at one particular wavelength (Kattawar and Plass, 1971).

Measurements at Blue Hill Meteorological Observatory over the whole

TABLE II

Downward Longwave Radiation at
Different Surface Vapor Pressures[a]

Vapor pressure (mbar)	Flux density $(W\ m^{-2})$
2	200
6	270
12	340
20	380

[a] Data from Barashkova (1960),
4-yr record, wind speed 5 m sec^{-1}
or higher. Karadag 45°N.

longwave spectrum (Brooks, 1952) found the zenithal flux to be 4 W m^{-2} sr^{-1} less than the hemispherical average. The flux from a band of the sky 15° above the horizon was 7 W m^{-2} sr^{-1} more than the average, which is in accord with computations (Kondratyev, 1969, p. 598). The flux from the band 38° above the horizon approximates the mean flux from the whole hemisphere (Dines and Dines, 1927; Dalrymple and Unsworth, 1978a). The suggestion that the flux be considered as two components—one from an isotropic and one from a directional source (Unsworth and Monteith, 1975)—is similar to the one noted in Chapter IV for a two-source geometry of scattered shortwave radiation.*

SOURCES OF VARIATION UNDER CLOUDS

Longwave radiation increases when water vapor condenses into droplets that radiate continuously over the whole longwave spectral range, including the 8–12 μm window in which water vapor does not radiate. A cloud base near the ground radiates virtually as a blackbody, but the window tends to reappear if the cloud base is at 3 km (Kondratyev, 1969, Fig. 9.22, p. 623), and hence colder. If the cloud base is at 9 km, the window is only half-filled because in the adjacent bands, around 7 μm and 14+ μm, the vapor and CO_2 below cloud level are radiating strongly. The lower and warmer the cloud base, the less likely that its radiation will be obscured by the intervening atmosphere (Niilisk, 1972).

* During much of the day, the area of the sky generating the least scattered solar radiation is also near the zenith, and a number of microclimatic practices utilize this cold hole for a kind of passive cooling (Brooks, 1959). It is particularly effective for ecosystems under dry, cloudless atmospheres.

Middle clouds, for example, increase the cloudless-sky flux density by 0.15 (or 40 W m^{-2}) when the surface vapor pressure is 6 mbar, but by only 0.09 (or 20 W m^{-2}) when vapor pressure is 12 mbar (Barashkova, 1960). Modeling shows the smaller effect in humid low latitudes (Kuhn, 1978), and night observations at Calcutta (De and Gupta, 1964) display an increase with overcast over clear skies in dry December (64 W m^{-2} change) that is twice as large as the increase in the moist season (June).

Considering the height and low temperature of cirrus clouds, they do not contribute a great deal to the longwave flux arriving at the earth's surface—about 0.05 in Barashkova's (1960) measurements. The effect would probably be relatively most important in cold, dry weather when warmth and vapor concentration in the intervening atmosphere are lacking. It was not uncommon in the Sierra Nevada to note reduced nocturnal cooling when a nearby airways station reported clear skies through the hours of darkness, but "high thin overcast" as soon as the sun made a cirrus deck visible. The effects of radiation from this deck were significant in these low-flux conditions and made up about half of the net loss of heat by exchange of radiation between snow and sky. Cirrostratus in storm systems is denser and appears to have a significant emissivity (Platt and Dilley, 1979).

MEASUREMENT AND ESTIMATION

The vertical component of incoming longwave radiation is measured by the difference between two hemispherical sensors open respectively to solar and all-wave radiation, not a precise procedure. An alternative is a sensor dome that screens out solar wavelengths below 3.6 μm, such as the one developed by Schulze (1970, p. 196) in the early 1950s. These hemispherical pyrgeometers need to be protected from solar overheating (Enz et al., 1975; Gaevskii, 1960). Other means of eliminating solar radiation have been proposed to make daytime measurements more accurate (e.g., Kozik, 1957).

Calibration is more difficult than for shortwave instruments, which are not always well calibrated either. Whether for this reason or because downward longwave radiation has not appeared as a separate flux in many energy-budget studies, few routine observations are made even in countries with good radiation networks. The Hamburg Observatory, reporting daily sums, has the only long published record known to me.

The lack of observations and the steadiness of the flux have encouraged development of calculation methods to determine long-term means. Empirical formulations for cloudless-sky radiation that have been used for 50 years include shelter temperature and vapor pressure with varying numerical coefficients, as described in Brunt (1934), Budyko (1974), and in many other places.* In some formulations the association of air temperature and vapor pressure is employed to fold the latter term into the former (Swinbank, 1963) as noted earlier, although in certain conditions vapor pressure is explicitly needed (Idso, 1974). Downward longwave radiation can be expressed as a fraction of blackbody radiation at shelter temperature, applying an apparent emissivity term that depends on humidity (Brutsaert, 1975; Marks and Dozier, 1979; Satterlund, 1979) or aerosol concentrations.

Employment of 1.5-m air temperature to represent the temperatures of the radiating layers of the atmosphere is another approximation because at different wavelengths the flux is generated by layers of different depths. The implied shelter–troposphere relationship is also disturbed when inversions occur. Kanemasu and Arkin (1974) ascribe a systematic diurnal deviation between observed and computed flux densities in Kansas to the fact that the shelter thermometer underestimates the temperature of the radiatively active depth of air at night and overestimates it in the daytime.

Paulsen and Torheim (1964) point out that "these empirical formulae will take different forms in the different climates...." This problem is approached by graphical formulations such as those of Elsasser or Shekhter, using temperature and vapor concentrations of separate layers from specific radiosonde ascents and various ways of integrating over wavelength (Charlock et al., 1976; Haltiner and Martin, 1957, Chap. 8). Cloudiness enters the graphical solutions by a blackbody assumption [criticized by Goody (1964)] and the algebraic solutions by relations like those used for shortwave radiation.

A cloud coefficient can be derived from the ratio of observed to clear-day total solar radiation (Anderson and Baker, 1967). One may be derived as in solar radiation formulas, perhaps as cloud cover n to some power, say 2.5 (Kreitz, 1954).

Graphical and algebraic expressions for downward longwave radiation are useful for large-scale work, perhaps in modeling, and in reconstructions of the past. It seems desirable, however, to measure locally at least long enough to assess errors in coefficients.

---

* These formulas are one of several combinations of air temperature and humidity, developed for such purposes as equivalent temperature and equivalent potential temperature in meteorology and a plethora of "effective" temperatures in physiology and air-conditioning engineering. Longwave radiation was felt to be implicit in the physiological expressions used in a study of sultry weather in the central United States (Havlik, 1976, p. 33).

VARIATIONS OVER TIME

The salient characteristic of the longwave radiation coming into eco-systems is its steadiness, compared with shortwave radiation fluxes and indeed with energy fluxes yet to be discussed. Paulsen and Torheim (1964) in noting that on a winter day in Bergen the flux is only 0.25 less than that on a summer day, refer to "this well-known, however re-markable, feature . . . ." Variations, though subdued in a relative sense, do, however, have ecological significance because of their absolute mag-nitude. The spectrum of time scales over which longwave radiation var-ies displays peaks associated with the occurrence of the two principal radiators—clouds and boundary-layer gases. Adding the two astronom-ical cycles of boundary-layer variation, day and year, give us a total of five time scales:

(1) minute to minute, caused by passing individual clouds;
(2) the day;
(3) variation from day to day, caused by passing cloud systems;
(4) the year;
(5) variation from year to year, caused by changes in the frequency of synoptic systems in different years, and in concentration of radiating gases.

Sources of the variation in longwave radiation are the variations in atmospheric heat and vapor at scales 2 and 4 and in cloud cover and aerosol concentration at scales 1, 3, and 5. The correlation with short-wave radiation is positive at scales 2 and 4, negative at scales 1, 3, and 5 (see Chapter VII).

Momentary Variations (Scale 1)

Microscale variations occur as a radiating cloud coming over the sky raises the flux density by 25 W m^{-2} or more. These have less impact than cloud-induced changes in the solar beam, but afford a degree of compensation in total energy input. Their effects on dew formation are familiar.

Diurnal Variations (Scale 2)

The difference between night and day values of flux density runs parallel with, not counter to, the diurnal march of solar radiation and are large in summer, small on winter days (Table III). Details of the daily march reflect the familiar lag in boundary-layer conditions behind the surface heat fluxes. A slow decrease occurs at night, sometimes expo-

TABLE III

Low and High Flux Densities of Downward Longwave Radiation in the Diurnal March

Station and season	Hourly values (W m^{-2})		Range (W m^{-2})	Source
	Low	High		
Bergen, summer clear day	—	—	75	Paulsen and Torheim (1964)
Bergen, June mean	327	375	48	Paulsen and Torheim (1964)
Bergen, winter clear day	—	—	~0	Paulsen and Torheim (1964)
Bergen, December mean	302	311	9	Paulsen and Torheim (1964)
Wien, July	325	350	25	Dirmhirn (1964)
Eastern Alps (3 km), summer				
clear days	—	—	50	Wagner (1980)
all days	—	—	35	Wagner (1980)
Karadag, July	280	345	65	Barashkova (1960)
Tashkent, June clear day	345	395	50	Lopukhin (1957)
Tashkent, June all days	350	400	50	Lopukhin (1957
Tashkent, December clear day	260	280	20	Lopukhin (1957)
Tashkent, December all days	280	310	30	Lopukhin (1957)
Central Asia, desert summer				
day	305	445	140	Aizenshtat and Zuev (1952)
Central Asia, desert July days	303	406	103	Aizenshtat (1960)
Central Asia, Pamir Mts. (4 km),				
July	195	237	42	Muminov (1959)
Windsor, autumn	290	360	70	Brazel and Osborne (1976)

nentially. At Tashkent (in a large oasis), the decrease from flux densities of the summer evening amounts to 25–30 W m^{-2} by dawn (Aizenshtat, 1957). It is about 20 W m^{-2} on May nights at Calcutta (De and Gupta, 1964). Brooks (1952) measured it in dry country as about 0.07 of the initial flux density.

The ecological significance of this flux at night lies in its effect on leaf-cell temperatures, especially useful in cold weather. In summer it is less welcome since it increases the respiration level of vegetation. Radiation (360–380 W m^{-2}) on July and August nights in Tashkent (and other oasis or low-latitude cities) slows the cooling of houses and adds to the discomfort of sleeping.*

Interdiurnal Variations (Scale 3)

Longwave radiation fluctuates from day to day as a result of changes in atmospheric warmth, moisture, and especially cloud content. Cold

* The death toll of the 1965 earthquake was less than might have been expected because many people were sleeping outside their houses, as they used to do in Arizona before air conditioning became common.

advection brings days when radiation from dry, cold, often cloudless air is very small. For example, in November 1977, a northerly stream of cool and very dry air (specific humidity 0.08 g kg^{-1} in the 80-60 kPa layer) over southeastern Wisconsin emitted a downward flux calculated as 190 W m^{-2} and measured as 195 W m^{-2}. Still lower fluxes occur in colder air later in the winter.

Warm advection, transporting millions of megajoules of energy poleward, supports rapid radiation of much of this heat to the underlying surface, and the downward flux might at times exceed the upward flux. The large excess (65 W m^{-2} or 0.26 of the mean) of radiation on cloudy days in December at Hamburg over that on clear days exemplifies a major avenue by which Atlantic heat reaches ecosystems of western Europe.

Radiation on clear and cloudy days in summer differs by only 20 W m^{-2}, or 0.05 of the mean. Interdiurnal variations at Hamburg average 19 W m^{-2} in December (0.06 of the mean, smaller than the 65 above because clear or cloudy days tend to come in clusters, and 13 W m^{-2} in June (0.03 of the mean).

Annual Variations (Scale 4)

Like the diurnal march, the annual march follows the rhythmic heating and moistening of the lower troposphere. Differences in June and December means are shown in Table IV, as well as the range between highest and lowest months which lag the solstices. (Maxima in the annual march typically occur from mid-July to mid-August.) The annual variation in Europe is small near the coast, larger inland (Wien) and still larger in the central chernozem zone [95 W m^{-2} at Kursk (Rudnev, 1980)].

The range in monthly mean values at Tashkent (vertical transect in Fig. 1) over the year is not much more than the range in hourly means from night to day (horizontal transect). All three forms of variation, the diurnal and annual associated with boundary-layer warmth and moisture and the interdiurnal associated with cloud systems, appear to display about the same magnitude.

Year-to-Year Variation (Scale 5)

Differences in synoptic weather systems from year to year affect cloud cover and air temperature and cause variation in longwave radiation. The standard deviation of mean monthly values of the flux in the same winter or summer month (data for 15 yr available at Hamburg) is about

TABLE IV

Flux Densities of Incoming Longwave Radiation in the Annual March[a]

Station	December	June	Difference June–December	Range between high and low months	Source
Barrow, 71°N	210[b]	270[b]	60	70	Maykut and Church (1973)
Bergen, 60°	305	350	45	75	Paulsen and Torheim (1964)
Hamburg, 54°	300	345	45	80	Germany, Wetterdienst (1954–1974)
Wien, 48°	282	353	71	95	Dirmhirn (1964)
Karadag, 45°					
night, clear weather	245	330	85	—	Barashkova (1960)
all days	295	365	70		
Tashkent, 41°					
night, clear	230	340	110	—	Aizenshtat (1957)
broken clouds	285	370	85		
overcast	300	360	60		
Tashkent, 41°					
all hours	291	370	79	—	Lopukhin (1957)
New Delhi, 29°					
early night	370	450	80	—	Mani and Chacko (1963)
Calcutta, 23°					
night, clear	320	418	98	—	De and Gupta (1964)
overcast	383	448	65		
Poona, 19°					
early night	345	375	30	—	Mani and Chacko (1963)
Sydney, Montana, 49°					
clear days	190[c]	265[c]	75	—	Aase and Idso (1978)
Phoenix, 33°					
clear days	315	450	—	135	Idso (1974)

[a] Data given in watts per square meter.
[b] Estimated.
[c] Winter and summer means.

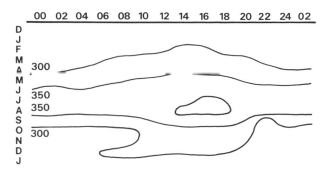

Fig. VI.1. Isopleth diagram of the downward flux of longwave radiation at Tashkent (41°N latitude) (W m^{-2}). [Data from Lopukhin (1957).] A horizontal transect across the diagram shows the restrained diurnal march of this flux in a given month; a vertical transect shows the annual march at a given time of day, which has a wider range. The pattern is repeated indefinitely as days and years follow one another.

4% of the mean flux. The frequency of cloud systems in different summers affects both the monthly mean and the day-to-day variance.

Although the role of radiating aerosols is not fully understood, it seems likely that their increase in the urban and global atmospheres will cause a systematic rise in the flux density of downward longwave radiation. The effect of increased CO_2 concentration has been much discussed on the basis of *calculated* radiation fluxes, and it is unfortunate that our baseline data on this radiation flux are so scanty.

SPATIAL VARIATIONS IN INCOMING LONGWAVE RADIATION

Micro- and Mesoscale

The longwave radiation entering an ecosystem is usually augmented downward by emission from the upper leaves. The vertical profiles of energy fluxes will be discussed in Chapter XX, but here we can note an example: A flux of 335 W m^{-2} entering a red pine stand was increased to 395 W m^{-2} at the ground (Reifsnyder, 1967). The environment of the lower leaves is one of isotropic radiation, quite different from the environment of the outer leaves.

It was early recognized that frost is a radiative phenomenon, and on still nights the important factors are incoming longwave radiation flux and that emitted by leaves and fruit. What could be more obvious than to supply a radiator above an orchard? At the time of the 1937 freeze in California it was believed that a soot cloud was such a radiator, and growers made their orchard heaters smoke as much as possible and burned old tires by the ton. The smoke from hundreds of citrus orchards drifted into Los Angeles. Although city people were months in cleaning up the gummy deposits, they did so cheerfully because they sympathized with the growers, but research by Brooks and others, sparked by this massive frost, determined that the smoke had less radiative benefit than the heaters did. Smokeless heaters with hot radiating stacks were subsequently developed.

The radiation principle was later applied by fogging systems that produce water droplets between 10 and 40 μm in diameter. A trial (Brewer *et al.*, 1974) showed downward longwave radiation was about 240 W m^{-2} under clear skies and 280 W m^{-2} under the fog (the heat demand for the orchard at 5°C was 335 W m^{-2}). The fog emitted 40 W m^{-2}, much less than was expected, and met 0.4 of the net loss of heat by radiative exchange. Changes in drift "disrupted the fog coverage and pushed the fog uphill to normally warmer areas—where it was not needed as badly as it was in the lower basin" (Brewer *et al.*, 1974). This tendency of artificial fogs or smokes to drift away has also been noted in European practice (Bagdonas *et al.*, 1978, pp. 97–99). Moreover, chemical smokes are hazardous.

Radiation from a ridge or stand of trees to an adjacent low ecosystem can be evaluated for various geometrical situations. Trees displace the horizon sectors of the sky from which, because of their warmth and optical thickness, the radiation flux to an ecosystem is relatively great, though at a low angle. Tree stands and hills do not cover the central area of the sky that is important by the cosine relation and usually the coldest radiator. Radiation input to slopes can be calculated as shown by Kondratyev (1969) or Unsworth and Monteith (1975), or it can be modeled (Zdunkowski and McDonald, 1966; Marks and Dozier, 1979). Greenland (1973) ascribes the winter surplus (instead of deficit) of all-wave radiation in a New Zealand valley at 43°S to radiation emitted by air heated by adjacent hillsides.

Urban atmospheres emit more longwave radiation than do rural, e.g., those of Windsor (Brazel and Osborne, 1976), Hamilton (Rouse *et al.*, 1973), and Montreal (Oke and Fuggle, 1972). The added load of excess longwave radiation in summer heat waves has lethal potential and should be monitored.

Larger Scales

Cloudy coastal regions presumably receive more longwave radiation in winter than interior regions as a result of heat advection. This is obvious in western Europe, where the Bergen and Hamburg stations cited lie on the oceanic side of a major air-stream boundary in January (Klaus, 1979). However, we need to separate the aspects of advection relevant to longwave radiation: (a) the warmth of the air itself, (b) advected clouds, and (c) conversion of latent heat into radiation when clouds form over the receiving region. For example, does summer cold advection onto the coast of California reduce the longwave flux more than the advected stratus clouds increase it?

Altitude. Warm moist lowland atmospheres contrast with the thin, usually dry, cold atmosphere above mountain ecosystems. The vertical difference in relatively dry air in central Asia in summer is about 30 W m^{-2} for a kilometer change in altitude (Zuev, 1960; Aizenshtat, 1960). The same variation occurs in summer in the eastern Alps, somewhat less in winter (Sauberer, 1954). The summer gradient is confirmed from Wagner's (1980) measurement at 3 km altitude.

Global Extremes. The incomplete station coverage in Table IV suggests annual means of about 240 W m^{-2} in Alaska,* 320 W m^{-2} at the European sites, more at Tashkent, still more in India, over the Indian Ocean 380 W m^{-2} (Portman and Ryznar, 1971, p. 46), and in regions under equatorial air 400 W m^{-2}. If we postulate air temperature near the ground at 33°C and vapor pressure near 45 mbar for air in equilibrium with a surface evaporating at its maximum rate in a high-energy climate, incoming longwave radiation from the usual formulas (Brutsaert, 1975; Staley and Jurica, 1972; Kondratyev, 1969; Swinbank, 1963) approaches 450 W m^{-2} from a cloudless sky and slightly more from thick low clouds.

The smallest fluxes come to ecosystems at high altitudes and high latitudes, and under cold dry atmospheres that possess and emit little energy. A classic investigation of atmosphere–surface interaction during a long cooling period in the Yukon valley (Wexler, 1936) determined a downward flux of only 85 W m^{-2} and was confirmed in a later study when a low temperature record for North America was set (Wexler, 1948). Fluxes in shorter interstorm periods on the Greenland ice cap (70°N) diminished to 125 W m^{-2} (Miller, 1956; Schlatter, 1972; measured fluxes in cloudless winter weather at Resolute (75°N) averaged 125 W m^{-2} also (Suckling and Wolfe, 1979).

Zonal means of downward longwave radiation in the northern hemisphere, calculated by Niilisk (1960) in comparing several methods, display poleward decline from 400 W m^{-2} in equatorial latitudes to 170 W m^{-2} at 60–70°N latitude. This range over the ecosystems of the globe

---

* Nevertheless, even here at 71°N latitude the flux is large enough in summer to play an important role in models of soil thawing and of tundra productivity (Miller *et al.*, 1976).

contrasts with the relatively small time variation to a given ecosystem. The comparatively steady and large input of energy to any particular ecosystem buffers fluctuations in its receipts of solar energy and plays a sustaining role in its total energy budget.

REFERENCES

Aase, J. K., and Idso, S. B. (1978). A comparison of two formula types for calculating longwave radiation from the atmosphere. *Water Resour. Res.* **14**, 623–625.

Aizenshtat, B. A. (1957). Radiatsionnye balans i temperatura poverkhnosti pochvy v Tashkent. *Tr. Tashk. Geofiz. Obs.* **13**, 3–74.

Aizenshtat, B. A. (1960). "The Heat Balance and Microclimate of Certain Landscapes in a Sandy Desert" (transl. by G. S. Mitchell). U.S. Weather Bur., Washington, D.C.

Aizenshtat, B. A., and Zuev, M. V. (1952). Nekotorye cherty teplovogo balansa peschanoi pustyni. *Tr. Tashk. Geofiz. Obs.* **6**, (7).

Anderson, E. A., and Baker, D. R. (1967). Estimating incident terrestrial radiation under all atmospheric conditions. *Water Resour. Res.* **3**, 975–987.

Bagdonas, A., Georg, J. C., and Gerber, J. F. (1978). Techniques of Frost Prediction and Methods of Frost and Cold Protection. *WMO Tech. Note* No. 157 (WMO No. 487).

Barashkova, E. P. (1960). Dlinnovolnovyi balans podstilaiushchei poverkhnosti po nabliudeniiam v Karadage. *Tr. Gl. Geofiz. Obs.* **100**, 141–153.

Brazel, A. J., and Osborne, R. (1976). Observations of atmospheric thermal radiation at Windsor, Ontario, Canada. *Arch. Meteorol., Geophys. Bioklimatol., Ser. B* **24**, 189–200.

Brewer, R. F., Burns, R. M., and Opitz, K. W. (1974). Evaluation of man-made fog for frost protection of citrus in California. *Calif. Agric.* **28**(5), 3–5.

Brooks, F. A. (1952). Atmospheric radiation and its reflection from the ground. *J. Meteorol.* **9**, 41–52.

Brooks, F. A. (1959). "An Introduction to Physical Microclimatology." Associated Student Store, Davis, California.

Brunt, Sir D. (1934). "Physical and Dynamical Meteorology." Cambridge Univ. Press, London.

Brutsaert, W. (1975). On a derivable formula for long-wave radiation from clear skies. *Water Resour. Res.* **11**, 742–744.

Budyko, M. I. (1974). "Climate and Life" (Engl. transl. ed. by D. H. Miller). Academic Press, New York.

Charlock, T., Herman, B. M., and Zdunkowski, W. G. (1976). Comments on "Discussion of the Elsasser formulation for infrared fluxes." *J. Appl. Meteorol.* **15**, 1317–1319.

Dalrymple, G. J., and Unsworth, M. H. (1978a). Longwave radiation at the ground: III. A radiometer for the "representative angle." *Q. J. R. Meteorol. Soc.* **104**, 357–362.

Dalrymple, G. J., and Unsworth, M. H. (1978b). Longwave radiation at the ground: IV. Comparison of measurement and calculation of radiation from cloudless skies. *Q. J. R. Meteorol. Soc.* **104**, 989–997.

De, A. C., and Gupta, P. K. (1964). Measurements of nocturnal radiation made at Dum Dum Airport, Calcutta. *Indian J. Meteorol. Geophys.* **15**, 439–446.

Deacon, E. L. (1970). The derivation of Swinbank's long-wave radiation formula. *Q. J. R. Meteorol. Soc.* **96**, 313–319.

Dines, W. H., and Dines, L. H. G. (1927). Monthly means of radiation from various parts of the sky at Benson, Oxfordshire. *Mem. R. Meteorol. Soc.* **20**(11), 1–8.

Dirmhirn, I. (1964). "Das Strahlungsfeld im Lebensraum." Akad. Verlagsges., Frankfurt a.M.

Donner, L., and Ramanathan, V. (1980). Methane and nitrous oxide: their effects on the terrestrial climate. *J. Atmos. Sci.* **37**, 119–124.

Enz, J. W., Klink, J. C., and Baker, D. G. (1975). Solar radiation effects on pyrgeometer performance. *J. Appl. Meteorol.* **14**, 1297–1302.

Gaevskii, V. L. (1960). Issledovanie dlinnovolnovogo izlucheniia atmosfery. *Tr. Gl. Geofiz. Obs.* **100**, 86–92.

Gates, D. M. (1963). The energy environment in which we live. *Am. Sci.* **51**, 327–348.

Germany, Wetterdienst (1954–1974). Beilage zum Medizin-meteorologischen Bericht. (Hamburg Observatorium, monthly.)

Goody, R. M. (1964). "Atmospheric Radiation. I: Theoretical Basis." Oxford Univ. Press, London and New York.

Greenland, D. (1973). An estimate of the heat balance in an alpine valley in the New Zealand Southern Alps. *Agric. Meteorol.* **11**, 293–302.

Haltiner, G. J., and Martin, F. L. (1957). "Dynamical and Physical Meteorology." McGraw-Hill, New York.

Havlik, D. (1976). Untersuchungen zur Schwüle im Kontinentalen Tiefland der Vereinigten Staaten von Amerika. *Freiburg. Geogr. Hefte,* **15**.

Idso, S. B. (1974). On the use of equations to estimate atmospheric thermal radiation. *Arch. Meteorol., Geophys. Bioklimatol., Ser. B* **22**, 287–299.

Kanemasu, E. T., and Arkin, G. F. (1974). Radiant energy and light environment of crops. *Agric. Meteorol.* **14**, 211–225.

Kattawar, G. W., and Plass, G. N (1971). Influence of aerosols, clouds, and molecular absorption on atmospheric emission. *J. Geophys. Res.* **76**, 3437–3444.

Klaus, D. (1979). Wärmemangel- und Trockengrenzen der Vegetation in ihrer Beziehung zu den Luftmassengrenzen. *Erdkunde* **33**, 258–266.

Kondratyev, K. Ya. (1969). "Radiation in the Atmosphere." Academic Press, New York.

Kozik, E. M. (1957). Absoliutny pirgeometr. *Tr. Tashk. Geofiz. Obs.* **13**, 133–137.

Kreitz, E. (1954). Registrierungen der langwelligen Gegenstrahlung in Frankfurt. *Geofis. Pura Appl.* **28**, 292–300.

Kuhn, P. M., Suomi, V. E., and Darkow, G. L. (1959). Soundings of terrestrial radiation flux over Wisconsin. *Mon. Weather Rev.* **87**, 129–135.

Kuhn, W. R. (1978). The effects of cloud height, thickness and overlap on tropospheric terrestrial radiation. *J. Geophys. Res.* **83**, 1337–1346.

Lopukhin, E. A. (1957). Izluchenie atmosfery v Tashkente. *Tr. Tashk. Geofiz. Obs.* **13**, 90–112.

Manabe, S., and Wetherald, R. J. (1980). On the distribution of climatic change resulting from an increase in CO_2 content of the atmosphere. *J. Atmos. Sci.* **37**, 99–118.

Mani, A., and Chacko, O. (1963). Studies of natural radiation at Poona and Delhi. *Indian J. Meteorol. Geophys.* **14**, 196–204.

Marks, D., and Dozier, J. (1979). A clear-sky longwave radiation model for remote alpine areas. *Arch. Meteorol., Geophys. Bioklimatol., Ser. B* **27**, 159–187.

Maykut, G. A., and Church, P. E. (1973). Radiation climate of Barrow, Alaska, 1962–66. *J. Appl. Meteorol.* **12**, 620–628.

Miller, D. H. (1956). The influence of snow cover on the local climate of Greenland. *J. Meteorol.* **13**, 112–120.

Miller, P. C., Stoner, W. A., and Tieszen, L. L. (1976). A model of stand photosynthesis for the wet meadow tundra at Barrow, Alaska. *Ecology* **52**, 411–430.

Muminov, F. A. (1959). Radiatsionnyi i teplovoi balansy Alaiskoi Doliny v Raione Sary-Tasha. *Tr. Sredneaziat. Nauchno-Issled. Gidrometeorol. Inst.*, **2**, 165–174.

Niilisk, Kh. (1960). K voprosu o raschetakh teplovogo izlucheniia atmosfery. *Izv. Akad. Nauk Est. SSR, Inst. Fiz. Astron., Issled. Fiz. Atmos., Tartu* **2**, 67–114.

Niilisk, Kh. Yu. (1972). Cloud characteristics in problems of radiation energetics in the earth's atmosphere (AGU transl.). *Izv. Atmos. Ocean. Phys.* 8(3), 154–159.

Oke, T. R., and Fuggle, R. F. (1972). Comparison of urban/rural counter and net radiation at night. *Boundary-Layer Meteorol.* **2**, 290–308.

Paulsen, H. S., and Torheim, K. A. (1964). Atmospheric radiation in Bergen, December 1957–June 1959. *Arbok Univ. Bergen, Mat.-Naturvitensk. Ser.* No. 11.

Platt, C. M. R., and Dilley, A. C. (1979). Remote sounding of high clouds; II. Emissivity of cirrostratus. *J. Appl. Meteorol.* **18**, 1144–1150.

Portman, D. J., and Ryznar, E. (1971). "An Investigation of Heat Exchange." East–West Center Press, Honolulu, Hawaii.

Ramanathan, V., Lian, M. S., and Cess, R. D. (1979). Increased atmospheric CO_2: Zonal and seasonal estimates of the effect on the radiation energy balance and surface temperature. *J. Geophys. Res.* **84**, 4949–4958.

Reifsnyder, W. E. (1967). Radiation geometry in the measurement and interpretation of radiation balance. *Agric. Meteorol.* **4**, 255–265.

Reiter, R., and Kanter, H. J. (1980). First results of simultaneous recordings of the CO_2-concentration from a valley station and a neighboring mountain station at an altitudinal difference of about 1 km. *Arch. Meteorol. Geophys. Bioklimatol. Ser. B* **28**, 1–13.

Rouse, W. D., Noad, D., and McCutcheon, J. (1973). Radiation temperature and atmospheric emissivities in a polluted urban atmosphere at Hamilton, Ontario. *J. Appl. Meteorol.* **12**, 798–807.

Rudnev, N. I. (1980). O raschete protivoizlucheniia atmosfery. *Meteorol. Gidrol. 1980*, **3**, 107–108.

Satterlund, D. R. (1979). An improved equation for estimating long-wave radiation from the atmosphere. *Water Resour. Res.* **15**, 1649–1650.

Sauberer, F. (1954). Zur Abschätzung der Gegenstrahlung in den Ostalpen. *Wetter Leben* **6**, 53–56.

Schlatter, T. W. (1972). The local surface energy balance and sub-surface temperature regime in Antarctica. *Am. Meteorol. Soc., Proc. Conf. Atmos. Radiat.*, pp. 239–246.

Schulze, R. (1970). "Strahlenklima der Erde." Steinkopf, Darmstadt.

Staley, D. O., and Jurica, G. M. (1972). Effective atmospheric emissivity under clear skies. *J. Appl. Meteorol.* **11**, 349–356.

Suckling, P. W., and Wolfe, M. E. (1979). Empirical methods for estimating incoming longwave radiation for cloudless winter days at Resolute, N.W.T. *McGill Univ. Climatol. Bull.* **25**, 1–11.

Swinbank, W. C. (1963). Long-wave radiation from clear skies. *Q. J. R. Meteorol. Soc.* **89**, 339–348.

Unsworth, M. H., and Monteith, J. L. (1975). Long-wave radiation at the ground I. Angular distribution of incoming radiation. *Q. J. R. Meteorol. Soc.* **101**, 13–24.

Wagner, H. P. (1980). Strahlungshaushaltsuntersuchungen an einen Ostalpengletscher wahrend der Hauptablationsperiode. Teil II: Langwellige Strahlung und Strahlungsbilanz. *Arch. Meteorol., Geophys. Bioklimatol. Ser. B* **28**, 41–62.

Wang, W.-C., Pinto, J. P., and Yung, Y. L. (1980). Climatic effects due to halogenated compounds in the earth's atmosphere. *J. Atmos. Sci.* **37**, 333–338.

Wang, W.-C., Yung, Y. L., Lacis, A. A., Mo, T., and Hansen, J. E. (1976). Greenhouse effects due to man-made perturbations of trace gases. *Science* **194**, 685–690.

Wexler, H. (1936). Cooling in the lower atmosphere and the structure of polar continental air. *Mon. Weather Rev.* **64**, 122–136.

Wexler, H. (1948). A note on the record low temperature in the Yukon Territory, January–February 1947. *Bull. Am. Meteorol. Soc.* **29**, 547–550.

Zdunkowski, W. G., and McDonald, D. V. (1966). The distribution of infrared radiation flux densities and their divergence along the symmetry line of an idealized valley. *Geofis. Pura Appl.* **65**(3), 185–195.

Zuev, M. V. (1960). "On the Heat Balance of the Kara-Kul' Lake (Ozero) Valley in the East Pamirs." U.S. Weather Bur., Washington, D.C. (Transl. by N. A. Stepanova of O teplovom balanse doliny ozera Kara-Kul'. *In* "Sovremmenye Problemy Meteorologii Prizemnogo Sloia Vozduka," M. J. Budyko, ed., pp. 61–66. Gidrometeorol. Izd., Leningrad, 1958.)

Chapter VII

RADIANT ENERGY ABSORBED BY ECOSYSTEMS

Photons of different wavelengths emerge at different angles from the hemisphere above an ecosystem and are variously received by it. Some are reflected by surfaces in the upper layers, some are absorbed there, others penetrate to deeper layers where they are reflected, absorbed, or transmitted farther. The net effect for the ecosystem as a whole is reflection of a fraction of the incident photons and absorption of the others. The pattern of these two processes with depth will be discussed in Chapter XX; here we look at ecosystems as radiation-absorbing, energy-transforming entities.

The interactions of photons with ecosystem surfaces, like their interactions with molecules and particles in the atmosphere, are a complex case of the general question of interaction of radiation and matter that occupied much of 19th century physics. Our task is to recognize the multiplicity of photons of specific frequency and direction striking the outer surfaces of an ecosystem that can be highly selective in rejecting or absorbing these photons and to generalize these events into time and spatial patterns of the flux density at which absorbed radiant energy becomes available to power ecosystem functions.

ECOSYSTEM CHARACTERISTICS THAT AFFECT ABSORPTION OF RADIANT ENERGY

Characteristics of an ecosystem that influence its interaction with incoming radiation include its depth and openness, roughness, moisture status, and the molecular composition of its surfaces.

Depth

The porosity or transparency of the receiving surface, whether water, vegetation, or snow, controls penetration and backscattering within an optically active layer, which might be 10 m deep in a forest, 1 ? mm in sand, or 0.4 mm in clay soil (Tolchel'nikov and Komiakov, 1980). Muddy water scatters incoming solar radiation back from its top layers, allowing little absorption in the deeper layers. In contrast, a *Pinus taeda* stand at Duke University (Gay, 1966) absorbed 0.89 of the incoming solar energy in spring and 0.87 in fall. The complex structure of a forest has to be known if we are to determine radiation absorption and conversion to biomass (Hutchison, 1979). Absorptivity of Nigerian ecosystems increases from 0.77 in those 0.3 m deep to 0.81 in those 3 m deep and 0.87 in those 30 m deep (Oguntoyinbo and Oguntala, 1979). Leaf area, which is associated with ecosystem depth, can be sensed by reflectivity in the 0.7–1.1 μm band (Thomas et al., 1977).

Overgrazing reduces ecosystem depth and roughness and tends to reduce absorptivity, with serious effects in many semiarid regions of the world. Exclosure of an area in the Sinai in 1974 increased its vegetation density and its absorptivity by 0.06 in the 0.55–0.9 μm spectral region (Otterman, 1977); logging in Quebec decreased biomass and shifted absorptivity from 0.93 to 0.82 (McCaughey, 1978), drought changes solar absorptivity over large areas (Norton et al., 1979), and Australian pastures in need of fertilization can be mapped by remote sensing (Vickery et al., 1980).

As snow ages, its active depth increases because fewer crystal facets scatter radiation and more solar radiation penetrates deeply, not to emerge again; the result is the familiar low albedo of old snow cover (Dirmhirn and Eaton, 1975). With snow grains larger than 1.5 mm, size appears to be less important than density (Bergen, 1975).

Roughness

A rough surface tends to absorb more radiation than a smooth one, partly because photons reflected downward by taller elements of the surface may be captured at their second contact. Roughened loam absorbs 0.03–0.04 more solar radiation than a smooth loam (Idso et al., 1975). Roughness tends to enhance midday absorption and to increase the dependence of absorptivity on sun height (Arnfield, 1975).

Longwave absorptivities in the 7–15 μm range are 0.94 for sand and 0.97 for lawn, a rougher surface (Lorenz, 1966). The high absorptivities of ecosystem canopies are consistent with their structure as an agglomeration of blackbody cavities or Hohlraum radiators (Reifsnyder, 1967b), especially in orchards and row crops (Sutherland and Bartholic, 1977). Tree crowns absorb more solar energy than individual leaves or needles (Miller, 1955, p. 97), and the stand canopy still more. Rudnev (1974) shows an increase from absorptivity of 0.90 of incident photosynthetically active radiation by hardwood leaves to 0.93 by crowns and to 0.944 by forest canopy; similar increases occur in solar infrared radiation.

Surface Wetness

A water film on leaves and particles of soil or snow also increases the likelihood of multiple internal reflection of photons. Wet sand is appreciably darker and more absorptive than dry sand. Wet snow during midday melting absorbs more solar radiation, and this positive feedback accelerates melting in sunny weather.

Solar absorptivity of drying loam at Phoenix (Idso et al., 1975) decreased from 0.83–0.86, depending on sun angle, to 0.67–0.71 in dry soil, the moisture of a thin top layer (2-mm thick or less) being most important. The change from wet to 'dry in such a thin layer is rapid; in a few hours "the value of the soil color is visibly changing from dark to light" (Idso et al., 1975). Ratios of reflectivity in six spectral bands from 0.449–0.484 up to 0.953–1.031 μm added no information on soil moisture and so can be used to identify soil types (Reginato et al., 1977). Wet leaves can be identified at radar as well as solar wavelengths, in which a wet pine forest displayed absorptivities several hundredths greater than a dry forest (Kasperavichene, 1965). Wet soils absorb about 0.01 more than dry in the longwave region.

Molecular Composition

Quartz absorbs less than 0.90 of incident longwave radiation, less than most other minerals or soils (Rudnev, 1974). Leaf chlorophyll absorbs in the blue and red regions of the spectrum and not in the green or solar infrared regions; water is highly selective in transmission and absorption. Most leaves absorb 0.90–0.98 of longwave radiation with some specific differences, such as 0.94 ± 0.04 for *Zea mays* and 0.995 ± 0.04 for *Saccharum officinarum* (Ross, 1975a, p. 173).

Determining reflectivity or absorptivity of a substance at different wavelengths is a standard laboratory measurement for leaves and other ecosystem components, and a spectrophotometer can be used in the field. Measurements have been most common in the visual range, but the growing employment of airborne or satellite-borne scanners has extended the range of wavelengths as well as the spatial scale. Ecosystem absorption over the solar spectrum is measured by a pair of pyranometers, one upright and the other inverted above the upper surface.

Determining Absorption

Albedo (1.0 − absorptivity in the solar spectrum; from the Latin *alba* meaning white) is measured (Fig. 1) and published at all stations in some

Fig. VII.1. Sensor for reflected solar radiation (left) registering data to be subtracted from measurements of incoming solar radiation to give the amount of solar energy absorbed by the grass, here on the observation field of the Glavnaia Geofizicheskaia Observatoriia at Voeikovo near Leningrad. Beam radiation is measured by pyrheliometer on post at right.

meteorological networks, usually at standard sites that are grassy in summer and snow-covered in winter. Thornthwaite's plea (1961) that albedo is a fundamental characteristic of the surface of the earth and should be intensively studied preceded a growth of interest on the part of agricultural, forest, radiation, and satellite meteorologists that has made it a familiar word, and its measurement is now more common. Dirmhirn (1968) shows that a reflectance measurement at 0.725 μm represents the solar spectrum, and this sampling is used in some satellite analyses.

Airborne measurements have been found to be in agreement with those at standard observation sites in central Asia (Sitnikova, 1963), but with snow cover and wave-roughened water the occurrence of small angles of incidence can generate enough specular (mirrorlike) reflection to bias the airborne readings (Dirmhirn and Eaton, 1975).

Longwave radiation reflected from the surface is virtually indistinguishable in the field from radiation emitted by the surface. Absorptivities or emissivities of ecosystems in this spectral range are usually built up from laboratory determinations for components, but data are "sparse and sporadic" (American Meteorological Society, 1972). Remote sensing studies have occasioned many attempts to estimate ecosystem and surface emissivity, often from scanning in a particular wavelength region (see, e.g., Sutherland et al., 1979).

GEOMETRIC COMPONENTS OF RADIANT ENERGY

In preceding chapters we distinguished between beam and diffuse radiation from the sun and also saw that solar radiation scattered in the atmosphere and longwave radiation emitted by it display geometric source patterns. Clearly, the shifting azimuth and altitude of the sun, the circumsolar source of much diffuse solar radiation, and the low-altitude bands of the sky that generate much diffuse solar and longwave radiation affect the angles at which photons strike ecosystem surfaces and hence ecosystem absorptivity.

Beam Radiation

The significance of direction is obvious when a receiving surface reflects the solar beam like a mirror, as snow and water may do under a low sun, or when the midday sunshine penetrates deep between leaves, individual plants, or crop rows. The glare from a water body at low sun angles indicates low absorptivity, contrasting with the high absorptivity (0.95 or more) when the sun is high, although values do not follow Fresnel's law precisely (Robinson, 1966, p. 203). Specular reflection from snow occurs only at sun heights lower than 45° (Middleton and Mungall, 1952) and is small in new snow (Dirmhirn and Eaton, 1975), but it is partly responsible for a diurnal variation in absorptivity, as with water bodies. Sun altitude affects the absorptivity of many terrestrial ecosyo tems, and midday values may be 0.05–0.10 greater than those at the ends of the day.

Diffuse Radiation

The absorptivity of grass is 0.04 more for diffuse than for direct short-wave radiation, the difference decreasing as the grass grows older and dies (Tooming, 1960); at Tbilisi, however, a small reduction in absorptivity sometimes occurred when diffuse radiation made up a larger share of total solar radiation (Tsutskiridze, 1961). Dirmhirn (1953) found that meadows and forests absorb more of the blue-sky diffuse radiation than of that emerging from cloud bases. The relative penetrations of beam and diffuse solar radiation depend not only on leaf area, but also on leaf angles and whether leaves were distributed in space in a random or clumped manner (Ross, 1975b).

Relative Contributions

How much incoming energy is directional, and how much comes from a hemispherical source? Since much of the diffuse shortwave comes from the circumsolar source, most of the radiant-energy intake incident on an ecosystem in daytime is directional, and sun height affects total solar energy absorbed (Ross, 1975a, p. 305).

The direct-beam component accentuates effects of slope in the terrain and the shape factor in biological organisms. Branches and animal bodies expose less extensive absorbing surfaces to the direct beam than to hemispherical-source radiation. Sun pits or points (Fig. 2) that follow the sun occur only in weak diffuse solar and longwave fields (cold, dry, very clear air) and strong beam radiation, as in the high Andes. Situations of beam dominance are, however, more the exception than the rule, judging by flux-density data presented in the foregoing chapters, particularly when one adds area-source longwave radiation to the area-source fraction of the diffuse solar flux.

SPECTRAL COMPOSITION OF RADIANT ENERGY ABSORBED BY ECOSYSTEMS

Absorbed radiation in the ultraviolet and visible ranges of the spectrum has photochemical and biological significance, dissociating molecules and fixing carbon, and all wavelengths have a heating value.

Fig. VII.2.   Sun-facing pits and points of a city snow remnant in late February 1978 on a dry, cold day when beam radiation was dominant over area-source solar and longwave radiation. Points move to follow the sun, and the next morning still indicate where it set on the day before.

Ultraviolet Radiation

Differently absorbed by the parts of a flower, ultraviolet radiation gives a signal to potential pollinators that helps them minimize the energy costs of foraging. In the black-eyed susan (*Rudbeckia hirta*) three glucosides with strong absorption in the 0.34–0.38 μm band, serve as nectar guides to orient insects (Thompson *et al.*, 1972). Other flowers use the strategy of reflecting petals against UV-absorbing foliage, or conversely, of UV-absorbing flowers against UV-reflecting backgrounds (soils or glaucous leaves) (Frolich, 1976).

Photosynthetically Active Radiation

In the growing state the absorptivity of most ecosystems, particularly aquatic ones, for photosynthetically active radiation is about 0.9 and increases with increasing chlorophyll concentration (Bray *et al.*, 1966). At the 55th parallel in the western Soviet Union 0.6 of the annual total comes during the months when mean air temperature exceeds 10°C (Efimova, 1965) and of this 0.9 is absorbed, i.e., about 130 W m^{-2} during the growing season.

In latitudes above 40° the winter flux densities of photosynthetically active radiation are small. They fall, in continental interiors, upon dormant vegetation, desert, or snow, and absorption is of the order of 10–20 W m^{-2}. Vegetation in coastal regions is active in winter and absorbs more energy than do the dormant ecosystems inland or at higher altitudes.

Solar Infrared Radiation

Absorptivity of foliage for radiation of wavelengths greater than 0.7 μm is much less than for photosynthetically active radiation. By using data from Dirmhirn (1964), we estimate that in June 120 W m^{-2} in the solar infrared reaches a grain field near Wien, of which 85 W m^{-2} is absorbed, and that 100 W m^{2} arrives as visible radiation, of which 90 W m^{-2} is absorbed. When this field is snow-covered, the situation is reversed: more solar infrared radiation is delivered, and much more is absorbed, 15 W m^{-2} of infrared compared with 5 W m^{-2} of visible. Absorptivity is less in deep ecosystems than in thin ones because leaves absorb less than soil. Corn with a leaf area index of 1.0 absorbs 0.73 of the solar infrared, but corn with an index of 6.0 absorbs only 0.63 (Ross, 1975a, p. 303).

Solar infrared radiation has only a heating effect when absorbed, like longwave radiation, but, unlike longwave, it is intermittent, and its daytime pulses change in size and spacing through the year. This intermittency is important in melting snow cover, giving meltwater generation an on-and-off character that eases the burden on drainage networks. Foliage absorbs relatively little solar infrared but does so at hours when it is absorbing a great deal of photosynthetically active radiation, so that the solar infrared only augments daytime heating.

Longwave Radiation

The regimes of longwave radiation absorbed by ecosystems are approximately expressed by the regime of the downward flux (Chapter VI), less a few percent reflected during periods of snow cover and full leaf, depending on species, and somewhat more in periods of bare soil, if any. These variations are a restrained diurnal variation of approximately 0.1, an annual variation of 0.2–0.3, and an interdiurnal variation nearly as large.

TABLE I

Absorptivity of a Field of Barley[a]

	Wavelength (μm)							
	0.4	0.5	0.55	0.6	0.65	0.7	0.75	0.8
Green field	0.97	0.96	0.90	0.89	0.91	0.59	0.50	0.58
Yellow field	0.96	0.85	0.78	0.75	0.74	0.56	0.60	0.68
Dead plants	0.92	0.81	0.74	0.72	0.70	0.64	0.58	0.62

[a] Data from Ross (1957).

SPECTRAL PATTERNS OF ABSORPTIVITY COEFFICIENT

At the ecosystem scale spectral absorptivity changes from time to time (Table I). As the barley matures, less red light (0.60–0.65 μm) is absorbed, and when it dies, less of all the visible wavelengths is absorbed. As an oak–hickory system at Oak Ridge, Tennessee, leafed out, its absorptivity of photosynthetically active radiation rose from 0.93 to 0.97, but that of infrared declined from 0.80 to 0.66 (Matt et al., 1979).

Substantial differences in spectral absorptivity exist among species. For example, over the 0.32–2.6 μm range upper leaf surface reflectivities of 0.45 for oak and 0.50 for plantains (MacBryde et al., 1971) led to oak-leaf temperatures of 45°C in the heat wave of July 1966, while plantain leaf temperatures were mostly lower than 40°C; a difference of this size is ecologically important at levels so near the lethal temperature.

Absorption by leaves depends on their state of health and moisture balance. Disease, moisture stress, insect attack, and other misfortunes change the reflectance in certain wavelengths to such an extent that the afflicted plants can be spotted by remote sensing. Multichannel scanners

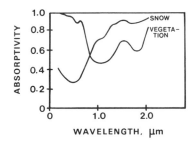

Fig. VII.3. Spectral absorptivity patterns of snow cover and green vegetation to 2.0 μm wavelength [from Sivkov (1968, p. 98)]. The blackness of snow in infrared wavelengths shows up clearly here.

are most effective because the ratios between reflectivity at two or more wavelengths can have diagnostic value and also serve to identify particular crops to be mapped by crop forecasters (MacDonald and Hall, 1980). Such differences are evident in the contrasting, almost reversed, spectral patterns of a typical green ecosystem and a snow cover (Fig. 3). The snow reflects strongly and absorbs little in the visible wavelengths, and the plants vice versa. The snow is dark in the solar infrared, especially between 1.6 and 2.5 μm (O'Brien and Munis, 1975), and the foliage is again the opposite.* The absorptivity of most inorganic materials changes more gradually with wavelength than is true of snow or leaves; however, the same effect can be obtained when a film that reflects solar radiation strongly and is transparent in the 8–13 μm longwave window is stretched above a surface with large longwave emissivity (Addeo et al., 1980); the space covered is cooled below outside air by 5–7°K.

Radiation Absorbed in Different Spectral Regions

The fluxes of radiant energy absorbed by an ecosystem vary with the season (Table II). While the winter sum (280 W m^{-2}) is about half the summer sum, its content of photosynthetically active radiation is an order of magnitude smaller.

The lag of the longwave component after the solar component tends

TABLE II

Spectral Components of Absorbed Radiation at Wien[a,b]

	Photosynthetically active radiation		Solar infrared		Longwave		Total
	W m^{-2}	Fraction absorbed[c]	W m^{-2}	Fraction absorbed[c]	W m^{-2}	Fraction absorbed[c]	W m^{-2}
June	90	0.17	85	0.16	350	0.67	525
December	5	0.02	15	0.05	260	0.93	280

[a] Latitude 48°N.
[b] Data from Dirmhirn (1964, pp. 131, 133, 141).
[c] Fraction of total absorbed radiation.

* However, solar flux densities per micrometer are much smaller in the infrared than in the visible wavelengths, so that the net heating effect is smaller than the curves suggest. Because the plant–snow difference is larger in the visible than in the solar infrared, plants absorb more total solar energy, as is well known and is illustrated by a zonal absorptivity of all land areas in the latitude band 48–50°N of 0.69 in December against 0.81 in June (Kukla and Robinson, 1980).

TABLE III

Components of Annual Absorbed Radiation along 55°N Latitude in Western USSR[a,b]

| | Longwave | Solar infrared | Photosynthetically active radiation at air temperature | | Sum |
			<10°C	>10°C	
Density (W m^{-2})	210	50	18	32	310
Fraction	0.68	0.17	0.05	0.10	1.00

[a] Longitude 20 to 100°E.
[b] Data from Niilisk's (1960) calculations and maps by Efimova (1965) and Budyko (1963).

to produce a lower-quality mix of forms of radiative energy in fall than at the same sun height in spring. However, the comparison of the amounts absorbed also depends on ecosystem absorptivity in spring and fall, which is a function of growth habit, i.e., whether the ecosystems are green at these seasons or present only bare branches, bare soil, or dying foliage.

Annual mean flux densities of the spectral components of radiant-energy intake can be estimated from published data, say along the 55th parallel north (Table III).

It is noteworthy that only 0.15 of the radiation has photochemical value and that a third of this small quantity is absorbed when low temperature hinders photosynthetic energy conversion.

As noted in Chapter VI, longwave and solar regimes of radiation tend

TABLE IV

Components of Radiant-Energy Intake at Hamburg on Two Days in January 1957[a,b]

	Clear 11 Jan 57	Cloudy 12 Jan 57[c]	Difference
Absorbed shortwave radiation	42	6	−36
Absorbed longwave radiation	260	327	+67
Radiant-energy intake	302	333	+31
Fraction as shortwave	0.14	0.02	
Fraction in direct beam	0.11	0.003	

[a] Data from Germany, Wetterdienst (1957).
[b] Data in watts per square meter.
[c] January 12, though optically dark, was bathed in invisible longwave radiation and enjoyed more radiant-energy intake than January 11.

TABLE V

Components of Radiant-Energy Intake at Hamburg in June 1954[a,b]

	Clear days	Cloudy days	Difference
Absorbed shortwave radiation	264	125	−139
Absorbed longwave radiation	314	335	+21
Radiant-energy intake	578	460	−118
Fraction as shortwave	0.46	0.27	

[a] Data in watts per square meter.
[b] Data from Schulze (1970, p. 106) (grass ecosystem).

to be associated—positively correlated in the daily and annual regimes and negatively in those variations caused by clouds. In the case of the negative association the passing of a cloud shifts the spectral mix of radiation delivered to an ecosystem with an enrichment of photosynthetically active radiation in the diffuse flux. Solely thermal effects may be either increased via longwave radiation emitted by the cloud or decreased as the solar infrared in the beam flux is cut off.

Whether or not the shading effect of a cloud or a cloud system outweighs the emissive effect depends on sun height, and this depends on season. Arrival of a cloud deck in winter means an increase in downward longwave radiation of 50 W m^{-2} or more, exceeding the decrease in the shortwave input. It brings a change in spectral composition of the radiant-energy intake and a net increase in its overall flux density of 31 W m^{-2} (Table IV).

The shading effect of clouds in summer, on the other hand, usually exceeds the radiation they emit (Table V). More solar radiation is lost to the grass ecosystem than is gained as longwave, for a net decrease of 118 W m^{-2} (or about 0.2) in all wavelengths, of which perhaps 50 W m^{-2} lies in the photosynthetically active region.

VARIATIONS IN RADIANT-ENERGY INTAKE OVER TIME

Short-Term and Diurnal Variations

Rapid fluctuations in radiant energy absorbed in an ecosystem are usually caused by the passing of a cloud or a break in an extensive cloud deck and have been noted in connection with the changed geometry or spectral distribution of the absorbed energy.

The diurnal march of absorbed radiant energy is the resultant of the regime of sun height in its effects on beam and diffuse radiation fluxes,

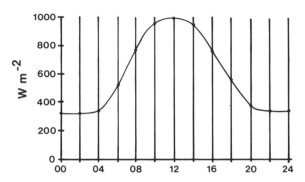

Fig. VII.4.   Diurnal march of radiant-energy intake (absorption of solar and incoming longwave radiation, W m^{-2}) by grass at Hamburg Observatory on 5 June 1954 [from Fleischer (1954)]. Note that the curve never is lower than 300 W m^{-2}.

modified by small changes in ecosystem absorptivity. All combine to cause a rapid increase in radiant-energy intake during the morning.

Midday flux densities on a clear summer day in the Sierra Nevada (1.6 km altitude) were nearly 1300 W m^{-2}, from which a decline to less than 400 W m^{-2} at night occurred. A complete day at a sea-level station (Fig. 4) shows a range of 675 W m^{-2}. Diurnal ranges at radiation sites for which values of absorbed solar and longwave radiation flux density could be computed generally exceed 500 W m^{-2} (Table VI). These large changes in available energy within the daily period bring about the familiar changes in ecosystem temperature and in all the other fluxes in its energy budget.

Aperiodic Variations

Aperiodic and interdiurnal variations in radiant-energy intake by ecosystems have been discussed in regard to cloud effects but not absorptivity. The net change in the two cases cited from Hamburg shows that overall flux density on the cloudy day in winter was 31 W m^{-2} more (Table IV) and in summer 118 W m^{-2} less (Table V) than on the clear day. Advection of heat, moisture, and clouds in winter storms usually increase the radiant-energy intake of the underlying ecosystems.

Some changes occur suddenly when rain wets ecosystems or snowfall covers them, or gradually when soil dries out or becomes smoother, when leaves unfold or fall, or snow cover ages. The surface with the most changeable albedo is snow cover. A new-snow value of 0.15–0.20 (depending on snowstorm conditions) is followed by an increase (Miller, 1955, pp. 128, 133), which brings a rise in absorbed radiant energy after

TABLE VI

Daily Ranges of Radiant-Energy Intake in Different Seasons[a]

Season	Location and site	Range (W m^{-2})	Observer
Winter	Krawarree, N.S.W., native grass	600	D. H. Miller (unpublished 1971)
Equinox	Central Sierra Snow Laboratory,[b] melting snow	500	Miller (1955)
Equinox	Columbia, Md., new city		
	field	635	Landsberg and Maisel (1972)
	paved area	805	Landsberg and Maisel (1972)
Equinox	Yale Forest, *Pinus resinosa* stand	765	Reifsnyder (1967a)
Summer	Hamburg, grass	675	Fleischer (1954)
Summer	El Mirage, Calif., dry-lake site	985	Vehrencamp (1953)
Summer	Blue Canyon, Calif.,[c] dry, sparse vegetation	1020	D. H. Miller and R. A. Muller (unpublished 1963)
Summer	Shafrikan, Uzbekistan		
	nearly bare soil	848	Aizenshtat (1960)
	irrigated cotton	935	Aizenshtat (1960)
	saxaul shrub	1020	Aizenshtat (1960)
Summer	Kara-Kul' Valley, Pamir Mts.[d]	525	Zuev (1960)

[a] Calculated by author from sets of solar and longwave radiation measurements.
[b] 2.2 km altitude.
[c] 1.6 km altitude.
[d] 4.0 km altitude.

a storm in spring equal to 14 W m^{-2} additional energy in the ecosystem every day. The increase is most rapid when alternate thawing and freezing and an abundance of free energy produce rapid weathering of snow; the feedback is positive.

A negative feedback occurs when plowed soil becomes smoother: absorptivity decreases with time (Table VII).

Seasonal Variations

The most obvious change in absorptivity with season is found in regions of winter snow in ecosystems. A different regime of ecosystem absorptivity is displayed by crop ecosystems in which the soil is bare, wet, and rough in winter and spring and is rapidly dried and warmed by reason of its high absorptivity then. Annual crops emerge and cover bare soil; native ecosystems put out new leaves and extend their canopies in spring, absorbing more photosynthetically active radiation and less of the solar infrared. Leaves grow old, lose chlorophyll, color, turn brown, and drop—all processes that affect ecosystem absorptivity. The

TABLE VII[a]

Type of surface	Fraction absorbed
Newly plowed soil	0.83
Large lumps of dirt	0.80
Small lumps	0.75
Dust cover	0.72
Graded, smooth surface	0.70

[a] Data from Kondratyev (1969, p. 424).

daily fall-color maps in Wisconsin newspapers suggest how quickly some of these seasonally induced changes take place, even in forest. Rudnev (1974) reports solar absorptivities in successive phenological periods in a mixed forest as 0.74, 0.80, 0.91, 0.88, 0.87, and 0.88.

In general, the phenological changes in ecosystems and the coming and going of snow in them reinforce the seasonal changes in incident solar and longwave radiation to produce a marked increase in radiant-energy absorption in the growing season. Radiant-energy intake in the yearly cycle of the grass ecosystem at Hamburg varies from a 24-hr mean of 300 W m^{-2} in winter to 500 in early summer. This variation is smaller than the range between winter (290 W m^{-2}) and summer (525 W m^{-2}) at Wien, which is inland and lower in latitude (48°N latitude) (from data of Sauberer, 1956). The contribution by absorbed solar radiation at Hamburg is minute in the dark winter when longwave radiation from the cloudy, warm marine air is the principal source of energy that helps the ecosystem to survive. Solar radiation absorbed in the bright, long days of summer comprises about a third of the total radiant-energy intake at that season.

Seasonality also appears in the day-to-day variation (Table VIII). Although we consider winter the season of storms and clouds, the standard deviation of daily values in summer is twice that in winter and makes up a larger fraction of the mean (coefficient of variation in December is 0.08, in June 0.12).

Absorption of radiant energy by the grass ecosystem increases rapidly in spring. The changes in mean value from month to month are as follows:

from mid-February to mid-March	+60 W m^{-2}
from mid-March to mid-April	+60 W m^{-2}
from mid-April to mid-May	+50 W m^{-2}
from mid-May to mid-June	+30 W m^{-2}

TABLE VIII

Monthly Means and Standard Deviations of Daily Values of the Components of Radiant-Energy Intake at Hamburg [a]

	Means and standard deviations[b]			Coefficients of variation		
	Absorbed shortwave	Absorbed longwave	Radiant-energy intake	Absorbed shortwave	Absorbed longwave	Radiant-energy intake
December 1957[c]	15 ± 10	300 ± 30	315 ± 25	0.67	0.10	0.08
June 1958[d]	165 ± 70	350 ± 22	515 ± 60	0.38	0.06	0.12

[a] Source: Miller (1973).
[b] Data given in watts per square meter.
[c] Coefficient of correlation between daily values of absorbed shortwave and longwave radiation is 0.62.
[d] Coefficient of correlation between daily values of absorbed shortwave and longwave radiation is 0.68.

These are substantial additions of energy each month to a situation in which relatively small rates of supply and demand had previously been in balance. If all the added energy had to be removed solely by increased emission from the radiating grass surface, grass temperature would have to rise, according to the fourth-power law, by 10°K to 12°K in each month to keep the energy budget in balance. However, the increased energy input evokes such other responses as reversals in soil-heat and sensible-heat fluxes and increases in the biological processes of photosynthesis and transpiration. These changes in flux strength produce the familiar rise of soil and air temperature, evaporation and atmospheric humidity, and biological production. Change in the radiant-energy intake of an ecosystem clearly expresses the seasonality of its forcing functions.

Variation from Year to Year

Extensive analyses at the Voeikov Geophysical Observatory in Leningrad of long series of radiation measurements (Budyko, 1971, pp. 137–143) show considerable stability in most radiation fluxes from year to year, and hence in the absorption of radiant energy. We can evaluate such variations in the growing season from the Hamburg record (Table IX). Sunny summers tend to have less longwave input. Variance in total radiant-energy intake among years is reduced, as is shown by a co-efficient of variation that is smaller than that of either component. Long-term changes unassociated with year-to-year differences in the frequency of cloud systems may be caused by change in absorptivity for

TABLE IX

Statistical Parameters of Radiant-Energy Intake for 16 Junes at Hamburg Observatory[a]

	Flux density (W m^{-2})			
	Mean	Range	Standard deviation	Coefficient of variation
Absorbed shortwave radiation	175	72	19	0.109
Absorbed longwave radiation	345	54	14	0.041
Radiant-energy intake	520	80	17	0.033

[a] Source: Miller (1973).

solar radiation that results from changes in biomass, depth, or roughness of an ecosystem. Some are caused by drought (e.g., solar absorptivity in the Sahel). The English drought in 1976 reduced solar absorptivity by an average of 0.10 (Henderson-Sellers, 1980), equivalent to 20 W m^{-2}. Overgrazing and logging also reduce absorptivity, as does the mulch residue left in crop ecosystems when no-till farming practices are adopted, which reduce radiant-energy absorption; the lower soil temperature in some areas slows the decay of organic residues in the soil, delays seed germination, and inhibits root growth (Gersmehl, 1978). Increasing utilization of passive solar energy may be expected to increase the radiation absorbed by cities about 1 W m^{-2}; active solar collectors in the numbers foreseen by Häfele (1980) can be calculated to increase it by an additional 1–2 W m^{-2}.

SPATIAL PATTERNS OF ABSORBED RADIANT ENERGY

Small Scale

Radiation absorbed by the upper part of a forest canopy is different in spectral and geometric composition and in quantity from that absorbed by snow on the forest floor or understory vegetation. Differences exist among the diverse microenvironments in a city, in which vertical surfaces are important in themselves and in their effect on adjacent systems, and in which green, paved and heated surfaces are closely mingled.

Large differences in absorbed radiant energy are more a matter of ecosystem-scale differences in shortwave absorptivity than they are of differences in the diffuse radiation fluxes, which show small spatial contrasts; differences in solar-beam incidence are important in dissected topography. Differences in ecosystem absorptivity, small in the long wavelengths, are large in the solar wavelengths, with snow cover forming a separate class. Absorptivity increases in this order: dry soil, low vegetation, forest stands, aquatic ecosystems.

Many contrasts reflect differences in terrain as it influences one or another radiation flux, individually noted in preceding chapters. Others follow from differences in ecosystem moistness or depth: Irrigated cotton in central Asia takes in approximately 100 W m^{-2} more energy than desert, and a deep shrub ecosystem (*Haloxylon sp.*) 150 W m^{-2} more than the native ecosystem (from data of Aizenshtat, 1960). Pine forest in eastern Oregon took in 580 W m^{-2} (24-hr mean) while a pumice surface took in only 530 (Gay, 1979).

Large Scale

The range of values of radiant-energy absorption by ecosystems can be visualized from some cases near the low and high extremes. The *lower limit* for a growing ecosystem lies around 300 W m^{-2} because a green cover at the freezing point must emit 315 W m^{-2} and must acquire this much energy from its environment, the most effective way being by radiative transport. We therefore can think of absorption of 300 W m^{-2} as a threshold of energy conversion, biologically related, like the 6°C temperature threshold.

Values smaller than 300 W m^{-2} occur in winter, at night, at high latitudes, and at high altitudes. Measurements at the Central Sierra Snow Laboratory (altitude 2 km) found about 240 W m^{-2} on spring nights (U.S. Corps of Engineers, 1956), evidenced in thick crusting of the snow mantle. On the Greenland icecap at 3 km altitude in summer, in the realms of endless day, similar values (Miller, 1956) are experienced, and winter values are about 125 W m^{-2}.

At low altitudes also the polar night brings small values of radiant-energy intake, as small as 80 W m^{-2} determined by Wexler (1936) in a classic study. Insulation by the snow mantle is a true survival factor.

The occurrence of *high values* depends on sun height and type of surface cover. Sparse vegetation on dry surfaces under dry, clear atmospheres experiences midday rates of radiant-energy intake in summer that are roughly the same in many places—about 1200 W m^{-2}. The noon maximum was 1230 W m^{-2} at El Mirage Dry Lake in California (Vehrencamp, 1953); Miller and Muller measured 1290 W m^{-2} in the Sierra Nevada; at a higher latitude in central Asia Aizenshtat's (1960) data for sparse camelthorn shrub yield a rate of 1175 W m^{-2}. Vegetation deepens the zone of radiation absorption, and Aizenshtat's data indicate absorption of 1265 W m^{-2} by a cotton field and 1350 W m^{-2} in a deep mantle of saxaul.

Data on absorptivity can be combined with data on longwave and solar radiation to yield zonal means of absorbed radiant energy:

60°N	290 W m^{-2}	20°N	510 W m^{-2}
40°N	410 W m^{-2}	Equator	510 W m^{-2}

These accord roughly with the few measurements we have. Equatorial clouds produce a flux of low photosynthetic value, as is evident in low rice yields. The principal difference between low and middle latitudes reflects the long period of high radiation at low latitudes, where "winter never comes" to ecosystems.

THE TEMPERATURE-INDEPENDENT INTAKE OF ENERGY

We have integrated the major geometric and spectral forms of radiant energy that ecosystems absorb. This is the energy that they convert into other forms, and so this quantity—although it lumps together direct and diffuse, ultraviolet and longwave, and all frequencies in between— has ecological, physical, and environmental significance (Miller, 1972). The resultant is not an energy flux per se, but a mix that sums up all the fluxes that an ecosystem must take in, regardless of its temperature.

The energy intakes described so far do not, by and large, depend on the temperature of the receiving ecosystem. Their absorption depends on such ecosystem properties as depth, leaf inclination to the sun, and albedo, but not to any important degree on ecosystem temperature: the immediate thermal response of an ecosystem to increased or decreased absorption affects the incoming fluxes not at all; it is essentially passive, with no feedback opportunity.

Clearly, the story cannot stop here. No ecosystem is so helpless in an often hostile universe as might be inferred from the preceding statement; rather, when it is heated by absorbing additional radiant energy, it has a means of accelerating heat-removing fluxes. These are in one way or another functions of its surface temperature (as well as of environmental conditions), such that rise in surface temperature in the energy budget is therefore central, as is recognized in energy-budget models. It is a means of balancing the temperature-dependent energy fluxes to be described against the temperature-independent fluxes we have considered up to now and summarized in this chapter.

REFERENCES 133

REFERENCES

Addeo, A., Nicolais, L., Romeo, G., Bartoli, B., Coluzzi, B., and Silvestrini, V. (1980). Light selective structures for large scale air conditioning. *Sol. Energy* **24**, 93–98.

Aizenshtat, B. A. (1960). "The Heat Balance and Microclimate of Certain Landscapes in a Sandy Desert" (transl. by G. S. Mitchell). U.S. Weather Bur., Washington, D.C.

American Meteorological Society, Atmospheric Radiation Working Group (1972). Major problems in atmospheric radiation: An evaluation and recommendations for future efforts. *Bull. Am. Meteorol. Soc.* **53**, 950–956.

Arnfield, A. J. (1975). A note on the diurnal, latitudinal and seasonal variation of the surface reflection coefficient. *J. Appl. Meteorol.* **14**, 1603–1608.

Bergen, J. D. (1975). A possible relation of albedo to the density and grain size of natural snow cover. *Water Resour. Res.* **11**(5), 745–746.

Bray, J. R., Sanger, J. E., and Archer, A. L. (1966). The visible albedo of surfaces in central Minnesota. *Ecology* **47**, 524–531.

Budyko, M. I., ed. (1963). "Atlas Teplovogo Balansa Zemnogo Shara." Akad. Nauk SSSR, Mezhduvedomstvennyi Geofiz. Komitet, Moscow.

Budyko, M. I. (1971). "Klimat i Zhizn." Gidrometeorol. Izd., Leningrad. (Engl. transl., "Climate and Life." Academic Press, New York, 1974.)

Dirmhirn, I. (1953). Einiges über die Reflexion der Sonnen- und Himmelsstrahlung an verschiedenen Oberflächen (Albedo). *Wetter Leben* **5**, 86–94.

Dirmhirn, I. (1964). "Das Strahlungsfeld im Lebensraum." Akad. Verlagsges., Frankfurt a.M.

Dirmhirn, I. (1968). On the use of silicon cells in meteorological radiation studies. *J. Appl. Meteorol.* **7**, 702–707.

Dirmhirn, I., and Eaton, F. D. (1975). Some characteristics of the albedo of snow. *J. Appl. Meteorol.* **14**, 275–279.

Efimova, N. A. (1965). Raspredelenie fotosinteticheski aktivnoi radiatsii na territorii Sovetskogo Soiuza. *Tr. Gl. Geofiz. Obs.* **179**, 118–130.

Fleischer, R. (1954). Der Jahresgang der Strahlungsbilanz und ihrer Komponenten. *Ann. Meteorol.* **6**, 357–364.

Frolich, M. W. (1976). Appearance of vegetation in ultraviolet light: Absorbing flowers, reflecting backgrounds. *Science* **194**, 839–841.

Gay, L. W. (1966). The radiant energy balance of a pine plantation. Ph. D. Thesis, Sch. For., Duke Univ., Durham, North Carolina.

Gay, L. W. (1979). Radiation budgets of desert, meadow, forest, and marsh sites. *Arch. Meteorol. Geophys. Bioklimatol. Ser. B* **27**, 349–359.

Germany, Wetterdienst (1957). Beilage zum Medizin-meteorologischen Bericht. (Hamburg Observatorium, monthly.)

Gersmehl, P. J. (1978). No-till farming: The regional applicability of a revolutionary agricultural technology. *Geogr. Rev.* **68**, 66–79.

Häfele, W. (1980). A global and long-range picture of energy developments. *Science* **209**, 174–182.

Henderson-Sellers, A. (1980). Albedo changes—surface surveillance from satellites. *Clim. Change* **2**, 275–281.

Hutchison, B. A. (1979). Forest meteorology, research needs for an energy and resource limited future, 28–30 August 1978, Ottawa, Ontario, Canada. *Bull. Am. Meteorol. Soc.* **60**, 331.

Idso, S. B., Jackson, R. D., Reginato, R. J., Kimball, R. A., and Nakayama, F. S. (1975). The dependence of bare soil albedo on soil water content. *J. Appl. Meteorol.* **14**, 109–113.

Kasperavichene, G. I. (1965). Dnevnoi khod al'bedo sosnovogo lesa. *Akad. Nauk Lit. SSR Ser. B* **2**(41).

Kondratyev, K. Ya. (1969). "Radiation in the Atmosphere." Academic Press, New York.

Kukla, G., and Robinson, D. (1980). Annual cycle of surface albedo. *Mon. Weather Rev.* **108**, 56–68.

Landsberg, H. E., and Maisel, T. N. (1972). Micrometeorological observations in an area of urban growth. *Boundary-Layer Meteorol.* **2**, 20–25.

Lorenz, D. (1966). The effect of the long-wave reflectivity of natural surfaces on surface temperature measurements using radiometers. *J. Appl. Meteorol.* **5**, 421–430.

MacBryde, B., Jefferies, R. L., Alderfer, R., and Gates, D. M. (1971). Water and energy relations of plant leaves during period of heat stress. *Oecol. Plant.* **6**, 151–162.

MacDonald, R. B., and Hall, F. G. (1980) Global crop forecasting. *Science* **208**, 670–679.

McCaughey, J. H. (1978). Estimation of net radiation for a coniferous forest, and the effects of logging on net radiation and the reflection coefficient. *Can. J. For. Res.* **8**, 450–455.

Matt, D. R., McMillen, R. J., and Hutchison, B. A. (1979). Spectral radiation balances above an oak-hickory stand during the spring leafing season. *In* "Symposium on Forest Meteorology, Proceedings" (W. E. Reifsnyder, ed.) pp. 217–219. World Meteorological Organization (WMO No. 527), Geneva.

Middleton, W. E. K., and Mungall, A. G. (1952). The luminous directional reflectance of snow. *J. Opt. Soc. Am.* **42**, 572–579.

Miller, D. H. (1955). "Snow Cover and Climate in the Sierra Nevada California," Publications in Geography, No. 11. Univ. of California Press, Berkeley.

Miller, D. H. (1956). The influence of snow cover on the local climate of Greenland. *J. Meteorol.* **13**, 112–120.

Miller, D. H. (1972). A new climatic parameter: radiant-energy intake. *In* "International Geography 1972, La géographie internationale" (W. P. Adams and F. M. Helleiner, eds.), pp. 164–166. Univ. of Toronto Press, Toronto.

Miller, D. H. (1973). On the variations of radiant-energy intake over time, with some notes on the responses of evapotranspiration and other energy fluxes as functions of the temperature of the surface of the earth. *Publ. Climatol.* **25**(3), 47–67.

Niilisk, Kh. (1960). K voprosu o raschetakh teplovogo izlucheniia atmosfery. *Izv. Akad. Nauk Est. SSR, Inst. Fiz. Astron., Issled. Fiz. Atmos., Tartu* **2**, 67–114.

Norton, C. C., Mosher, F. R., and Hinton, B. (1979). An investigation of surface albedo variations during the recent Sahel drought. *J. Appl. Meteorol.* **18**, 1252–1262.

O'Brien, H. W., and Munis, R. H. (1975). Red and near-infrared spectral reflectance of snow. *In* "Operational Applications of Satellite Snowcover Observations" (A. Rango, ed.), pp. 345–360. Natl. Aeronaut. Space Adm., Washington, D.C.

Oguntoyinbo, J. S., and Oguntala, A. B. (1979). Aspects of the forest climate in southern Nigeria. *In* "Symposium on Forest Meteorology, Proceedings." (W. E. Reifsnyder, ed.), pp. 198–212. World Meteorological Organization (WMO No. 527), Geneva.

Otterman, J. (1977). Monitoring surface albedo change with Landsat. *Geophys. Res. Lett.* **4**, 441–444.

Reginato, R. J., Vedder, J. F., Idso, S. B., Jackson, R. D., Blanchard, M. B., and Goettelman, R. (1977). An evaluation of total solar reflectance and spectral band ratioing techniques for estimating soil water content. *J. Geophys. Res.* **82**, 2101–2104.

Reifsnyder, W. E. (1967a). Forest meteorology: the forest energy balance. *Int. Rev. For. Res.* **2**, 127–179.

Reifsnyder, W. E. (1967b). Radiation geometry in the measurement and interpretation of radiation balance. *Agric. Meteorol.* **4**, 255–265.

Robinson, N., ed. (1966). "Solar Radiation." Am. Elsevier, New York.

Ross, Iu. K. (1957). O korotkovolnovom radiatsionom rezhime poverkhnosti, pokrytoi rastitel'nost'iu. *Liet. TSR, Mokslu Akad., Geol. Geogr. Inst. Moksliniai Pranesimai* **5**, 41–60.

Ross, Iu. K. (1975a). "Radiatsionnyi Rezhim i Arkhitektonika Rastitel'nogo Pokrova." Gidrometeorol. Izd., Leningrad.

Ross, J. (1975b). Radiative transfer in plant communities. In "Vegetation and the Atmosphere. Vol. 1: Principles" (J. L. Monteith, ed.), pp. 13–55. Academic Press, New York.

Rudnev, N. I. (1974). Issledovanie radiatsionnogo rezhima lesnoi rastitel'nosti. In "Issledovaniia Genezisa Klimata," for B. L. Dzerdzeevskii, pp. 214–239. Akad. Nauk Inst. Geogr., Moscow.

Sauberer, F. (1956). Über die Strahlungsbilanz verschiedener Oberflächen und deren Messung. *Wetter Leben* **8**, 12–26.

Schulze, R. (1970). "Strahlenklima der Erde." Steinkopff, Darmstadt.

Sitnikova, M. V. (1963). Al'bedo nekotorykh podstilaiushchikh poverkhnostei. *Tr. Sredneaziat. Nauchno-Issled. Gidrometeorol. Inst.* **16**(31), 37–40.

Sivkov, S. I. (1968). "Metody Rascheta Kharakteristik Solnechnoi Radiatsii." Gidrometeorol. Izd, Leningrad.

Sutherland, R. A., and Bartholic, J. F. (1977). Significance of vegetation in interpreting thermal radiation from a terrestrial surface. *J. Appl. Meteorol.* **16**, 759–764.

Sutherland, R. A., Bartholic, J. F., and Gerber, J. F. (1979). Emissivity correction for interpreting thermal radiation from a terrestrial surface. *J. Appl. Meteorol.* **18**, 1165–1171.

Thomas, D. A., Rebella, C., and Chartier, P. (1977). An analysis of the vertically reflected radiation from a maize crop as a possible means of determining biomass and water content. *Agric. Meteorol.* **18**, 101–114.

Thompson, W. R., Meinwald, J., Aneshansley, D., and Eisner, T. (1972). Flavonols: pigments responsible for ultraviolet absorption in nectar guide of flower. *Science* **177**, 528–530.

Thornthwaite, C. W. (1961). The task ahead (a presidential address). *Ann. Assoc. Am. Geogr.* **51**, 345–356.

Tolchel'nikov, Iu. S., and Komiakov, A. K. (1980). Fotometricheskie metody opredeleniia opticheski aktivnykh sloev rykhlykh porod i pochv. *Izv. Vses. Geogr. Obshch.* **112**, 64–68.

Tooming, Kh. (1960). Dnevnye i sezonnye izmeneniia al'bedo nekotorykhestest vennykh poverkhnostei. *Izv. Akad. Nauk Est. SSR, Inst. Fiz. Astron., Issled. Fiz. Atmos., Tartu* **2**, 115–163.

Tsutskiridze, Ya. A. (1961). Al'bedo nekotorykh kul'turnykh rastenii i drugikh estestvennykh poverkhnostei. *Tr. Tbilis. Nauchno-Issled. Gidrometeorol. Inst.* **8**, 34–41.

U.S. Corps of Engineers (1956). "Snow Hydrology." Portland, Oregon. (Reprinted by U.S. Gov. Print. Off., Washington, D.C., 1958.)

Vehrencamp, J. E. (1953). Experimental investigation of heat transfer at an air-earth interface. *Trans. Am. Geophys. Union* **34**, 22–30.

Vickery, P. J., Hedges, D. A., and Duggin, M. J. (1980) Assessment of the fertilizer requirement of improved pasture from remote sensing information. *Remote Sens. Environ.* **9**, 131–148.

Wexler, H. (1936). Cooling in the lower atmosphere and the structure of polar continental air. *Mon. Weather Rev.* **64**, 122–136.

Zuev, M. V. (1960). "On the Heat Balance of the Kara-Kul' Lake (Ozero) Valley in the East Pamirs." U.S. Weather Bur., Washington, D.C. (Transl. by N. A. Stepanova of O teplovom balanse doliny ozera Kara-Kul'. In "Sovremmenye Problemy Meteorologii Prizemnogo Sloia Vozdukha," pp. 61–66. Gidrometeorol. Izd., Leningrad, 1958.)

Chapter VIII

SURFACE TEMPERATURE OF ECOSYSTEMS

The reaction of an irradiated ecosystem to an increase in the flux of radiation it absorbs is an immediate rise in the temperature of its outer surface T_0. This increase accelerates outgoing energy fluxes and maintains the balanced energy state of the ecosystem. Surface temperature, one of the most interesting and variable of all climatic elements, is the mediator in this balancing process.

The place of T_0 in individual outflows will be discussed in the following chapters, and the final chapters of the book will show its role in the whole energy budget. This chapter is therefore a hinge between our earlier consideration of energy flows that are independent of surface temperature and those that are dependent on it. We have come halfway through our inquiry into the energetics of ecosystems.

Surface temperature is also a basic environmental condition of the processes of photosynthesis and respiration in ecosystems and is "at once a consequence and a control of the radiative exchanges and the chemistry of green tissues" (Hare, 1966)—those tissues upon which all life depends. Its significance is therefore both physical and biological.

DEFINING SURFACE TEMPERATURE T_0

Differentiation from Air Temperature

Veselovskii observed in 1857 [cited in Budyko (1960)] that atmospheric warmth is not the same as surface warmth, which he expresses as the "heating power of the sun's rays or the temperature obtained from the sun by bodies at the earth's surface." Air temperature* and surface tem-

---

* Or shelter temperature, unfortunately sometimes called "surface temperature" by meteorologists as distinct from air temperature aloft.

perature are generated differently and are not parallel in their distributional patterns. The difference between them is often substantial and varies with time, as well as with such factors as moisture status. "Plant temperature may be a valuable qualitative index to differences in plant water regimes" (Tanner, 1963) because drying leaves must heat up to stay in energy balance, and the leaf–air difference becomes an indicator of moisture stress (Jackson et al., 1977).

"Don't believe the ice in your driveways; the National Weather Service says the temperature has been above freezing in Milwaukee since 10 p.m. Wednesday," reads an item in an evening newspaper in February. It explains that the ground is still cold (in fact it was frozen to a depth of 0.2–0.3 m), but "weathermen keep their thermometers in shelters 5 feet off the ground, and the air there may be 10 degrees [F] warmer than it is at ground level" (Milwaukee Journal, 1976). The 1.5-m temperature at 0900 was $+5°C$, and the surface temperature was estimated as $-4°C$. Similarly, the temperature of the air over a snow cover gives little indication of the energy conversion at the snow surface. I have seen snow remain frozen and unthawing when the shelter thermometer read $+6°C$, and I have seen it melting and wet when the thermometer read $-3°C$.

In the desert, midday surface temperatures averaging 57°C were 19°C above air temperature at 2.0 m (and 19°C above soil temperature at -0.2 m) (Dodd and McPhilimy, 1959). Because in daytime the 1.5-m temperature continuously seeks, through the exchange of radiation and sensible heat, a compromise between surface temperature and upper-air conditions, it seldom displays a unique relation with the temperature at the true active surface of the earth. Recognition of the inadequacies of 1.5-m temperature in practical questions of plant survival and growth led Geiger (1927, 1965) and others to seek information on the true environment and in the process to develop the discipline of microclimatology.

The Locus of Surface Temperature

As we saw in Chapter II, the "surface of the earth" usually is not a neat two-dimensional plane but has substantial depth. One operational definition of this surface is the plane below which most heat moves by conduction (Miller et al., 1976). Yet the penetration of solar radiation into snow cover can produce net subsurface heating with melting while the top surface remains dry (Munn, 1966, p. 135). Bare soil is penetrated to shallow depths by solar radiation, and in it the locus of surface temperature is defined within a few millimeters and is to be distinguished from the temperature of the top layer of the soil column. Patches of bare soil among sparse plants (Fig. 1) exert radiative effects on them

Fig. VIII.1.   Bare soil reflecting and radiating upward into shrub canopies of sand-plain ecosystem in Western Australia. Sheep on wheat stubble in background.

(Reifsnyder and Lull, 1965) and make it hard to define the true ecosystem surface (Budyko, 1960).

In ecosystems of greater plant density the leaves "comprise the essential portion of the active surface" (Budyko, 1960) and display a temperature regime that differs from that at the soil surface just beneath (Fimpel, 1964). Leaf exposures range all the way from physiological full-sun leaves to leaves in intermittent sun flecks to complete-shade leaves, and their temperatures range accordingly. Ecosystems searching for light and CO_2 reach up into the atmosphere and blur the earth/air interface by forming a medium for optimum leaf ventilation that is mostly air, but the temperature of the air within the space "is, by no means, equal to the temperature of the active surface" (Budyko, 1960). Leaves have different radiation exchanges than do air molecules.

There exist, in fact, several energy-converting loci, each at a temperature that balances its energy inputs and outputs, as will be discussed in Chapter XX. Operationally, it is convenient to define the surface where T_0 is being measured as the locus generating the upward flux of longwave radiation (see next chapter) within the field of view of a remote-sensing thermometer above the ecosystem, which tends to represent the field of absorption of radiation from the upper part of the sky, i.e., midday beam and circumsolar diffuse solar radiation.

SIGNIFICANCE OF SURFACE TEMPERATURE

Surface Temperature as an Indicator of State

The importance of leaf temperature in the functioning of plants was outlined years ago when Brown and Escombe (1905) found tight quantitative relationships among CO_2 uptake, respiration, transpiration, and leaf temperature. The energy budget that relates leaf temperature to external conditions came to be studied intensively in the resurgence of research on evapotranspiration as an energy conversion (Budyko, 1948; Penman, 1948; Thornthwaite, 1948), which led to a physically based biometeorology in which leaves held a central place (Raschke, 1956; Budyko, 1956; Gates, 1962, 1963; Tanner, 1963).* Subsequent work is based on "an understanding of the physical theory underlying most of the [biological] processes studied" (Murphy, 1974), and practical problems in agronomy and forestry that were conceptualized in Geiger's (1927) microclimatology focus on these physical processes.

Photosynthesis (without photorespiration) has been studied to determine the temperature dependency of all its subprocesses and allow "interfacing with equations describing the leaf energy budget" (Tenhunen et al., 1976b), "which is our most powerful tool for linking leaf and environment." Net photosynthesis displayed a shift in maximum rate to a lower leaf temperature when radiation flux density decreased (Tenhunen et al., 1976a). It appears that leaves perform well when strongly irradiated and warm—a parallel to the joint occurrence of both photosynthetically active and inactive (infrared) spectral components in solar radiation. Leaves of many desert plants are typically small and therefore less superheated, but large-leaved species represent a special adaptation with some advantages in photosynthesis (Smith, 1978).

The role of leaf temperature in biomass production led to the suggestion that growth units (the cumulated sum of daily temperatures above some threshold such as 10°C) would give better results if leaf

---

* Earlier work on leaf temperature was mostly oriented toward special environments like the timberline or to the leaf as environment of certain insects like needle miners. A few studies paid attention to sun and wind (Ehlers, 1915; Michaelis, 1932; Henson and Shepherd, 1952) and reported data that could be used in a leaf-temperature submodel in an energy-budget study of pine canopy (Miller, 1955, pp. 97–104, 1956).

temperature were substituted for air temperature (Davitaia, 1960; Budyko, 1960). For comparison, annual sums are as follows at stations at about the 30th meridian east:

	(a)	(b)	Ratio (b)/(a)
Murmansk (69°N)	780	1090	1.40
Leningrad (60°N)	1820	2140	1.18
Kiev (50°N)	2600	2790	1.07

where (a) represents heat units calculated from air temperature (2-m shelter height) (kelvin-days), and (b) heat units calculated from surface temperature; values are substantially greater on south slopes. The large discrepancy in northern regions partly explains why vegetation growth is better than the air temperature would indicate; tundra ecosystems are usually warmer than standard climatic data indicate, and surface temperature is employed in modelling them.

Other plant organs also experience surface temperatures that differ from air temperature. Pine bark absorbing 746 W m^{-2} of radiation reached 27°C while air temperature was -2°C, and other surfaces at the north edge of a pine stand in the Netherlands were as warm at a time when solar radiation was 560 W m^{-2} on a flat surface (Stoutjesdijk, 1977). The surface temperature of bare soil is important to seedlings; if rain comes so late in the pastoral zone of northern Australia that sorghum does not germinate until early summer, the emerging plants encounter, in sunny weather, soil-surface temperatures as high as 65°C, and few survive (Wadham *et al.*, 1964, p. 75). Winter wheat sown into stubble left after the harvest of spring wheat on the northern Great Plains exhibits T_0 that is 10°K lower on hot days than if the wheat is sown into bare soil (Aase and Siddoway, 1980). In mountains pine seedlings are vulnerable to heat damage at and just above the hot soil surface. A more favorable situation is found in porous rocks of the dry valleys of Antarctica, which harbor an indigenous microbial flora 1–2 mm beneath their surface (Friedman and Ocampo, 1976). This near-surface zone absorbs meltwater and warmth that permit brief periods of metabolic activity.

Surface Temperature and Evapotranspiration.   The relation of the temperature of an evaporating surface and the rate at which liquid or solid molecules are converted to vapor derives from the nature of the process: The warmest molecules have the most kinetic energy and are most likely to break free of their bonds. The warmer the surface, the greater is the number of escaping molecules, each of which removes energy from the surface and tends to depress its temperature unless the energy is replaced. Evaporation, as an energy conversion, is most fruitfully investigated in an energy framework [see Miller (1977, Chapters 10–12) for examples] in which T_0 is an index of energy supply. Surface temperature is remotely measured for evaporation determination from cotton in different stages of moisture stress [see, e.g., Brown (1974)] and wheat (Jackson et al., 1977), as well as over larger, nonuniform landscapes (Soer, 1980). Moisture stress resulting from disease or fungal attack also raises T_0, e.g., root rot of sugar beets and cotton (Pinter et al., 1979). Soil moisture can be estimated from soil temperature under barley canopy, which can be derived from canopy temperature T_0 (Heilman and Moore, 1980).

Surface Temperature and Sensible-Heat Flux

The sensible-heat flux is largely a daytime phenomenon because it depends on a surface that is warmer than the air. It is familiar to us in a qualitative way, yet its effect on the atmosphere is sometimes poorly understood because of the lack of data on surface temperature.

High leaf temperatures warm the air, and this heat augments snow melting on adjacent meadows (Miller, 1955, p. 166) or creates a warmth in the local climate that belies the textbook examples of the climate of snowy regions (Miller, 1956). A synoptic meteorologist, B. L. Dzerdzeevskii, started an extensive research program in forest meteorology because forest covers much of the USSR and its surface temperature controls air-mass transformation.

Surface Temperature and Substrate Conditions

Soil temperatures at different depths are related to surface temperature, but the relation is a shifting one, to be understood only by analyzing the flow of heat, which in natural soil is not simple. However, since there is little opportunity for lateral movement of heat in the soil, the long-term relations of ecosystem soil and surface temperature are usable.

The temperature of bare rock used to be considered important in its weathering, but measurements in the Mojave Desert found only small substrate temperature gradients, too small to produce rock disintegration (Roth, 1965). The surface temperature of roofs and walls in winter is measured in order to assess the fossil-energy input necessary to maintain a steady internal temperature in human microenvironments. A hot roof means a heat leak, and an aerial survey over 100 km^{-2} in Houston served as a basis for planning urban rehabilitation (Coiner and Levine, 1979).

Surface Temperature and Emission of Longwave Radiation

Integration of the equation for Planck's law over wavelength into the Stefan–Boltzmann relation defines emitted radiation E,

$$E = \epsilon \sigma T_0^4,$$

in which ϵ is surface emissivity and σ is the Stefan–Boltzmann constant. A rise in surface temperature accelerates the removal of energy by radiation. We shall look at one example of this effect: its value in ridding the merino sheep in the sunny pastoral lands of Australia of what would otherwise be a crushing heat load. An energy-budget analysis (Priestley, 1957) gives data from which we can calculate that out of a radiant-energy loading of 860 W m^{-2}, averaged over the entire surface of the sheep, only 25 W m^{-2} penetrates to the skin. All the rest is dumped back into the environment by radiative (650 W m^{-2}) and convective fluxes driven by a surface temperature that has been measured as high as 85°C (Macfarlane, 1964). Ability to raise T_0 is one factor in the success of this animal in a harsh thermal environment.

DETERMINING SURFACE TEMPERATURE

Because surface temperature fluctuates constantly in its task of keeping the temperature-dependent energy fluxes in balance with absorbed radiant energy, it is hard to measure. In contrast with such surface properties as albedo, episodic measurements of T_0 cannot be taken as representing time variations and cannot be adequately interpreted without information on the concomitant energy fluxes.

Thermometers and Thermocouples: Measurement by Contact

The first measurements of T_0 were made in an ecosystem in which difficulties are smallest—a plot of mowed grass at night. The radiative, convective, and conductive energy flows to and from the thermometer bulb approximate those to and from the short-grass sward, and the site is easily standardized. Grass minimum thermometers have been used since the mid-19th century in the Russian and other meteorological networks and provide an antidote to the tendency to rely on shelter temperature as an all-purpose measure of energy level. The temperature of the top layer of bare soil is sometimes measured but only approximates T_0 since it is influenced by the thermal conductivities of the layers below it.

The discrepancy between the two-dimensional nature of many active surfaces and a large thermometer bulb led to use of thermocouple junctions, which can be made very small. These work well in leaves, in tent canvas, and in soil. Connecting several junctions in a network meets problems of microscale spatial variations (Budyko, 1960; Davitaia, 1960).

Radiation Temperatures

Bolometers have been used as radiation thermometers for some time, e.g., the Linke–Feussner directional radiometer (Monteith and Szeicz, 1962), and can be improved by excluding reflected shortwave radiation and viewing reference surfaces alternately with the target. If a particular channel in the longwave spectrum is employed, Planck's law at the appropriate wavelength is used rather than the spectrally integrated Stefan–Boltzmann relation. The field of view may be as large as 30° or more, or as small as 2 to 4°; measurements can be recorded or digitized and time-integrated over the appropriate response period to smooth out fluctuations. Finally, infrared thermometers define an acceptable "active surface," i.e., the radiators that the sensor sees. In most crops the viewing angle is no great problem (Fuchs et al., 1967), but the problem of emissivity remains since the convenient assumption of $\epsilon = 1.00$ is, in general, not true. Fuchs and Tanner (1966) suggest ways to correct infrared thermometers for this effect.

Scanning infrared thermometers have brought a new era in surface-temperature measurement, especially because most scanners are designed to be airborne or satellite-borne and to view ecosystems from above. This remote view of the surface provides spatial integration of the microscale temperature pattern, and resolution often corresponds to the areal dimensions of many ecosystems, e.g., 72 m in the Skylab scanner. Scanners can also resolve flux density, hence surface temperature, to $2^8 - 1$ or 255 levels, which permits enhancing spatial contrasts, filtering, compressing data, calibrating, and displaying, as well as comparisons between wavelength channels.

Atmospheric constituents between ecosystem and sensor are not perfectly transparent, even in the 8–12-μm "window," and ground-based temperature measurements of such uniform surfaces as a lake are used for calibrating the sensor signal. Corrections for atmospheric water vapor, particulates, and gaseous pollutants have also been developed. In regions where few atmospheric constituents other than water vapor absorb and emit radiation in a particular channel, the vertical temperature distribution of vapor is used to correct the signal and obtain surface temperature (Cogan and Willand, 1976). The enormous volume of data from airborne and satellite scanners demands mechanized archiving, and this computer connection is useful in analysis (Sutherland *et al.*, 1979).

Modeling Surface Temperature

The role of T_0 in the radiation and energy budgets can be utilized to determine its numerical value. By one method, the radiation-budget equation is solved for upward longwave radiation. An error analysis from data at Coshocton, Ohio, recommended that the method not be used as a "substitute for the use of infrared thermometers" (DeWalle and Parmele, 1974), but primarily as a check on radiation-budget measurements. The temperature of the top layer of soil can be modeled (e.g., Lin, 1980), sometimes assuming an often unrealistic homogeneity in the soil column, but in any case it differs from T_0 itself (Bhumralkar, 1975).

Models gave good reconstitutions of T_0 at flat and sloping land in Alaska (Brazel and Outcalt, 1973) and also of leaf temperature at different depths in an ecosystem (Waggoner and Reifsnyder, 1968), along with the exchanges of energy, CO_2, and water. (Other models are described in Chapter XXIII). Coarser submodels for surface temperature have been included in atmospheric models, for example, to reproduce the land-sea temperature contrast important in predicting summer weather in Australia (McBride, 1975).

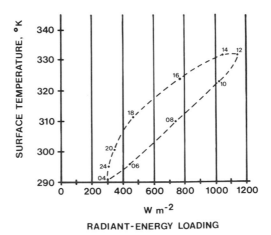

Fig. VIII.2.   Diurnal march of the surface temperature (°K) of a desert ecosystem through a day in summer as a function of its radiant-energy intake or loading (W m^{-2}). The hysteresis shown in the curve suggests the decline of heat-removing fluxes in the late afternoon hours, due to warmth of both media—soil and air—adjacent to the interface. [From measurements at Shafrikan, 40°N latitude, by B. A. Aizenshtat, (1960).]

VARIATIONS OVER TIME

Momentary Fluctuations

When a leaf is shaded, there is an immediate response in its temperature because it has little thermal inertia (Knoerr and Gay, 1965). Soil surface-temperature fluctuations due to passing cumulus clouds were greatest at a sky cover of 0.6; the coefficient of variation over a period of time was 0.55 for solar radiation and 0.35 for T_0 (Pyldmaa, 1978). The autocorrelation coefficient went to zero at 200 sec at a dry soil surface, 420 sec at a moist soil.

The Diurnal Regime

Except where nonradiative energy fluxes can become very large, as at a water surface, T_0 usually responds to the regime of radiant-energy intake, which is dominated by the solar beam, and surface temperature peaks about an hour after solar noon (Fig. 2). The T_0 range during the 24 hr depends chiefly on the midday flux density of absorbed radiation and the heat-removing potential of the temperature-dependent energy fluxes, that is, the couplings of surface with air and substrate, which can be estimated if diurnal range of T_0 is remotely sensed (see Chapter XVII).

Daily regimes of soil and plant surface temperatures in England (Table I) show bare soil to be far hotter than the air during the daytime and cooler during the night. Grass leaves display about half as much midday superheating above air temperature as the soil and are coldest at night. Taller leaves are subjected to less extreme conditions than the shorter ones, and at night remained within 3°K of the air temperature. A tree branch is also closely coupled with the air but still cools to 3–4°K below air temperature on a clear night (Häckel, 1978).

TABLE I

Surface and Air Temperatures at Rothamsted Experimental Station, England, 29–30 June 1961[a]

	09	12	15	18	21	24	03	06	09	Max.	Min.	Mean	Range
Bare soil (powdery mulch)	34	43	43	30	14	10	9	18	36	44	9	29	35
Short grass	30	34	33	20	12	8	7	17	27	36	7	19	29
Barley	20	24	24	20	13	10	10	14	26	25	9	19	16
Air	18	23	24	23	19	15	12	16	22	25	12	19	13

[a] Data from Monteith and Szeicz (1962).

TABLE II

Morning Increase in Radiant-Energy Intake and Surface Temperature at Two Desert Sites[a]

Site	Increase (W m^{-2})	T_0 rise (°K)
Shafrikan, Uzbekistan (Fig. 2)	848	41
El Mirage Dry Lake, California	985	48

[a] Data from Aizenshtat (1960) and Vehrencamp (1953).

In an aerial survey of diversified farmland in south-central Wisconsin, Lenschow and Dutton (1964) found the greatest diurnal variation of T_0 at the sandy glacial outwash plains, where substrate and latent-heat fluxes are small. The smallest ranges were found in hilly wooded areas and lakes, which are closely coupled with the atmosphere and substrate, respectively. The diurnal range of desert surface temperature is related to the characteristically large increase in radiant-energy absorption as shown in Table II. Concrete shows a smaller range in T_0 than desert, but with a similar response to radiation: A range of 9°K on days with a mean solar flux density of 50 W m^{-2} increases to 32°K on days when flux density is 300 W m^{-2} (Mahringer, 1962). The desert response is 0.05°K for each flux increase of 1 W m^{-2} and that of the concrete surface, which has closer substrate coupling, is about 0.04°K. Smaller heating coefficients would occur at moist surfaces.

Interdiurnal Variations

Variations in surface temperature from day to day follow variations in radiant-energy intake (Chapter VII) and in the temperature-dependent fluxes, which can vary even in the relative stability of a desert climate. Williams (1967) found no relation of incoming solar radiation and T_0 at Yuma in a sample of 97 hot days (shelter temperature exceeding 42°C) because warm air aloft suppressed the sensible-heat flux to a varying degree on different days.

Day-to-day variability in a humid climate is shown by surface-temperature frequencies in a clearcut in the southern Appalachians (see Table III). The highest surface temperatures were infrequent after vegetation began to shade the soil in June.

Annual Regime

The annual regime of T_0 is less marked than the daily because radiant-energy intake changes less. At Hamburg, for instance, the energy in-

TABLE III

Frequencies of Maximum Surface
Temperature on Summer Days in a 3-
hectare Clearcut in the Southern
Appalachians[a]

Max. Surface Temperature T_0 (°C)	Frequency
>43	0.27
>49	0.19
>54	0.10
>60	0.03

[a] Data from McGee (1976).

crease from winter to summer is 190 W m^{-2}, and the increase in grass mean surface temperature is about 20°K. The heating coefficient in the annual cycle is thus 0.11°K for an increase of 1 W m^{-2} in mean daily radiant-energy intake. My calculations from measurements made at Tashkent by Aizenshtat (1957) give an almost identical heating coefficient.

Spatially integrated midday T_0 at rolling farmland in south-central Wisconsin on sunny days from May to September runs about 30–32°C, drops to 20°C in October, and to 5°C in November (Lenschow and Dutton, 1964); it remains below 0°C in the snow-cover season, and in clear weather may reach −20°C; its range over the year is about 50°K, double that at the ice and water of Lake Mendota.

A plotting of monthly maxima of T_0 and radiant-energy intake at Tashkent (Aizenshtat, 1957) displays hysteresis like that in the daily march shown in Fig. 2. At the same value of radiation, T_0 in the fall is 15°K higher than in the spring (Fig. 3), presumably because the warmer soil in fall accepts less heat from the surface.

SPATIAL PATTERNS

Ecosystem-Scale Patterns

Transpiring biological surfaces tend toward a restrained variation in T_0, from about 0°C to 30–35°C, for, except in strong heat advection, moist surfaces seldom exceed 32–33°C. Contrasts in T_0 at adjacent active ecosystems are therefore not extreme and usually are due to differences in turbulent heat removal rather than in radiant-energy intake, except where exposure to the solar beam is different. Midday values of T_0 were

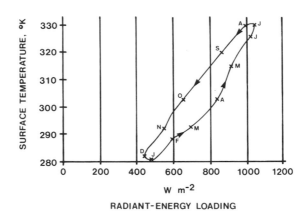

W m⁻²

RADIANT-ENERGY LOADING

Fig. VIII.3. Annual march of the surface temperature (°K) of a sparse unirrigated grass ecosystem at the radiation site of the Tashkent Geophysical Observatory, as a function of its radiant-energy intake (W m⁻²) on the maximum day in each month [measurements from Aizenshtat (1957)]. Hysteresis is like that in the diurnal curve.

6°K higher on a slope than flat land, for example, in both observation and model (Brazel and Outcalt, 1973).

Gay (1972) measured the ecosystem temperatures from a helicopter on a hot day in the Willamette Valley (see Table IV). Radiant-energy intake can be calculated as being 1050-1150 W m⁻² in the different eco- systems, and T_0 differences between ecosystems were larger than within ecosystems; the group of dry ecosystems was by far the hottest.

Rock areas at treeline on the Sonnblick (3.1 km altitude) reached tem- peratures 17°K above air temperature on sunny, windy days (>10 m

TABLE IV

Surface Temperatures at Moist and Dry Ecosystems in the Willamette Valley, Oregon[a]

Site	Temperature (°C)	Site	Temperature (°C)
Moist Ecosystems		Dry Ecosystems	
River	19	Dried grass	42
Deciduous forest	28	Stubble fields	38
Coniferous forest	30	Burned fields	51
Clover	32	Plowed fields	56
Unirrigated corn	30	City	52
Irrigated corn	24		

[a] Data from Gay (1972). Midday in summer; shelter temperature 35°C and incoming solar radiation 860 W m⁻².

sec^{-1}) and 28°K excess on calm days (1.5 m sec^{-1}) (Dirmhirn, 1952). Dry urban surfaces are also hot: In Bonn midday pavement temperature of 47°C exceeded grass temperature by 15–16 degrees (Kessler, 1971); artificial grass on a baseball diamond in St. Louis was 19°K above air temperature.

Differences among ecosystem temperatures at night result primarily from substrate contrasts and mesoscale advective effects like cold-air drainage or shelterbelt effects, rather than from differences in incoming radiation. Transects of a landscape of bare soil, citrus orchards, and shelterbelts in south Texas (Nixon and Hales, 1975) found contrasts of 3–5°K. Shelterbelts and heated orchards were slightly warmer than other ecosystems, as were fields recently irrigated. Low values of T_0 distinguish winter crops on drained organic soils in the Florida Everglades, as much as 5°K colder than nearby soils (Chen et al., 1979).

Large Scale

Surface temperatures of ecosystems on the east side of Mt. Fuji rose rapidly during the morning, and 3 hours after sunrise reached 32°C at all altitudes, while west-side surfaces were not only colder (by 10–20°K) but also varied more with altitude (Fujita et al., 1968). Formation of clouds above the hotter side suggests the large scale of the influence of T_0. Cold (-35°C) valleys or "parks" in the Rocky Mountains are delineated in more detail on GOES images (Maddox and Reynolds, 1980) than by observations at weather stations.

Midday aerial measurements of T_0 over large regions (Bogdanova and Lebedev, 1961) exceeded air temperature at 0.5 km by 8–10°K above mixed forest, by 10–15°K above tundra, and by 30–40°K above desert. If air temperature at flight level was 3–5°K lower than shelter temperature, the surface-to-shelter differences become approximately 5, 8, and 25°K in the three biomes. Kessler (1974) found in a winter traverse of West Africa that foliage temperatures in most places were in the low 30s, but bare ground showed a marked increase southward, reaching values above 50°C.

Extreme surface temperatures among all the earth's ecosystems occur on sun-facing slopes at high altitudes (Turner, 1958) and on sandy ground, where temperatures exceeding 80°C have been reported (Cloudsley-Thompson, 1977, p. 8) and lichens exist at 70°C (Cloudsley-Thompson, 1977, p. 93). Day-active "insects tend to leave the sand when its temperature reaches about 50°C (122°F). Some climb grasses, some dive into holes, while others fly about above the ground making hurried landings to enter their burrows" (Cloudsley-Thompson, 1977, p. 72).

Global low extremes are found at the surface of snow cover under a cold, dry atmosphere in the polar night, when T_0 is commonly $-70°C$ or colder. This surface is as hostile to life as the hot sand, and deeper layers of the snow are the refuge of many small mammals (Pruitt, 1960).

Although surface temperature is a condition, not an energy flux, it plays a pivotal role in this book on energy fluxes because it mediates the ways that energy made available from absorption of radiation fluxes by an ecosystem is converted into energy-removing outward fluxes. Preceding chapters have dealt with the T_0-independent fluxes summed up as radiant-energy intake; later ones will deal with the fluxes that depend on T_0.

REFERENCES

Aase, J. K., and Siddoway, F. H. (1980). Stubble height effects on seasonal microclimate, water balance, and plant development of no-till winter wheat. *Agric. Meteorol.* **21**, 1–20.

Aizenshtat, B. A. (1957). Radiatsionnye balans i temperatura poverkhnosti pochvy v Tashkente. *Tr. Tashk. Geofiz. Obs.* **13**, 3–74.

Aizenshtat, B. A. (1960). "The Heat Balance and Microclimate of Certain Landscapes in a Sandy Desert" (transl. by G. S. Mitchell). U.S. Weather Bur., Washington, D.C.

Bhumralkar, C. M. (1975). Numerical experiments on the computation of ground surface temperature in an atmospheric general circulation model. *J. Appl. Meteorol.* **14**, 1246–1248.

Bogdanova, N. P., and Lebedev, A. N. (1961). Sviaz' pogodnykh i klimaticheskikh kharakteristik s radiatsionnoi temperaturnoi podstilaiuschei poverkhnosti. *Tr. Gl. Geofiz. Obs.* **109**, 38–52.

Brazel, A. J., and Outcalt, S. J. (1973). The observation and simulation of diurnal evaporation contrast in an Alaskan alpine pass. *J. Appl. Meteorol.* **12**, 1134–1143.

Brown, H. T., and Escombe, F. (1905). Researches in some of the physiological processes of green leaves, with special reference to the interchange of energy between leaf and its surroundings. *Proc. R. Soc. London Ser. B* **76**, 29–111.

Brown, K. W. (1974). Calculations of evapotranspiration from crop surface temperature. *Agric. Meteorol.* **14**, 199–209.

Budyko, M. I. (1948). "Isparenie v Estestvennykh Usloviiakh." Gidrometeorol. Izd., Leningrad.

Budyko, M. I. (1956). "Teplovoi Balans Zemnoi Poverkhnosti." Gidrometeorol. Izd., Leningrad. (Transl. by N. A. Stepanova as "The Heat Balance of the Earth's Surface." U.S. Weather Bur., Washington, D.C., 1958.)

Budyko, M. I. (1960). "The Temperature of the Active Surface and Its Bioclimatic Significance." U.S. Weather Bur., Washington, D.C. (Transl. by N. A. Stepanova and G. S. Mitchell from "Sovremennye Problemy Meteorologii Prizemnogo Sloia Vozdukha" (M. I. Budyko, ed.), pp. 201–211. Gidrometeoizd., Leningrad, 1958.)

Chen, E., Allen, L. H., Jr., Bartholic, J. F., Bill, R. G., Jr., and Sutherland, R. A. (1979). Satellite-sensed winter nocturnal temperature patterns of the Everglades agricultural area. *J. Appl. Meteorol.* **18**, 992–1002.

Cloudsley-Thompson, J. L. (1977). "Man and the Biology of Arid Zones." University Park Press, Baltimore.

Cogan, J. L., and Willand, J. H. (1976). Measurement of sea surface temperature by the NOAA 2 satellite. *J. Appl. Meteorol.* **15,** 173–180.

Coiner, J. C., and Levine, A. L. (1979). Applications of remote sensing to urban problems. *Urban Systems* **4,** 205–219.

Davitaia, F. F. (1960). "An Appraisal of Microclimatic Peculiarities in the Distribution of Cultivated Plants and the Specialization of Agriculture." U.S. Weather Bur., Washington, D.C. (Transl. by G. S. Mitchell and N. A. Stepanova of Uchet mikroclimaticheskikh osobennosti v razmeshchenii kul'turnykh rastenii i spetsializatsiia sel'skogo khoziaistva. *In* "Sovremennye Problemy Meteorologii Prizemnogo Sloia Vozdukha" (M. I. Budyko, ed.), pp. 192–200. Gidrometeoizd., Leningrad, 1958.)

DeWalle, D. R., and Parmele, L. H. (1974). Application of error analysis to surface temperature determination by radiation budget techniques. *J. Appl. Meteorol.* **13,** 430–434.

Dirmhirn, I. (1952). Oberflächentemperaturen der Gesteine im Hochbirge. *Arch. Meteorol., Geophys. Bioklimatol., Ser. B* **4,** 43–50.

Dodd, A. V., and McPhilimy, H. S. (1959). Yuma summer microclimate. *U.S. Quartermaster Res. Eng. Cent., Tech. Rep.* **EP-120.**

Ehlers, J. H. (1915). The temperature of leaves of *Pinus* in winter. *Am. J. Bot.* **2,** 32–70.

Fimpel, H. (1964). Messungen der Temperatur einer Grasoberfläche mit einen Gesamtstrahlungspyrometer. Univ. München Meteorol. Inst. *Wiss. Mitt.* **9,** 98–106.

Friedmann, E. I., and Ocampo, R. (1976). Endolithic blue–green algae in the dry valleys; Primary producers in the Antarctic desert ecosystem. *Science* **193,** 1247–1249.

Fuchs, M., Kanemasu, E. T., Kerr, J. P., and Tanner, C. B. (1967). Effect of viewing angle on canopy temperature measurements with infrared thermometers. *Agron. J.* **59,** 494–496.

Fuchs, M., and Tanner, C. B. (1966). Infrared thermometry of vegetation. *Agron. J.* **58,** 597–601.

Fujita, T., Baralt, G., and Tsuchiya, K. (1968). Aerial measurement of radiation temperatures over Mt. Fuji areas and their application to the determination of ground- and water-surface temperatures. *J. Appl. Meteorol.* **7,** 801–816.

Gates, D. M. (1962). "Energy Exchange in the Atmosphere." Harper, New York.

Gates, D. M. (1963). The energy environment in which we live. *Am. Sci.* **51,** 327–348.

Gay, L. W. (1972). Radiative temperatures in the Willamette Valley. *Northwest Sci.* **46,** 332–335.

Geiger, R. (1927). "Das Klima der bodennahen Luftschicht." Vieweg, Braunschweig.

Geiger, R. (1965). "The Climate Near the Ground." Harvard Univ. Press, Cambridge, Massachusetts. (Transl. of "Das Klima der bodennahen Luftschicht," 4th ed. Vieweg, Braunschweig, 1961.)

Häckel, H. (1978). Modellrechnungen über die Temperaturen von Pflanzen in winterlichen Strahlungsnächten. *Agric. Meteorol.* **19,** 497–504.

Hare, F. K. (1966). The concept of climate. *Geography* **51,** 99–110.

Heilman, J. P., and Moore, D. G. (1980). Thermography for estimating near-surface soil moisture under developing crop canopies. *J. Appl. Meteorol.* **19,** 324–328.

Henson, W. R., and Shepherd, R. F. (1952). The effects of radiation on the habitat temperatures of the lodgepole needle miner, *Recurvaria milleri* Busck (Gelechiidae: Lepidoptera). *Canad. J. Zool.* **30,** 144–153.

Jackson, R. D., Reginato, R. J., and Idso, S. B. (1977). Wheat canopy temperature: A practical tool for evaluating water requirements. *Water Resour. Res.* **13,** 651–656.

Kessler, A. (1971). Über den Tagesgang von Oberflächentemperaturen in der Bonner Innenstadt an einem sommerlichen Strahlungstag. *Erdkunde* **25,** 13–20.

Kessler, A. (1974). Infrarotstrahlungsmessungen auf einer Reise durch Westafrika und die

Sahara, 1. Mitteilung: Effektive Strahlungstemperaturen verschiedener Oberflächen. *Arch. Meteorol., Geophys. Bioklimatol., Ser. B* **22,** 135–147.

Knoerr, K. R., and Gay, L. W. (1965). Tree leaf energy balance. *Ecology* **46,** 17–24.

Lenschow, D. H., and Dutton, J. A. (1964). Surface temperature variations measured from an airplane over several surface types. *J. Appl. Meteorol.* **3,** 65–69.

Lin, J. D. (1980). On the force-restore method for prediction of ground surface temperature. *J. Geophys. Res.* **85,** 3251–3254.

McBride, J. L. (1975). The effect on land-sea temperature contrast on short-term numerical forecasts. *Austral. Meteorol. Mag.* **23**(4), 75–98.

Macfarlane, W. V. (1964). Terrestrial animals in dry heat: Ungulates. *In* "Handbook of Physiology–Environment" (D. B. Dill, E. F. Adolph, and C. G. Wilber, eds.), Vol. 4, pp. 509–539. Am. Physiol. Soc., Washington, D.C.

McGee, C. E. (1976). Maximum soil temperature on clearcut forest land in western North Carolina. *U.S. Dep. Agric., For. Serv. Res. Note* **SE-237.**

Maddox, R. A., and Reynolds, D. W. (1980). GOES satellite data maps areas of extreme cold in Colorado. *Mon. Weather Rev.* **108,** 116–118.

Mahringer, W. (1962). Das Temperaturregime einer Betonoberfläche. *Arch. Meteorol., Geophys. Bioklimatol., Ser. B* **11,** 533–559.

Michaelis, P. (1932). Ökologische Studien an der alpinen Baumgrenze. I. Das Klima und die Temperaturverhältnisse der Vegetationsorgane im Hochwinter. *Ber. Dtsch. Bot. Ges.* **50,** 31–42.

Miller, D. H. (1955). "Snow Cover and Climate in the Sierra Nevada California," Publications in Geography, 11. Univ. of California Press, Berkeley.

Miller, D. H. (1956). The influence of open pine forest on daytime temperature in the Sierra Nevada. *Geogr. Rev.* **46,** 209–218.

Miller, D. H. (1977). "Water at the Surface of the Earth." Academic Press, New York.

Miller, P. C., Stoner, W. A., and Tieszen, L. L. (1976). A model of stand photosynthesis for the wet meadow tundra at Barrow, Alaska. *Ecology* **57,** 411–430.

Milwaukee Journal (1976). It's only the ground that's cold. Feb. 12, Part 2, p. 4.

Monteith, J. L., and Szeicz, G. (1962). Radiative temperature in the heat balance of natural surfaces. *Q. J. R. Meteorol. Soc.* **88,** 496–507.

Munn, R. E. (1966). "Descriptive Micrometeorology" Academic Press, New York.

Murphy, C. E., Jr. (1974). Biological application of atmospheric physics. (Review of J. L. Monteith, "Principles of Environmental Physics," 1973.) *Ecology* **55,** 677–678.

Nixon, P. R., and Hales, T. A. (1975). Observing cold-night temperature of agricultural landscapes with an airplane-mounted radiation thermometer. *J. Appl. Meteorol.* **14,** 498–505.

Penman, H. L. (1948). Natural evaporation from open water, bare soil, and grass. *Proc. R. Soc. London, Ser. A* **193,** 120–145.

Pinter, P. J., Jr., Stanghellini, M. E., Reginato, R. J., Idso, S. B., Jenkins, A. D., and Jackson, R. D. (1979). Remote detection of biological stresses in plants with infrared thermometry. *Science* **205,** 585–587.

Priestley, C. H. B. (1957). The heat balance of sheep standing in the sun. *Austral. J. Agric. Res.* **8,** 271–280.

Pruitt, W. O., Jr. (1960). Animals in the snow. *Sci. Am.* **202**(1), 60.

Pyldmaa, K. V. (1978). Temperatura podstilaiushchei poverkhnosti pri kuchevoi oblachnosti. *In* "Izmenchivost' Oblachnosti i Polei Radiatsii" (Variability of Cloudiness and Radiation Field) (Iu. Ross *et al.*, eds.), Issledovaniia po Fizike Atmosfery, Tartu, No. 21, pp. 105–112. Akad. Nauk Eston, SSR, Inst. Astron. Atmos. Fiz., Tartu.

Raschke, K. (1956). Über die physikalischen Beziehungen zwischen Wärmeübergangszahl, Strahlungsaustausch, Temperatur und Transpiration eines Blattes. *Planta* **48**, 200–238.

Reifsnyder, W. E., and Lull, H. W. (1965). Radiant energy in relation to forests. *U.S. Dep. Agric., Tech. Bull.* No. 1344.

Roth, E. S. (1965). Temperature and water content as factors in desert weathering. *J. Geol.* **73**, 454–468.

Smith, W. K. (1978). Temperatures of desert plants: Another perspective on the adaptability of leaf size. *Science* **201**, 614–616.

Soer, G. J. R. (1980). Estimation of regional evapotranspiration and soil moisture conditions using remotely sensed crop surface temperatures. *Remote Sens. Envir.* **9**, 27–45.

Stoutjesdijk, P. (1977). High surface temperature of trees and pine litter in the winter and their biological importance. *Int. J. Biometeorol.* **21**(4), 325–331.

Sutherland, R. A., Langford, J. L., Bartholic, J. F., and Bill, R. G., Jr. (1979). A real-time satellite data acquisition, analysis and display system—a practical application of the GOES network. *J. Appl. Meteorol.* **18**, 335–360.

Tanner, C. B. (1963). Plant temperatures. *Agron. J.* **55**, 210–211.

Tenhunen, J. D., Weber, J. A., Yocum, C. S., and Gates, D. M. (1976a). Development of a photosynthesis model with an emphasis on ecological applications. II. Analysis of a data set describing the PM surface. *Oecologia (Berlin)* **26**, 101–119.

Tenhunen, J. D., Yocum, C. S., and Gates, D. M. (1976b). Development of a photosynthesis model with an emphasis on ecological applications. I. Theory. *Oecologia (Berlin)* **26**, 89–100.

Thornthwaite, C. W. (1948). An approach toward a rational classification of climate. *Geogr. Rev.* **38**, 55–94.

Turner, H. (1958). Maximaltemperaturen oberflächennaher Bodenschichten an der alpinen Waldgrenze. Messungen in der Schönwetterperiode Juni/Juli 1957. *Wetter Leben* **10**, 1–12.

Vehrencamp, J. E. (1953). Experimental investigation of heat transfer at an air–earth interface. *Trans. Am. Geophys. Union* **34**, 22–30.

Wadham, S., Wilson, R. K., and Wood, J. (1964). "Land Utilization in Australia," 4th ed. Melbourne Univ. Press, Melbourne.

Waggoner, P. E., and Reifsnyder, W. E. (1968). Simulation of temperature, humidity and evaporation profiles in a leaf canopy. *J. Appl. Meteorol.* **7**, 400–409.

Williams, L. (1967). Occurrence of high temperatures at Yuma Proving Ground, Arizona. *Ann. Assoc. Am. Geogr.* **57**, 579–592.

Chapter IX

LONGWAVE RADIATION EMITTED BY ECOSYSTEMS

One means by which a radiation-loaded ecosystem stays in energy balance when its surface temperature rises is to increase its emission of longwave radiation. This outward flux of energy increases immediately and inevitably when the surface heats up and is the temperature-dependent energy flux most closely related to variation in surface temperature. It is, on the average, the largest of all the energy fluxes and bears a large part of the task of preventing overheating of ecosystems in high-energy conditions. In low-energy conditions, on the other hand, it is a remorseless drain on an ecosystem's energy resources.

THE LONGWAVE FLUX

A price for the existence of any object is that the kinetic energy of its molecules gives off bursts of energy that are radiated away. A liquid or solid nonmetallic surface at 0°C radiates 300–315 W m^{-2}, and this loss of energy can be reduced only if the surface can be cooled.

Constraints on the free motion of molecules in solids and liquids cause their emission of radiation to be spread over a continuous series of wavelengths rather than being concentrated in narrow spectral bands as is true of gas molecules. Therefore the formulation of a single expression for emission of radiation as a function of wavelength (E_λ) is possible—Planck's law with its three universal constraints, noted earlier, which form a connected tissue among the dimensions of length, mass, time, and temperature. Differentiation of this expression identifies the wavelengths at which peak energy flow per unit wavelength (micrometer) is given off (Wien's displacement law). Maximum intensity of radiation from a low-temperature source like cold snow cover ($T_0 = 253°K$) occurs

TABLE I

Flux Densities[a] in Different Spectral Bands of the Radiant Energy Emitted by a Surface at $T_0 = 300°K$

Spectral band (μm)	Flux density (W m^{-2})
3–8	64
8–12	122
12–16	93
16–20	58
20–24	35
>24	70
Sum	442

[a] Data from Smithsonian Institution (1966, Table 1290).

at a wavelength of 12 μm, from hot soil at 8.5 μm, and from active ecosystems at intermediate wavelengths. Integration of the expression for Planck's law yields the total energy flux from the emitting surface at wavelengths within any spectral band. From a surface at 300°K, the spectral distribution of blackbody radiation is shown in Table I. The substantial upward flux of longwave radiation in the 8–12 μm "window" contrasts with the small downward flux from a cloudless atmosphere in this spectral band. Over the entire spectrum the energy flux (442 W m^{-2}) is given by the Stefan–Boltzmann law

$$E = \sigma T_0^4$$

for a surface that radiates as an ideal blackbody.* The constant σ is named for the two physicists; it is equal to 56.7×10^{-9} W m^{-2} deg^{-4}. For natural surfaces this expression is reduced by a dimensionless emissivity coefficient ϵ, noted in Chapter III. Thus

$$E = \epsilon \sigma T_0^4.$$

* This law was discovered before Planck's law was formulated. Stefan was an experimental physicist in Wien and Boltzmann was a theoretician, who worked out a general explanation in the 1880s after long research into radiant energy and its relation to matter. Planck's law and Einstein's work on the photoelectric effect led to the quantum theory and the 20th century revolution in physics. Only in recent decades have ecologists and climatologists had available even a coarse network of radiation instruments, but it is no longer necessary to stop with a description of thermal regimes; we can now explain ecosystem phenomena in terms of energy flows, conversions, and storages.

ROLE IN ECOSYSTEM ENERGY BUDGETS

High-Energy Conditions

In conditions of large energy intake, emission of longwave radiation by foliage helps shed excess heat. Other outgoing energy fluxes also help keep leaf temperature down, but the prompt response of emission, its size (about 5 W m^{-2} increase for a 1°K rise in temperature), and particularly its independence of environmental conditions make it effective.

Movement of heat from an ecosystem by nonradiative means depends on outward gradients from the ecosystem into the environment, so that heat, vapor concentration or turbulence in the environment act as governors on the fluxes. This is not true of radiation. The ecosystem radiates by the Stefan–Boltzman law, into which only emissivity and temperature enter.[*]

Low-Energy Conditions

In low-energy periods, a system that can allow its surface to chill to a low temperature adjusts its radiative heat loss to fit a current small intake of energy. The fleece of a sheep serves such an insulating role in winter, as does a snow cover. When aquatic ecosystems in continental middle and high latitudes freeze over in winter, their ice and snow cover reduce T_0 and radiative heat loss; system energy loss is minimized, and plants and fish survive (Fig. 1).

Short low-energy periods occur every night of the growing season. Foliage radiates away its heat and reduces respiration, conserving the energy content of the photosynthate produced during the preceding daylight hours. Nocturnal cooling of foliage and soil surfaces cools the local air. Threads of chilled air descend from the forest canopy; cold air piles up on the ground, and cold-air drainage allows replacement warm air to sink into an ecosystem. These local circulations of air tend to compensate for radiative cooling that might otherwise be excessive. A similar technique is used to reduce heat loss from solar energy collectors (Lehane, 1980).

Artificial reduction by 0.04–0.12 of the emissivity of grapefruit trees, effected by sprinkling the canopy with aluminum powder, decreased the longwave flux by several watts per square meter and increased leaf temperature by 0.6–0.9°K (Hales, 1974). These manifestations of upward

---

[*] The atmosphere also radiates independently, and in Chapter X we shall look at the net difference between these two independent fluxes of energy.

Fig. IX.1. Frozen aquatic ecosystems insulate themselves from their outer surface, which is allowed to cool to a temperature at which little energy is lost to the atmosphere; the sacrifice allows the main ecosystem to remain relatively warm.

longwave radiation are so familiar that Weischet (1979, p. 83) has to remind us that this flux is not just confined to night or winter.

DETERMINATION OF UPWARD LONGWAVE RADIATION

The measurement of upward longwave radiation as an individual energy flux has benefited by the advent of infrared thermometry for measuring surface temperature, as discussed in Chapter VIII. Methods of determining the flux also include direct measurement by an inverted hemispherical radiation sensor (Fig. 2), paired with an inverted short-wave sensor.

Separating reflected longwave from emitted longwave radiation is done through the coefficient of reflectivity or its complement, absorptivity or emissivity, integrated over the longwave region of interest. The separation should be kept in mind if the reflected flux, which is independent of surface temperature, follows a different regime than the emitted flux (Brooks, 1952, 1959).

Fig. IX.2. When seen from below, replicated all-wave Schulze radiometers at Hamburg Meteorological Observatory show the domes on sensors that measure upward radiation fluxes (reflected solar plus longwave from grass). A nearly unique long record has been maintained at this station.

Emissivity determinations are sometimes made over only a restricted spectral region, e.g., 8–13 μm (Maxwell, 1971) or 9–12 μm (Kondratyev, 1969, p. 43), in which the values range from 0.95 for sand and sandy soil to 0.98 for grass. "The problem of the measurement of the relative emissivity of natural underlying surfaces is far from being satisfactorily solved" (Kondratyev, 1969, p. 43) because the extent to which readings in a particular spectral band represent spectrally-integrated emissivity is not always clear.

Emissivity values over the 2–3 μm to 15–30 μm range vary with species (Ross, 1975, p. 173) and are higher for ecosystems than for bare ground or single leaves (Fuchs and Tanner, 1966). Multiple reflection of longwave radiation tends to increase the resemblance of an ecosystem to a set of blackbody cavities. In orchard rows the effect of bare soil diminishes as the ratio of the height to spacing increases (Sutherland and Bartholic, 1977).

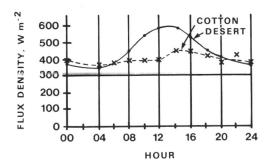

Fig. IX.3.   Diurnal marches of longwave radiation emitted by two ecosystems in central Asia in summer (W m^{-2}) [data from Aizenshtat (1960)]. The lower surface temperature of the cotton field results from irrigation and reduces the emission of radiation substantially.

VARIATIONS IN THE UPWARD FLUX OF LONGWAVE RADIATION

Variations in Time

Figure 3 shows the diurnal variations in the flux density of longwave radiation emitted by a desert ecosystem of widely spaced plants, dormant at the time observations were made, and an irrigated field of cotton. The fluxes are similar during the night, but not during the day when the desert was hotter than the cotton by 25–30°. Although the cotton has the greater emissivity, the lower temperature of its transpiring leaves causes it to emit less energy than the dry system; more of the energy available to it goes into the biological conversions of photosynthesis and transpiration.

Averaging flux densities through the 24-hr cycle eliminates a large part of the total variation over time. Mean daily values of upward longwave radiation (emitted + reflected*) from grass at Hamburg Observatory (Fig. 4) vary by a relatively small amount from day to day. The mean interdiurnal variation is 5.5 W m^{-2}; no day-to-day change exceeded 16 W m^{-2}.

The seasonal decline, hinted at in Fig. 4, actually occurred during a few periods of rapid change in weather, for example, from the 17th to the 19th. The mean decrease during autumn at this station is about 20 W m^{-2} per month, in both the 1954–1956 averages (Fleischer and Gräfe, 1955–1956) and the 1955–1964 averages (Schulze, 1970, p. 100). The

---

* The flux of reflected longwave radiation here is small—only a few percentage points of the values graphed.

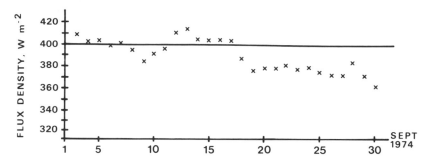

Fig. IX.4.   Daily mean flux densities of longwave radiation emitted by grass at Hamburg Observatory on successive days in fall, 1974 [data from Germany, Wetterdienst (1974)]. Note the synoptic-scale clustering of days.

mean increase in spring is at about the same rate. The total winter-to-summer change is approximately 85 W m^{-2} (Schulze, 1970), with more than a month's lag after the solstices, a range not much greater than the night-to-day range. The phase is affected by the heat-storage capacity available to the irradiated surface; upward radiation from an aquatic ecosystem or a city reaches its peak later than that from a terrestrial ecosystem.

Long-term variations in emitted radiation might reflect a change in radiant-energy absorption or of nonradiative factors in T_0. Peak monthly means in eight summers at Hamburg (Schulze, 1970, p. 108) showed a standard deviation of 10 W m^{-2}.

Spatial Contrasts

Microscale exchanges of longwave radiation within ecosystems smooth out variations in leaf temperature, help support evaporation from shade leaves, melt out rings in snow cover around tree trunks (Mazzarella, 1979), and provide information for animals (e.g., for infrared-sensing snakes) about the thermal field of their environment (Gamow and Harris, 1973).

Aquatic and terrestrial ecosystems present contrasts, varying with the season, that resemble those between the desert and irrigated ecosystems described earlier in this chapter. Urban systems generate greater outward fluxes than do their environs, and these will become better known as more thermal scans of cities are analyzed.

The upward component of longwave radiation is always large, the growing-season flux density being around 400 W m^{-2}. Independent of the environment, it is a reliable means of relieving excess heat loading on an ecosystem, responding quickly to any increase in surface temperature. In cold conditions, however, it is a continuing leakage from the energy stored by an ecosystem.

REFERENCES

Aizenshtat, B. A. (1960). "The Heat Balance and Microclimate of Certain Landscapes in a Sandy Desert" (transl. by G. S. Mitchell). U.S. Weather Bur., Washington, D.C.

Brooks, F. A. (1952). Atmospheric radiation and its reflection from the ground. *J. Meteorol.* **9**, 41–52.

Brooks, F. A. (1959). "An Introduction to Physical Microclimatology." Associated Student Store, Davis, California.

Fleischer, R., and Gräfe, K. (1955–1956). Die Ultrarotstrahlungsströme aus Registrierung des Strahlungsbilanzmessers nach Schulze. *Ann. Meteorol.* **7**, 87–95.

Fuchs, M., and Tanner, C. B. (1966). Infrared thermometry of vegetation. *Agron. J.* **58**, 597–601.

Gamow, R. I., and Harris, J. F. (1973). The infrared receptors of snakes. *Sci. Am.* **228**(5), 94–100.

Germany, Wetterdienst (1974). Beilage zum Medizin-meteorologischen Bericht. (Hamburg Observatorium, monthly.)

Hales, T. A. (1974). Alteration of nocturnal radiation balance of "Red Blush" grapefruit trees by application of aluminum powder suspensions to upper leaves. *Agric. Meteorol.* **13**, 59–67.

Kondratyev, K. Ya. (1969). "Radiation in the Atmosphere." Academic Press, New York.

Lehane, R. (1980). Towards industrial use of solar energy. *Ecos (Melbourne)* **23**, 18–22.

Maxwell, F. D. (1971). A portable IR system for observing fire thru smoke. *Fire Technol.* **7**, 321–331.

Mazzarella, D. A. (1979). Snow-free rings around trees. *Weatherwise* **32**, 247–250.

Ross, Iu. K. (1975), "Radiatsionnyi Rezhim i Arkhitektonika Rastitel'nogo Pokrova." Gidrometeorol. Izd., Leningrad.

Schulze, R. (1970). "Strahlenklima der Erde." Steinkopff, Darmstadt.

Smithsonian Institution (1966). "Smithsonian Meteorological Tables" (R. J. List, ed.), 6th ed., Smithsonian Institution Publ. No. 4014. Washington, D.C.

Sutherland, R. A., and Bartholic, J. F. (1977). Significance of vegetation in interpreting thermal radiation from a terrestrial surface. *J. Appl. Meteorol.* **16**, 759–763.

Weischet, W. (1979). "Einführung in die allegemeine Klimatologie. Physikalische und Meteorologische Grundlagen." 2te. Aufl. B. G. Teubner, Stuttgart.

Chapter X

RESULTANTS OF THE UPWARD AND DOWNWARD RADIATION FLUXES

The net resultant of the fluxes of radiant energy at the surface of an ecosystem is a useful quantity that sums up all forms of a particular mode of energy transfer by adding the vertical components of both short-wave and longwave fluxes. It is not a physical flux but a mixture, consideration of which can be justified because it helps us see interactions between an ecosystem and its environment. We begin with the simple case of a radiating ecosystem and the radiating atmosphere, i.e., the net exchange of energy between them by longwave radiation: land and sky confronting each other. Later in the chapter we consider land, sun, and sky.

THE NET EXCHANGE OF LONGWAVE RADIATION BETWEEN ECOSYSTEM AND ATMOSPHERE

One of these radiators, the ecosystem, approaches being an ideal blackbody. The other radiates only in certain spectral bands. Therefore, neglecting any difference in their temperatures, the net exchange of energy by radiation in the long wavelengths is the difference between the large upward flux emitted by an ecosystem and the downward flux emitted by the overlying atmosphere. For example, the mean difference at Hamburg in December is between $+305$ W m^{-2} downward and -325 W m^{-2} upward, equal to -20 W m^{-2} [from data in Schulze (1970, p. 100)]. The negative sign indicates a loss of energy by the grass surface.

Because it is most easily measured in the absence of solar radiation, this net difference has been called nocturnal radiation. However, both the fluxes continue day and night without ceasing, and the word "nocturnal" is a misnomer; in fact, the net exchange is usually greater by day because the surface warms up more than the atmosphere.* The significance of this resultant quantity is best seen at night, however. If it is large, say -50 W m^{-2}, it has to be made up by large nonradiative energy inputs into the ecosystem from its soil and local air. Removal of heat from biomass, soil, and air produces a "radiation night" that favors heavy dew formation and in winter the chilling and perhaps freezing of plant tissue. When the deficit is small, as it usually is on a cloudy night, dew may not form, and the threat of frost recedes. The ecosystem retains more of its daytime warmth.

A contrasting case, a system to apply the net difference for cooling a house, stresses isolation of the collector from the atmosphere. While solar energy is easier to exploit than the small cooling power of the exchange of longwave radiation, cooling of a house by use of a 27-m^2 radiator produced "cold" equivalent to 125 MJ, which, stored in buried rocks, served to cool the living space the next day (Daniels, 1964, p. 242). This exploitation of net longwave radiation is most feasible where downward longwave radiation is small, i.e., under a dry atmosphere. In such conditions a collector can reject 100 W m^{-2} to the sky at a cooling of 11°C (Meinel and Meinel, 1976, p. 453), which is about 0.15 of the input of a solar collector.

The net difference between downward and upward fluxes is maximized if it can be confined to the window region of the spectrum, in which the atmosphere produces little downward radiation. A collector that radiates in the 8–12-μm region is working in a region in which the atmosphere emits little for it to absorb. If it also reflects radiation from the atmosphere in the 3–8-μm and 12–20-μm bands, it attains a substantial cooling effect and exemplifies an extreme degree of uncoupling of a system from its atmospheric environment.

The Net Exchange as a Local Recycling of Energy

The atmosphere is an emanation of the earth, and its temperature and other characteristics are governed to a large degree by the earth's surface, but not necessarily in the same part of the world. It is logical to ask whether or not the exchange of longwave radiation between an ecosystem and the overlying air represents a local recycling of energy. Does the ecosystem emit radiation *only* insofar as it is warmed by absorbing radiation emitted by the atmosphere? Does the air radiate *only*

---

* The term "effective radiation" is also open to objection, like its sisters "effective precipitation," "effective temperature," and others that do not specify the adjustment made in some original datum.

insofar as it absorbs energy from the upward flux emitted by the ecosystem? These questions ask about interaction between ecosystems and air.

In some conditions the downward flux emitted by the atmosphere expresses primarily a recycling of energy within the local landscape; we can conceive of the earth's surface being heated by the sun *and then* heating the local air by radiative and nonradiative means; the air *then* radiates energy back to the surface. In this "symbiosis" of ecosystems and local air the quotient of net longwave radiation and the emitted flux expresses leakage of energy through the atmospheric window and out of the symbiotic system, caused by the incomplete emissivity of atmospheric gases as compared to ecosystem surfaces, and is called the Ångström ratio by Lettau (1969). It is large under dry air because of its smaller emissivity: 0.23 over the United States in January but 0.18 in July (Sellers, 1965, p. 53). It is small where a cold surface underlies a warm low-level isothermal layer, as in Antarctica (Schwerdtfeger, 1977), in warm advection over snow covered landscapes, and at night (0.17 at Shafrikan at midnight and 0400 as compared with 0.32 in the early afternoon). When atmospheric emissivity increases because of cloud formation and surface and air temperatures tend to converge, the Ångström ratio approaches zero.

In other conditions, however, no such "symbiotic" sequence takes place, and we can think of the flux of downward longwave radiation as at least partially independent of local surface energy fluxes. An obvious case is a storm, which draws upon a vast reservoir of atmospheric energy augmented by high rates of condensation in the invading moist airstream, much of which is radiated downward by the vapor and clouds. At a precipitation intensity of 5 mm hr^{-1}, for instance, latent heat is made available at an areal rate of 3500 W m^{-2}. In the initial days of the transformation of maritime air into polar continental air over the dark snowfields of the Yukon (Wexler, 1936), the invading warm, moist air radiates to a cold passive surface below and is clearly the initiator. Only later do the air and snow approach radiative equilibrium, in which fluxes from the atmosphere and surface do resemble a local recycling of energy.

Local recycling occurs in conditions in which a body of air is isolated from the rest of the free air and stagnates. Sierra valleys in quiet weather harbor air that is a little warmer and moister than the free air and radiates more strongly to the underlying snow cover, thereby maintaining the energy budget of the snow at a higher temperature than would otherwise be the case. An urban atmosphere likewise contributes to a warm nocturnal urban surface and is reciprocally supported by it. The increase

TABLE I

Daily Regime of Net Exchange of Longwave Radiation, Karadag, 1953–1956[a] W m⁻²

	0030	0630	0930	1230	1530	1830	Mean
June	−58	−93	−160	−185	−160	−93	−110
December	−45	−35	−45	−45	−35	−45	−42

[a] Data from Barashkova (1960). Data in watts per square meter.

in both upward and downward fluxes in Montreal as compared with rural areas is about 40 W m⁻² (Oke and Fuggle, 1972). On a global scale this protective role is called the "atmospheric effect."*

The daytime intake of atmospheric carbon dioxide by an ecosystem and its outputs of water vapor and sensible heat, as well as radiation, bring about a diurnal variation in the radiating power of the overlying air. These local influences also operate in the annual cycle, but are countered by large-scale advection, especially in winter. The downward longwave flux from tropical air that invades continents is the last stage in the delivery of energy from the low-latitude oceans to terrestrial ecosystems of middle or high latitudes. At this season these ecosystems can pay back relatively little energy to the air by radiation or other means.

FACTORS IN THE NET EXCHANGE OF ENERGY BY LONGWAVE RADIATION

Formulas for estimating the difference of the longwave fluxes utilize the parameters of vapor pressure and air temperature as measured near the ground for general averaging. It is preferable, however, to calculate the downward flux from a radiosonde profile taking account of the layer structure of the lower air, then to calculate separately the upward flux from the temperature and emissivity of the ecosystem surface, and then to take their difference.

This procedure reveals the large effect of surface temperature on radiation, especially if the soil is bare. The net exchange is about the same at night in June and December in the Crimea (Table I), but by day in June (as compared with December) the surface warms up, and the net loss is three times as large as during the night.

This example illustrates a situation in which the two flows of longwave

---

* This is a more correct term than "greenhouse effect," which it does not resemble (Fleagle and Businger, 1963, 1975).

TABLE II

Net Loss of Energy by Longwave Radiation at Different Values of Cloud Cover[a]

Cloud cover, tenths	0	2	5	8	10
Net longwave radiation (W m^{-2})	−83	−78	−58	−34	−13

[a] Data from Barashkova (1961).

radiation are to a degree independent of one another. The warming of the underlying surface by day does not evoke a commensurate change in the downward flux, and the net difference becomes large. This temperature-jump effect invalidates the traditional formulas for net loss unless corrected, e.g., by a method (Budyko, 1974, p. 60) based on the energy budget.

The net exchange of longwave radiation decreases nonlinearly (Sauberer, 1936; Budyko, 1974, p. 57) with an increase in cloudiness and by 10–20 W m^{-2} with haze (Table II). This represents a discharge of a part of the latent heat released when clouds form, especially if they precipitate water, and to a degree it results from closing the 8–12-μm window in which few atmospheric gases radiate. A continual variation of the net flux occurs as a cyclone moves over an ecosystem.

With a clear sky and surface and air temperatures between 10 and 15°C net longwave radiation is related to vapor pressure as shown in Table III. Similar findings were made in Japan (Takahashi *et al.*, 1960).

Variations in Net Longwave Radiation

The diurnal regime (Table I) shows effects of a large increase in the upward flux, caused by the rise in surface temperature, accompanied by a small increase in the downward flux, caused by a smaller rise in atmospheric temperature. The large net loss at midday helps prevent solar overheating of an ecosystem; the smaller net loss at night fits the lower energy conditions then.

The loss is small in winter also (Table I). We tend to think of a large net loss of energy as characteristic of the chill nights of the Arctic, but

TABLE III

Net Loss of Energy by Longwave Radiation at Different Values of Vapor Pressure[a]

Vapor pressure (mbar)	6–8	12–14	18–20
Net longwave radiation (W m^{-2})	−80	−70	−60

[a] Data from Barashkova (1961).

in reality it is not particularly large because protective snow cover can attain a low surface temperature and thereby reduce energy outflows.

In dissected land the angle between slopes affects radiation exchanges. For example, the net loss from an ecosystem at an angle of 15° is 0.97 of that from a horizontal surface, and at 30° it is 0.90 (Kondratyev, 1965, p. 305). At exposed substations of the Lunz Biological Station in the eastern Alps, the net loss was double that at sheltered substations (Lauscher, 1937, p. 20).

Local deposits of dew in areas of strong cooling show up in reflected light (Mattsson, 1974). Three ecosystems in northern Germany displayed large contrasts in nocturnal net radiation in July: sandy upland -29 W m^{-2}, upland peat -19, and low peat bog -10 (Miess, 1968, p. 57), suggesting differences in stored heat.

On a larger spatial scale the net longwave quantity tends to follow cloudiness patterns. The monthly mean at Niigata in December, under heavy snowstorm clouds in Siberian air that has crossed the relatively warm Sea of Japan, is only -46 W m^{-2}—half the value at Tokyo (-89 W m^{-2}), which lies to the lee of the mountain chain (Kondo, 1967)— this regional difference is the consequence of a massive transport of energy from the sea to terrestrial ecosystems near the Sea of Japan. Zonal distributions of net longwave radiation are represented by a 20-layer atmospheric model (Falconer and Peyinghaus, 1975) as -70 W m^{-2} at latitudes up to 40° in summer and -45 W m^{-2} from 50°N to 70°N.

COMBINING ALL THE RADIATION FLUXES: THE NET ALL-WAVE DEFICIT OR SURPLUS

We have seen that the resultant of the upward and downward fluxes of longwave radiation usually represents a net loss of energy from an ecosystem. The resultant of the upward and downward fluxes of radiation of *all* wavelengths, the all-wave net quantity or radiation budget,[*] is in deficit status when solar fluxes are weak or absent and ecosystems must call upon nonradiative sources of energy. When it is in surplus, ecosystems are well-endowed, sometimes over-endowed, with energy and have a surplus to partition among the nonradiative modes of heat removal.

The seven individual streams of radiant energy that we have separately examined up to this point differ with respect to wavelength, paths, and direction toward or away from an ecosystem. For meteorological

---

[*] Not "radiation balance," because the equilibrium state is transient.

and biological research each stream should be separately measured; in fact, forest meteorologists recommend that the incoming direct and diffuse fluxes of photosynthetically active radiation ought to be included also. However, they all transport energy by the same basic process, which is markedly different from the other means by which energy is transported in the environment of ecosystems, and on this basis can be lumped into one quantity: net all-wave radiation. The "net" means that incoming and outgoing fluxes are compared to see whether the ecosystem gains or loses energy. "All-wave" means that we combine those regions of the electromagnetic spectrum in which appreciable amounts of energy are transmitted in nature, specifically, solar and longwave radiation (Table IV). Seven fluxes that have been discussed are recapitulated in the combinations shown diagrammatically in Table IV.

The Role of the Sun

The sun is usually the most important factor, especially in view of the large variation of solar radiation over time. Sun height is closely related to flux density in the direct beam and less closely related to the diffuse flux. It is related to surface and atmospheric temperature, hence to the fluxes of longwave radiation.

The fact that longwave fluxes change slowly and that absorbed shortwave radiation is more or less a constant fraction of the incoming shortwave flux means that the radical change of the solar beam from zero at night to 700–900 W m^{-2} in midday overshadows the daily regimes of the other radiation fluxes and dominates net all-wave radiation. The radiation surplus at Hamburg at noon and the deficit at night (Table V) show the solar dominance but also the stabilizing influence of the downward flux of longwave radiation.

The Role of Clouds

Clouds cut off the solar beam but have compensating effects on the diffuse fluxes of solar and longwave radiation. This compensation, however, is incomplete at midday hours during the growing season (Table VI). Loss of 592 W m^{-2} in the beam was much greater than the area-source increase of 139 W m^{-2}. Concurrent changes in the upward fluxes modify the net further: A greater fraction of solar radiation is absorbed, and less longwave radiation is emitted by the grass. Overall, the all-wave radiation surplus decreases from +373 to +93 W m^{-2}.

Monthly means, standardized as to sun height (40°) and surface absorptivity (0.8–0.9) (Barashkova et al., 1961, p. 135) show that net all-wave radiation decreased about 60 W m^{-2} (about 0.18) when cloudiness

TABLE IV *Schematic Diagram Showing All the Radiation Fluxes Involved in Ecosystem Energy Budgets*

DIRECTION	SHORTWAVE		LONGWAVE		ALL-WAVE
DOWNWARD FLUXES (+)	(a) Direct solar beam	(b) Diffuse solar radiation	(d) Radiation emitted by atmospheric gases	(e) Radiation emitted by clouds	(a + b + c + e) Total incoming radiation
	(a + b) Total solar radiation		(d + e) Downward longwave radiation		
UPWARD FLUXES (−)	(c) Reflected solar beam	(c¹) Reflected diffuse solar radiation	(f) Reflected longwave radiation		(c + c¹ − f) Total reflected radiation
			(g) Radiation emitted by the surface		(g) Emitted radiation
					(c + c¹ + f + g) Total upward radiation
ALGEBRAIC SUMS	(a + b + c + c¹) Absorbed (net) solar radiation		(d + e + f) Absorbed longwave radiation		(a + b + c + c¹ + d + e + f) Absorbed radiant energy
			(d + e + f + g) Net longwave radiation		(a + b + c − c¹ + d + e + f + g) Net all-wave radiation

TABLE V

Schematic Presentation of All Radiation Fluxes at Grass Surface at Hamburg Observatory at Noon and Midnight of a Day in May[a,b]

	1200			2400			WHOLE-DAY AVERAGE		
	SHORT-WAVE	LONG-WAVE	ALL-WAVE	SHORT-WAVE	LONG-WAVE	ALL-WAVE	SHORT-WAVE	LONG-WAVE	ALL-WAVE
	+777	+360	+1137	0	+372	+372	+326	∓319	+645
	−140	−441	−581	0	−417	−417	−62	−395	−457
NET	+637	−81	+556	0	−45	−45	+264	−76	+188

[a] Data given in watts per square meter. The range in solar energy absorbed, 637 W m^{-2}, contrasts with the 41 W m^{-2} range in incoming longwave radiation.

[b] Data from Fleischer and Gräfe (1955/56).

TABLE VI

Schematic Presentation of All Radiation Fluxes at a Meadow[a] in Wien under Cloudless and Overcast Skies in Summer[b]

	CLOUDLESS				OVERCAST			
	SHORTWAVE DIRECT	DIFFUSE	LONGWAVE	ALL-WAVE	SHORTWAVE DIRECT	DIFFUSE	LONGWAVE	ALL-WAVE
	+592	+105	+324	+1021	0	+161	+407	+568
	−128	−10	−510	−648	0	−23	−452	−475
NET	+464	+95	−186	+373	0	+138	−45	+93

[a] Albedo of meadow is 0.22 for direct beam, 0.12 for diffuse solar radiation. Sun altitude is 45°. Radiant-energy intake dropped from 883 W m^{-2} under the cloudless sky (assuming longwave absorptivity = 1.0) to 545 W m^{-2} under the overcast sky.

[b] Data from Sauberer (1956).

increased from scattered to full overcast. A solar eclipse reduces direct and diffuse shortwave fluxes in proportion to the obscuring of the sun's disk. When both fluxes decreased to 0.76 in one eclipse but the longwave fluxes changed only slowly, the net all-wave surplus decreased to 0.71 of its previous value (Pruitt et al., 1965).

Other Factors

Systems of low absorptivity are not closely coupled with the sun, and changes in solar radiation have less effect on the all-wave radiation surplus. The surplus at high-altitude snow (absorptivity 0.16) changed from 9 W m^{-2} in clear weather to 32 W m^{-2} in cloudy (Ambach, 1974).

Linacre (1968), in examining some of the many attempts to relate the net quantity of all-wave radiation to such single factors as insolation, found that other factors cannot be neglected. Solar radiation is modified by ecosystem absorptivity, and temperature and cloudiness as indexes to net longwave radiation must be included also. Net longwave cannot be taken as a constant fraction (Gay, 1971; Nkemdirim, 1973).

Determination

Net all-wave radiation is determined by algebraic addition of the densities of the individual fluxes. In Tables IV–VI, the net quantity is either the sum of incoming and outgoing all-wave values in the right-hand column or of shortwave and longwave net values in the bottom row.

Direct measurement of this quantity is done by thermoelectric "balance" meters, thermopiles with absorbent top and bottom surfaces facing the upper and lower hemispheres, under domes of some material like polyethylene ("Lupolen-H" in Europe) that more or less uniformly (about 0.83 transmissivity) transmits radiation of both solar and longwave regions and shields the sensors from erratic nonradiative fluxes (Funk, 1963). Net radiometers suffer from their apparent simplicity and ease of use and do not always receive necessary maintenance and calibration. Many published data on net all-wave radiation should not be used without verification. A network of net radiation sensors has been deployed in the USSR, Canada, and other countries for years. There are at least 100 in the USSR (Budyko, 1974, p. 26), and their data have been used in developing methods for estimating monthly or long-term mean values.

VARIATIONS IN NET ALL-WAVE RADIATION OVER TIME

The rotation of the earth produces in all except polar regions a 24-hr cycle of alternating darkness and illumination, which is expressed as an alternation of deficit and surplus in net all-wave radiation. At the two times of crossover the algebraic sum is zero.

Equilibria in the Diurnal Radiation Budget

Crossovers are interesting because the nonradiative fluxes have either just changed direction (substrate-heat flux) or are about to change (sensible-heat flux) and all of them are small, indicating more or less isothermal profiles in the soil and air. The morning crossover marks the end of the inward-looking night period, during which the ecosystem drew heat from soil and air to meet the net deficit of radiation and sustain its temperature in readiness for the day to come. Absorption of solar radiation in the morning usually becomes equal to the net longwave loss about an hour after sunrise when the sun is 9–10° above the horizon (higher for a snow-covered ecosystem).

The second equilibrium occurs in late afternoon when the diminishing quantity of absorbed solar radiation reaches the level of the net loss by longwave radiation, a loss that is greater in the afternoon than morning, and so the equilibrium occurs more than an hour before sunset. The times at Tashkent are shown in Table VII. Many textbooks to the contrary, this equilibrium in the radiation budget does not coincide with the hour when temperature in a thermometer shelter reaches its maximum, but almost always comes several hours later. The persistence of this plausible but erroneous idea in textbooks attests to the long neglect of observations of radiative energy. At the evening crossover the contrasts between layers in and above an ecosystem have died out and with them most of the violent thermal activity of the day; the ecosystem is "going to sleep," as it were.

The hours of crossover separate the active, solar-powered day from the quiet night, a difference that also appears in the different kind of boundary layer an ecosystem produces in the overlying atmosphere.

TABLE VII

Time of Afternoon Equilibrium in the
Radiation Budget at Tashkent[a]

Month	Time of equilibrium
December	16:02
March	17:31
June	18:58
September	17:42

[a] Data from Berliand (1970, Table 4).

The morning radiative equilibrium at Hamburg (56°N) in June comes at 0445, 1 hr after sunrise, and thermal equilibrium in the gradient of potential temperature at a height of 6 m occurs at 0600. The order is reversed in the evening: thermal equilibrium at 1800, radiative at 1950, and sunset at 2110 (Frankenberger, 1955). The occurrence of radiative equilibrium after sunrise and before sunset shortens the period of radiative surplus. Over the year at St. Paul it averages 9.3 hr in length (Blad and Baker, 1971).

Equilibria in the Annual Regime

Fall. The daytime surplus period is so short in winter in middle and high latitudes that it can deliver only a small amount of energy. Long nights of radiative deficit result in a large net loss and the mean 24-hr budget is then in deficit. The daytime surplus and nighttime deficit come to equality at a date in fall and one in spring.

These dates occur earlier in autumn at high latitudes than at middle, as is shown by a meridional profile in the western Soviet Union (Table VIII). Equilibrium in the radiation budget at high latitudes occurs nearly a hundred days before the winter solstice. In the following months the mean daily isarithm of zero net all-wave radiation moves south at an average speed of a half-degree of latitude per day. Its movement southward encounters higher angles of the noonday sun, but this effect is

TABLE VIII

Mean Dates of 24-Hr Radiative Equilibrium in Autumn[a]

Station	Latitude (°N)	Average date	No. of days before winter solstice	Height of noonday sun
Karadag	45	1 Dec	20	24°
Kiev	51	15 Nov	36	21°
Voeikovo	60	18 Oct[b]	60	20°
Kola	69	30 Sept	82	19°
Bukhta Tikhaia	80	15 Sept	97	13°
Drifting stations[c]	80–85	12 Sept	100	13°
North Pole	90	5 Sept	107	≈7°

[a] Data from Barashkova et al. (1961), Gavrilova (1963), and Vowinckel and Orvig (1965).

[b] At Palmer, Alaska (62°N), the equilibrium data is 15 Oct (Branton et al., 1972, p. 7).

[c] Summary of observations at several stations on the ice of the Arctic Ocean.

TABLE IX

Mean Dates of Radiative Equilibrium in Spring[a]

Station	Latitude (°N)	Average date	No. of days after winter solstice	Height of noonday sun
Karadag	45	5 Jan	15	23°
Kiev	51	5 Feb	46	24°
Voeikovo	60	8 Mar[b]	78	19°
Kola	69	13 Apr	114	30°
Bukhta Tikhaia	80	25 Apr	126	23°
Drifting stations	80–85	18 Apr	119	18°
North Pole	90	25 Apr	126	13°

[a] Data from Barashkova *et al.* (1961), Gavrilova (1963), and Vowinckel and Orvig (1965).

[b] 15 March at Palmer, Alaska (62°N) (Branton *et al.*, 1972, p. 6).

countered by the later, shorter days. Radiative equilibrium occurs in days 12–15 hr long in the high latitudes, 9 hr long in the south. The product of sun height and day length at this date is about the same at all stations on the profile.

Winter snow cover usually is not established until after the date of radiative equilibrium, and once established its low solar absorption tends to retain the deficit status. Warm soil or invading warm airstreams help determine the date when winter snow cover is established, as also does the depth of snowfall in the early storms. Big snowstorms usually establish the mantle in the Sierra Nevada at a date when noon sun height is still 30°, higher than when snow cover is established at lowland stations in Russia.

Spring. In the season of lengthening days and increasing sun height the average date when radiative equilibrium is attained at places on the meridional profile occurs longer after the winter solstice than the date of fall equilibrium occurs before it (Table IX). (This phase shift is the opposite of that of radiative equilibrium in the daily cycle.) The delay of the date of radiative equilibrium in spring is large where a winter snow cover exists, which prolongs the winter deficit in the radiation budget and is itself prolonged by the deficit.

The Long, Cold Night and the Long, Cold Winter: Deficits in the Radiation Budget

The period of deficit is longer than the period of darkness, and at St. Paul varies from 11.5 hr in June to 17.0 hr in December (Blad and Baker,

1971). Deficits occur in 0.6 of the hours of the year, but, fortunately, the rates of energy loss seldom exceed 50 W m^{-2} from any ecosystem without access to a large reserve of stored energy. Weekly means at Palmer, Alaska (Branton et al., 1972), lie close to -30 W m^{-2} for 10 consecutive weeks; no week exceeds -36 W m^{-2}.

The isarithm that bounds the 24-hr deficit in midwinter crosses the North American Plains at about 42°N, a latitude near the edge of persistent snow cover. It lies north of 45°N in western Europe, and at about 42° in central Asia. Snow cover prolongs the period of deficit and remains 30–40 days even after radiative equilibrium is attained; this lag depends on the energy required to melt it. The deep snow of the Sierra Nevada (39°N) requires 250–300 MJ m^{-2} heat supply, and melting takes about 2 months after the date in late March when radiative equilibrium occurs, 90 days after the winter solstice. At lowland stations (Table IX) a delay this long is found only in high latitudes.

The winter budgets of radiation of ecosystems in lands poleward of 45° are in true deficit status; more radiative energy is going out than coming in. Except at the transient equilibria, *there is no "balance."* The deficit, nocturnal or winter-long, is real and by the law of conservation of energy must be made up from nonradiative sources.

Surpluses in the Radiation Budget

Periods of radiation surplus are times of copious energy supply, growth and renewal, and intense biological, chemical, and physical activity in ecosystems.

Regimes of the net surplus of all-wave radiation at different ecosystems display midday flux densities exceeding 800 W m^{-2} at cotton, sugar cane, and forest on days of strong sunshine. Such a midday rate suggests an increasing input of energy to ecosystems in the morning hours, producing a rapid rise of surface temperature and of soil and air temperatures that follow it. It can be represented by the height of a triangle whose base is represented by the length of daytime radiation surplus, 12–14 hr in summer. The area of this triangle, minus the nocturnal deficit, is equal to the 24-hr surplus of all-wave radiation.*

In most summer days at all latitudes the 24-hr means are concentrated between 100 and 150 W m^{-2}. Average 24-hr means in June over grass

---

* Because the nocturnal deficit does not change a great deal from winter to summer while the midday surplus increases sharply, the daily range becomes larger more or less in proportion to the rising 24-hr mean. For example, at Tashkent the night-to-midday range increases 3 W m^{-2} for each 1 W m^{-2} rise in the 24-hr mean (Aizenshtat, 1957).

at stations near 60°N latitude in western Europe, the hearth of radiation research (Table X), are about 120–125 W m^{-2}; coastal cloudiness reduces solar absorption more than it increases the longwave input. Values are slightly larger at stations near the 40th parallel.

The contrast between the radiation surplus of summer and the deficit of winter (Table X) indicates seasonal changes at middle and high latitudes. For example, the following changes occur from December to June at Hamburg:

167 W m^{-2} increase in absorbed solar energy

40 W m^{-2} increase in downward longwave radiation gives

207 W m^{-2} increase in radiant-energy intake, which

less −65 W m^{-2} greater emission of longwave radiation,

nets a 142 W m^{-2} change in net all-wave radiation.

The duration of the radiation surplus in middle latitudes exceeds that of the growing season, which does not begin until soil and air temperatures have responded to spring radiation. The accumulated surplus in these thawing and warming-up days is about equal to the winter deficit (Budyko, 1974, p. 344), and as a result the radiation-surplus sums over the whole year and over the growing season are about the same. The principal exception is found in high latitudes in continental interiors, where warming up is brief.

Longer-Term Variations

Variations in solar and longwave radiation from year to year, usually caused by differences in numbers of cloud systems, affect the surplus

TABLE X

Net Daily Surplus or Deficit of All-wave Radiation in June and December[a]

	June	Dec.	Range	Source
Stations near 60°N				
Lerwick	+117	−32	149	Berliand (1970)
Uppsala	+120	−21	141	Rodskjer (1979)
Helsinki	+121	−10	131	Berliand (1970)
Leningrad	+124	−9	131	Berliand (1970)
Copenhagen (56°)	+133	−11	144	Aslyng and Jensen (1966)
Stations near 40°N or S				
Tashkent	+147	+4	143	Barashkova et al. (1961)
Tbilisi	+146	+3	143	Berliand (1970)
Argonne	+173	−8	181	Moses and Bogner (1967, p. 22)
Aspendale	+17	+157	140	Funk (1963)

[a] Data given in watts per square meter.

or deficit of all-wave radiation, perhaps offset by precipitation-related changes in surface absorptivity and temperature. For example, the mean surplus in two dry summers (1946 and 1954) at Kiev averaged 6 W m^{-2} greater than in two wet summers (1945 and 1950) (Danilova, 1960). Stronger solar radiation in the drought years (12–15 W m^{-2} greater) was not wholly countered by increased longwave radiation from the hotter ecosystems.

The general magnitude of the year-to-year variation in a specific month is given from June measurements at Voeikovo [in Barashkova *et al.*, (1961, pp. 452–453)] with an 8-year mean of 126 W m^{-2} and a standard deviation of 16 W m^{-2} (about 0.13). At other long-record stations the standard deviations average about 0.10 (Budyko, 1974, p. 145).

SPATIAL CONTRASTS IN NET ALL-WAVE RADIATION

At the Ecosystem Scale

Contrasts in the diurnal march of net all-wave radiation occur at juxtaposed ecosystems such as desert and cotton fields and snow in meadows and forest. Irrigated ecosystems usually enjoy surpluses a third or a half greater than adjacent desert ecosystems, which reflect and emit large upward fluxes.

Forest ecosystems that have surface properties similar to irrigated fields also enjoy larger radiation surpluses than lower vegetation. With good access to moisture and ample ventilation the cool foliage minimizes the emission of longwave radiation, and the deep absorbing layer minimizes reflection of solar radiation. Several cases of forest versus low-ecosystem pairing show a difference of about 15 W m^{-2} (Rauner, 1965). Logging a coniferous stand in Quebec resulted in a smaller deficit at night and a much smaller surplus by day, with a net 24-hr reduction to 0.88 of the forest mean (McCaughey, 1978). Aquatic ecosystems display superiority to terrestrial because they minimize the outward fluxes of both solar and longwave radiation.

Contrasts at Large Scales

The annual net surplus of radiation changes little with altitude, short of the zone of snow cover (Budyko, 1974, p. 192), because the altitudinal increase in absorbed solar energy more or less counterbalances the decrease in incoming longwave radiation. Lowland ecosystems average 65 W m^{-2} over the year near the Caucasus, and at 3-km altitude the surplus is still 55 W m^{-2}; only on Mt. Kazbeg does the surplus decline to 40 W m^{-2}. The range in yearly values over Tasmania is from 75 W m^{-2} in the cloudy west to 95–105 W m^{-2} in sunny parts of the east coast (Nunez, 1978).

The large radiation surpluses of ecosystems in well-watered regions of the low latitudes (90 W m^{-2} or more) produce a cellular pattern in which "zonality is abruptly broken by the effect of differences in moisture conditions" (Budyko, 1974, p. 156). A zonal pattern is displayed at middle and high latitudes in the Soviet radiation network (snow cover in winter):

68°N	31 W m^{-2}	50°N	57 W m^{-2}
60°N	41 W m^{-2}	40°N	75 W m^{-2}

Surpluses at ecosystems of Europe, Asia, and North America average around 60 W m^{-2} and in Africa, Australia, and South America 90 W m^{-2}. Land, sky, and sun have a different relation, and the availability of half again as much radiant energy to ecosystems of the low-latitude continents produces large nonradiative energy fluxes that influence all terrestrial physical, chemical, and biological phenomena.

The quantity expressing the net effect of all the radiation fluxes to and from an ecosystem is equal and opposite to the net sum of all the non-radiative conversions and fluxes and exhibits frequent reversals in direction: it is in deficit status almost every night and in surplus almost every day. Except in winter at high and middle latitudes the daytime surplus is larger than the nocturnal deficit, and over the year all ecosystems everywhere enjoy a net surplus. The net radiative quantity can now be equated to the sum of the nonradiative energy fluxes and the on-site energy transformations, which will be taken up in chapters that follow.

REFERENCES

Aizenshtat, B. A. (1957). Radiatsionnye balans i temperatura poverkhnosti pochv v Tashkente. Tr. Tashk. Geofiz. Obs. **13**, 3–74.

Ambach, W. (1974). The influence of cloudiness on the net radiation balance of a snow surface with high albedo. J. Glaciol. **13**, 73–84.

Aslyng, H. C., and Jensen, S. E. (1966). Radiation and energy balances at Copenhagen 1955–1964. R. Vet. Agric. Coll. (Copenhagen), Yearbk. 1965, pp. 22–40.

Barashkova, E. P. (1960). Dlinnovolnovyi balans podstilaiushchei poverkhnosti po nabliudeniiam v Karadage. Tr. Gl. Geofiz. Obs. **100**, 141–153.

Barashkova, E. P. (1961). Dlinnovolnovyi balans v nekotorykh punktakh SSSR. Tr. Gl. Geofiz. Obs. **109**, 25–37.

Barashkova, E. P., Gaevskii, V. L., D'iachenko, L. N., Lugina, K. M., and Pivovarova, Z. I. (1961). "Radiatsionnyi Rezhim Territorii SSSR." Gidrometeorol. Izd., Leningrad.

Berliand, T. G. (1970). "Solar Radiation and Radiation Balance Data (The World Network)." Hydrometeorol. Publ. House, Leningrad.

Blad, B. L., and Baker, D, G, (1971). A three-year study of net radiation at St. Paul, Minnesota. *J. Appl. Meteorol.* **10**, 820–824.

Branton, C. I., Shaw, R. H., and Allen, L. D. (1972). Solar and net radiation at Palmer, Alaska/1960–71. Univ. Alaska, Inst. Agr. Sci. Tech. Bull. 3.

Budyko, M. I. (1974). "Climate and Life" (transl. ed. by D. H. Miller). Academic Press, New York.

Daniels, F. (1964). "The Direct Use of the Sun's Energy." Yale Univ. Press, New Haven, Connecticut.

Danilova, N. A. (1960). Radiatsionnyi balans v lesostepnoi i stepnoi zonakh Evropeiskii chasti SSSR v zasushlivye i vlazhnye gody. *In* "Gidroklimaticheskii Rezhim Lesostepnoi i Stepnoi Zon SSSR v Zasushlivye i Vlazhnye Gody" (B. L. Dzerdzeevskii, ed.), pp. 90–101. Izd. Akad. Nauk, Moscow.

Falconer, P. D., and Peyinghaus, W. (1975). Radiative balance in the atmosphere as a function of season, latitude and height. *Arch. Meteorol., Geophys. Bioklimatol., Ser. B* **23**, 201–223.

Fleagle, R. G., and Businger, J. A. (1963). "An Introduction to Atmospheric Physics." Academic Press, New York.

Fleagle, R. G., and Businger, J. A. (1975). The "greenhouse effect." *Science* **190**, 1042–1043.

Fleischer, R., and Gräfe, K. (1955–1956). Die Ultrarotstrahlungströme aus Registrierung des Strahlungsbilanzmessers nach Schulze. *Ann. Meteorol.* **7**, 87–95.

Frankenberger, E. (1955). Über vertikale Temperatur-, Feuchte- und Windgradienten in den untersten 7 Dekametern der Atmosphäre, den Vertikalaustausch und den Wärmehaushalt an Wiesenboden bei Quickborn/Holstein 1953/1954. *Ber. Dtsch. Wetterdienstes* **3**, No. 20.

Funk, J. P. (1963). Radiation observations at Aspendale, Australia, and their comparison with other data. *Arch. Meteorol., Geophys. Bioklimatol., Ser. B* **13**, 52–70.

Gavrilova, M. K. (1963). "Radiatsionnyi Klimat Arktiki." Gidrometeorol. Izd., Leningrad.

Gay, L. W. (1971). The regression of net radiation upon solar radiation. *Arch. Meteorol., Geophys. Bioklimatol., Ser. B* **19**, 1–14.

Kondo, J. (1967). Analysis of solar radiation and downward long-wave radiation in Japan. *Sci. Rep. Tohoku Univ., Ser. 5* **18**, 91–124.

Kondratyev, K. Ia. (1965). "Aktinometriia." Gidrometeorol. Izd., Leningrad.

Lauscher, F. (1937). Grundlagen des Strahlungsklimas der Lunzer Kleinklimastationen. Wien. Zentralanst. Meteorol. Geodyn., Publ. 146 (Beiheft Jahrbuch 1931).

Lettau, H. (1969). Evapotranspiration climatonomy. I. A new approach to numerical prediction of monthly evapotranspiration, runoff, and soil moisture storage. *Mon. Weather Rev.* **97**, 691–699.

Linacre, E. T. (1968). Estimating the net-radiation flux. *Agric. Meteorol.* **5**, 49–63.

McCaughey, J. H. (1978). Estimation of net radiation for a coniferous forest, and the effects of logging on net radiation and the reflection coefficient. *Can. J. For. Res.* **8**, 450–455.

Mattson, J. O. (1974). Climatic information in night-recorded aerial photographs, with special regard to registration made in retroflected light. *Lund. Univ. Naturgeogr. Inst., Rapp. Not.* No. 23.

Meinel, A. B., and Meinel, M. P. (1976). "Applied Solar Energy: An Introduction." Addison-Wesley, Reading, Massachusetts.

Miess, M. (1968). Vergleichende Darstellung von meteorologischen Messergebnisse und Wärmehaushalts-untersuchungen an drei unterschiedlichen Standorten in Norddeutschland. *Tech. Univ. Hannover, Inst. Meteorol. Klimatol., Ber.* **2**.

Moses, H., and Bogner, M. A. (1967). "Fifteen-Year Climatological Summary, January 1, 1950—December 31, 1964," ANL-7084. Argonne Natl. Lab., Argonne, Illinois.

Nkemdirim, L. C. (1973). Radiative flux relations over crops. *Agric. Meteorol.* **11**, 229–242.

Nunez, M. (1978). The radiation index of dryness in Tasmania. *Austral. Geogr. Stud.* **16**, 126–135.

Oke, T. R., and Fuggle, R. F. (1972). Comparison of urban/rural counter and net radiation at night. *Boundary-Layer Meteorol.* **2**, 290–308.

Pruitt, W. O., Lourence, F., and Crawford, T. V. (1965). Radiation and energy balance changes during the eclipse of 30 July 1963. *J. Appl. Meteorol.* **4**, 272–278.

Rauner, I. L. (1965). O gidrologicheskoi roli lesa. *Izv. Akad. Nauk SSSR, Ser. Geogr.* No. 4, 40–53.

Rodskjer, N. (1979). Net long-wave radiation at Uppsala, Sweden. *Arch. Meteorol., Geophys. Bioklimatol., Ser. B* **27**, 189–192.

Sauberer, F. (1936). Messungen des nächtlichen Strahlungshaushaltes der Erdoberfläche. *Meteorol. Z.* **8**, 296–302.

Sauberer, F. (1956). Über die Strahlungsbilanz verschiedener Oberflächen und deren Messung. *Wetter Leben* **8**, 12–26.

Schulze, R. (1970). "Strahlenklima der Erde." Steinkopff, Darmstadt.

Schwerdtfeger, W. (1977). Temperature regime of the South Pole: Results of 20 years' observations at Amundsen–Scott Station. *Antarct. J. U.S.* **12**(4), 156–159.

Sellers, W. D. (1965). "Physical Climatology." Univ. of Chicago Press, Chicago, Illinois.

Takahashi, K., Katayama, A., and Asakura, T. (1960). A numerical experiment of the atmospheric radiation. *J. Meteorol. Soc. Jpn.* **38**, 175–181.

Vowinckel, E., and Orvig, J. (1965). Energy balance of the Arctic. II. Long wave radiation and total radiation balance at the surface in the Arctic. *Arch. Meteorol., Geophys. Bioklimatol., Ser. B* **13**, 452–479.

Wexler, H. (1936). Cooling in the lower atmosphere and the structure of polar continental air. *Mon. Weather Rev.* **64**, 122–136.

Chapter XI

FIXING OF CARBON BY ECOSYSTEMS

Radiant energy absorbed by ecosystems is transformed in ways that add up to keep the energy budget in balance and include transformations into chemical energy. A variety of photoelectric and photochemical phenomena are carried on by bacteria or algae,* but we will here consider only photosynthesis, energy conversion by ecosystems fixing atmospheric carbon†, the major basis of life.

In contrast to such energy exchanges as the absorption of longwave radiation that are typified by large flux density but yield low-quality energy, the fixing of carbon transforms high-frequency radiant energy into high-quality chemical energy in the form of food, fiber, and environmental amenity. The products of photosynthesis, moreover, can be stored and transported, hence have a time and place utility not, in general, characteristic of radiant energy. The central role of photosynthesis in ecosystem studies is illustrated by a model constructed in order to identify gaps in the knowledge of a widespread desert species, *Larrea tridentata*; the gain and allocation of carbon formed "a logical choice for investigation because it is so intimately tied to the procurement and allocation of energy, mineral nutrients and water" (Cunningham and Reynolds, 1978).

GROSS PRIMARY PRODUCTION

The fixing of radiant energy by photosynthesis is studied at several scales. Below the scale of the ecosystem are levels within the leaf canopy,

---

* Certain enzyme systems and green algae in a radiation field can photolyze water and produce hydrogen, for example (Bishop *et al.*, 1977), a possible base for an efficient hydrogen fuel economy but not now widespread.
† Ecosystem cycling of such other elements as sulfur and nitrogen also involves energy conversions, which we must pass over.

the much-studied individual leaf, and the biochemical level within the leaf cells. Energy transfer processes at the microscale are outside the scope of our inquiry and have not always been found to be well-correlated with ecosystem dry-matter production. Time scales also must be recognized; short-period assimilation rates in an environment of changing sun angle and cloud passage do not necessarily indicate long-term production or ecosystem survival. Downstream from photosynthetic production, the yield of crop ecosystems also depends on the partitioning of the products into other organs and energy storages.

The Supply of Radiant Energy to Photosynthesizing Ecosystems

Selectivity. The rate at which radiant energy is fixed in plant cells forms a small component in the total energy budget of an ecosystem, in part because only certain spectral bands are absorbed by chlorophyll (centered at 0.4 and 0.6 μm). Radiation in other wavelengths, however, is absorbed by and warms water and other leaf components and keeps chloroplasts at working temperature. The fraction of the total absorbed radiation (radiant-energy intake or loading, discussed in Chapter VII) fixed photosynthetically is of the order of 1% or less but constitutes a larger fraction of the solar radiation absorbed and a still larger fraction of the photosynthetically active radiation, typically up to 2% (Wassink, 1975). Most models assume absorption of 10 quanta per molecule of CO_2 fixed (Loomis and Gerakis, 1975).

Because no radiation having a wavelength greater than 0.7 μm is absorbed by chlorophyll molecules, the solar infrared spectrum, conveying nearly half the total flux density, enters the carbon fixing process only by heating leaf cells and countering the evaporative cooling as water leaks out of them. Even this function is not particularly needed because certain visible solar wavelengths are absorbed by water and can warm the leaf cells. Solar infrared radiation, in fact, can overheat cells and is usually not strongly absorbed by leaves, as was seen in Chapter VII.

The peculiar on-and-off regime of solar radiation and the directional energy flux that shifts from minute to minute present problems to a photosynthesizing ecosystem. Beam azimuth and angle above the horizon change rapidly, and absorption depends on leaf inclination. Vertical leaves do well when the sun is low in the sky; CO_2 conversion in rice, for example, is greater with erect than with horizontal leaves, especially in the upper layers (Uchijima, 1976).

Because leaves grow at many angles, modeling the absorption of the solar beam within an ecosystem requires consideration of the frequency distribution of angles between leaves and beam at different sun altitudes.

Height growth of trees in the Alps doubles on slopes exposed to stronger radiation (Schönenberger, 1975), and the highest rates of ecosystem photosynthesis are reached in direct-beam radiation, not because it more efficiently penetrates the foliage or is more efficiently absorbed but because its flux density is usually greater than that of diffuse solar radiation. Mean rates over longer periods of time are smaller because cloudy days are included. Absorptivity is reduced in some arid-land plants by leaf hairs or other obstacles that reduce transpiration as well as photosynthesis (Ehleringer et al., 1976).

The diffuse flux of solar radiation is richer in photochemical value than the direct beam and penetrates ecosystems deeply. A modest increase in the diffuse fraction, even if accompanied by some decrease in total solar radiation such as might occur under a polluted atmosphere, tends to increase net photosynthesis—more so for C_3 carbon-pathway plants like sugar beets than for C_4 plants like maize* (Lister and Lemon, 1976). Specifically, increasing the fraction of diffuse radiation from 0.1 of the total to 0.3 puts more leaves in the energy flux range of 250–300 W m^{-2}, reduces the number of leaves in the higher range of 350–450 W m^{-2}, and slightly increases flux density on shaded leaves, which make up two-thirds of the total leaf area of the system studied (Lister and Lemon, 1976, Fig. 6). The small size of the photosynthesizing unit in a leaf (a few hundred molecules) is appropriate for diffuse light or shade because it makes possible the concentration of 8–10 quanta in a very short time "(less than 10 μs) and practically in one place" (Heath, 1969, p. 257), so that the leaf utilizes small flux densities of radiation "by collecting together the energy from relatively scattered quanta" (Heath, 1969, p. 257).

Light Saturation. The flux density of diffuse solar radiation falls within the light-utilization capacity of most leaves. While species differ, most of the length of the curve relating their CO_2 assimilation to radiative flux density is concave downward and tends to level off at a light-sat-

---

* C_3 and C_4 plants differ in the number of carbon atoms in the molecule of the first detectable product formed from CO_2. C_4 plants in general display little or no photorespiration, usually photosynthesize at higher leaf temperatures and higher radiation flux densities than do C_3 plants, and show less reaction to changes in CO_2 concentration (Zelitch, 1975; Lister and Lemon, 1976). A variant pathway in photosynthesis is crassulacean acid metabolism (CAM), in which the two major processes involving CO_2 "occur in the same cells but at different times" (Solbrig and Orians, 1977) so that stomata are open at night and closed by day.

uration value beyond which further increases in flux density produce no increase in rate of assimilation. Optimal flux densities for photosynthesis in several species in Alpine ecosystems under strong radiation in summer lie between 200 and 400 W m^{-2} (Aulitzky, 1963). In arid climates net assimilation is greatest if flux density is near 370 W m^{-2} (Warren Wilson, 1967). Higher intensities, in fact, increase respiration and damage leaf tissue by overheating. Sun leaves typically display a high value of light saturation, as do some species of low-latitude origin, but this capacity may come at the cost of higher protein content and consequent expenditure of energy to construct and maintain the leaf (Berry, 1975). The plant, in effect, allocates energy and nutrients to create additional photosynthetic apparatus to exploit the strong solar beam, a strategy common in early successional plants (Bazzaz, 1979).

Some full-sun plants in deserts can use high flux densities; an evening primrose in Death Valley showed little falling off in its utilization of solar energy at full midday sun, the rate being equivalent to 0.10 of the energy in the 0.4–0.7 μm range (Mooney et al., 1976). This rate is at least as great as that assumed for solar cells that produce electricity, 20 W m^{-2} (Häfele, 1980).

In contrast, high light saturation would not be worth its cost to forest systems in cloudy western Oregon, where, moreover, moderate leaf temperature and high atmospheric vapor pressure do not often call for stomatal closure and consequent restriction of CO_2 flux and favor very large leaf areas—25–50 m^2 m^{-2} (Waring et al., 1978).

As a result of light saturation of photosynthetic apparatus, the midday hours of high energy flux density might not display a peak, but rather a plateau, in CO_2 assimilation. Geometric considerations relating sun height and leaf angle may also account for midday flattening (Turitzen, 1978); Nichiporovich (1966) shows a truncated midday maximum lasting 10 hr in a one-layer leaf canopy as compared with a slightly flattened curve for a deep canopy.

Association and Analogy with Evapotranspiration

The radiant-energy sink of chlorophyll deployed in ecosystems is small in the energy budget but unusual in that it keeps energy out of the contemporaneous thermal arena. In this respect it differs from absorption that raises ecosystem temperature. Initial absorption is followed immediately by electron changes of a nonthermal nature, and the energy thus sequestered does not reenter the thermal arena until decomposition or combustion occurs. This temporary withdrawal of energy resembles the effect of evaporation.

In evaporation turbulent processes remove the end product, i.e., molecules of water vapor; in photosynthetic conversion turbulent processes supply the raw material, i.e., molecules of carbon dioxide. The same eddies that bring CO_2 to a photosynthesizing, transpiring ecosystem carry off water vapor; similar physical laws govern both fluxes, and biological controls operate on both. When leaf stomata are open to allow CO_2 molecules to diffuse inward to the wet cell walls where they are dissolved and carried into the cells, molecules of water vapor from the cell walls are free to leak out. The two energy conversions are associated though not identical and can be analyzed together. In this chapter we take note of the similarities of their regimes through the day, responses to day-to-day fluctuations in solar energy and soil moisture, variations as plants undergo phenological aging, and as succession takes place.

Gross assimilation of CO_2 (i.e., without respiration) can convert about 0.05 of total solar radiation (Budyko, 1974, equations 8.1–8.13). This energy budget approach uses the analogy between the mass exchanges of carbon and water to divide the daytime net surplus of all-wave radiation R among the nonradiative fluxes and transformations. The energy fixed in assimilation A can be expressed in terms of conversion constants of CO_2 and H_2O, quasi-constants a, b, and c, and daytime differences between temperature T and specific humidity q at leaf-cell surfaces and in the ambient air. The expression [Budyko, 1974, Eq. (8.13)] is as follows:

$$lA = lcR/[(L/a)(q_0 - q_e) + lc + b(T_0 - T_e)],$$
where

$$l = 10.4 \times 10^3 \ \text{kJ kg}^{-1}$$
$$L = 2.5 \times 10^3 \ \text{kJ kg}^{-1}$$
$$a \approx 0.8$$
$$c = 0.46 \times 10^{-3}$$
$$b = (c_p/a)(1 + D''/D') = 2.6 \ \text{kJ kg}^{-1} \text{K}^{-1}$$

where the specific heat of air $c_p \approx 1 \ \text{kJ K}^{-1} \text{kg}^{-1}$, D'' is a coefficient of external diffusion, and D' a coefficient of internal diffusion; subscripts 0 and e refer to surface and air, respectively.

At a temperature difference between leaf and air of 5° and a specific humidity difference of 1.3×10^{-2}, typical of daytime conditions in summer, the equation above becomes

$$lA = 0.08 R.$$

The rate of assimilation is 0.08 of the net surplus of all-wave radiation. This analogy does not offer a precise mode of determining photosynthetic conversion because it lumps together such details as the varying spectral composition of sunlight and differing plant responses. Moreover, plants using the C_4 biochemical pathway absorb CO_2 more effectively at low concentrations than C_3 plants do and can alter inward diffusion of CO_2 and hence outward diffusion of water vapor. The existence of the C_4 pathway in plants of different taxonomic groups suggests its survival value.

Another means of escaping the syndrome of vapor loss and restricted CO_2 intake lies in exploiting the cycle of day and night by the CAM route. The pineapple and many succulents store CO_2 at night and release it within cells the next day for use in the regular C_3 cycle (Solbrig and Orians, 1977). Their stomata stay closed during the daytime hours, when vapor pressure in the sun-heated leaf mesophyll far exceeds atmospheric vapor pressure.

Both these deviations, achieved at considerable cost to the plant, illustrate the analogy between photosynthesis and evapotranspiration. Because rain clouds intercept the radiation flux that powers photosynthesis, there is a possibility for trade-offs. Irrigation supplies water without cutting off radiation and is employed on sugar plantations of the lee sides of the Hawaiian Islands where solar radiation is strong; windward plantations get frequent rain but less radiation. For example, sugar yield was 1.2 kg m^{-2} when solar radiation averaged 205 W m^{-2}, 2.2 kg m^{-2} when it averaged 240 (Chang, 1968, p. 65). A crop–weather model for corn in Indiana uses solar radiation and evapotranspiration (Nelson and Dale, 1978).

The ratio between production and transpiration varies from place to place and expresses a kind of efficiency of water use. It increases with leaf temperature excess over air temperature and with wind speed (Rauner, 1973), as well as with favorable moisture conditions (Rauner, 1972). Improved corn management at Coshocton Experiment Station in Ohio increased evapotranspiration by 0.11 but productivity by 0.44 (from 0.82 to 1.19 kg m^{-2}) (Dreibelbis, 1963). On a global scale the relation can be expressed (Lieth and Box, 1972) as

$$\Pi = 3 - 3 \exp[-0.00097(E - 20)],$$

where Π is net production and E is evapotranspiration, both in kilograms per square meter.

Evapotranspiration can use low-grade energy converted from all wavelengths of radiation as well as abstracted from the warmth of air and soil. Photosynthesis is selective as to energy source but produces high-grade products that have almost infinite possibilities for conversion into other forms of energy and matter.

Other Limiting Factors

Few ecosystems approach the rates of photosynthesis expressed by ideal leaf radiation because there are at least a hundred steps in the whole chain and they "require a very high degree of organization and are very fragile" (Berry, 1975). A limiting factor operating on a single step can throw it out of balance with other steps. In so complex a system such factors are not always easy to identify, whether they are the atmospheric concentration of CO_2, the supply of photosynthetically active radiation, cell moisture, stomatal opening, leaf-cell temperature, leaf-cell concentrations of other elements taking part in one stage or another of the photosynthetic sequence, or internal controls that the plant exerts if the facilities to translocate or store photosynthate are inadequate. This complexity is not surprising if we consider that the basic problem is to extract a trace concentration of an atmospheric gas by means of energy supplied in a peculiar regime and partly at a continually shifting angle, and at the same time to assemble the necessary chemical compounds at the right aqueous concentrations in a medium at optimal temperature. The system must deploy its photosynthesizing equipment for maximum ventilation by the air, and simultaneously obtain optimum illumination by the sun and sky, hold a certain temperature range, and provide security for the displayed surfaces against static and dynamic loading by wind, rain, snow, and ice.

Ventilation. The CO_2 flux is usually downward to an ecosystem by day and away from it at night and can be calculated from CO_2-concentration measurements or eddy correlation. It may reach 10 g CO_2 m^{-2} hr^{-1} above maize on a clear windy day (Uchijima, 1976) (energy equivalent of $\approx$30 W m^{-2} flux), supplemented by soil respiration by day, which may add as much as 0.2 of the daytime assimilation in wheat (Denmead, 1976) and even more in low-latitude forest (Kira, 1978). Ventilation also helps prevent excessive temperature in leaves. Alpine trees survive best in windy sites (Schönenberger, 1975).

Leaf Temperature. The influence of leaf temperature on photosynthesis differs with species. A desert plant, *Tidestromia*, exhibits maximum CO_2 uptake at about 45°C (assuming no moisture stress), while a coastal plant, *Atriplex*, exhibits maximum uptake at temperatures from 15°C to 30°C (Berry, 1975). As we saw in Chapter VIII, leaf temperature is in continual fluctuation as a consequence of its role in mediating and balancing the leaf energy budget, and the broad range of photosynthetic response to leaf temperature expresses an accommodation of ecosystems to the variability of radiation flux densities.

C_4 plants display sharper peaks in their temperature curves than C_3 plants (Lister and Lemon, 1976), and these peaks often occur at higher temperatures, with no negative effect appearing until temperature exceeds 30°C, certainly a benefit to a plant growing in strong sunshine and often dry soil.

Low environmental temperatures early in the growing season reduce utilization of solar radiation approaching its annual peak and decrease the use of assimilates in Arctic plants, which slows assimilation rates (Warren Wilson, 1966).

Scale Factors. The initial advantages at the cellular level in the C_4 pathway are reduced at the leaf scale by resistances to the inward diffusion of CO_2 and dark respiration.* They are reduced further at the ecosystem scale by slow downward flow of atmospheric CO_2 into the ecosystem, self-shading, and cloud interception of solar radiation (Evans, 1975b, p. 340). For example, low-latitude C_4 crops have recorded daily growth rates as high as 60 g m^{-2} of crop area (Evans, 1975b), but midlatitude C_3 crops also approach similar rates. The C_4 advantage decreases at higher levels of organization (Milthorpe and Moorby, 1979, p. 128). The CAM pathway has rates about half as large.

A further reduction in mean assimilation is found when we look at the whole growing season. In early growth an annual ecosystem has not deployed optimum leaf area and cannot intercept all the incoming radiant energy. This effect is marked in potatoes, soybeans, and maize; the loss in potential energy conversion is of concern in Chinese research on food production. It is less important in such long-season crops as sugar cane, in which mean rates of dry-matter production can average, if water is not excessive or short, as high as 40 g m^{-2} day^{-1} (Bull and Glasziou, 1975). Duration of the growing season accounts for much of the difference among global ecosystems and can be allowed for by the concept of leaf-area duration; one leaf-area index-month accounts for about 0.085 kg m^{-2} of grass production, for example (Kira, 1975).

NET PRIMARY PRODUCTION

Respiration

Ordinary, that is, "dark," respiration (as distinguished from photorespiration in C_3 plants at an earlier stage in the sequence) reduces gross

---

* Dark respiration is ordinary respiration, which can be measured only at night.

photosynthesis by a third to a half, depending on the time period under consideration. It is less than 0.1 of gross assimilation in strong sunlight, but a larger fraction totaled over a day (Zelitch, 1975).

This process appears to be parasitic on production but in reality is a part of the operating and maintenance processes that keep an ecosystem running. It can be divided into two energy transformations: the one is often associated with accumulated dry weight of the system, representing the replacement or repair of essential structures, whether photosynthesizing apparatus, stems, or roots; the other is associated with photosynthetic rate, representing the building up of new structure or compounds (Evans, 1975b, p. 341), especially in young plants (Milthorpe and Moorby, 1979, p. 124).

The daily maintenance cost is equivalent to 10–25 g of photosynthate consumed per kilogram of plant biomass and tends to increase with temperature. When air and leaf temperature rise during the day, it is an inescapable drain on current photosynthesis; by night it is a loss unassociated with production. On warm nights when downward longwave radiation from a humid atmosphere has a high flux density, respiration is large. This effect, year long, limits the potential value of the humid low latitudes (Chang, 1968) for meeting world food needs. Respiration loss in cane is 0.15–0.20 (Bull and Glasziou, 1975) and perhaps still higher in fields lying under the heat plume of a city as Chang (1970) has noted over sugarcane fields downwind from Honolulu. The effect of longwave radiation occurs even on Arctic shores; Miller et al. (1976) show a decline in net photosynthesis when incoming longwave radiation is large.

An ingenious means of determining ecosystem respiration was demonstrated by Woodwell and Dykeman (1966) for the local air in and above a 10-m oak forest on Long Island outwash sands. Atmospheric sampling on towers 125 m tall identified nocturnal temperature inversions strong enough to isolate the oak–local-air system from the main body of the atmosphere, and showed that most of the accumulation of CO_2 during the night displayed a temperature dependence. This was marked in summer when respiration R_s in grams per square meter per night was

$$R_s = 0.5°K - 128,$$

where °K is absolute temperature, varying from 278 to 293. This study illustrates one approach to problems of determining forest productivity, a complex area.

The yearly sum respired amounts to the breakdown of 2.1 kg m^{-2} dry matter and includes below-ground respiration as well as the more commonly measured above-ground amount. Since net production of dry matter by the forest was measured as 1.15 kg m^{-2}, respiration was about two-thirds the gross primary photosynthesis of 2.1 + 1.15 = 3.25 kg m^{-2}. In energy terms it averaged approximately 1 W m^{-2} over the year and 2 W m^{-2} in summer.

The growth component of total respiration in sugar cane may be as large as 0.4 of gross photosynthesis when roots are being extended to search for water, especially into resistant soil, and new leaves put out (Bull and Glaszious, 1975). In other crops it is also large for similar purposes, for example, growing the nitrogen-fixing nodules on the roots of peas. Nitrogen fixation is estimated to cost 2.5% of primary photosynthate (Gutschick, 1978).

The building component of respiration is associated with ecosystem mass and indirectly with leaf-area index. Photosynthesis and respiration in wheat crops at different leaf areas show the leaf-area influence (Table I). High respiration in forest is an energy cost of the mechanical structure that raises leaves high in the air, increases leaf area, and accounts for the typically high gross production (Kira, 1975, 1978) (Fig. 1). An extreme case is a 450-year Douglas fir forest in Oregon, in which respiration consumes 0.9 of the gross primary production (Grier and Logan, 1977), equivalent in energy terms to 8 W m^{-2}.

The heat released in respiration in a few species is important in other ways. Skunk cabbage melts its way through spring snow and maintains temperatures 15–35° above air temperature (Knutson, 1974) with respiration rates equivalent to $\frac{1}{3}$ W.

Net Photosynthesis

The daily curve of net photosynthesis may be flattened by respiration losses in the hot part of the day, as well as by the effects of light sat-

TABLE I

Mass Transformation in Photosynthesis and Respiration of Wheat[a]

Leaf-area index (LAI)	In photosynthesis	In respiration
2	2.5	0.9
6	4.0	1.2
10	4.5	1.3

[a] Data from Evans *et al.*, 1975, p. 112. Data given in grams per square meter per hour.

Fig. XI.1. Late-afternoon sun on strongly photosynthesizing forest of several age classes in the Oregon Coast Ranges in June 1978; the high values of the biomass production are shown by the steepness of slopes that have been clear-cut.

uration or leaf geometry on gross photosynthesis, or by restricted moisture uptake that reduces CO_2 uptake. Interactions among several factors may account for a steep morning rise of net photosynthesis to a plateau as long as 12 hr (Lommen *et al.*, 1971). Carbon dioxide uptake in a C_4 plant growing in summer in Death Valley, at midday air temperatures around 45°C and solar radiation flux densities of 900 W m^{-2}, displays a plateau of 6–8 hr (Björkman *et al.*, 1972). This behavior contrasts with midday depressions in other arid-land plants and is ascribed to the C_4 pathway and high light saturation.

Flexibility of net photosynthesis is shown by measurements on wheat (Evans *et al.*, 1975, p. 111), in which midday canopy photosynthesis at 800 W m^{-2} solar radiation was 5.5 g m^{-2} hr^{-1} on 14 October, 7.5 on 15 October, and 3.0 on 16 October. Day-to-day variations in tundra near the Arctic Ocean result from changes in wind direction from over the land to off the ice-filled ocean (Miller *et al.*, 1976) because net photosynthesis increased with solar radiation and air and soil temperatures (both low). The ecosystem operates so as to maximize carbon assimilation under those environmental conditions that occur most frequently during the growing season. Hellmers (1964) notes that the long summer days of moderate radiation in western Europe account for often-reported high efficiencies of forest photosynthesis. As much as 0.02 of radiant energy in the photosynthetically active region is converted into energy content of wood.

Translocations of photosynthate from the leaves to other parts of the plant for operations, maintenance, or storage are sometimes blocked. These downstream processes may then exert a restraining influence on photosynthetic activity (Warren Wilson, 1966; Evans, 1975b, p. 331).

Models of Productivity Factors

Canopy or ecosystem models show the whole energy-conversion* picture of carbon cycling. For example, a model of grassland combines three photosynthesis functions: Those related to solar radiation, to water potential in the leaves, and to leaf temperature (Ripley and Redmann, 1976); these also enter the model of Bikhele *et al.* (1980).

A much-studied plant, maize, presents to the sun and atmosphere a well-ventilated, absorptive, efficient canopy equipped with C_4 pathways that work well under strong solar intensities. Highest maize yields, of the order of 1.2 kg m^{-2}, are attained in densely planted systems that range up to 11 or 12 plants per square meter and intercept the solar flux completely (Duncan, 1975, p. 43); leaf area index is used to modify solar radiation in one growth model (Coelho and Dale, 1980).

This high variability is made to order for modeling techniques. These often operate in steps of individual daytime periods; each day's simulation starting with the results reached the day before, introducing "discontinuous and variable daily inputs into a system whose response is continually changing" (Duncan, 1975, p. 44). In one model the daily net production of photosynthate is a function of leaf area, solar radiation, moisture status, and air temperature and is allocated to roots, stems, leaves, and corn ear as a function of plant development. It describes ecosystem condition at any date and gives a means of predicting what would happen if environmental factors change. Using weather records, the model can " 'grow' any described variety of maize at any location in the world" (Duncan, 1975, p. 47), or at any growing season in the

---

* In centering our attention on productivity we are not unmindful of other significant dimensions. As Major (1969) warns, "Productivity is a simplifying measure which can ignore the ecosystem idea and which then gives the same numbers for Death Valley as for the northernmost coast of Alaska." Energy is only one of the dimensions of life, but the important one in this book.

past as if different varieties or cultural practices had then been in use. One model is essentially one leaf thick and combines a leaf-energy budget with CO_2 diffusion equations (Ehleringer et al., 1979); in contrast, the SPAM (soil-plant-atmosphere model) (Lemon et al., 1971) uses both leaf and community (e.g., light penetration) submodels.

Allocation of Product

Above and Below Ground. Models for simulating the growth of crop ecosystems allocate net photosynthetic product to system growth or to storage, i.e., in sugar beets a partition between tops and fibrous roots versus the storage root (Fick et al., 1975). This beet model begins by calculating hourly values of sun height and flux density of direct and diffuse solar radiation and proceeds through net photosynthesis as a function of radiation, air temperature, and moisture status; it pinpoints respiration as a process that still needs study. The model can determine scheduling of harvest of beets growing in each of the numerous local climates of California for best date of delivery to the beet-sugar factories (Fick et al., 1975, p. 290).

Translocation of biomass in such perennial species as wetland *Carex* (Bernard, 1974) produces a winter peak in below-ground biomass at 328 g m^{-2}, which, after a decline as above-ground biomass grows during spring and summer, then recovers after July. The ecosystem increase from winter to summer, 560 g m^{-2}, incorporates an internal shift of about a third of the mass.

Three species of lupine in different habitats of the central coast of California display three patterns of translocation, in which the percentages of annual energy devoted to the below-ground system and reproductive tissues are as shown in Table II. The annual species invests three times as much energy in formation of seeds and other reproductive tissues as the longer-lived perennial and shrub species do, and its below-ground growth is much less. These strategies of energy allocation are

TABLE II

Allocation of Energy to Below-Ground and Reproductive Parts of Lupine[a]

	Below-ground	Reproductive
Annual (*Lupinus nanus*)	3–4%	61%
Perennial (*L. varicolor*)	40%	18%
Shrub (*L. arboreus*)	25%	20%

[a] Data from Pitelka (1977).

related to habitat; the perennial lupine grows in windy, cold bluff sites. Below-ground allocation of energy in many species supports mycorrhizae that aid water and nutrient uptake.

Metabolites. Allocation of energy into such metabolites as hydrocarbons is important where photosynthesis is considered as an oil source (Nielsen *et al.*, 1977). These compounds are sometimes complex chemically, even beyond laboratory synthesis (e.g., jojoba wax). To the extent that they act as insect repellents, they keep balance among ecosystem members. Root exudates may govern spatial patterns in ecosystems.

Reproductive Tissue. Even during the reproductive phase only half the dry-matter production of cotton is allocated to reproductive growth (Stanhill, 1976), but in sunflower the seeds get all the photosynthates; no new leaves are produced after seeds are started (Saugier, 1976). Models that allocate photosynthate between vegetative and reproductive tissues may utilize two compartments (labile and nonlabile fractions) in each of several plant organs (leaves, stems, early reproductive buds, fruits, etc.) (Cunningham and Reynolds, 1978). Changes of energy in each compartment depend on the rates of maintenance respiration and growth and follow a scheme of translocation priorities.

A model of the widespread, successful desert plant, creosote bush (Cunningham and Reynolds, 1978) includes a photosynthesis submodel (moisture, temperature, biomass of each leaf age-class) and a respiration submodel, with high maximum rates per hr for reproductive tissues (152 mg CO_2 per gram dry weight of flowers, contrasted with 5 mg CO_2 per gram of leaves). It reproduces flushes of both vegetative and reproductive growth during a desert year. Genotype also affects this allocation: open-pollinated maize of the early 1920s yielded 0.24 of its dry weight, while hybrid maize of the late 1950s yielded 0.43 (Zelitch, 1975).

Energy Aspects of Pollination. Allocation of energy to the purpose of gene selection, i.e., information transfer, takes many forms, especially when a plant species and a pollinator species have evolved in mutual dependence. The energy manifestations of this process include the payment in nectar that provides operational support for the pollinator, which often has rapid metabolism with large energy requirements. Although pollination by wind transport works well in large populations of the same species, it is not so reliable when individuals of a species are spatially isolated (Regal, 1977).

In a dispersed plant population spatial considerations become important because the "animal pollinators require increased energy rewards per flower" (Heinrich and Raven, 1972) and must be large enough to travel long distances. While a bumblebee expends 5 mW, a hummingbird expends 750 mW or more. Ants, on the other hand, are low-energy pollinators of plants that have many small flowers with energy rewards too small to encourage flying insects (Hickman, 1974). Foraging costs of a hummingbird in Kenya, about 1.4 W, need to be minimized, as also do defense flights (about 3 W) (Wolf, 1975). These costs are modified in different species of bees by the costs of life in colonies (Schaffer et al., 1979).

Energy costs increase in cold conditions, and in consequence flowers that "open in early morning and in the evening are generally large" (Heinrich and Raven, 1972). Some Arctic flowers provide a small solar furnace that warms the pollinators, as noted in Chapter III.

Pollination, whether supported by atmospheric kinetic energy or photosynthetic conversion of solar radiation, is an information transfer that makes possible trial-and-error experimentation in the myriad processes, catalysts, enzymes, etc., in ecosystem productivity and does so on an enormous scale (Berry, 1975). The complicated C_4 solution has been tried and has succeeded in many distinct species. A person looking at the great variety of leaf shapes and canopy architecture in contemporary and prehistoric plants, as arranged perhaps in a paleontological sequence, is impressed with the numerous workable solutions that have been found to the problem of fixing radiant energy.

INFLUENCES ON PRODUCTIVITY AND TRANSLOCATION

Temperature and Moisture

Air temperature over a region to a degree reflects boundary fluxes at the underlying ecosystems and therefore indexes some processes (though not evapotranspiration) better than solar radiation does, for example, the storage rate of photosynthate in wheat or cotton (Evans, 1975b). Yields of rice and maize depend on temperature as well as radiation; rice displays a curvilinear relation to the mean temperature of the 40-day ripening period, which for maximum yield is 22°C (Uchijima, 1976). Water temperature is important earlier in its life cycle. Increased radiation always gives more productivity, and a radiation–temperature interaction also occurs. A low temperature gives the greatest productivity at low radiation, 15°C gives the greatest productivity at 200 W m^{-2}, and 20°C at 400 W m^{-2}.

Most of the time the temperature in low-latitude ecosystems remains within the range needed for rapid photosynthesis and can be omitted from production models. For example, a yield model of Townsville stylo, a legume important in northern Australian pastures, is based on antecedent phosphate application, a standard soil-moisture submodel, and current evaporation (Rose et al., 1976), which is a surrogate for radiant-energy input. The major temperature parameter is that of the topmost layer of the soil at the time of germination.

In high-energy summers of midlatitudes "water is the principal driving variable in shortgrass prairie ecosystems" (Lauenroth and Sims, 1976). Investigation of the influence of nitrogen on ecosystem functions at the Pawnee site of the IBP Grassland Biome Program found no nitrogen effects under natural rainfall in 3 years, but a positive response in irrigated grassland, described as an indirect effect "mediated by vegetation biomass." Irrigated ecosystems transpire more water not only because more is available but also because the size and productivity rate of the standing crop are greater. Wet meadows in central Europe outproduce drier ones substantially (Rychnovská, 1979).

Phenological Phases in the Annual Cycle

Ecosystem productivity is also influenced by the phases of ecosystem development and the need to prepare for the end of the growing season. Below-ground storage is built up in such perennial species as *Carex* and *Poa* or is accumulated in tubers, but tuberization is hindered if temperature remains high. One crop of potatoes is grown in the Murrumbidgee Irrigation Area in Australia in the spring and one in the fall, but none in the hot months (Moorby and Milthorpe, 1975, p. 233). The rate at which energy enters storage in wheat and cotton is diminished by high temperature (Evans, 1975b, p. 345) in a relation opposite to the effect of solar radiation, and the usual correlation of radiation and temperature results in an indeterminate effect on crop yield.

The length of the translocation period primarily determines wheat yield (Evans et al., 1975, p. 131). This period lasts longer if the wheat is not under moisture stress, a condition that can be determined by remotely sensing its surface temperature (Idso et al., 1977).

A tundra ecosystem does not begin to store net photosynthetic production until its leaves are grown, late in July (Miller et al., 1976). After this date it would be pointless to start leaves because too little time remains for the investment in them to be recovered. Allocation to leaves in oak canopy became less than that to branches after August (McLaughlin et al., 1979).

Annual Cycle.   Solar radiation is the source of energy for photosynthesis and the forcing function for phenological phases and the regimes of temperatures of leaves, soil, and air, all of which influence total energy conversion. The regime of soil moisture in many regions also displays a simple annual rise and fall, in which a maximum at the beginning of the growing season is followed by a long period of drawdown. In other regions late-summer rains shift production far behind the peak solar output. Elsewhere, irregular delivery of water causes sporadic periods of moisture stress, and daily values of soil moisture must be entered in growth models of maize (Duncan, 1975; Morgan et al., 1980). Soil moisture is exhausted in summer-dry climates while the energy supply is still strong, and California chaparral makes much of its growth before the equinox. Annual grasses and grains follow the same early pattern. Shrubs of the Mojave Desert must deal with two regimes: winter–spring rains dominate some years; summer rains from a different atmospheric circulation dominate others. Spring-active shrubs reach peak hourly assimilation rates of 30 g CO_2 fixed per kilogram of plant dry weight in March and decline to zero by June; summer-green plants, leafless in winter, reach lower, later peaks (10 g CO_2 per kilogram in late May) (Bamberg et al., 1975).

Length of the growing season is a major factor in annual production since peak rates do not differ greatly in different parts of the world. Where the short season of high latitudes is further shortened under slow melting patches of snow, lichen growth is greatly reduced (Koerner, 1980). Growing-season length is related to the annual surplus of net all-wave radiation, which in regions of low moisture stress is correlated with production as shown in Table III. Production is reduced about half in regions of frequent moisture stress (radiative index of dryness = 2). On the other hand, the year-long warmth of equatorial latitudes leads to excessive respiration and to disease and insect attacks, particularly damaging to nonforest agro-ecosystems (Chang, 1968; Janzen, 1973; Courtenay, 1978).

Variations in Production between Years. Year-to-year variations in ecosystem production usually cannot be expressed simply as results of fluctuations in seasonal total rainfall or mean solar radiation, but are more likely to sum up many events throughout the growing season, which can be detected only by modeling daily production. The moisture-energy model for Townsville stylo, for instance, "has contributed toward the interpretation of a very variable seasonal production" (Rose et al., 1976).

An increase in year-to-year variability of summer weather in the corn belt following 1974, coming after agricultural technologists had grown overconfident of the powers of technology as displayed in a period of stable years, created a shock that caused reevaluation of thinking about

TABLE III

Production during the Growing Season in Regimes of Low Moisture Stress[a] with Different Levels of All-Wave Radiation Surplus[b]

Surplus of net all-wave radiation $(W\ m^{-2})$	Annual production $(kg\ m^{-2})$
25	0.4
50	1.2
75	2.5

[a] Defined as $LP > R$, where $L = 2.5$ MJ kg^{-1}, P is precipitation in kilograms per square meter, and R is the yearly surplus of all-wave radiation (see Chapter X) in megajoules per square meter.
[b] Data from Budyko (1974, p. 426). Data are means over whole year.

the role of the environment in crop productivity. Variability was seen to be more important in grain production than a cooling trend, even one that would be cold as that of the 1830s (Thompson, 1975). It is interesting that each of three severe winters in Wisconsin (1976–1977, 1977–1978, and 1978–1979) was followed by record-breaking growing-season production.

Production of a forest ecosystem is usually greatest at the time of complete canopy closure (Kira, 1975; Rauner, 1976). It tends to decrease thereafter because maintenance respiration increases as the biomass grows, but some conifer stands in the Pacific Northwest continue to grow in height and diameter into their fifth century (Waring and Franklin, 1979).

Spatial Diversity

Photosynthetic production varies with inputs of radiation and CO_2, as well as water and nutrients, all of which vary spatially over the surface of the earth. In addition, neighboring ecosystems create their own smaller-scale patterns of diversity in age, fire history, and species composition.

Ecosystem Structure. Different layers of an ecosystem have different daytime energy-conversion rates and sometimes different annual cycles. The spring flowers of Wisconsin woodlots, for example, accomplish much of their annual energy conversion in the short period between the end of snow cover and the leafing out of the trees above them, after which they receive only a small quantity of solar energy. Flowering begins as soon as there is any chance of pollination and is completed before the tree canopy closes (Schemske et al., 1978). Biomass productivity of the herb structure was about 0.117 kg m^{-2} under aspen but 0.063 under denser maple-aspen-birch forest in northern Wisconsin (Zavitkovski, 1976), and radiation measurements under different forest canopies showed significant correlation with biomass production. A model of overstory density in the Adirondacks includes light supply to the understory, hence can estimate the production of browse for wildlife (Cooperrider and Behrend, 1980).

Many ecosystems display an intricate dovetailing of the energy-fixing regimes of their component species, reflecting their respective accommodations to moisture, light, ventilation, frost hazard, and temperature of each layer in the ecosystem (Waggoner, 1975). Xerophytes that coexist in many deserts exploit different soil zones or seasons of water availability (Solbrig and Orians, 1977). Such dovetailing represents an allocation of yearly inputs that can be analyzed in terms of the energy budgets of the individual components, animals as well as plant species. A forest ecosystem exemplifies this complexity in the growth in roots, stems, branches, twigs, and leaves, requiring difficult field measurement (Whittaker and Marks, 1975). In Townsville stylo pastures much depends on conditions during germination and establishment; legume and grass species and "their changes in space and time are . . . the result of interaction with . . . micrometeorological variables" (Rose et al., 1976). In most crop ecosystems "weeds have always been an integral part" (Furtick, 1978), though not necessarily recognized as being as costly as they are, for example, in accounting for half the energy applied in crop cultivation.

Intercropping increases species diversity and the complexity of conversions and allocations of solar energy, and in some cases raises total ecosystem production. Intercropping of peanuts and corn increases productivity because peanuts attract a predator of the corn borer (Wade, 1975). Monocultures that present extensive areas of vegetation at the same phenological phase to insects and disease vectors may be vulnerable to attack, and production may fall short of that of native ecosystems. Crops occupy 0.11 of the land of the earth, but their harvests account for only 0.01 of terrestrial photosynthetic production (Evans, 1975a).

Mesoscale Contrasts. Productivity varies from ecosystem to ecosystem, as all farmers know, and as is true also of native systems. In the

TABLE IV

Productivity of Forest-Steppe Ecosystems[a]

Association[b]	Terrain	Production (kg m^{-2})
Mountain sage	north slopes	0.25
Stipa-Poa	ridge tops	0.29
Tall grass	clearing	0.45
Oak forest-nettle	gully bottom	0.71
Feathergrass-sage	dry south slope	0.34

[a] Data from Utekhin (1972).
[b] Listed in order of increasing radiation supply.

Alekhin National Park of central Russia biometeorological research has identified hydrologic and production differences among the virgin prairie and oak ecosystems (Table IV). Greatest production is found in the relatively moist, moderate-radiation sites occupied by tall grass or oak-nettle associations; forest outproduces low vegetation on all comparable sites.

Dairy farms in Wisconsin typically include corn, oats, and hay fields, as well as highly productive wetlands and woodlots, which form a mosaic that extends, with local changes in the mix, over much of the state. In the high-yield summer of 1977 the agricultural ecosystems averaged the energy productivities shown in Table V. Growing periods differ, corn gets an energy subsidy, and alfalfa hay allocates photosynthates to fixing atmospheric nitrogen.

TABLE V

Production by Agricultural Ecosystems of Wisconsin in 1977[a]

System	Dry-weight production (kg m^{-2})	Energy production (MJ m^{-2})
Corn for grain	0.52	10
Oats	0.15	3
Hay	0.4	7[b]

[a] Data from Wisconsin Statistical Reporting Service (1977).
[b] Yield in this good summer from three, or in some places four, cuttings approximated 0.6 W m^{-2} over a long season.

TABLE VI

Annual Production Estimates from Ecosystems in
Different Zones of the Earth[a]

Zonal group	Annual production (kg m^{-2})
Low-latitude rainforests	3.2
Midlatitude broadleaf forests	1.3
Midlatitude steppes	1.1
Semidesert shrub	0.1
High-latitude tundra	0.1

[a] Data from Rodin and Basilevič (1968).

Regional Productivity. Where low ecosystems cover flat terrain like the Great Plains, productivity can be integrated by tower measurements of the CO_2 flux. Midday production rates of ecosystems of this terrain averaged 6 W m^{-2} in early June and 18–25 W m^{-2} in early August when ground cover was complete. Net rates over the daily cycle averaged 1.5 W m^{-2} in early June and 3.0 in early August (Verma and Rosenberg, 1976).

A comprehensive study in the White Mountains of New England reports net primary production of deciduous forest over a 4-month season as 2 W m^{-2} [data from Gosz et al. (1978)], of which about a quarter represented net increase in forest biomass.

When energy and moisture are used in mapping production of native ecosystems in the Soviet Union a value of 0.2 kg m^{-2} is found in the tundra of the Arctic Coast and the deserts of central Asia, 0.8 in most of central European Russia, and 2.0 kg m^{-2} on the warm humid coast of Gruziia on the Black Sea (ancient Colchis) (Budyko, 1974, p. 428).

A world survey of annual productivity (Rodin and Basilevič, 1968) which indicates the effects of water and solar energy inputs and their durations over the year in zonal groups of ecosystems is shown in Table VI.

A summary of many production studies under the International Biological Program (Wassink, 1975) shows forest as the most productive class of ecosystem, herbaceous less, and tundra and deserts least, as reflections of their brief seasons of growth. The energy equivalents are 0.6–1.7 W m^{-2} for forest and 0.2–0.7 for herbaceous ecosystems.

These values of net primary productivity are, in a steady-state ecosystem, equal to consumption by grazers and browsers and decomposition, which will be discussed in Chapter XII.

REFERENCES

Aulitzky, H. (1963). Bioklima und Hochlagenaufforstung in der subalpinen Stufe der Innenalpen. *Schweiz. Z. Forstwes.* **12,** 1–25.
Bamberg, S. A., Kleinkopf, G. E., Wallace, A., and Vollmer, A. (1975). Comparative pho-
tosynthetic production of Mojave shrubs. *Ecology* **56,** 732–736.
Bazzaz, F. A. (1979). The physiological ecology of plant succession. *Ann. Rev. Ecol. Syst.*
10, 351–371.
Bernard, J. M. (1974). Seasonal changes in standing crop and primary production in a
sedge wetland and an adjacent dry old-field in central Minnesota. *Ecology* **55,** 350–359.
Berry, J. A. (1975). Adaptation of photosynthetic processes to stress. *Science* **188,** 644–650.
Bikhele, Z. N., Moldau, Kh. A., and Ross, Iu. K. (1980). "Matematicheskoe Modelirovanie
Transpiratsii i Fotosinteza Rastenii pri Nedostatke Pochvennoi Vlagi." Gidrometeorol.
Izd., Leningrad.
Bishop, N. I., Frick, M., and Jones, L. W. (1977). Photohydrogen production in green
algae: Water serves as the primary substrate for hydrogen and oxygen production. *In*
"Biological Solar Energy Conversion" (A. Mitsui, S. Miyachi, A. San Pietro, and S.
Tamura, eds.), pp. 3–22. Academic Press, New York.
Björkman, O., Pearcy, R. W., Harrison, A. T., and Mooney, H. (1972). Photosynthetic
adaptation to high temperatures: a field study in Death Valley, California. *Science* **175,**
786–789.
Budyko, M. I. (1974). "Climate and Life" (transl. ed. by D. H. Miller). Academic Press,
New York.
Bull, T. A., and Glasziou, K. T. (1975). Sugar cane. *In* "Crop Physiology" (L. T. Evans,
ed.), pp. 51–72. Cambridge Univ. Press, London and New York.
Chang, J.-H. (1968). The agricultural potential of the humid tropics, *Geogr. Rev.* **58,** 333–361.
Chang, J.-H. (1970). Personal communication.
Coelho, D. T., and Dale, R. F. (1980). An energy-crop growth variable and temperature
function for predicting corn growth and development: planting to silking. *Agron. J.* **72,**
503–510.
Cooperrider, A. Y., and Behrend, D. F. (1980). Simulation of forest dynamics and deer
browse production. *J. For.* **78,** 85–88.
Courtenay, F. P. (1978). Agriculture in North Queensland. *Austral. Geogr. Stud.* **16,** 29–42.
Cunningham, G. L., and Reynolds, J. F. (1978). A simulation model of primary production
and carbon allocation in the creosotebush (*Larrea tridentata* [DC] Cov.). *Ecology* **59,** 37–52.
Denmead, O. T. (1976). Temperate cereals. *In* "Vegetation and the Atmosphere" (J. L.
Monteith, ed.), Vol. 2, pp. 1–30. Academic Press, New York.
Dreibelbis, F. R. (1963). Land use and soil type effects on the soil moisture regimen in
lysimeters and small watersheds. *Soil Sci. Soc. Am., Proc.* **27,** 455–460.
Duncan, W. G. (1975). Maize. *In* "Crop Physiology" (L. T. Evans, ed.), pp. 23–50. Cam-
bridge Univ. Press, London and New York.
Ehleringer, J., Björkman, O., and Mooney, H. A. (1976). Leaf pubescence: effects on ab-
sorptances and photosynthesis in a desert shrub. *Science* **192,** 376–377.
Ehleringer, J., Mooney, H. A., and Berry, J. A. (1979). Photosynthesis and microclimate
of *Camissonia claviformis*, a desert winter annual. *Ecology* **60,** 280–286.
Evans, L. T. (1975a). Crops and world food supply, crop evolution, and the origin of crop
physiology. *In* "Crop Physiology" (L. T. Evans, ed.), pp. 1–22. Cambridge University
Press, London and New York.
Evans, L. T. (1975b). The physiological basis of crop yield. *In* "Crop Physiology" (L. T.
Evans, ed.), pp. 327–355. Cambridge Univ. Press, London and New York.

Evans, L. T., Wardlaw, I. F., and Fischer, R. A. (1975). Wheat. In "Crop Physiology" (L. T. Evans, ed.), pp. 101–149. Cambridge Univ. Press, London and New York.

Fick, G. W., Loomis, R. S., and Williams, W. A. (1975). Sugar beet. In "Crop Physiology" (L. T. Evans, ed.), pp. 259–295. Cambridge Univ. Press, London and New York.

Furtick, W. R. (1978). Weeds and world food production. In "World Food, Pest Losses, and the Environment" (D. Pimentel, ed.), AAAS Selected Symposia Series, pp. 51–62. Westview Press, Boulder, Colorado.

Gosz, J. R., Holmes, R. T., Likens, C. E., and Bormann, F. H. (1978). The flow of energy in a forest ecosystem. Sci. Am. 238(3), 93–102.

Grier, C. C., and Logan, R. S. (1977). Old-growth Pseudotsuga menziesii communities of a western Oregon watershed:Biomass distribution and production budgets. Ecol. Monogr. 47, 373–400.

Gutschick, V. P. (1978). Energy and nitrogen fixation. BioScience 28, 571–575.

Häfele, W. (1980). A global and long-range picture of energy developments. Science 209, 174–182.

Heath, O. V. S. (1969). "The Physiological Aspects of Photosynthesis." Stanford Univ. Press, Stanford, California.

Heinrich, B., and Raven, P. H. (1972). Energetics and pollination ecology. Science 176, 597–602.

Hellmers, H. (1964). An evaluation of the photosynthetic efficiency of forests. Q. Rev. Biol. 39, 249–257.

Hickman, J. C. (1974). Pollination by ants: a low-energy system. Science 184, 1290–1292.

Idso, S. B., Jackson, R. D., and Reginato, R. J. (1977). Remote-sensing of crop yields. Science 196, 19–25.

Janzen, D. H. (1973). Tropical agroecosystems. Science 182, 1212–1219.

Kira, T. (1975). Primary production of forests. In "Photosynthesis and Productivity in Different Environments" (J. P. Cooper, ed.), International Biological Programme, Synthesis Ser. No. 3, pp. 5–40. Cambridge Univ. Press, London and New York.

Kira, T. (1978). Community architecture and organic matter dynamics in tropical lowland rain forests of Southeast Asia with special reference to Pasoh Forest, West Malaysia. In "Tropical Trees as Living Systems" (P. B. Tomlinson and M. H. Zimmerman, eds.), pp. 561–570. Cambridge Univ. Press, London and New York.

Knutson, R. M. (1974). Heat production and temperature regulation in Eastern skunk cabbage. Science 186, 746–747.

Koerner, R. M. (1980). The problem of lichen-free zones in Arctic Canada. Arct. Alpine Res. 12(1), 87–94.

Lauenroth, W. K., and Sims, P. L. (1976). Evapotranspiration from a shortgrass prairie subjected to water and nitrogen treatments. Water Resour. Res. 12, 437–444.

Lemon, E., Stewart, W., and Shawcroft, R. W. (1971). The sun's work in a cornfield. Science 174, 371–378.

Lieth, H., and Box, E. (1972). Evapotranspiration and primary productivity: C. W. Thornthwaite memorial model. Publ. Climatol. 25(3), 37–46.

Lister, R., and Lemon, E. (1976). Interactions of atmospheric carbon dioxide, diffuse light, plant productivity, and climate processes—model predictions. In "Atmosphere-Surface Exchange of Particulate and Gaseous Pollutants (1974)," pp. 112–135. ERDA, Washington, D.C.

Lommen, P. W., Schwintzer, C. R., Yocum, C. S., and Gates, D. M. (1971). A model describing photosynthesis in terms of gas diffusion and enzyme kinetics. Planta 98, 195–220.

Loomis, R. S., and Gerakis, P. A. (1975). Productivity of agricultural ecosystems. In "Photosynthesis and Productivity in Different Environments" (J. P. Cooper, ed.), International Biological Programme, Synthesis Ser. No. 3, pp. 145–172. Cambridge Univ. Press, London and New York.

Major, J. (1969). Historical development of the ecosystem concept. In "The Ecosystem Concept in Natural Resource Management" (G. M. van Dyne, ed.), pp. 9–22. Academic Press, New York.

McLaughlin, S. B., McConathy, R. K., and Beste, B. (1979). Seasonal changes in within-canopy allocation of ^{14}C-photosynthate by white oak. For. Sci. 25(2), 361–370.

Miller, P. C., Stoner, W. A., and Tieszen, L. L. (1976). A model of stand photosynthesis for the wet meadow tundra at Barrow, Alaska. Ecology 57, 411–430.

Milthorpe, F. L., and Moorby, J. (1979). "An Introduction to Crop Physiology" 2d ed., Cambridge University Press, London and New York.

Mooney, H. A., Ehleringer, J., and Berry, J. A. (1976). High photosynthetic capacity of a winter annual in Death Valley. Science 194, 322–323.

Moorby, J., and Milthorpe, F. L. (1975). Potato. In "Crop Physiology" (L. T. Evans, ed.), pp. 225–257. Cambridge Univ. Press, London and New York.

Morgan, T. H., Biere, A. W., Kanemasu, E. T. (1980). A dynamic model of corn yield response to water. Water Resour. Res. 16(1), 59–64.

Nelson, W. L., and Dale, R. F. (1978). A methodology for testing the accuracy of yield predictions from weather-yield regression models for corn. Agron. J. 70, 734–740.

Nichiporovich, A. A. (1966). Zadachi rabot po izucheniiu fotosinteticheskoi deiatel'nosti rastenii kak faktora produktivnosti. In "Fotosinteziruiushchie Sistemy Vysokoi Produktivnosti" (A. A. Nichiporovich, ed.), pp. 7–50. Izd. Nauka, Moscow.

Nielsen, P. E., Nishimura, H., Otvos, J. W., and Calvin, M. (1977). Plant crops as source of fuel and hydrocarbon-like materials. Science 198, 942–944.

Pitelka, L. F. (1977). Energy allocation in annual and perennial lupines (Lupinus: Leguminosae). Ecology 58, 1055–1065.

Rauner, Iu. L. (1972). Energeticheskie kharakteristiki produktsionnogo protsessa lugovo-stepnykh fitotsenozov na razlichnykh elementakh rel'efa. Izv. Akad. Nauk SSSR, Ser. Geogr. No. 4, 95–105.

Rauner, Iu. L. (1973). Energeticheskaia effektivnost' produktsionnogo protsessa rastitel'nykh soobshchestv. Izv. Akad. Nauk SSSR, Ser. Geogr. No. 6, 17–28.

Rauner, Iu. L. (1976). Deciduous forest. In "Vegetation and the Atmosphere" (J. L. Monteith, ed.), Vol. 2, pp. 241–262. Academic Press, New York.

Regal, P. J. (1977). Ecology and evolution of flowering plant dominance. Science 196, 622–629.

Ripley, E. A., and Redmann, R. E. (1976). Grassland. In "Vegetation and the Atmosphere" (J. L. Monteith, ed.), Vol. 2, pp. 349–398. Academic Press, New York.

Rodin, L. E., and Basilevič, N. I. (1968). World distribution of plant biomass. In "Functioning of Terrestrial Ecosystems at the Primary Production Level," (F. E. Eckardt, ed.) UNESCO Natural Resources Research, No. 5, pp. 45–52. UNESCO, Paris.

Rose, C. W., Begg, J. E., and Torssell, B. W. R. (1976). Townsville stylo (Stylosanthes humilis H.B.K.). In "Vegetation and the Atmosphere" (J. L. Monteith, ed.), Vol. 2, pp. 151–169. Academic Press, New York.

Rychnovská, M. (1979). Ecosystem synthesis of meadows. In "Grassland Ecosystems of the World: Analysis of Grasslands and Their Uses" (R. T. Coupland, ed.) pp. 165–180. Cambridge University Press, London and New York.

Saugier, B. (1976). Sunflower. In "Vegetation and the Atmosphere" (J. L. Monteith, ed.), Vol. 2, pp. 87–119. Academic Press, New York.

Schaffer, W. M., Jensen, D, B., Hobbs, D. F., Gurevitch, J., Todd, J. R. and Schaffer, M.

V. (1979). Competition, foraging energetics and the cost of sociality in three species of bees. *Ecology* **60**(5), 976–987.

Schemske, D. W., Willson, M. F., Melampy, M. N., Miller, L. J., Verner, L., Schemske, K. M., and Best, L. B. (1978). Flowering ecology of some spring woodland herbs. *Ecology* **59**, 351–356.

Schönenberger, W. (1975). Standortseinflüsse auf Versuchsafforstung an der alpinen Waldgrenze (Stillberg, Davos). *Mitt. Schweiz. Anst. Forstl. Versuchswes.* **51**, 357–428.

Solbrig, O. T., and Orians, G. H. (1977). The adaptive characteristics of desert plants. *Am. Sci.* **65**, 412–421.

Stanhill, G. (1976). Cotton. *In* "Vegetation and the Atmosphere" (J. L. Monteith, ed.), Vol. 2, pp. 121–150. Academic Press, New York.

Thompson, L. M. (1975). Weather variability, climatic change, and grain production. *Science* **188**, 535–541.

Turitzen, S. N. (1978). Canopy structure and potential light competition in two adjacent annual plant communities. *Ecology* **59**, 161–167.

Uchijima, Z. (1976). Maize and rice. *In* "Vegetation and the Atmosphere" (J. L. Monteith, ed.), Vol. 2, pp. 33–64. Academic Press, New York.

Utekhin, V. D. (1972). Rastitel'nyi pokrov territorii Kurskogo Statsionara i ego produktivnost'. *In* "Biogeograficheskoe i Landshaftnoe Izuchenie Lesostepi" (D. L. Armand, ed.), pp. 143–179. Izd. Nauka, Moscow.

Verma, S. B., and Rosenberg, N. J. (1976). Carbon dioxide concentration and flux in a large agricultural region of the Great Plains of North America. *J. Geophys. Res.* **81**, 399–405.

Wade, N. (1975). International agricultural research. *Science* **188**, 585–589.

Waggoner, P. E. (1975). Micrometeorological models. *In* "Vegetation and the Atmosphere" (J. L. Monteith, ed.), Vol. 1, pp. 205–228. Academic Press, New York.

Waring, R. H., Emmingham, W. H., Gholz, H. L., and Grier, C. C. (1978). Variation in maximum leaf area of coniferous forests in Oregon and its ecological significance. *For. Sci.* **24**(1), 131–140.

Waring, R. H., and Franklin, J. F. (1979). Evergreen coniferous forests of the Pacific Northwest. *Science* **204**, 1380–1386.

Warren Wilson, J. (1966). An analysis of plant growth and its control in Arctic environments. *Ann. Bot. (London)* **30**, 384–402.

Warren Wilson, J. (1967). Effects of seasonal variation in radiation and temperature on net assimilation and growth rates in an arid climate. *Ann. Bot. (London)* **31**, 41–57.

Wassink, E. C. (1975). Photosynthesis and productivity in different environments—conclusions. *In* "Photosynthesis and Productivity in Different Environments" (J. P. Cooper, ed.), International Biological Programme, Synthesis Ser. No. 3, pp. 675–687. Cambridge Univ. Press, London and New York.

Whittaker, R. H., and Marks, P. L. (1975). Methods of assessing terrestrial productivity. *In* "Primary Productivity of the Biosphere" (H. Lieth and R. H. Whittaker, eds.), pp. 55–118. Springer-Verlag, Berlin and New York.

Wisconsin Statistical Reporting Service (1977). Monthly Report, November, 1977. Madison.

Wolf, L. L. (1975). Energy intake and expenditures in a nectar-feeding sunbird. *Ecology* **56**, 92–104.

Woodwell, G. M., and Dykeman, W. R. (1966). Respiration of a forest measured by carbon dioxide accumulation during temperature inversions. *Science* **154**, 1031–1034.

Zavitkovski, J. (1976). Ground vegetation biomass, production, and efficiency of energy utilization in some northern Wisconsin forest ecosystems. *Ecology* **57**, 694–706.

Zelitch, I. (1975). Improving the efficiency of photosynthesis. *Science* **188**, 626–633.

Chapter XII

THE RELEASE OF CARBON FIXED IN ECOSYSTEMS

The biomass that represents the net primary production of an ecosystem is subsequently broken down by grazing animals, decomposers, or fire, which transform the fixed solar energy. These transformations have their own energy budgets, which may be compressed in time or concentrated in space. The detritus food chain is the most important in all except phytoplankton ecosystems, which are heavily grazed; it outweighs grazing even in grassland, according to International Biological Programme (IBP) research (Petrusewicz and Grodzinski, 1975).

BIOLOGICAL ENERGY CONVERSIONS

Grazers

It is convenient to distinguish large grazers from leaf-eating insects because size differences create different relations to the environment. For example, insects are dependent on heat exchanges with their immediate surroundings and are more of a problem in ecosystems that are continuously warm. They also convert leaves into animal tissue more efficiently (Coupland, 1979).

Insect Grazers. Pine needles, tunneled out by mining insects, become both food and winter shelter. Needle temperature determines the survival of miners and has been measured under different conditions of solar radiation and wind speed that determine the energy fluxes between needles and environment (Henson and Shepherd, 1952).

Pine bark beetles are less exposed than needle miners, and their activity is directly related to late-winter energy flow, indexed as potential evapotranspiration, and inversely related to summer energy flow (Kalkstein, 1974). The biomass loss of trees girdled in an outbreak in 1973 was equivalent to about 6 mW m^{-2}.

A model for insect holes in *Liriodendron* leaves that includes leaf growth after the attack, determined grazing to be 2.5% of net primary production (Reichle *et al.*, 1973). Over the 153-day growing season the energy flux density of herbivore consumption averaged 12 mW m^{-2}, and the loss of photosynthesizing surface amounted to 40 mW m^{-2} in energy-fixing capacity. Protection of rice in the Philippines raised yield from 5.3 MJ m^{-2} to 9.9 (Smith and Calvert, 1978). Many insect larvae find tree leaves a poorer food source than herbaceous leaves, which are less "apparent" in the landscape (Scriber and Feeny, 1979).

Ants daily cut approximately 108 cm^2 of leaves per square meter of forest area from wet-forest trees in Costa Rica (Lugo *et al.*, 1973) to serve as a medium for culturing fungus. The leaf mass removed daily has an energy flux density equal to 24 mW m^{-2}, and its loss results in a potential decrease of 63 mW m^{-2} or more in photosynthesis, but the total reduction is only a small fraction of the 6 W m^{-2} production of the forest and is, moreover, offset by recycling of phosphorus-rich ash to the forest soil, estimated to increase forest metabolism by 90 mW m^{-2}.

Grazing is not always favorable, however: insect grazing on eucalypts in Australia represents a large and continuous suppression of productivity (Morrow and LaMarche, 1978). In other lands, where most grazers are unable to bypass the tree's chemical defenses, plantation eucalypts often grow rapidly and produce large quantities of fuel.

Larger Animals. Benefits conferred by large grazers include opening of a grass system to light, especially if grazers remove inactive or dead leaves as lemmings do in early summer in Arctic tundra (Miller *et al.*, 1976). Grazing by sheep "improves the light supply to leaves, which would otherwise be starved for lack of light" (Myers, 1969), and can increase growth by 0.3. Grazing keeps the desired balance between legumes and grasses in pasture in northern Australia.

The energy subsidy of a continuously warm environment in the leaf-cutting ant system reduces the need for the insects to seek microclimates that would reduce body heat losses, but large animals that must maintain body temperature at night and through the cold season face an energy problem. Peccaries in Arizona, for example, lose 150 W of heat per square meter of skin surface in winter nights (Zervanos and Hadley, 1973), which is six times the loss in summer; 120 W m^{-2} represents the seasonal cost of night cold, even as reduced by basking on winter days. The food intake, prickly-pear cactus, that supports this heat loss as well as metabolic activities is about 0.27 mW m^{-2}, which is equal to 0.2 of the production of cactus.

White-tailed deer experience colder winters and respond by behavioral changes in level of activity and by exploiting diverse ecosystems: conifer stands for wind shelter, and level terrain and shallow snow cover for grazing. These activity patterns can be expressed in terms of energy exchanges; for example, in conditions of heavy clouds and little wind, a standing 60-kg doe loses approximately $140 - 4T$ watts, in which T is a mean antecedent air temperature (Moen, 1976). Reducing activity patterns from a mix of 25% of the day bedded, 30% standing, 30% foraging, 12% walking, and 3% running, to a pattern of 75% of the day bedded, 10% standing, 10% foraging, and 5% walking in 0.5-m snow cover reduces mean energy costs from 120 W to 85 W and makes possible a reduction in conversion of carbon compounds in food intake (Moen, 1973, p. 358).

To complete the trophic picture, grazing pressure in 20 large-mammal communities [compiled in Farlow (1976)] averaged 14 mW m^{-2}, about the same as the leaf-cutting ants. The secondary productivity of prey animals was 175 μW m^{-2}, and predation pressure was 79 μW m^{-2}. Lions are impressive animals, but grazers and the organisms in the soil convert far more of the net primary production of these grazed ecosystems.

Decomposers

Energy flows at prey–predator trophic levels have flux densities of the order of microwatts per square meter, as we have seen. Since net primary production in most ecosystems is of the order of 1 W m^{-2}, it is clear that other mechanisms are at work in transforming fixed carbon and energy—the decomposers.

The combinations of grazers and decomposers in an ecosystem vary through the year and in different summers. An outbreak of caterpillars in New Hampshire ate 44% of one year's production of foliage (Gosz et al., 1978). Respiration of grasshoppers on Larrea tridentata in successive years was equal to flux densities of 3.1, 7.2, 3.2, 0.8, and 4.2 μW m^{-2} (Mispagel, 1978), a substantial variation typical of herbivore grazing of forest (Olson, 1975).

Decomposition of Litter.   Net primary production by second-growth maple, beech, and yellow birch in the basin of Hubbard Brook (1.90 W m^{-2} over 4 months) was partitioned as follows:

0.49 W m^{-2} to growth   (0.8 above ground, 0.2 below)
1.41 W m^{-2} to grazing and detritus.

Grazing was important only in summers of caterpillar population explosions. The detritus represented litter fall from the trees (1.17 W m^{-2})
and shrubs (0.03), plus 0.03 organic matter washed off leaves by intercepted rainwater (Gosz et al., 1978), plus the death of roots (0.18). Its
decomposition supported insect life that was a major food source for
bird metabolism (2 mW m^{-2}). Part of the transfer into the detritus pool,
60 mW m^{-2}, remained to build up the organic matter of the forest floor
and improve its valuable hydrologic properties, and 20 mW m^{-2} was
carried into the streams and served as an energy source for animal life
in the basin and downstream. Birds and fish were beneficiaries of the
decomposition of the detritus.

Grazing was even less important in a 450-year-old Douglas fir forest
in the Oregon Cascades, near the former Willamette Basin Snow Laboratory, where net primary production (0.6 W m^{-2}) was detritus, part
as litter fall but more than half as dead trees. Decomposition of litter
and logs converted about 0.3 W m^{-2} and soil respiration 0.1 W m^{-2},
leaving a small net ecosystem product "entirely as an accumulation of
woody detritus on the soil surface" (Grier and Logan, 1977).

Most litter in younger ecosystems falls as small fragments, including
half-eaten leaves clipped off by caterpillars (Heinrich, 1980) and twigs
with leaves cut by squirrels in Wisconsin or wild parrots in Australia.
These are rapidly decomposed. Its quantity can be appreciated from the
fact that the ants in the Costa Rican forest allocated three times as much
metabolic energy to clearing this rain of fragments from their paths* as
to carrying leaf fragments into their nest (Lugo et al., 1973). Magnitudes
of litter fall reflect ecosystem leaf production (Table I). Leaf and twig
litter fall in a dry forest in South Australia, at about 0.15 W m^{-2}, was
five times greater than the fall of larger material (Lee and Correll, 1978)
and reached a summer maximum of 0.3 W m^{-2}. Standard deviation in
six summers was 0.1 W m^{-2}. In northeastern Australia, litter fall in both
rainforest and planted Araucaria averaged 0.5 W m^{-2} (Brasell et al., 1980),
substantially more than in the south.

The Soil.   As litter decomposes it is gradually incorporated in the organic fraction of the soil, except in such ecosystems as the wet jungle

---

* The energetic cost of trail maintenance works out to 4 μW m^{-2} of forest floor area and
might help limit extension of the nest system and possible overgrazing.

TABLE I

Annual Litter Production as Leaves and Skeletal Biomass[a,b]

	Leaf litter	Nonleaf litter
Arctic-alpine ecosystems	70	40
Cool midlatitude forests	250	90
Warm midlatitude forests	360	190
Equatorial forests	680	350

[a] Data given in grams per square meter.
[b] Data from Bray and Gorham (1964).

of snags and logs in Oregon. Death of roots supplies perhaps 0.2–0.3 of the total amount, directly within the soil.

Though higher forms of life may serve, as earthworms do, to reduce particle sizes, most of the decomposition in the soil is microbial, exclusively so for cellulose and lignin (Dagley, 1975). The soil of the long-studied grain plots at Rothamsted, England, harbors bacteria in concentration of 1 to 4 billion per gram, which convert six times as much fixed carbon as is removed as human food. "Much of our agricultural effort goes to sustaining a large and varied population of living things in the soil; we get only the by-products of their activity" (Russell, 1961, p. 172). Organic matter in forest soil provides energy for ectomycorrhizal activity and fixing of nitrogen (Harvey et al., 1979).

Conditions for Decomposition.   Decomposition is slow in cold litter and soils, and organic material accumulates to many times the annual detritus production (Table II). Accumulation attains steady state only

TABLE II

Annual Conversion and Accumulation of Litter in Forest Ecosystems[a]

	Annual conversion		Energy equivalent of accumulation (MJ m^{-2})
	MJ m^{-2}	W m^{-2}	
Ghana and other equatorial wet forests	15	0.5	6
Southeastern U.S. pines	6	0.2	30
Minnesota pines	3	0.1	35
Sierra Nevada	3	0.1	90

[a] Data from Olson (1963).

slowly in cold environments and in wetlands, regardless of warmth (Chamie and Richardson, 1978). Decomposition is so fast in warm, moist environments that little stored energy accumulates. Equatorial wet forests annually produce five times as much biomass and organic material in the soil as midlatitude forests, but decomposition is also five times as fast (Sanchez and Buol, 1975).

An index combining temperature and moisture conditions for decomposition is actual evapotranspiration, which requires moisture and moderate to high temperatures (Chapter XV). Inflow–outflow budgeting of water thus defines the environment of decomposition (D)

$$D = 0.166E - 36.3,$$

in which D is the percentage of organic matter decomposed in a year, and E is annual evapotranspiration in kilograms per square meter (Meentemeyer, 1974, p. 59). Where E is 600 mm yr^{-1} (600 kg m^{-2} yr^{-1}), characteristic of a subhumid climate or one with a cold season, 64% of organic matter decomposes. Departures from this relation appear to be associated with the lignin fraction of the litter (Meentemeyer, 1978) because this complex substance is resistant to chemical and biochemical attack (Dagley, 1975). The relation crosses a curve relating leaf production to annual evapotranspiration at about 850 mm yr^{-1}, and equatorial ecosystems with larger evapotranspiration experience rapid turnover of leaf litter and its nutrients as Olson (1963) found.

Temperature and moisture conditions that determine decomposition rates vary with depth in the soil. The below-ground part of the ecosystem in a model of decomposer dynamics in Colorado grassland was stratified into litter and three soil layers, in which energy conversions were 105 in the litter, 40 (0–4 cm), 16 (4–15 cm), and 4 mW m^{-2} (15–60 cm) (Hunt, 1977).

Ecosystem Disturbances

Net primary production in excess of decomposition and grazing in most perennial systems accumulates over the years into a standing crop of biomass and a litter accumulation, which is sometimes abruptly cut short by windfall, fire, or human intervention to harvest a crop. Harvesting of the yields noted in the preceding chapter removes biomass from the ecosystem* which, if annual, must start anew from seed and, if perennial, must regroup and build up its production machinery and

---

* Decomposition does not necessarily end with harvesting: stored food is subject to all kinds of spoilage. Even in good storage conditions grain decomposes at a rate (Smith, 1969, p. 21) that is equivalent to 4 W m^{-3}.

skeletal and storage material. In preparation for these processes careful practices in harvesting need to be followed. Windfall will be discussed in Chapter XIX and fire at the end of this chapter.

ECOSYSTEMS IN STEADY STATE

Natural Ecosystems

Steady-state ecosystems still exist on this teeming earth, especially where they play a protective function or provide on-site consumption of vegetation amenity. For example, natural areas set aside by Wisconsin (Tans, 1974) and the Forest Service (Moir, 1972) serve as monitors for the productivity or health of ecosystems operating in degraded environments. Carbon compounds remain in balance and biological energy is released at the same overall rate as solar energy is fixed; human activities are limited. In national parks "man does not supply or divert significant amounts of materials and energy to or from the park ecosystem, and he is not a significant part of the food chain" (Houston, 1971). Otherwise, disturbance results; for instance, food import and the disposal of 7000 tons of garbage every year in Yellowstone Park affect the spatial distribution of bears. Park policy allows wildfires in the Sierra parks to burn and attacks of herbivorous insects to occur (as in Yosemite), so that episodic processes exist alongside the quieter decomposition processes.

Many cases cited in this book are from research sites in the biosphere reserve program: Coweeta Experimental Forest (N. C.), Hubbard Brook Experimental Forest (N. H.), Andrews Experimental Forest (Oregon), San Dimas Experimental Forest (California), and Pawnee National Grassland (Colorado) (Franklin, 1977). The UNESCO biosphere program includes these and others all over the world, e.g., in the Soviet Union (Gerasimov, 1978), Africa, Australia (Costin and Groves, 1973), and elsewhere.

Alekhin National Park in the chernozem region of European Russia preserves virgin steppe and provides a baseline of information on water and energy processes under natural conditions and various degrees of human impact, such as mowing and grazing (Akad. Nauk, 1967). Energy-budget studies (Rauner *et al.*, 1974) are integrated with hydrologic research; for example, twice as much water needed for biological production infiltrates virgin steppe as infiltrates adjacent stubble fields, and soil loss is one-fiftieth (Chernyshev, 1972, pp. 109, 111). In such natural areas it is possible to see how much ecosystem energy is allotted to maintenance functions that enable systems to respond to external perturbations. Net primary production of the ecosystem is larger than heterotrophic respiration (or decomposition): 0.065 more in the 60-year-old deciduous forest in New England (Gosz et al., 1978) and only 0.008 more in the 450-year Douglas-fir stand in Oregon (Grier and Logan, 1977). Energy conversions at different trophic levels (Lindeman, 1942) are complex, but provide a means of approaching the true operational nature of ecosystems.

Intellectual, esthetic, or atavistic satisfactions are found in native and slightly modified ecosystems by humans coming in small numbers and not bringing or removing energy or materials. Protected ecosystems serve this function better in a steady-state condition than do ecosystems in earlier stages of succession; the complexity of a woodlot with its spring flowers is more satisfying than the neighboring cornfield. In winter there is cross-country skiing in the shelter of trees, and in autumn the enjoyment of color and hunting in the forest. Recreational use of the national forests has increased about 0.1 annually for a long time, and the consumption of forage by big game nearly as fast (Clawson, 1979). Off-site values are also important to the human love of nature, ". . a manifestation of almost religious feeling; most never expect to see the Alaskan wilderness, but they are heartened to realize that it exists and is protected" (Wright, 1974).

Native ecosystems also correct problems generated by abused cultural ecosystems; for example, a floodplain forest in Georgia is calculated to provide groundwater storage, purification of water polluted in cultural systems, and other services worth 19.5¢ m^{-2} yearly. Forest ecosystems on the mountains east of Los Angeles reduce downslope movement of soil that would otherwise require control costs that are equivalent to 68¢ m^{-2} of forest area (Westman, 1977), a function that is threatened by ozone coming out of the city. In contrast, Zurich has maintained a city forest for 5 centuries to protect its water supply, to provide amenity onsite, and to yield fuel and timber off-site (Vaux, 1980).

Grazed Ecosystems

Decomposition in natural ecosystems is forestalled by grazing and browsing by wildlife and domesticated animals. Conversion rates are usually small and the enterprise extensive in scale with a minimum of human labor.

Deer in the upper Great Lakes region exemplify a wild animal population for which vegetation is managed to assure sufficient browse from trees and shrubs, especially in the understory, early in the ecological succession following logging. Management endeavors to provide a mix

Fig. XII.1.   Heat loss can be lethal to these shorn sheep in winter weather on the western slopes of New South Wales.

of ecosystems within a larger landscape, depending on the spatial range of deer seeking forage and winter shelter, the one to provide energy and the other to reduce its loss, considering the carrying capacity of the landscape (Moen, 1973, p. 365).

Much of Australia is occupied by low-intensity ecosystems that have small rates of production, which can be garnered only by an animal that fits the energy and moisture environment. The merino sheep, an Iberian animal that is well insulated, thrifty with water, and mobile, is the means of harvesting "protein from the wasteland"* (Macfarlane, 1968).

Direct manipulation of the ecosystems by fertilization or species introduction is not feasible; "the state and productivity of the whole ecosystem must be controlled by adjustments in the animal factor alone" (Perry, 1969). One period of mismanagement, like the catastrophic overstocking of some areas in the 1890s, degrades vegetation and soil to a degree that is "virtually irreversible under most conditions" (Perry,

* "Wasteland" is not being used pejoratively; these ecosystems supported millennia of successful low-density human occupation. The hunting of kangaroos (highly efficient grazers) and the gathering of small animals and plant products supported a complex social organization and a unique synthesis of culture and religion with the environment.

1969). The ecosystem destruction of the 1890s was relived in the 1970s in the Sahel and other regions of desertification.

The energy balances of the grazers affect their efficiency as converters of biomass energy into wool or meat. Heat losses in wind and rain storms from uninsulated bodies of newly shorn sheep can become lethal (Fig. 1); when fleece thickness is reduced to 5 mm, sheep in a 5 m sec^{-1} wind at 0°C experience sensible-heat loss of -185 W m^{-2}, their skin temperature drops to 13°C, and heart rate accelerates to 160 per minute. Heat production to meet the convective loss is stimulated by the release of adrenalin, "the last line of defence" (Hutchinson and Bennett, 1966), and can continue for several hours until the sheep suddenly breaks down.

The sun-baked shrub lands of the interior present a different problem with surface temperatures above 50°C, air temperatures above 40°C, and solar radiation flux densities exceeding 1000 W m^{-2} (Table III). This

TABLE III

Energy Budget of Pastoral Ecosystem in Interior Australia[a]

A. Radiant-energy loading[b]		
Solar beam absorbed		695
Diffuse solar radiation absorbed		100
Incoming longwave radiation		400
	Total	1195
B. Disposition[c]		
Longwave radiation emitted		-645
Photosynthesis		~ -10
Latent-heat flux (evapotranspiration)		-100
Soil-heat flux		-140
Sensible-heat flux to air		-300
	Total	-1195
	Balance	± 0

[a] Data given in watts per square meter. The environment is midday in summer.

[b] Sun: Direct beam, vertical component 925 W m^{-2}. Sky: Diffuse solar radiation 135 W m^{-2}; longwave radiation 400 W m^{-2}; atmospheric temperature 40°C, vapor pressure 13 mbar, and movement 2 m sec^{-1}.

[c] Lower hemisphere: Soil albedo 0.25, surface temperature 55°C; reflected solar radiation 270 W m^{-2}, emitted longwave 645 W m^{-2}.

energy budget shows that absorbed radiant energy (Chapter VII) totals 1195 W m^{-2} and raises surface temperature to a level at which the same quantity of energy is disposed of through emission of longwave radiation (Chapter IX), photosynthesis (Chapter XI), and other nonradiative fluxes of energy to be discussed in Chapters XV–XVII. The survival strategy is based in isolating the animal from the energy fields of the hostile environment; the merino sheep "withstands summer desert heat by physical rejection of most of the energy falling on it" (Macfarlane,

TABLE IV

Three Energy Budgets of the Merino in Interior Australia[a,b,c]

1. The fleece–air interface		
A. Radiant-energy loading (considering appropriate		
shape factors):		
Direct solar beam		280
Diffuse solar radiation		70
Solar radiation reflected from ground		45
Longwave radiation from air and ground		465
	Total	860
B. Disposition of radiant-energy loading:		
Emission of longwave radiation by fleece		−650
Sensible-heat flux to air from fleece		185
Conduction into the fleece[d]		−25
	Total	−860
2. The skin surface		
Conduction through the fleece[d]		+25
Evaporation from skin		−10
Conducted into the body[d]		−15
	Balance	±0
3. The body core		
Heat conducted from skin[d]		+15
Metabolic heat generated in rumen and tissues		+45
Evaporation from nasal membranes		−60
	Balance	±0

[a] Data given in watts per square meter.

[b] Fleece: albedo 0.4, thickness 40 mm, emissivity 1.0, area 1 m^2, air content 0.9, thermal conductivity 4×10^{-2} W m^{-1} deg^{-1}, thermal admittance 80 J m^{-2} deg^{-1} sec$^{-1/2}$. Body: core temperature 40°C, metabolic heat generation by protozoa, muscular activity, etc. 45 W m^{-2}.

[c] Data from Priestley (1957) and Macfarlane (1964).

[d] Flux carried over from one energy budget to another.

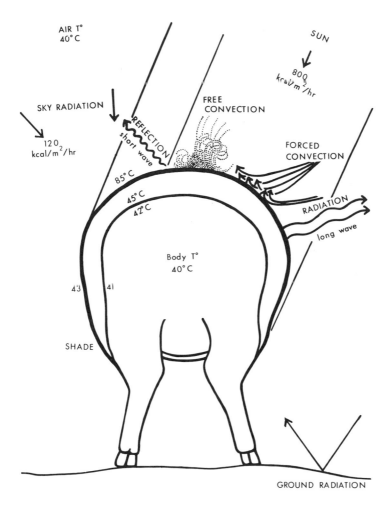

Fig. XII.2. A merino in her environment, a pastoral ecosystem in central Australia. [From Macfarlane (1964). Reproduced with permission from the American Physiological Society.]

1968), accomplished in part by high albedo (about 0.4), but mostly, as noted earlier, by allowing the outer surface of the fleece to attain temperatures exceeding 80°C (Table IV). The weak coupling of this environment to the ewe is shown by casting energy budgets at three sites: the outer surface, the skin, and the body core [Fig. 2; Macfarlane (1971)], which show how the merino can reject enormous heat loading without recourse to much evaporative cooling, which would exact large energy costs for travel to distant watering points.* In contrast to the heat load of 25 W m^{-2} on the skin of sheep, that on the skin of an unclothed human being would be 160 W m^{-2} (Auliciems and Kalma, 1979), an intolerable quantity.

Energy flux densities are small when distributed over the whole ecosystem. In western New South Wales, at a density of 1 sheep per 20 ha (Gruen, 1970), the areal metabolic flux density is 200 μW m^{-2} and in more humid range at 1 sheep per 3 ha (Heathcote, 1975, p. 99) 1.5 mW m^{-2} flux density, to which should be added similarly small energy costs of growth and reproduction and of daily movement between water and forage. The total human energy flux is 5–10 μW m^{-2} in the more intensive systems [calculated from data in Heathcote (1975)], to which imported energy and energy-rich materials might add 5 μW m^{-2}. The product is wool of energy equivalent 50 μW m^{-2} and unique energy insulation value.

Grazing camels, cattle, sheep, and goats of a typical Touareg nomad group consume 7 mW m^{-2} of herbage and produce 0.03 mW m^{-2} of young animals and 0.21 mW m^{-2} of milk and meat (de Felice-Katz, 1980). This food, a larger output than in the Australian case, supports the human components of the system which expends 11 mW m^{-2} labor in milking and caring for the animals.

Silviculture

Silvicultural plans that extend over decades or a century attempt to assess the biomass that can be logged from a forest as sawlogs, pulpwood, or fuel. One approach utilizes a computer model to simulate growth of individual trees in stands of mixed species and uneven ages in the Hubbard Brook basins of the White Mountains (Botkin et al., 1972) at existing suboptimal levels of light and moisture. This population dynamics model is centered on a growth submodel with a temperature dependence, occurrence of soil-moisture stress on some sites, and leaf area to represent collection of solar energy. "Growth" is converted into dimensions of height and trunk diameter. Birth and death also enter the demographic picture. Certain species, like yellow birch, that are overshadowed in the early decades of a simulation, become more dominant later; perseverance in adversity is important, here expressed as "the minimum growth that a tree can sustain without significantly increasing its chance of death" (Botkin et al., 1972).

Careless logging can affect later production. Clear-cutting that allows

---

* From data of Schmidt-Nielsen (1972) I calculate travel cost of a merino as about 30 J m^{-1}. Moen (1973, p. 351) gives costs of other activities.

full sun for regeneration of shade-intolerant species may degrade the site by soil erosion and rapid mobilization of nutrients into soluble forms. An increase in runoff and percolation (because of reduced evaporative removal of soil moisture) might carry off so much N, Ca, K, and other nutrients as to hamper subsequent production (Likens *et al.*, 1978). Small logged tracts on sites with recuperative capacity make for quick entry by successional species that can rapidly cover the soil, start evapotranspiration, and reduce soil temperature and loss of nutrients by too rapid decomposition (Marks and Bormann, 1972). It is necessary to avoid mechanical damage to the forest floor, minimize road building, and leave uncut strips along streams, both for hydrologic benefits and to ensure productivity of succeeding ecosystems.

Self-Sufficient Agriculture

Some food-producing ecosystems in which humans are numerous continue to operate with only small injections of external energy. Traditional hunting and gathering economies are self-sufficient in energy and exhibit a turnover in the microwatt per square meter range. Agricultural systems based on human labor without draft animals illustrate energy costs of directing photosynthetic production into ways that produce human food and still maintain the system. These systems are still to be found although some have a small energy subsidy in the form of tools. Isolation has largely been responsible; hundreds of energy-self-sufficient groups in highland New Guinea, unknown to the rest of the world until the 1920s, retain intensive land and water management systems (Brookfield, 1968; Rappaport, 1971).

Swidden, or crop-forest rotation, in low-latitude agriculture maintains equilibrium by transferring nutrients and energy from the forest years of the rotation period into the cropping years. Manpower studies make it possible to estimate energy flux densities, as was done by Conklin (1957) for a group in the Philippines. The sequence begins in careful selection of a piece of village land to be planted, preparing and burning it, planting it with upland rice and other crops, cultivating and weeding out wild grasses, guarding the rice plants from birds and animals, harvesting the rice, sometimes planting aftermath root and tree crops, and finally returning the land to forest fallow. Direct labor expenditure on the area of the crop itself in all these operations totals approximately 20 mW m^{-2}, and a similar figure was found in a New Guinea group (Rappaport, 1971). To this figure should be added such indirect costs as making tools, loss of stored rice to insects, loss in food preparation, household operation, the support and education of children, overhead costs of the village social organization that directs the whole system, and so on. When this larger number is allocated to the whole area of village lands under the 12-year rotation in the Philippines and added to nutrient recovery, additional food items, and fuel from the forest, the result is on the order of 5 mW m^{-2} over the community area. In more densely settled rural nations (Indonesia, Bangladesh, and Nigeria) per capita energy consumption averages for food 100 W, fuel wood 250 W, crop residue and dung 70 W, for a total of 420 W (Smil, 1979). The energy flux density at a population density of 5 ha^{-1} is 210 mW m^{-2}, with little margin. Nonlocal energy inputs are approximately 50 mW m^{-2}, and further growth of such imports might be minimized by more effective use of animals (Ward et al., 1980).

Energy costs rise in systems requiring human operation of a moisture input because water is heavy and much is needed. Wilken (1977a) measured energy costs of scoop irrigation and pot irrigation, the former lifting water only a fraction of a meter from canals to plants, the latter several meters from shallow wells scattered through a field. The large labor requirement of pot irrigation at a rate of 10–20 kg water per minute,* limits the crop area that can be cultivated by one man to 1000–2000 m². Uncertainty of crop yield is reduced but the margin between food energy grown and human energy consumed is smaller than in some hunting and gathering economies.

A steady-state agriculture prevailed for millennia in Île-de-France, a transformed landscape in use since Neolithic times without losing productivity, and, until a few hundred years ago, without external energy subsidy (Dubos, 1973, 1976). The present mosaic of forest, aquatic, meadow, urban, and field-crop ecosystems "can be maintained only by individualized ecological management" (Dubos, 1973), often by methods long antedating present-day inputs of fossil energy.

ENERGY CONVERSION BY FIRE

Fire in Ecosystems: On-Site Combustion

Where decomposition is slow, its place in maintaining equilibrium may be taken by wildfire, an episodic and dramatic conversion of energy.

Fuel Volume in Ecosystems. The amounts of energy released by fire can be very large, from fuel loadings as great as 100 MJ m^{-2} (5 kg m^{-2}).

---

* Lifting the water and running with it to the plants is equivalent to work of 200–300 W, which, on a unit-area basis of the watered soil, is 200–300 mW m^{-2} during irrigation hours and a quarter as much when averaged over the 24 hours.

Heaviest loadings among North American ecosystems are found in the slash or residue after logging of western conifer stands, which leaves half in large pieces with a 100-hr time lag in drying (due to a small ratio of surface area to volume, 100 m^{-1}), a quarter in smaller pieces of 10-hr lag, and a quarter in fine fuels (Deeming and Brown, 1975). Loadings half as large occur in stagnating stands of conifers like lodgepole pine and are about 16 MJ m^{-2} in hardwood stands.

Fuel in chaparral ecosystems includes 10 MJ m^{-2} of live plants, which dry out during the long summer, and 50 MJ m^{-2} of litter, forming a fuel bed 2 meters or more deep with a large fraction of fine fuels. Dryness that slows decomposition permits litter to accumulate at a rate of 80–100 g m^{-2} each year until episodic consumption occurs in wildfires that seem to "compensate for the lag in decomposition" (Odum and Odum, 1959, p. 374). Fuel models begin with the age of chamise (*Adenostoma fasciculatum*), for instance, the number of years since the last burn, and determine the depth of fuel bed and fuel loading, fraction of dead material, and the size-class distribution of living and dead material. Chamise has many dead branches with a high surface-to-volume ratio, perhaps as an aid to flammability (Biswell, 1974, p. 338), and this is included in the model, along with seasonal changes in moisture and heat content of living fuel, which depend on antecedent weather (Rothermel and Philpot, 1973). The model yields data on rate of fire spread—0.03 m sec^{-1} in 10-year-old chamise under given wind conditions, and 0.9 m sec^{-1} in 40-year-old; it was noted that "there is a fuel age beyond which fires often become uncontrollable." It also calculates reaction intensity or energy flux density—55 kW m^{-2} for 10-year chamise at the most severe part of the season, 660 kW m^{-2} for 40-year-old. Such a model has value in planning fire control, and the rates of predicted heat release determine the rise of the smoke plume and downwind concentrations of smoke (U.S. Forest Service, 1976, pp. 33, 49, 52).

Grass forms a continuous bed of fine fuels of small loading (6 MJ m^{-2}) but rapid drying. Even before all the snow in Wisconsin is melted, grass fires are reported along railroad rights-of-way. Before sap rises, shrub and sapling stands reach low fuel moisture after only a few days of dry, sunny anticyclonic weather. The spring fire season in New England and the Lake States, which is unexpected to a Westerner who associates wildfire with late summer and autumn, may come so early that the soil is too wet to allow off-road use of control equipment. Grassland fires were 15–20 times as numerous as usual in the English drought of 1976 (Doornkamp and Gregory, 1980).

A long period of drying converts large pieces of wood into ready fuel. The anticyclonic summer of 1871 in the Lake States, like that of 1967 in

Idaho, had this effect on slash from logging and land clearing. Enormous volumes of fuel were at hand by October 1871, when strong flow ahead of a trough* fanned and spread escape fires (Haines and Kuehnast, 1970) over large areas (15,000 km^2) of Wisconsin [1300–1500 died in Peshtigo and Door County (Wells, 1968, p. 169)] and Michigan (Manistee), and the city of Chicago.

Terrain Effects. Ecosystem site affects production and decomposition, hence fuel volume, and also fuel moisture and the combustion process. Nocturnal air drainage patterns in northern Idaho determine daily regimes of fuel moisture (Furman, 1978); fires burn uphill more easily than down; the spread rate on a 10° slope is twice as great as on the level (Daubenmire, 1968, p. 214). Fire is less frequent in ecosystems in concave topography in northern Sweden than on convex, and less frequent in north than south slopes (108 year spacing versus 70), due in part to the attractiveness of south slopes for agricultural burning (Zackrisson, 1977). Terrain-associated fire enhances the "diversity and mosaic structure" of this landscape, as it does that of the boreal forest of northeastern Minnesota (Wright, 1974), where a long fire history shows average intervals of 80–100 years.

Combustion Models. Fuel models incorporating changing moisture become inputs for models for spreading (with a slope and wind factor) and rate of energy release. Model outputs include the potential of a fire to crown, i.e., to generate flames high enough to reach the tree crowns, where additional fine fuel and higher wind speeds increase the rate of spread.

As the combustion zone moves through the fuel bed, the following general relation† between line intensity (watts per meter of fire line) and areal intensity (watts per square meter) holds true:

Line intensity (W m^{-1})
= spread (m sec^{-1}) × residence time (sec) × areal intensity (W m^{-2})

Line intensities of 2–3 MW m^{-1} mark "fairly hot" fires, where flames are 3-m high and uncomfortably warm to observers at 10–12 m distance. At a spreading rate of 0.1 m sec^{-1} and a residence time of 3 min, the flux density of energy conversion is 10^5 W m^{-2}, i.e., about two orders

---

* Such flow is often the culmination of increasing dry, hot advection and, in addition, has the kinetic energy to carry sparks and brands. This pattern occurred in major fires in the southeastern United States in 1966 and in Tasmania in 1967 (Brotak, 1980).
† See Roussopoulos and Johnson (1975).

of magnitude greater than the midday flux density of solar energy. Such intensities can be reached even in light fuel on days of extreme fire danger in Australia (Lumbers, 1979).

Grass fires in central Washington exhibit line intensities up to 0.5 MW m^{-1} and spreading rates up to 0.3 m sec^{-1} that could be predicted with high accuracy from wind speed, relative humidity, air temperature, and grass dryness (Sneeuwjagt and Frandsen, 1977).

Line intensities reached 80 MW m^{-1} in the 1967 Sundance fire in Idaho, and overall energy releases peaked three times at 1.52, 3.76, and 5 × 10^8 kW (Anderson, 1968). The areal growth rate was 6700 m^2 sec^{-1} in strong winds (Finklin, 1973, p. 1). Areal growth probably approached its maximum in the Tillamook holocaust in western Oregon: 90,000 acres went in 30 hr on 25–26 August 1933 (Weaver, 1974, pp. 279, 309), a rate equivalent to 8300 m^2 sec^{-1}. The author saw the smoke of this fire 200 km north; Weaver (1974) saw it 250 km south.

Energy budgets of these intense conversions of biomass show that the largest fluxes go into the atmosphere as radiation and convection and that little heat moves into the underground parts of the ecosystem during the short residence time of combustion at any spot. Most soil fauna are spared [e.g., pocket gophers in fire-climax ecosystems in Georgia (Mendez, 1980)]; roots of sprouting species in chaparral ecosystems escape, and within a few years the site is reoccupied. Perennial prairie grasses are little affected by fall burning of the thin above-ground parts that have usually completed their year's work, and perennating buds deeper than a few centimeters usually are not damaged (Daubenmire, 1968, p. 231). Decomposer organisms in the soil of fields where grain straw is burned do not suffer in a single fire although repeated fires reduce the bacteria cohort (Biederbeck *et al.*, 1980). Time-response curves of biomass, habitat, and wildlife after fires in Arizona show a variety of benefits and losses and can be used to evaluate total impact (Lowe *et al.*, 1978).

Prescribed Burning. The violence of wildfires tends to obscure their sometimes beneficial role in landscapes where ecosystems have good survival mechanisms, as chaparral, prairie, eucalyptus, and some pines do. Prescribed burning in the Lake States is done principally to control forest encroachment on ecosystems valuable for wildlife food and habitat, but acreages average only 60 km^2 annually (Donoghue and Johnson, 1975). Hazard-reduction burning is common where litter accumulations

---

* This line intensity is the same as that in a line of oil burners at an airport in a model for fog dissipation (Tag, 1979).

might become hazardous, and aerial incendiary bombs in Australian forests result in line intensity less than 0.4 MW m^{-1} (Kayll, 1974, p. 500).

Fire of both natural and human origin is connected with landscape diversity where ecosystems display contrast in age, species composition, density, height, and therefore in energy budgets (Chapter XXI). Diversity at a human scale affords amenity and other benefits for human activities (Miller, 1978), and prescribed burning produces edge and diversity with high wildlife and recreational values (Kayll, 1974, p. 485).

Grasslands that accumulate large volumes of litter are benefited by fires that remove this "dominating mantle that stifles growth" (Vogl, 1974, p. 158). Crop residues are burned to reduce diseases, but pollution occurs in the weak air circulation of such valleys as the Willamette and Sacramento. Prescribed burning in the southeastern United States maintains the subclimax pine ecosystems and maintains habitat of deer, quail, and turkey (Komarek, 1974, p. 257).

The ancient practice of burning to clear land for cultivation in swidden rotation is a means of managing forest ecosystems, and fire in the right circumstances improves soil fertility more than plowing under biomass of high carbon/nitrogen ratio (Igbozurike, 1974). It causes little erosion in the Nigerian terrain, perhaps because underground biomass is favored at the expense of the above-ground component. Local control practices seem to be adequate; "Most fires are set late in the afternoon or early in the evening, and any that burns on until later is put out by the substantial dewfall of dry-season nights" (Igbozurike, 1974), when weak longwave radiation allows leaves to cool to the dew point. Upslope fires slow or stop at ridges, and the relative ease of clearing slopes may be an important factor in the large population densities of many highlands of low-latitude America (Aschmann, 1955). Fire in prehistory in Mediterranean lands mobilized "nutrients tied up in the highly lignified wood and dead and slowly decomposing litter" (Naveh, 1974, p. 414) and removed phytotoxins, but in later generations the synergistic effect of fire with overgrazing and overcutting (Naveh, 1974, p. 431) caused loss of productive capacity. Permanent loss occurs if the swidden rotation period is shortened under population pressure, reducing control of water and soil erosion (Brookfield, 1968) so that nutrients are "taken out of the ecosystem faster than they can be replaced." The release of carbon from forest and forest soil in the low latitudes brings an unexpected biological dimension into the atmospheric CO_2 problem that is emerging at the present time (Woodwell, 1978; Woodwell et al., 1978; Seiler and Crutzen, 1980).

Fire in Human Systems: Off-Site Combustion

Fire has long been a major tool in human systems. Sauer (1961) feels that early humans in East Africa, in a diversity of microenvironments and in the presence of frequent fire along the volcanic Rift Valley, were among the first to appropriate this energy conversion mechanism.* This "innovation and its elaboration was mainly done by woman, keeper of the hearth and provider of food" (Sauer, 1961), enlarged the number of sources of metabolic energy for humans and overcame the "limitations of environment."

Food Preparation. Fuel wood carried from forests, and crop residues, supply fuel for cooking and space heating in energy equivalents that in India are, respectively, 37 and 23 mW m^{-2} of area of the crop landscape (Revelle, 1976), half again the size of the food harvest from that

Fig. XII.3. The biomass production of forest on the slopes of Malinche in Tlaxcala is en route to lowland villages.

* Fire is a protection from animals at night; such use of wood in West Africa is said to rival the enormous per-capita use of fossil energy by industrial societies.

landscape. Forest litter is carried out of the mountains (Fig. 3) to serve as fuel and fertilizer in the Mediterranean world and the Americas.* Charcoal is a common medium of energy storage (30 MJ kg^{-1}) that reduces the energy costs of distance, 100–200 J kg^{-1} m^{-1} by human carriage, 50–60 by cart (Bronson, 1978).

A household in developing countries burns more than a ton of fuel-wood per year† and may require a quarter of a laborer's efforts because it must be carried from greater and greater distances. The use of cattle dung for fuel increases, with consequent loss of nutrients and organic matter inputs into crop ecosystems (Eckholm, 1975, 1977). Reforestation is difficult under continuous pressure from animals and humans, solar cookers are expensive, and the 1973 price-rise of kerosene eliminated that fuel for many rural people. "The firewood crisis . . . is also forcing governments and analysts back to the basics of man's relationship with the land" (Eckholm, 1977). Deforestation for fuel and stock fodder is pushing Nepal "toward environmental ruin" (Hoffpauir, 1978). Burning of forest for fuel and clearing releases nearly an order of magnitude more energy and CO_2 than wildfires do, worldwide.

Charcoal production and fuel-wood cutting, as well as forest burning to clear land for cultivation, seem to account for the deforestation of northern Morocco after the 7th century (Mikesell, 1960). When forest cutting was followed by overgrazing, regeneration became impossible, and steppe and desert ecosystems took the place of oak and pine. Mayan fuel supply may have required 3 ha of forest per person (Wiseman, 1978).

Industrial Fuel. To meet large demands for fuel, eucalyptus and other plantations were widely planted in lands that had little coal in the years before cheap oil came on the market; remnants of the eucalyptus boom of the 1880s still exist in the Sacramento Valley. Plantations around Addis Ababa provide fuel for that city, and those planted in southern Brazil provided fuel for the railroads. Plantations of coppicing hard-woods, to be harvested on very short rotations (5 years) are under current study in the U.S.

Smelting iron by wood charcoal has an ancient origin, and charcoal shortages in East Africa led, 1500 years ago, to an advanced technology

---

* Wilken (1977b) estimates that this transfer of the forest floor to vegetable plots amounts to about 4 kg m^{-2} on cropped land (2 W m^{-2}). At an annual litter production of 0.4 kg m^{-2}, the neighboring pine–oak stands support cropland at a ratio of ten to one.

† In a town in the Sudan Digernes (1979) found per capita consumption of wood to be 400 W and of charcoal nearly 300 W; the annual mass totals 1.1 tonnes, and self-collection is diminishing.

that made carbon steel (Schmidt and Avery, 1978). Iron furnaces were widely dispersed in Sweden during the 17th and 18th centuries, "each being assigned a territory from which it could draw wood and charcoal" (Alexandersson, 1971). One of the few forested regions not occupied by iron making, inland from Kalmar, was taken over for glass making and is still the center of the art-glass industry.

The energy economy of North America as late as the third quarter of the 19th century (Clawson, 1979) was based on wind-powered ocean and lake vessels and charcoal and wood-fueled* lake boats and railroads. Most railroads in Michigan were wood-burning as late as 1872 (Dunbar, 1969, p. 215). Logging camps and sawmills were powered off the land; small water mills converted wheat to flour, and textiles came from the great hydropower installations of New England. Mill residues supported such energy-intensive industries as salt-boiling.

Power production in some places now uses sawmill and logging residues. Pine plantations in the lower southeast of South Australia supply softwood construction lumber; the residues, converted into electricity at Mt. Gambier, help power the region. Several power stations in Vermont operate on chipped residues from logging and mills, and crop residues in California have this potential (Becker et al., 1978). Residues can meet the respective energy requirements of manufacturing such commodities as lumber and plywood, but fall short of the requirements for such reconstituted products as particleboard. Railroad cars loaded with residue are a common sight in western Oregon. Mill residues now utilized in the wood industry (Hammond, 1977) are equivalent to a flux density of 6 mW m^{-2} over the U.S.

The use of forest biomass for fuel offers certain benefits, which include the small input of fossil fuel needed in forestry compared with crop ecosystems (Alich and Inman, 1975). Wood is a dense fuel and can be stored with little loss for many years—until the tree is cut and some years thereafter. Nutrient demands on the soil are reasonably small; the fuel is clean with low sulfur content (Burwell, 1978). The collectible yield has an energy flux density of 30 mW m^{-2} over the U.S., but estimates have many uncertainties (Burgess, 1978). In a Swedish plan (Lönroth et al., 1980) residues would provide 15% and energy plantations 75% of a total quantity of biomass energy equivalent to 90 mW m^{-2} of the area of the country. About a third would be converted to methanol, largely to power transportation.

---

* Since the forests were not replanted, fuel was available only while farmers were clearing their land. The energy economy depended on one-time consumption of stored photosynthetic energy and replenishable wind and water energy.

Crop and animal residues, more bulky to transport, offer such possibilities as fermentation to ethanol or anaerobic digestion to methane.*
Like the use of wood for a space-heating fuel, this application seems
best suited to on-site uses that reduce the costs of collecting and transporting (Pimentel et al., 1978). Of the potential harvest in the U.S. (Hammond, 1977) (energy equivalent 20 mW m^{-2}) some could be used to
heat dairy-barn water or dry grain in years like 1972 or 1979 when its
field moisture content is too high for it to go into storage. "Perhaps it's
in country areas that fuels made from plants will have the most impact"
in Australia (Anonymous, 1977), and China (Anonymous, 1979), providing on-site production of fuels used by farm machinery. Centralized
production from cane, cassava, or pineapple in dry areas (Marzola and
Bartholomew, 1979) offers yet uncertain energy gains.

The dispersed nature of these conversions of forest or crop-residue
biomass is a consequence of the ubiquitous gentle flow of solar energy
itself. For many purposes this quality is not a disadvantage if fuel production is integrated with fiber or food production (Lipinsky, 1978).
Energy is fixed at a broad scale in the cycling of carbon into ecosystems;
much of the return part of the cycle also may operate at a broad scale.

REFERENCES

Akademiia Nauk SSSR, Institut Geografii (1967). "Geofizika Landshafta" (D. L. Armand,
    ed.). Izd. Nauka, Moscow.
Alexandersson, G. (1971). Regional planning in Sweden: an example of government in-
    fluence in a developed country. In "Government Influence and the Location of Economic
    Activity" (G. J. R. Linge and P. J. Rimmer, eds.), Res. Sch. Pac. Stud., Dept. Hum.
    Geogr., Publ. HG/5, pp. 53–76. Austral. Natl. Univ. Press, Canberra.
Alich, J. A., Jr., and Inman, R. E. (1975). Utilization of plant biomass as an energy feed-
    stock. In "Energy, Agriculture, and Water Management" (W. J. Jewell, ed.), pp. 453–466.
    Ann Arbor Sci. Publ., Ann Arbor, Michigan.
Anderson, J. E. (1968). The Sundance fire. U.S. For. Serv., Res. Pap. INT-56.
Anonymous (1977). Growing fuel—a future option? Ecos (Melbourne) 14, 24–31.
Anonymous (1979). Popularizing use of methane in rural areas. Beijing Rev. 22(29), 5.
Aschmann, H. (1955). Hillside farms, valley ranches. Landscape 1955–1956, Winter, 18–24.
Auliciems, A., and Kalma, J. D. (1979). A climatic classification of human thermal stress
    in Australia. J. Appl. Meteorol. 18, 616–626.
Becker, C. F., Jenkins, B. M., Horsfield, B., and Goss, J. R. (1978). Attitudes of farmers
    toward using crop residues as fuel. Calif. Agric. 32(12), 8–10.

* In order for the methane reaction to go, a process temperature must be maintained
at an energy cost, except in the low latitudes, of about 0.2 of the methane produced. In
contrast, decomposition in the soil occurs in an environment warmed by absorption of
parts of the solar spectrum that are not photosynthetically active.

Biederbeck, V. A., Campbell, C. A., Bowren, K. E., Schnitzer, M., and McIver, R. N. (1980). Effect of burning cereal straw on soil properties and grain yields in Saskatchewan. *Soil Sci. Soc. Am. J.* **44**, 103–111.

Biswell, H. H. (1974). Effects of fire on chaparral. *In* "Fire and Ecosystems" (T. T. Kozlowski and C. E. Ahlgren, eds.), pp. 321–361. Academic Press, New York.

Botkin, D. B., Janak, J. F., and Wallis J. R. (1972). Rationale, limitations, and assumptions of a northeastern forest growth simulator. *IBM J. Res. Dev.* **16**, 101–116.

Brasell, H. M., Unwin, G. L., and Stocker, G. C. (1980). The quantity, temporal distribution and mineral-element content of litterfall in two forest types at two sites in tropical Australia. *J. Ecol.* **68**, 123–139.

Bray, J. R., and Gorham, E. (1964). Litter production in forests of the world. *Adv. Ecol. Res.* **2**, 101–157.

Bronson, B. (1978). Angkor, Anuradhapura, Prambanan, Tikal: Maya subsistence in an Asian perspective. *In* "Pre-Hispanic Maya Agriculture" (P. D. Harris and B. L. Turner, II, eds.), pp. 255–300. Univ. of New Mexico Press, Albuquerque.

Brookfield, H. C. (1968). New directions in the study of agricultural systems in tropical areas. *In* "Evolution and Environment" (E. T. Drake, ed.), pp. 413–439. Yale Univ. Press, New Haven, Connecticut.

Brotak, E. A. (1980). A comparison of the meteorological conditions associated with a major wildland fire in the United States and a major bush fire in Australia. *J. Appl. Meteorol.* **19**, 474–476.

Burgess, R. L. (1978). Potential of forest fuels for producing electrical energy. *J. For.* **76**, 154–157.

Burwell, C. C. (1978). Solar biomass energy: an overview of U.S. potential. *Science* **199**, 1041–1048.

Chamie, J. P. M., and Richardson, C. J. (1978). Decomposition in northern wetlands. *In* "Freshwater Wetlands" (R. E. Good, D. F. Whigham, R. L. Simpson, and G. J. Crawford, Jr., eds.), pp. 115–130. Academic Press, New York.

Chernyshev, E. P. (1972). Struktura vodnogo balansa i protessy erozii v usloviiakh razlichnykh landshaftov lesostepi. *In* "Biogeograficheskoe i Landshaftnoe Izuchenie Lesostepi" (D. L. Armand, ed.), pp. 107–121. Izd. Nauka, Moscow.

Clawson, M. (1979). Forests in the long sweep of American history. *Science* **204**, 1168–1174.

Conklin, H. C. (1957). Hanunoo agriculture: A report of an integral system of shifting cultivation in the Philippines. *FAO For. Dev. Pap.* No. 12.

Costin, A. B., and Groves, R. H. (1973). "Nature Conservation in the Pacific." Austral. Natl. Univ. Press, Canberra.

Coupland, R. T. (1979). Conclusion. *In* "Grassland Ecosystems of the World: Analysis of Grasslands and Their Uses" (R. T. Coupland, ed.), International Biological Programme, Synthesis Series No. 18, pp. 335–355, Cambridge Univ. Press, London and New York.

Dagley, S. (1975). Microbial degradation of organic compounds in the biosphere. *Am. Sci.* **63**, 681–689.

Daubenmire, R. (1968). Ecology of fire in grasslands. *Adv. Ecol. Res.* **5**, 209–266.

Deeming, J. E., and Brown, J. K. (1975). Fuel models in the national fire-danger rating system. *J. For.* **73**, 347–350.

de Felice-Katz, J. (1980). Analyse éco-énergétique d'un élevage nomade (Touareg) au Niger, dans la région de l'Azawak. *Ann. Géogr.* **89**, 57–73.

Digernes, T. H. (1979). Fuelwood crisis causing unfortunate land use—and the other way around. *Norsk geogr. Tidsskr.* **33**, 23–32.

Donoghue, L. R., and Johnson, V. J. (1975). Prescribed burning in the north central states. *U.S. For. Serv., Res. Pap.* **NC-111.**

Doornkamp, J. C., and Gregory, K. J. (1980). The Great Drought recorded. *Geogr. Mag.* **52,** 297–301.

Dubos, R. J. (1973). Humanizing the earth. *Science* **179,** 769–772.

Dubos, R. J. (1976). Symbiosis between the earth and humankind. *Science* **193,** 459–462.

Dunbar, W. F. (1969). "All Aboard! A History of Railroads in Michigan." Eerdmans, Grand Rapids, Michigan.

Eckholm, E. P. (1975). Deterioration of mountain environments. *Science* **189,** 764–780.

Eckholm, E. P. (1977). The other energy crisis: firewood. *Focus (New York)* **27,** 9–16.

Farlow, J. O. (1976). A consideration of the trophic dynamics of a late Cretaceous large-dinosaur community (Oldman formation). *Ecology* **57,** 841–857.

Finklin, A. I. (1973). Meteorological factors in the Sundance fire run. *U.S. For. Serv., Gen. Tech. Rep.* **INT-6.**

Franklin, J. F. (1977). The biosphere reserve program in the United States. *Science* **195,** 262–267.

Furman, R. W. (1978). Wildfire zones on a mountain ridge. *Ann. Assoc. Am. Geogr.* **68,** 89–94.

Gerasimov, I. P. (1978). Biosfernye stantsii-zapovedniki, ikh zadachi i programma deiatel'nosti. *Izv. Akad. Nauk SSSR, Ser. Geogr.* No. 2, 5–17.

Gosz, J. R., Holmes, R. T., Likens, G. E., and Bormann, F. H. (1978). The flow of energy in a forest ecosystem. *Sci. Am.* **238,** 93–102.

Grier, C. C., and Logan, R. S. (1977). Old-growth *Pseudotsuga menziesii* communities of a western Oregon watershed: Biomass distribution and production budgets. *Ecol. Monogr.* **47,** 373–400.

Gruen, F. A. G. (1970). Grassland production. *In* "Australian Grasslands" (R. M. Moore, ed.), pp. 401–411. Austral. Natl. Univ. Press, Canberra.

Haines, D. A., and Kuehnast, E. L. (1970). When the Midwest burned. *Weatherwise* **23,** 112–119.

Hammond, A. L. (1977). Photosynthetic solar energy: Rediscovering biomass fuels. *Science* **197,** 745–746.

Harvey, A. E., Jurgensen, M. F., and Larsen, M. J. (1979). Role of Forest Fuels in the Biology and Management of Soil. *U.S. For. Serv., Gen. Tech. Rep.* **INT-65.**

Heathcote, R. L. (1975). "Australia." Longmans, London.

Heinrich, B. (1980). Artful diners. *Nat. Hist.* **89**(6), 42–51.

Henson, W. R., and Shepherd, R. F. (1952). The effects of radiation on the habitat temperatures of the lodgepole needle miner, *Recurvaria milleri* Busck (Gelechiidae: Lepidoptera). *Can. J. Zool.* **30,** 144–153.

Hoffpauir, R. (1978). Deforestation in the Nepal Himalaya: A village perspective. *Assoc. Pac. Coast Geogr. Yearb.* **40,** 79–89.

Houston, D. B. (1971). Ecosystems of National Parks. *Science* **172,** 648–651.

Hunt, H. W. (1977). A simulation model for decomposition in grasslands. *Ecology* **58,** 469–484.

Hutchinson, J. C. D., and Bennett, J. W. (1966). The effect of cold on sheep. *Wool Technol. Sheep Breed.* **9,** 11–16.

Igbozurike, M. U. (1974). Vegetation burning and forest reservation in a segment of the forest-savanna mosaic of Nigeria: a preliminary investigation. *Landscape Plann. (Amsterdam)* **1,** 81–103.

Kalkstein, L. S. (1974). The effect of climate upon outbreaks of the southern pine beetle. *Publ. Climatol.* **27**(3), 1–65.

Kayll, A. J. (1974). Use of fire in land management. *In* "Fire and Ecosystems" (T. T. Kozlowski and C. E. Ahlgren, eds.), pp. 483–511. Academic Press, New York.

Komarek, E. V. (1974). Effects of fire on temperate forests and related ecosystems: Southeastern United States. In "Fire and Ecosystems" (T. T. Kozlowski and C. E. Ahlgren, eds.), pp. 251–277. Academic Press, New York.

Lee, K. E., and Correll, R. L. (1978). Litter fall and its relationship to nutrient cycling in a South Australian dry sclerophyll forest. Austral. J. Ecol. 3, 243–252.

Likens, G. E., Bormann, F. H., Pierce, R. S., and Reiners, W. A. (1978). Recovery of a deforested ecosystem. Science 199, 492–496.

Lindeman, R. L. (1942). The trophic-dynamic aspect of ecology. Ecology 23, 399–418.

Lipinsky, E. S. (1978). Fuels from biomass: integration with food and materials systems. Science 199, 644–651.

Lönroth, M., Johanssen, T. B., and Steen, P. (1980). Sweden beyond oil: Nuclear commitments and solar options. Science 208, 557–563.

Lowe, P. O., Ffolliott, P. F., Dieterich, J. H., and Patton, D. R. (1978). Determining potential wildlife benefits from wildfire in Arizona ponderosa pine forests. U.S. For. Serv., Gen. Tech. Rep. RM-52.

Lugo, A., Farnworth, E. G., Pool, D., Jerez, P., and Kaufman, G. (1973). The impact of the leaf cutter ant Atta colombica on the energy flow of a tropical wet forest. Ecology 54, 1292–1301.

Lumbers, J. (1979). Our changing bushfire scene. Rural Res. (Melbourne), 105, 8–10.

Macfarlane, W. V. (1964). Terrestrial animals in dry heat: Ungulates. In "Handbook of Physiology—Environment" (D. B. Dill, E. F. Adolph, and C. G. Wilber, eds.), Vol. 4, pp. 509–539. Am. Physiol. Soc., Washington, D.C.

Macfarlane, W. V. (1968). Weather, climate and domestic animals. WMO Agric. Meteorol. Symp., 1966; Bur. Meteorol., (Melbourne) 1, 119–157.

Macfarlane, W. V. (1971). Personal communication.

Marks, P. L., and Bormann, F. H. (1972). Revegetation following forest cutting: Mechanisms for return to steady-state nutrient cycling. Science 176, 914–915.

Marzola, D. L., and Bartholomew, D. P. (1979). Photosynthetic pathway and biomass energy production. Science 205, 555–559.

Meentemeyer, V. (1974). Climatic water budget approach to forest problems. I. The prediction of forest fire hazard through moisture budgeting. II. The prediction of regional differences in decomposition rate of organic debris. Publ. Climatol. 27(1), 1–74.

Meentemeyer, V. (1978). Macroclimate and lignin control of litter decomposition rates. Ecology 59, 465–472.

Mendez, R. A. (1980). The pocket gopher mound probe. Nat. Hist. 89(6), 36–41.

Mikesell, M. W. (1960). Deforestation in Northern Morocco. Science 132, 441–448.

Miller, D. H. (1978). The factor of scale: Ecosystem, landscape mosaic, and region. In "Sourcebook of the Environment" (K. A. Hammond, G. Macinko, and W. B. Fairchild, eds.), pp. 63–88. Univ. of Chicago Press, Chicago, Illinois.

Miller, P. C., Stoner, W. A., and Tieszen, L. L. (1976). A model of stand photosynthesis for the wet meadow tundra at Barrow, Alaska. Ecology 57, 411–430.

Mispagel, M. E. (1978). The ecology and bioenergetics of the acridid grasshopper, Bootettix punctatis on creosotebush, Larrea Tridentata, in the northern Mojave desert. Ecology 59, 779–788.

Moen, A. N. (1973). "Wildlife Ecology: An Analytical Approach." Freeman, San Francisco, California.

Moen, A. N. (1976). Energy conservation by white-tailed deer in the winter. Ecology 57, 192–198.

Moir, W. H. (1972). Natural areas. Science 177, 396–400.

Morrow, P. A., and LaMarche, V. C., Jr. (1978). Tree ring evidence for chronic insect suppression of productivity in subalpine *Eucalyptus*. *Science* **201**, 1244–1246.

Myers, L. F. (1969). The effect of the grazing animal on plant growth—through grazing. In "Intensive Utilization of Pastures" (B. J. F. James, ed.), pp. 26–29. Angus & Robertson, Sydney, Australia.

Naveh, Z. (1974). Effects of fire in the Mediterranean region. In "Fire and Ecosystems" (T. T. Kozlowski and C. E. Ahlgren, eds.), pp. 401–434. Academic Press, New York.

Odum, E. P., and Odum, H. T. (1959). "Fundamentals of Ecology." 2d ed. Saunders, Philadelphia, Pennsylvania.

Olson, J. S. (1963). Energy storage and the balance of producers and decomposers in ecological systems. *Ecology* **44**, 322–331.

Olson, J. S. (1975). Productivity of forest ecosystems. In "Productivity of World Ecosystems," pp. 33–43. Natl. Acad. Sci., Washington, D.C.

Perry, R. A. (1969). Rangelands research and extension. In "Arid Lands of Australia" (R. O. Slayter and R. A. Perry, eds.), pp. 291–302. Austral. Natl. Univ. Press, Canberra.

Petrusewicz, K., and Grodzinski, W. L. (1975). The role of herbivore consumers in various ecosystems. In "Productivity of World Ecosystems," pp. 64–70. Natl. Acad. Sci., Washington, D.C.

Pimentel, D., Nafus, D., Vergara, W., Papaj, D., Jaconetta, L., Wulfe, M., Olsvig, L., Frech, K., Loye, M., and Mendoza, E. (1978). Biological solar energy conversion and U.S. energy policy. *BioScience* **28**, 376–382.

Priestley, C. H. B. (1957). The heat balance of sheep standing in the sun. *Austral. J. Agric. Res.* **8**, 271–280.

Rappaport, R. A. (1971). The flow of energy in an agricultural society. *Sci. Am.* **224**, 116–122, 127–132.

Rauner, Iu. L., Ananeva, L. M., and Samarina, N. N. (1974). Radiatsionno-teplovoi rezhim lugovo-stepnykh fitotsenozov na razlichnykh elementakh rel'efa. In "Issledovaniia Genezisa Klimata". Akad. Nauk SSSR, Inst Geografii, Moscow.

Reichle, D. E., Goldstein, R. A., Van Hook, R. I., Jr., and Dodson, G. J. (1973). Analysis of insect consumption in a forest canopy. *Ecology* **54**, 1076–1084.

Revelle, R. (1976). Energy use in rural India. *Science* **192**, 969–975.

Rothermel, R. C., and Philpot, C. W. (1973). Predicting changes in chaparral flammability. *J. For.* **71**, 640–643.

Roussopoulos, P. J., and Johnson, V. J. (1975). Help in making fuel management decisions. *U.S. For. Serv., Res. Pap.* **NC-112**.

Russell, Sir E. J. (1961). "The World of the Soil." Collins, London.

Sanchez, P. A., and Buol, S. W. (1975). Soils of the tropics and the world food crisis. *Science* **188**, 598–603.

Sauer, C. O. (1961). Fire and early man. *Paideuma, Mitt. zur Kulturkunde* **7**, 399–407.

Schmidt, P., and Avery, D. H. (1978). Complex iron smelting and prehistoric culture in Tanzania. *Science* **201**, 1085–1089.

Schmidt-Nielsen, K. (1972). Locomotion: Energy cost of swimming, flying, and running. *Science* **177**, 222–228.

Scriber, J. M., and Feeny, P. (1979). Growth of herbivorous caterpillars in relation to feeding specialization and to the growth form of their food plants. *Ecology* **60**, 829–850.

Seiler, W., and Crutzen, P. J. (1980). Estimates of gross and net fluxes of carbon between the biosphere and the atmosphere from biomass burning. *Clim. Change* **2**, 207–247.

Smil, V. (1979). Energy flows in the developing world. *Am. Sci.* **67**, 522–531.

Smith, C. V. (1969). Meteorology and grain storage. *WMO Tech. Note* No. 101.

Smith, R. F., and Calvert, D. J. (1978). Insect pest losses and the dimensions of the world food problem. *In* "World Food, Pest Losses, and the Environment" (D. Pimentel, ed.), AAAS Symposium, No. 13, pp. 17–38. Westview Press, Boulder, Colorado.

Sneeuwjagt, R. J. and Frandsen, W. H. (1977). Behavior of experimental grass fires vs. predictions based on Rothermel's fire model. *Can. J. For. Res.* **7**, 357–367.

Tag, P. M. (1979). A numerical simulation of fog dissipation using passive burner lines. Part I: Model development and comparison with observations. *J. Appl. Meteorol.* **18**, 1442–1454.

Tans, W. (1974). Priority ranking of biotic natural areas. *Mich. Bot.* **13**, 31–39.

U.S. Forest Service, Southern Forest Fire Laboratory and Southeastern Forest Experiment Station (1976). Southern forestry smoke management guidebook (H. E. Mobley, senior compiler). *U.S. For. Serv., Gen. Tech. Rep.* **SE-10**.

Vaux, H. J. (1980). Urban forestry/Bridge to the profession's future. *J. For.* **78**, 260–262.

Vogl, R. J. (1974). Effects of fire on grasslands. *In* "Fire and Ecosystems" (T. T. Kozlowski and C. E. Ahlgren, eds.), pp. 139–194. Academic Press, New York.

Ward, G. M., Sutherland, T. M., and Sutherland, J. M. (1980). Animals as an energy source in Third World agriculture. *Science* **208**, 570–574.

Weaver, H. (1974). Effects of fire in temperate forests: Western United States. *In* "Fire and Ecosystems" (T. T. Kozlowski and C. E. Ahlgren, eds.), pp. 279–319. Academic Press, New York.

Wells, R. W. (1968). "Fire at Peshtigo." Prentice-Hall, Englewood Cliffs, New Jersey.

Westman, W. E. (1977). How much are nature's services worth? *Science* **197**, 960–964.

Wilken, G. C. (1977a). Manual irrigation in Middle America. *Agric. Water Manage.* **1**, 155–165.

Wilken, G. C. (1977b). Integrating forest and small-scale farm systems in Middle America. *Agro-Ecosystems* **3**, 291–302.

Wiseman, F. M. (1978). Agricultural and historical ecology of the Maya lowlands. *In* "Pre-Hispanic Maya Agriculture" (P. D. Harrison and B. L. Turner, II, eds.), pp. 63–115. Univ. of New Mexico Press, Albuquerque.

Woodwell, G. M. (1978). The carbon dioxide question. *Sci. Am.* **238**, 34–43.

Woodwell, G. M., Whittaker, R. H., Reiners, W. A., Likens, G. E., Delwiche, C. C., and Botkin, D. B. (1978). The biota and the world carbon budget. *Science* **199**, 141–146.

Wright, H. E., Jr. (1974). Landscape development, forest fires, and wilderness management. *Science* **186**, 487–495.

Zackrisson, O. (1977). Influence of forest fires on the North Swedish boreal forest. *Oikos* **29**, 22–32.

Zervanos, S. M., and Hadley, N. F. (1973). Adaptational biology and energy relationships of the collared peccary (*Tayassu tajacu*). *Ecology* **54**, 759–774.

Chapter XIII

BROAD-SCALE TRANSFORMATIONS OF FOSSIL ENERGY

The preceding chapter described energy conversions of biomass produced more or less currently by fixing solar energy. This chapter discusses their control and support by fossil energy, in which the biomass produced over a million years is consumed in one year—an enormous concentration of energy transformations in time. We will consider extensive systems, mostly agricultural but also including rural circulation systems outside urban nodes.

Nonfossil-energy conversions of forest and crop residues into fuel, feed, and food crops for animal and human activity, and passive utilization of solar energy in well-designed homes, together with hydropower and windmills, sustained the economies and societies of the past. The first transition from this energy economy in North America began with European settlement, which was founded on exploitation of accumulated biomass. New England sold white pines for ship masts and converted whole forests to potash during its first clearing for farming. Midwestern forests powered lake shipping and railroads. The forests around Puget Sound were logged and "agriculture was relegated to the background" (U.S. Writers' Program, 1950, p. 64) in the 1850s; water-powered and steam-powered saw mills provided lumber for the frequent rebuildings of San Francisco after fire. Farms in the Southeast exploited the accumulated organic matter in the topsoil, often to its disappearance in one generation. Biomass that had accumulated over a hundred years or more was converted into working capital of the invading economy.

The second transition, which occurred from the mid-19th century in North America, drew upon biomass that had been stored much longer. Anthracite coal replaced wood and charcoal, coal replaced New England hydropower and wood in midwestern industries and cities, and oil replaced horsepower on midwestern farms. The time scale of accumulation jumped to millions of years instead of a century and represented the liquidation in a few human lifetimes of energy fixed over millions of years.

FOSSIL ENERGY IN THE MANAGEMENT OF EXTENSIVELY EXPLOITED ECOSYSTEMS

Wilderness and Wildlands

By definition energy is not exported from or imported into true wilderness. The food a backpacker at activity levels of 150 W every square kilometer carries in amounts to 0.1 mW m^{-2}, a flux density three orders of magnitude smaller than even the low rates of biomass production in these generally infertile, dry, or cold terrains.

Harvesting game animals usually involves some mechanization. The Eskimos of the east coast of Baffin Island (Foote, 1967) make the summer hunt of the ringed seal from a boat driven by an outboard motor in order to cover the maximum area of sea during the period of light. The average area observed is 15 km^2 at an energy cost of 1–2 J m^{-2} plus the energy cost of the motor and guns and human effort. A household hunts 20 to 30 times over a summer at a fuel cost of 40 J m^{-2}, and a human cost of 15 over the whole season to capture a food yield of 30 J m^{-2} plus lamp oil for space heating and dog food for winter hunting. Flux densities of fuel input and food yield are both about 3 μW m^{-2}. Net primary production of these waters is, of course, much larger since seals occupy the third or fourth trophic level above phytoplankton. Fishing can be similarly analyzed (Foote and Greer-Wootten, 1968).

Rangeland

Rangelands of small productive capacity because of shortages of energy (Arctic tundra) or moisture (interior Australia) are managed chiefly by manipulating the numbers of grazing animals that harvest primary production. The small cash flow of the operation purchases only small inputs of fossil energy. Rangeland of greater production repays more expenditures for management and harvesting. Grazing animals are controlled by fencing, water points, salt licks, mechanized roundup for inspection and medication, and sometimes cultivation of the soil–plant resource. The energy expenditure, largely fuel for pickup trucks, is about 0.8 MJ for each MJ of animal tissue growth (Ward et al., 1977). Making hay on flood-irrigated valley floors in summer increases the fossil-energy input, as does sowing or other attempts to upgrade species composition.

Range in Utah (360 mm annual rainfall) (Pimentel *et al.*, 1975) is grazed by sheep at a rate of 4 mW m^{-2}, a small yield that has to be supplemented by feed in order to produce a lamb yield of 0.38 mW m^{-2}. Fossil-energy costs of range management (0.29 mW m^{-2}) and those associated with growing and bringing in the supplementary feed (0.06 mW m^{-2}) total about the same as the energy content of the lamb output. This small secondary product can buy only a small fossil-energy subsidy.

Forest

Managing forest often requires control of cattle, sheep, or deer, reduction of fire hazard, perhaps prescribed burning, and thinning (Fig. 1). The greatest fossil-energy input comes at harvest time: building or opening roads, felling trees, bucking trunks into logs, hauling them to the mill, burning or scattering the slash or residue, and doing whatever else is felt to be necessary to secure regeneration for continued biomass production. Many of these operations are mechanized since logs are heavy and dangerous to handle, and energy inputs are large: 5 mW m^{-2}

Fig. XIII.1. Fossil energy moves water from lakes of roadless western Quebec to suppress fires, thus serving a gate function in fostering biological energy conversions.

in Texas forests (LePori and Coble, 1975). Helicopter logging fuel costs 0.2 of the energy content of the logs (Dykstra et al., 1978).

Forest stands harvested more frequently, as for pulpwood, generally lie in flatter terrain and consist of smaller stems or coppice sprouts, which can be harvested by a mowing process. Proposed *Populus* tree farms in Wisconsin, which in a 10-year growth period accumulate 322 MJ m^{-2} of biomass, would have incurred an energy cost of 24 mW m^{-2} for irrigation, 38 for fertilizer, and 23 for operations (Zavitkovski, 1979); the harvest is chipped and dried at a further cost of 150 mW m^{-2}. Yield is 1.02 W m^{-2}, somewhat larger than Fege et al. (1979) computed for a 6-year cropping. In some stands the fertilizer cost might be reduced by interplanting of a nitrogen-fixing species like red alder, which also increases production (Bergstrom, 1979). Foresters learned long ago in German spruce plantations that the humus capital of such intensively cropped sites could be depleted, and the soils scientist Hans Jenny (1980) warns us of this possibility in biomass cropping. Transportation costs remain a barrier to high-volume use of wood as an energy source in the Rockies (Sampson, 1979), just as in methanol or other liquid-fuel conversion from biomass, but forest biomass may have advantages over grain as a feedstock.

FOSSIL-ENERGY CONVERSIONS IN FARM OPERATIONS

Except in a few isolated areas, American farming has never been subsistence-oriented but has produced cash crops for off-farm sale, often at great distances, like the pork of the Ohio Valley in the 1830s and the wheat of California in the 1870s. With this cash flow fossil-energy inputs were feasible and began earliest in cultivation and harvesting.

Mechanical Power

Land preparation and management is back-breaking work, and new machines were introduced all through the 19th century; each one represented an input to the farm of urban labor and industrial energy that increasingly came from coal. These machines were horse-drawn* for several decades, and the inputs of fossil energy to agriculture consisted largely of that incorporated in the steel used. External energy could hardly have exceeded 1 to 2 mW m^{-2} of crop land by the turn of the century, far less than 1% of net primary production. The only petroleum

---

* Or were powered by crop residues, e.g., straw that fueled threshing machines (Anonymous, 1978).

products used on a Kansas farm in the mid-1920s were kerosene for light and gasoline for the car and an occasionally used silage shredder. Draft animals on farms in 1915 supplied energy (Heichel, 1973) equivalent to about 4 mW m^{-2}, and human labor was about 0.2 mW m^{-2} as the farmer worked side by side with his team.* Much of the farm production went to the grazers and other primary consumers—horses, cows, hogs. After 1910, 30 × 10^4 km^2 of land in the U.S. that had been committed to producing feed for horses (urban as well as rural animals) went into other production (Landsberg, 1964, p. 150).

Energy inputs via machinery include the energy content of the metals and repairs and maintenance: 0.3 mW m^{-2} in 1940, 1.4 mW m^{-2} in 1970 (C. E. Steinhart and Steinhart, 1974) over the entire U.S., and higher for maize ecosystems alone: 6 mW m^{-2} in 1945, up to 14 mW m^{-2} in 1970 (Pimentel et al., 1973). The work of tractors can be described as the density of power they make available to do mechanical work, and this potential on farms in the U.S. in the mid-1960s (Giles, 1967) was 75 mW m^{-2},† quite a lot more than mean use year-round.

Direct costs of fuel alone, most of which powers mechanical work on crop fields, were 25 on Illinois maize for grain, 34 on Iowa maize for silage, 19 on Missouri alfalfa, and 13 mW m^{-2} on Minnesota oats (Heichel, 1973). These rates are 7–8 times the rates of power applied by horses 55 years earlier and are to be added to machinery investment (1.4 mW m^{-2}) and to human labor (0.04 mW m^{-2}).

Mass Budgets of Crop Ecosystems

Ecosystems that receive outside material inputs that contain fossil energy are receiving an energy subsidy.

Fertilizers. Plant nutrients in the past were generally cycled within the crop ecosystem, or the farm, as manure was returned to the fields. Nitrogen likely to be lost by leaching or removal of harvested material was made good by local fixing; alfalfa rotated with maize in the Midwest, for example. Clover is sown with grass in New Zealand pastures, in a highly productive system that receives a fossil-energy input of only 8 mW m^{-2} (Pearson and Corbet, 1976), of which nitrogen is only 0.4 mW m^{-2}. Curry (1962) felt that the competitors of New Zealand farmers

---

* For comparison, draft animals in India provide 9 mW m^{-2} and humans 13 (Revelle, 1976).

† In more intensive agriculture in Japan the potential tractor power density is 154 mW m^{-2}, mostly from garden tractors sized to paddy cultivation of rice. Mean power is supplemented by 7 mW m^{-2} of animal and 11 mW m^{-2} of human power.

"cannot possibly afford to pour as much nitrates into the soil as the clover components of New Zealand pastures do".

These practices were gradually displaced in the United States by application of chemical fertilizers after World War II; Pimentel *et al.* (1973) show that the use of nitrogen fertilizers (which comprise three-quarters of the energy content of almost all fertilizers applied) on maize increased 16 times in 25 years; and represent a substantial part of the entire fossil-energy subsidy to maize and meat agriculture, as they do to mechanized farms in developing countries. In fact, nitrogen use serves "as a marker for all technology" (Nelson and Dale, 1978) in separating technological from weather effects in yield models. Nitrogen was applied in 1972 to 98% of the maize ecosystems in Wisconsin (Wisconsin Statistical Reporting Service, 1973) at $17–20$ mW m^{-2}, plus the costs of transportation* and spreading.

Less fertilizer is used on other crops of the dairy system. Whole-farm averages come down to approximately 10 mW m^{-2} in Wisconsin and 1 mW m^{-2} or less on Amish farms (Johnson *et al.*, 1977). Substitution of animal manures for purchased fertilizer seems to be most feasible within 5 km (Heichel, 1976), with only a small increase in fuel use. Preventing damage to ground water and aquatic ecosystems in lakes and streams, 5 times as much nutrient loading as from cities (U.S. Council on Environmental Quality, 1978, p. 119), should also be included in the energy cost of fertilizers (Evans, 1980). Accelerating biological fixing of atmospheric nitrogen (Evans and Barber, 1977), although an energy cost to the plant, would reduce the fossil-energy input, and research is active in this direction (Wharton, 1980).

Biocides. Insecticides and herbicides are used at relatively small energy flux densities; those on maize ecosystems cost about 0.7 mW m^{-2} plus the cost of spreading. Externalities follow these information-bearing compounds, as they do after fertilizers, and downstream costs should properly be charged to the crop ecosystem receiving them. The problems caused by off-site movement of DDT spread so far that it would be difficult to calculate their energy effects in such remote ecosystems as Lake Michigan.

Organic farming, in which commercial biocides and nitrogen fertilizers are replaced by biological means of cycling nutrients and controlling insects and disease may yield less biomass, but reduces fossil-fuel inputs by up to a half (Carter, 1980). Depending on the price of inputs,

* The costs of transport became clear in spring of 1973 when fertilizer shortages through the Midwest were caused by interruption of barge traffic by the Mississippi River flood and problems in allocating trucks to haul it from barge terminals.

these farms show net profits similar to those of conventional farms (Lockeretz *et al.*, 1978). Organic dairy farms in Australia used fertilizer less and machines more and had the same ratio of milk output to fossil input as conventional farms (Dornom and Tribe, 1976).

Irrigation. Maize ecosystems that are irrigated require 30–50 mW m^{-2}, about the same as the fertilizer and machine-operation expenditures. The crop becomes more vulnerable to shocks from the outside world: increases in the price of gas for the pumps (High Plains), shortages of water (California in 1977), exhaustion of ground-water aquifers (southern Great Plains), increased lift caused by overpumping (central Arizona). In the Gila basin each added meter of lift requires about 20 × 10^6 kWh electric energy (U.S. Water Resources Council, 1978, p. 27); water shortages in the California drought of 1976–1977, on the other hand, resulted in a substantial decrease in the use of electricity (Ritschard and Tsao, 1980), which was also in short supply. It appears that shortages in either energy or water tend to bring about more realistic evaluations of our resources.

Maize irrigated from ground water at 90 m depth requires 7 kJ per mm water on one square meter of field (Pimentel *et al.*, 1975). To meet summer evapotranspiration of 5 mm per day, the energy input is 35 kJ m^{-2}, or 0.4 W m^{-2}, a figure that approaches the rate at which solar energy is being fixed by the ecosystem. Surface water may be nearly as costly: alfalfa in San Diego County irrigated with water pumped over the mountains from the Colorado River involves an energy subsidy of about 0.2 W m^{-2} (Knutson *et al.*, 1977).

Use of sprinkler irrigation (Fig. 2) saves on costs of land preparation and leveling and saves water; however, the pressure needed in the pipes has an energy cost of about 1 kJ per kilogram of water conveyed to the crop, which might come from solar energy (Matlin and Katzman, 1979). This mass of water is 1 mm in depth over a square-meter unit area, and its pressurizing cost of 1 kJ has to be added to the 7 kJ mentioned earlier for a lift of 90 m, which is the average lift in the 80,000 gas-powered wells on the high plains of western Texas (LePori and Coble, 1975). This rate continues 2700 hr per year. At the low natural-gas prices of the past, about 1 cent per ton of water, this subsidy of 100–150 mW m^{-2} plus indirect costs of well casing and pumps and other farm operations could be financed by crops of only moderate returns, like sorghum and wheat. A rise in energy prices is a more immediate threat to this economy than the certain exhaustion of the aquifers; farmers shift to crops using less water, cut back on acreage, or revert to dryland farming (U.S. Department of Agriculture, 1977). "Abandonment of large acreages of ir-

Fig. XIII.2. This energy-intensive sprinkler irrigation of corn on hilly land not otherwise irrigable near Wild Rose, Wisconsin, saves water but requires more energy than gravity irrigation.

rigated areas began recently when the required high pumping lifts combined with sharp increases in prices for natural gas used to run the pumps . . ." (U.S. Water Resources Council, 1978, p. 20). Proposals to pump water from the Mississippi River to the High Plains (more than 1 km lift) (Graves, 1971, p. 78) would multiply this energy-cost problem ten times.

Great Plains irrigation in Colorado and Nebraska, which also taps a limited source of water, experiences transient fuel shortages, during which irrigation systems (increasing in the early 1970s at a rate of 2000 per year) compete for fuel with wheat combines. A rainstorm is as welcome as it was during dry-farming days because it means the irrigation machinery, operating at a cost of 300–400 mW m^{-2} (Splinter, 1976), can be turned off for a few days. Fossil energy used for irrigation is about 0.4 of the metabolizable energy in alfalfa, pasture, and silage and 0.6 of that in maize grain (Ward *et al.*, 1977). Annual energy costs, including capital costs, of center-pivot irrigation run about 150 mW m^{-2} as compared with 90 for surface irrigation (Batty *et al.*, 1975). However, flood irrigation may leach nitrogen fertilizers equal in energy cost to the cost of pressurizing water for sprinkler application (Rawlins and Raats, 1975) and should also include the energy costs of dams, canals, and the corrective works necessary to control salting and waterlogging that almost without exception follow a large irrigation project (Manners, 1978).

Frost Protection

The direct application of thermal energy to crop ecosystems is uncommon but of long standing in some high-value crops, particularly winter vegetables and fruit orchards. Orchard heaters are sometimes paired with, and sometimes replaced by, direct mechanical energy generated by wind machines that force warmer air from aloft into a near-surface, cold, radiating layer.

The direct energy flux of these conversions in California, where they are most used, is about half the total fossil-energy input to the area in fruits and nuts, and yearly flux density averages 180–200 mW m^{-2} over the area protected. Flux densities during operation reach 600–700 W m^{-2} (Bagdonas et al., 1978, p. 100), double the natural intake of energy. In one budget analysis (Crawford, 1964) the fossil-energy input of 225 W m^{-2} was found to go in equal parts to meet the deficit in longwave radiation exchange, to heat inflowing cold air from upslope, and to support thermal convection above the orchard (chimney effect). Only a few percent enters the soil (Fritton and Martsolf, 1980). Total input of fossil energy far exceeds the transitory shortfall in downward longwave radiation that creates the frost hazard, and wind machines are equally inefficient in energy terms (Bagdonas et al., 1978).

FOSSIL ENERGY IN ECOSYSTEMS AND MOSAICS

Individual Crops

There are substantial differences between the fossil-energy inputs used in major crops. Table I sets out the mean energy inputs into several crops along with biomass yields. Wheat fields are not frequently worked between planting and combining, both of which are efficient operations on the plains, and wheat receives less fertilizer than many other crops; its moderately low input is about half its yield. Soybeans receive more input, but have not responded well to efforts to force higher yields. Rice grown in the U.S. receives heavy inputs, including aerial sowing and fertilization.

TABLE I

Fossil Energy Inputs to Types of Crops[a]

Crop	Inputs	Yields
Wheat, U.S.	50	99
Wheat–sheep, Australia	5	19
Rice, U.S.	216	278
Rice, Philippines	8	80
Soybeans, U.S.	70	100
Maize for grain, U.S.	88	237
Maize, Mexico	0.7	90
Sugar cane for alcohol, Brazil	55[b]	290

[a] Data from Pimentel et al. (1975), Gomes da Silva et al. (1978), Handreck and Martin (1976), and Pimentel and Pimentel (1980). Data given in milliwatts per square meter.

[b] A further input, at least 140 mW m^{-2}, is necessary in converting cane to alcohol.

Conversion of biomass to other fuels to replace those derived from fossil deposits appears to present many advantages (Hall, 1980).* Diverse ecosystems adapted to many environments of the earth are, as we saw in Chapter XI, "well designed for the collection and storage of solar energy" (Boardman, 1980), and also produce hydrocarbons and other special compounds now derived from petroleum (Coffey and Halloran, 1979). A survey in Australia (Stewart et al., 1979) shows that some biomass conversions are not at present competitive with those from coal, which of course is more concentrated to begin with, and entail possible loss of erosion-control benefits (Watson, 1980; Stewart et al., 1980). A critical point, considering the objective of generating liquid fuels, hinges on the quantities of fossil liquid fuels needed in the fermentation, distillation or other conversion processes, and in field production, especially irrigation. Any permanent arrangement depends on developing a close interaction of technological and agronomic processes (Weisz and Marshall, 1979), on careful system design (Sachs, 1980), and on the selection of the best crops or residues in the light of their alternative values.

* One attraction is an alleviation of concern about rising atmospheric CO_2 if we cease to dig up and burn fossil carbon. Assessment of this factor depends on the degree to which ecosystems can increase their fixing and storing of atmospheric CO_2, which is a matter for argument. One critical factor is the stimulation of ecosystem production by increased CO_2 concentration, a relation that turns out to be more important in ecosystems under moisture stress than in well-watered ones (Gifford, 1979).

Maize Ecosystems

Maize, an ancient American crop, has become a case study in agricultural history and environmental energy budgets. Table II shows changes over a critical quarter-century of cheap fuel. Fossil-energy inputs of fuel in 1945, when draft animals had largely been replaced, increased as machines grew still more numerous and much larger. The big change was in application of fertilizer, especially nitrogen with its high energy density (77 MJ kg^{-1}). Insecticides came into use in the 1950s and herbicides in the 1960s. The use of electric power increased by an order of magnitude when farm electrification came to power all kinds of on-site machines, like milking machines and milk coolers on dairy farms. Energy used to dry grain artificially increased to a rate equal to the energy content of phosphorus and potassium fertilizers. Human on-site labor shrank to 160 μW m^{-2}, half what it was in 1945, but more off-site labor was imported into the ecosystem in the form of machinery.

Minimum tillage methods reduce fuel needs but increase biocides, for a net reduction from 95 mW m^{-2} by conventional methods to 90 mW m^{-2} (Heichel, 1976). Fertilization by animal manures (within 5 km) reduces the energy input of purchased fertilizers but increases machine use, reducing net input to 80 mW m^{-2}. The practice of irrigating maize has grown further since 1970 and has intensified mean energy inputs far beyond the 1 mW m^{-2} average given in Table II.

In a cornfield we see an ecosystem that receives 600–800 W m^{-2} of radiant energy in midsummer, reflects a part, radiates away 400 or so, helps to warm and moisten the regional atmosphere (that responds in severe storms and even tornadoes), and still fixes approximately 1 W

TABLE II

Energy Inputs into Maize Production in 1945 and 1970[a]

	1945	1970
Machinery, transportation and fuel	25.0	42.0
Fertilizer	2.5	35.0
Seeds and biocides	1.3	3.0
Irrigation	0.6	1.0
Electricity	1.3	10.0
Drying	0.3	4.0
Human labor	0.3	0.16
	31.3	95.0
Maize yield as grain	121.0	270.0

[a] Data from Pimentel et al. (1973). Data given in milliwatts per square meter.

m^{-2} in maize kernels. The carbon-fixing mechanism is steered and supported by the summertime import of 300 mW m^{-2} energy equivalent in machinery, fertilizer, and fuels; this expenditure is guided by on-site human activity at rates of 150–200 μW m^{-2}. The range of energy transactions extends over six orders of magnitude.

Vulnerability. Coming on top of earlier increases, the tripling of the fossil-energy inflow between 1945 and 1970 increased the dependence of ecosystem managers. Fluctuations in inflow cause crises in farm operations, as noted earlier with regard to irrigation. For example, the fall of 1972, although prior to the oil embargo, was a time of dislocation in fuel distribution systems; late summer was wet, September was cool and cloudy, and the moisture content of maize in Wisconsin in mid-October was 0.325 (Wisconsin Statistical Reporting Service, 1972), much too high for it to go into storage. Driers were operating at full capacity around the clock* until fuel began to run out, and propane supplies for winter house-heating in rural Wisconsin were seen to be threatened. Frantic efforts brought in enough fuel for the approaching winter. A fortunate outcome was that the state government was alerted to the lack of information about the complex networks by which fuels are brought into and distributed through the state and began to grapple with the situation, a concern that alleviated fuel crises in the following spring and during the oil embargo the next fall.

A Mosaic of Ecosystems

A student who has learned that each crop has an optimum set of environmental conditions and expects to find only cotton in one place, only wheat in another, and so on, is baffled by the real heterogeneous landscapes that are seemingly not dominated by any one ecosystem. What accounts for this diversity? Sometimes there are obvious topographic reasons, but often a deterministic explanation fails. A mix of crops may be seen to extend over fairly homogenous terrain (excepting woodlots and tree strips on land obviously too rough or wet to plow), and in much of Wisconsin this consists of maize, oats, and hay, in proportions that vary according to markets, soil, climate, availability of feed, and other factors. In the 1950s the mix was 0.40 maize, 0.45 hay, and 0.15 oats, but differed locally (Weaver, 1954). It has changed to 0.49, 0.39, and 0.12, respectively (Wisconsin Agricultural Reporting Service,

* The quantity of energy can be estimated on the basis of 0.03 excess moisture in 5 × 10^9 kg of grain corn in Wisconsin: to evaporate 150 × 10^6 kg of water requires 375 × 10^6 MJ plus losses. If actual fuel needs were 600 × 10^6 MJ, they are equivalent over a month to 230 MW, and over the crop area of 8000 km^2 the flux density is 30 mW m^{-2}.

TABLE III

Fossil-Energy Inputs and Yields of Ecosystems of the North American Dairying Landscape[a,b]

	Maize for grain	Oats	Maize for silage	Hay
Inputs				
Wisconsin, Ozaukee Co.	55	35	58	35
New York State	16	17	27	17 (20[c])
U.S.	88	40	73	37
Yields				
Wisconsin, Ozaukee Co.	590	320	850	850
U.S.	237	98	319	151

[a] Inputs and yields of woodlots that are commonly integrated into dairying systems are not shown. Power calculations are made on a whole-year basis. Individual growing seasons differ, being quite short for oats and longer and in high-energy months for corn; alfalfa is perennial. Data given in milliwatts per square meter.

[b] Wisconsin data from Kobriger and Stearns (1972), New York State data (fuel use only) from Price et al. (1975), U.S. data from Pimentel et al. (1975).

[c] Silage hay.

1980), as direct fertilizer displaced nitrogen fixing by legumes in a crop rotation. Summer energy inputs, 130 mW m^{-2} in the corn–oats–alfalfa rotation, increase to 210 mW m^{-2} in continuous corn (Heichel, 1978). Soil loss has exceeded tolerable levels (Brink et al., 1977) where the maize component reaches 0.64.

Energy inputs in the component crop ecosystem (Table III) are of the order of 30 mW m^{-2} to alfalfa and oats and 50–60 mW m^{-2} to maize. Crop yields are of the order of 300–800 mW m^{-2}, but little of this goes to feed human beings; it passes through the grazing trophic level before becoming food (Fig. 3).

Dairying. These three ecosystems are integrated by the milk cow into a rather self-contained unit that produces an ancient product—milk.* At a higher trophic level than that of ecosystem primary production,

* Milk production also comes from mosaics that are almost entirely composed of grass ecosystems, as it does in the milder winter of New Zealand. This milk–butter–cheese economy accepts a slump in animal production during the winter decrease in plant production (Curry, 1962) and has no need for expensive barn storage of summer energy for winter feeding. Lewthwaite (1964) points out the paradox of a moderately even flow of product from a Wisconsin landscape that experiences a seasonal variation in natural energy input, and a variable product flow from a landscape that experiences less natural variation.

milk is produced in energy quantities about an order of magnitude smaller than crop yields (which are presented in Table IV). The Pennsylvania and Wisconsin cases show further comparisons in farming style: Amish farmers abjure certain kinds of dependence on the outer world (particularly the use of tractors) and are frugal in their use of consumer goods.

Electric power used for milking machines and cooling makes dairy farms vulnerable to interruptions in this flow of energy. California farms have "a low level of immunity" from such disruption (Fairbank *et al.*, 1978), and Wisconsin farms suffered large losses in the ice storm of 1976, not only in loss of milk, but still more from mastitis in unmilked cows.

Intensification of energy inputs in southwestern Wisconsin led to concentration of effort on soils that are best for moisture storage. Heavy fertilizing removed one soil factor as a potential limit to yield, but without irrigation the hydrologic characteristics became more limiting; "yield increases have been acquired at the cost of increased energy expenditures *and* requirements for specific site attributes" (Auclair, 1976).

Fig. XIII.3. A farmstead is an energy-concentrating and energy-converting node where biomass from a large producing area is stored as hay and grain to maintain a steady level of conversion by animals into milk and meat. The crib is still full as the new growing season commences; central Wisconsin in May.

TABLE IV

Overall Fossil-Energy Inputs and Yields of Dairying
Systems[a]

Location	Fossil inputs	Yields
New York State[b]		
Whole farm (105 ha)	21	19
Cropped area (49 ha)	45	41
Pennsylvania[c]		
Old Order Amish	35	42
"Nebraska" Amish	15	23
English	74	41
Southwestern Wisconsin[c]		
Amish	11	17
English (small)	81	22
Gippsland, Australia[d]	31	53

[a] Data given in milliwatts per square meter.
[b] Data from Williams et al. (1975) (40-cow, one-family farm). Direct energy inputs only.
[c] Data from Johnson et al. (1977).
[d] Data from Dornom and Tribe (1976).

The energy content of milk as the major off-farm yield is further concentrated in many dairying regions when the milk is made into cheese. The true output of a mosaic of crop ecosystems is focused, by way of several score milking barns, on one factory, from which come the principal products, cheese and whey,* of the mosaic.

Process temperatures are maintained in the cheese factory by fossil fuel that warms the vats and in the cow barns by metabolic energy, even in cold weather. Wash water for milking is heated by fossil energy to 90–95°C, but solar energy can provide economic preheating to 60°C (Currier and Westwood, 1976; Thompson and Clary, 1977). Process warmth of the photosynthesizing apparatus in the fields is derived from natural radiant energy.

Aside from process heat, the field ecosystems receive large external inputs and, indeed, could not yield heavily without them (Evans, 1980). While they do little to increase the rate of photosynthetic production, they enable plant breeders to select varieties that allocate more of the converted solar energy to harvestable components. They are more than gate or directive inputs of energy, but rather, in his phrase, "supportive

* Whey was formerly often dumped into streams, where its high nutrient value made it an unwelcome source of energy. Now it is more likely to be recycled into animal feed.

inputs." While such inputs in the past have come from human or animal labor, at present they come from fossil energy and in some cases have overshot the purpose Evans describes. They bring externalities in the form of sediment or pollutant outputs; crop ecosystems are the greatest polluters of water bodies in the United States (U.S. Council on Environmental Quality, 1978, pp. 119, 129). Rising costs of fossil energy in the U.S. that make fuels and fertilizer 0.2–0.3 more expensive each year may mean that the rates of energy conversion described here are near their maximum flux densities. Societal problems may, in fact, make developing countries cautious in wholesale application of fossil energy, as in China (Evans, 1980) and Papua New Guinea (Wood-Bradley et al., 1980).

ENERGY IN CIRCULATION SYSTEMS

Materials Flows

The economic and energetic integration of maize, oats, and alfalfa ecosystems into an operating farm system is made possible by physical integration of material flows. Green chop is carried from the alfalfa field to the cow barn for daily consumption; harvested hay and corn are collected at the farmstead and after processing (grain drying, silage shredding) are put into storage in corn cribs, silos, and hay barns. Hay baled and left in the field is eventually brought to the cows. Manure is removed from the central energy-converting node, the barn. All this transportation of bulky material takes energy, as illustrated by the following data for a New York State dairy farm (Williams et al., 1975):

trucking associated with crop systems requires 0.8 mW m^{-2} of farm area,
manure handling requires 1.6 mW m^{-2},
truck work around the farmstead requires 0.8 mW m^{-2},
total energy requirement is 3.2 mW m^{-2}.

Off-farm traffic includes the linkage of milking barn to cheese factory or bottling plant which must be made every day or every other day. Dairy regions usually have a dense net of all-weather roads, which however, permit long hauls (averaging 110 km in Wisconsin) and excessive overlap of hauling routes (Cropp and Graf, 1977). The cost of transport in southern Australia is about 6% of total farm energy use, i.e., 1.7 mW m^{-2} of farm area (Dornom and Tribe, 1976).

On the national scale, assuming that half the truck transport of food (J. S. Steinhart and Steinhart, 1974) is intercity, we compute a flux density of 2–2.5 mW m^{-2} for agricultural supplies (fuel, fertilizer) and products (wheat, feed grains*). To this should be added at least 0.5 mW m^{-2} expended in rail and water carriage of agricultural commodities. The food industry, responsible for 12% of the national energy budget, takes an even greater share of the conversion of fossil energy to operate intercity circulation systems. This is a result of the extensive, dispersed nature of agricultural production, which, in turn, is a result of the extensive, dispersed nature of the flux of solar energy. Fossil energy compensates, to a degree, for the small solar flux density, as well as for differences between the spatial distributions of the producing ecosystems and the consuming feed animals and humans in the U.S. and abroad.

Intercity freight traffic of all commodities is compared in fossil-energy flux densities in Table V.

Road traffic demands 0.6 of the fossil energy flux, but carries only 0.22 of the freight. Rail contributes 0.22 of the energy cost and carries 0.40 of the load—the long unit-trains of coal, the heavy pellet and ore trains, the trains of lumber climbing over the mountains from western Oregon, the cars of paper and machinery and chemicals shipped by industrial centers.

Rail freight, as a standard, has an energy cost of 0.44 J m^{-1} kg^{-1}, somewhat more than for barges and pipelines and much less than for trucks. A shipment that uses several modes incurs transfer costs; getting steel pipe from the Midwest to a geothermal energy project in California works out at 2.9 J kg^{-1} m^{-1} overall [data from Gilliland (1975)]. The highway mesh over the country consumes fossil energy at a flux density of 5 mW m^{-2} of the country's area.

Energy Moves Energy

If the freight is rich in energy, say 17 MJ kg^{-1}, as is true of much biomass, then the quotient of 17×10^6 and the kilogram meter cost is the number of meters the energy content of the substance can carry it if completely and frictionlessly converted. Such quotients are so large as to make plain the inefficiency of actual transport methods. A pipeline bringing natural gas into Wisconsin uses about 0.1 of its gas to run the

---

* Insofar as fed off the farm where they are grown. The discussion of dairying systems showed that farms that sell animal products rather than primary biomass production must devote energy to their internal circulation systems. For instance, few of the millions of tons of hay grown in this country appear on public highways. When they do, it usually indicates the supplying of an intensified form of energy conversion, like the dry-feed dairy lots of Los Angeles.

TABLE V

Freight[a] Traffic in the United States by Mode[b]

Mode	Energy flux density (mW m^{-2})	Traffic flux density (kg m sec^{-1} × 10^9)	Unit energy cost (J m^{-1} kg^{-1})
Road	5.1	25.0	2.4
Rail	1.9	43.0	0.44[c]
Water	0.6	17.0	0.35
Pipelines	0.7	24.0	0.30
Air	0.5	0.2	28.5
Total	8.8	109.2	Mean 0.8

[a] The energy costs of moving water, though not included in these figures, are substantial, as described earlier under irrigation.
[b] Data from C. E. Steinhart and Steinhart (1974, pp. 222–224.)
[c] The decrease in rail freight cost in Germany (on a slightly different basis) was from 1.6 in 1965 to 0.6 in 1976 (Bauermeister, 1979).

compressors that move the gas a few hundred kilometers. Other estimates are 0.04 per 100 km for natural gas, 0.05 for coal per 100 km by rail. Loss of 7 MJ in moving 235 MJ worth of coal from mine to power plant (C. E. Steinhart and Steinhart, 1974, p. 209) works out, for a typical distance of 300 km, to an energy cost of 23 J m^{-1} or 10^{-7} m^{-1}, that is, a ten-millionth of its initial size per meter traveled. The converted electric energy is transmitted at an energy cost of 70 J m^{-1}, equal to 0.8 × 10^{-6} m^{-1}, but does not include the energy equivalent of outages in this weather-exposed system. Buried systems encounter problems of soil-heat flux, i.e., "the inability of the earth to absorb heat that is produced by the transmission of power" at 230 kV (Metz, 1972). Cryoresistive transmission at very low temperatures generates little heat, but needs refrigeration to remove heat leaking from the soil into the cable (Metz, 1972).

Energy in thermal form is costly to move; atmospheric heat advection of low-latitude air over snow-covered land seldom survives distances greater than a few hundred kilometers (Lamb, 1955). Transmission of hot water or steam for space heating from a plant outside the district to be heated is possible to 30 or 40 km, but at a cost increase of 35–40% more than from a local heating plant (Karkheck et al., 1977). The transport of food as a form of energy was discussed earlier.

The conversion of energy in transit occasionally reaches catastrophic intensities. A truck with 36 tons of pressurized natural gas skidded on a Mexican highway in 1978, turned over and burned, igniting a large area. The conversion was so fast that buses and cars following could not avoid the wreck and were destroyed; if we assume 300 sec, the rate of energy conversion over an area of 1 ha gives a flux density of 250 kW m^{-2}, which exceeds the energy release in a big forest fire. "The ground shook like an earthquake," said a witness who escaped. When the energy costs of dozens of dead and of restoring 150 burned people are considered, this mode of transport becomes very costly. A few days earlier a similar event in Spain killed several hundred people. Tankers of energy-dense materials are involved in several dozen accidents every year. Propane that burned in a Canadian wreck created thermal convection that removed half the chlorine in another car, a fortunate but not a predictable outcome. The vulnerability of electrical systems has already been noted.

Transportation of People

The energy costs of carrying people over the wide expanses of the United States (Table VI) are larger than those for freight and flux density is double that of freight (Table V), but vulnerability of this far-flung system is not prevented by its large energy expenditures; in fact, these

TABLE VI

Intercity Passenger Traffic in the United States[a]

Mode	Energy flux density $(mW\ m^{-2})$	Energy cost $J\ m^{-1}$ per passenger[b]	$J\ m^{-1}\ kg^{-1}$
Highway			
car	13.3	2410	40
bus	0.1	780	13
Rail	0.03	600	10
Air	4.3	7710	128
	17.7	Mean 2830	Mean 47[c]

[a] Data from C. E. Steinhart and Steinhart (1974, p. 223).

[b] By somewhat different definitions of long-distance travel, values per passenger-meter are cited as 2200 J by car, 670 J by bus, and 6700 J by air (Schipper and Lichtenberg, 1976).

[c] For comparison it will be recalled that the energy cost of moving upward against gravity is 9.8 J m^{-1} kg^{-1}. Costs of animal locomotion are 3.4 J m^{-1} kg^{-1} for a 40-kg sheep and 0.8 J m^{-1} kg^{-1} for a 600-kg horse (Schmidt-Nielsen, 1972).

expenditures suggest attempts to travel at times when the road–vehicle system cannot operate. Storms of 1977–1978 in Illinois injured 2000 people and killed 62, mostly on the roads (car accidents, people frozen or suffocated in cars trapped in snowdrifts) (Changnon and Changnon, 1978). Losses would be very large in energy terms and, when added to the energy costs of roads and vehicles, make a large total. This may be reduced by at least 0.1 by telecommunications (Kumar *et al.*, 1978) in a spread-out country like Canada or the United States in which the per capita cost of automobile travel is 6 times that of more compact Japan (Doernberg, 1978).

Quantities of fossil energy used in operating the areally extensive systems that circulate materials, energy, and humans over the countryside are comparable with the quantities applied to crop ecosystems. The circulation systems remain exposed to natural processes and suffer frequent disruption in bad weather. Moreover, since they are largely dependent on liquid fuels, they are highly vulnerable; even if there is no long-term energy shortage, liquid fuels in particular present "an awesome transitional problem" (Fri, 1978). They are limited in quantity and can be produced from coal only by methods that "will be constrained by a scarcity of freshwater" (Harte and El-Gasseir, 1978). Some indications [see, e.g., Wall Street Journal (1980), Marshall (1980)] are that this conversion of fossil energy to movement, whether purposeful or merely restless motion sensation, might not increase much above the 15–20 mW m^{-2} in Table VI.

Fossil energy, formerly inexpensive, has come to be applied in large quantities "to manipulate and manage the environment" (Pimentel and Pimentel, 1978), specifically in the mechanization, chemicalization, and irrigation of crop ecosystems and in powering extensive circulation systems that in part transport ecosystem production.

Quantities of fossil energy applied to agriculture vary from country to country. They are small (8–10 mW m^{-2}) in New Zealand's clover–grass ecosystems; in Australia, where fertilizer and irrigation inputs are small; and in India, where they merely supplement bullock labor and dung. The quantity in Canada is not much greater (16 mW m^{-2}) (Downing, 1975), but in the U.S. the number jumps to 42 mW m^{-2}, somewhat greater than biomass output, and to 76 mW m^{-2} in the United Kingdom (Gifford, 1976). Israel and the Netherlands (150–180 mW m^{-2}) represent cases of heavy irrigation and glasshouse culture, respectively. Large in terms of biological energy conversion, this application of biological product from past millennia seems to be "probably the most important factor responsible for the rapid growth of the world population" (Pimentel and Pimentel, 1978), which is the primary problem we all face.

REFERENCES

Anonymous (1978). Threshing day was a lively time. *Wis. Then Now* **25**(1), 2, 3, 7.

Auclair, A. N. (1976). Ecological factors in the development of intensive-management ecosystems in the midwestern United States. *Ecology* **57**, 431–444.

Bagdonas, A., Georg, J. C., and Gerber, J. F. (1978). Techniques of frost prediction and methods of frost and cold protection. *WMO Tech. Note* No. **157**.

Batty, J. C., Hamad, S. N., and Keller, J. (1975). Energy inputs to irrigation. *J. Irr. Div. Proc. Am. Soc. Civil Eng.* **101** No. IR4, 293–307.

Bauermeister, K. (1979). Consumption of energy in transport. *Rail Internat.* **10**, 893–906.

Bergstrom, D. (1979). Let's harness the energy of red alder. *U. S. For. Serv., Forestry Res-West,* **August,** 1–4.

Boardman, N. K. (1980). Energy from the biological conversion of solar energy. *Phil. Trans. R. Soc. London A.* **295**, 477–489.

Brink, R. A., Densmore, J. W., and Hill, G. A. (1977). Soil deterioration and the growing world demand for food. *Science* **197**, 625–630.

Carter, L. J. (1980). Organic farming becomes "legitimate." *Science* **209**, 254–256.

Changnon, S. A., Jr., and Changnon, D. (1978). Winter storms and the record-breaking winter in Illinois. *Weatherwise* **31**, 218–225.

Coffey, S. G., and Halloran, G. M. (1979). Higher plants as possible sources of petroleum substitutes. *Search* **10**, 423–428.

Crawford, T. V. (1964). Computing the heating requirements for frost protection. *J. Appl. Meteorol.* **3**, 750–760.

Cropp, R. A., and Graf, T. F. (1977). "Economic Analysis of Farm to Milk Plant Hauling," Res. Rep. Coll. Agric. Life Sci., Univ. of Wisconsin, Madison.

Currier, J. W. R., and Westwood, D. C. (1976). Solar energy in the farm dairy. *Search* **7**, 434–435.

Curry, L. (1962). The climatic resources of intensive grassland farming: The Waikato, New Zealand. *Geogr. Rev.* **52**, 174–194.

Doernberg, A. (1978). Energy use in Japan and the United States. In "International Comparisons of Energy Consumption" (J. Dunkerly, ed.), Res. Rep. R-10, pp. 56–81, Resources for the Future, Washington, D.C.

Dornom, H., and Tribe, D. E. (1976). Energetics of dairying in Gippsland. *Search* **7**, 431–433.

Downing, C. G. E. (1975). Energy and agricultural biomass production utilization in Canada. In "Energy, Agriculture and Waste Management" (W. J. Jewell, ed.), pp. 261–269. Ann Arbor Sci. Publ., Ann Arbor, Michigan.

Dykstra, D. P., Aulerich, D. E., and Henshaw, J. R. (1978). Prebunching to reduce helicopter logging costs. *J. For.* **76**, 362–364.

Evans, H. J., and Barber, L. E. (1977). Biological nitrogen fixation for food and fiber production. *Science* **197**, 332–339.

Evans, L. T. (1980). The natural history of crop yield. *Am. Sci.* **68**, 388–397.

Fairbank, W. D., Eide, R. N., Gurtle, G. G., and Etchegaray, H. S. (1978). Energy supplies for milking parlors. *Calif. Agric.* **32**(7), 10–11.

Fege, A. S., Inman, R. E., and Salo, D. J. (1979). Energy farms for the future. *J. For.* **77**, 358–361.

Foote, D. C. (1967). In "Baffin Island—East Coast. An Area Economic Survey" Rep. No. 66/4. Can. Dep. Indian Aff. North. Dev., Ottawa.

Foote, D. C., and Greer-Wootten, B. (1968). An approach to systems analysis in cultural geography. *Prof. Geog.* **20**, 86–91.

Fri, R. W. (1978). Energy imperatives and the environment. *In* "Resources for an Uncertain Future" (C. H. Hitch, ed.) pp. 43–58. Johns Hopkins Press, Resour. Future, Baltimore, Maryland.

Fritton, D. D., and Martsolf, J. D. (1980). Soil heat under an orchard heater. *Soil Sci. Soc. Am. J.* **44**, 13–16.

Gillord, K. M. (1976). An overview of fuel used for crops and national agricultural systems. *Search* **7**, 412–417.

Gifford, R. M. (1979). Carbon dioxide and plant growth under water and light stress: Implications for balancing the global carbon budget. *Search* **10**, 316–318.

Giles, G. W. (1967). Agricultural power and equipment. *In* "The World Food Problem," Vol. 3, pp. 175–216. U.S. Pres. Sci. Advis. Comm., Washington, D.C.

Gilliland, M. W. (1975). Energy analysis and public policy. *Science* **189**, 1051–1056.

Gomes da Silva, J., Serra, G. E., Moreira, J. R., Conçalves, J. C., and Goldemberg, J. (1978). Energy balance for ethyl alcohol production from crops. *Science* **201**, 903–906.

Graves, J. (1971). Texas: "You ain't seen nothing yet." *In* "The Water Hustlers" (R. H. Boyle, J. Graves, and T. H. Watkins, eds.), pp. 15–129. Sierra Club, San Francisco, California.

Hall, D. O. (1980). Bio-Energy '80. *Nature* **285**, 135.

Handreck, K. A., and Martin, A. E. (1976). Energetics of the wheat-sheep farming system in two areas of South Australia. *Search* **7**, 436–442.

Harte, J., and El-Gasseir, M. (1978). Energy and water. *Science* **199**, 623–634.

Heichel, G. H. (1973). Comparative efficiency of energy use in crop production. *Conn. Agric. Exp. Stn., New Haven, Bull.* No. 739.

Heichel, G. H. (1976). Agricultural production and energy resources. *Am. Sci.* **64**, 64–72.

Heichel, G. H. (1978). Stabilizing agricultural energy needs: Role of forages, rotations, and nitrogen fixation. *J. Soil Water Conserv.* **33**, 279–282.

Jenny, H. (1980). Alcohol or humus? *Science* **209**, 444.

Johnson, W. A., Stoltzfus, V., and Craumer, P. (1977). Energy conservation in Amish agriculture. *Science* **198**, 373–378.

Karkheck, J., Powell, J., and Beardsworth, E. (1977). Prospects for district heating in the United States. *Science* **195**, 948–955.

Knutson, J. D., Jr., Curley, R. G., Roberts, E. B., Hagan, R. M., and Cervinka, V. (1977). Energy for irrigation. *Calif. Agric.* **31**(5), 46–47.

Kobriger, N., and Stearns, F. (1972). Productivity and energy storage. *Univ. Wis. (Milwaukee), Field Stn. Bull.* **5**(2), 13–16.

Kumar, A., Langlois, J., and Watson, K. (1978). Communications–transportation substitution in the reduction of energy requirements: An exploration of the social and economic impact on human settlements. *Habitat Int.* **3**, 435–451.

Lamb, H. H. (1955). Two-way relationship between the snow or ice limit and 1,000–500 mb thickness in the overlying atmosphere. *Q. J. R. Meteorol. Soc.* **81**, 172–189.

Landsberg, H. H. (1964). "Natural Resources for U.S. Growth." Johns Hopkins Press, Resour. Future, Baltimore, Maryland.

LePori, W. A., and Coble, C. G. (1975). Assessment of energy inputs for Texas agricultural production. *In* "Energy, Agriculture and Waste Management" (W. J. Jewell, ed.), pp. 49–59. Ann Arbor Sci. Publ., Ann Arbor, Michigan.

Lewthwaite, G. R. (1964). Wisconsin and the Waikato: a comparison of dairy farming in the United States and New Zealand. *Ann. Assoc. Am. Geogr.* **54**, 59–87.

Lockeretz, W., Shearer, G., Klepper, R., and Sweeney, S. (1978). Field crop production on organic farms in the midwest. *J. Soil Water Conserv.* **33**, 130–134.

Manners, I. R. (1978). Agricultural activities and environmental stress. In "Sourcebook on the Environment" (K. A. Hammond, G. Macinko, and W. B. Fairchild, eds.), pp. 263–294. Univ. of Chicago Press, Chicago, Illinois.

Marshall, E. (1980). Energy forecasts: Sinking to new lows. Science 208, 1353–1356.

Matlin, R. W., and Katzman, M. T. (1979). Assessing solar photovoltaic energy systems for crop irrigation. Water Resour. Bull. 15, 1308–1317.

Metz, W. D. (1972). New means of transmitting electricity: a three-way race. Science 178, 968–970.

Nelson, W. L., and Dale, R. F. (1978). Effect of trend or technology variables and record period on prediction of corn yields with weather variables. J. Appl. Meteorol. 17, 926–933.

Pearson, R. G., and Corbet, P. S. (1976). Energy in New Zealand agriculture. Search 7, 418–423.

Pimentel, D., and Pimentel, M. (1978). Dimensions of the world food problem and losses to pests. In "World Food, Pest Losses, and the Environment" (D. Pimentel, ed.), AAAS Symposium No. 13, pp. 1–16. Westview Press, Boulder, Colorado.

Pimentel, D., and Pimentel, M. (1980). Counting kilocalories. Focus (New York) 30(3), 9–10.

Pimentel, D., Hurd, L. E., Bellotti, A. C., Forster, M. J., Oka, I. N., Sholes, O. D., and Whitman, R. J. (1973). Food production and the energy crisis. Science 182, 443–449.

Pimentel, D., Dritschilo, W., Krummel, J., and Kutzman, J. (1975). Energy and land constraints in food protein production. Science 190, 754–761.

Price, D. R., Gunkel, W. W., and Casler, G. L. (1975). Accounting of energy inputs for agricultural production in New York. In "Energy, Agriculture and Waste Management" (W. J. Jewell, ed.), pp. 61–76. Ann Arbor Sci. Publ., Ann Arbor, Michigan.

Rawlins, S. L., and Raats, P. A. C. (1975). Prospects for high-frequency irrigation. Science 188, 604–610.

Revelle, R. (1976). Energy use in rural India. Science 192, 969–975.

Ritschard, R. L., and Tsao, K. (1980). Energy and water conservation strategies in irrigated agriculture. Water Resour. Bull. 16, 340–347.

Sachs, R. M. (1980). Crop foodstocks for fuel alcohol production. Calif. Agric. 34, 11–14.

Sampson, G. R. (1979). Energy potential from central and southern Rocky Mountain timber. U.S. For. Serv., Res. Note RM-368.

Schipper, L., and Lichtenberg, A. J. (1976). Efficient energy use and well-being: the Swedish example. Science 194, 1001–1013.

Schmidt-Nielsen, K. (1972). Locomotion: Energy cost of swimming, flying, and running. Science 177, 222–228.

Splinter, W. E. (1976). Center-pivot irrigation. Sci. Am. 234(6), 90–99.

Steinhart, C. E., and Steinhart, J. S. (1974). "Energy; Sources, Use, and Role in Human Affairs," Duxbury Press, North Scituate, Massachusetts.

Steinhart, J. S., and Steinhart, C. E. (1974). Energy use in the U.S. food system. Science 184, 307–316.

Stewart, G. A., Gartside, G., Gifford, R. M., Nix, H. A., Rawlins, W. H. M., and Siemon, J. R. (1979). Liquid fuel production from agriculture and forestry in Australia. Search 10, 382–387.

Stewart, G. A., Gartside, G., Gifford, R. M., Nix, H. A., Rawlins, W. H. M., and Siemon, J. R. (1980). Reply. Search 11, 96–97.

Thompson, P., and Clary, M. G. (1977). Solar heating for milking parlors. U.S. Dep. Agric., Farmers' Bull. No. 2266.

U.S. Council on Environmental Quality (1978). "Environmental Quality," 9th Annual Report. U.S. Gov. Print. Off., Washington, D.C.

U.S. Department of Agriculture, Crop Reporting Board (1977). "Agricultural Situation," March. U.S. Gov. Print. Off., Washington, D.C.

U.S. Water Resources Council (1978). "The Nation's Water Resources. Second National Assessment," Part II. Water Manage. Probl. Profiles, Washington, D.C.

U.S. Writers' Program (1950). "Washington," rev. ed. Wash. State Hist. Soc. and Binfors & Mort, Portland, Oregon. (Orig. publ., 1941.)

Wall Street Journal (1980). Sputtering along; reduced gasoline use cuts into tax levies, hurts many businesses, 3 April 1980, pp. 1, 20.

Ward, G. M., Knox, P. L., and Hobson, B. W. (1977). Beef production options and requirements for fossil fuel. *Science* **198**, 265–271.

Watson, C. L. (1980). Fuel crops and the soil. *Search* **11**, 96.

Weaver, J. C. (1954). Crop-combination regions in the Middle West. *Geogr. Rev.* **44**, 175–200.

Weisz, P. B. and Marshall, J. F. (1979). High-grade fuels from biomass farming: Potentials and constraints. *Science* **206**, 24–29.

Wharton, C. R., Jr. (1980). Food, the hidden crisis. *Science* **208**, 1415.

Williams, D. W., McCarty, T. R., Gunkel, W. W., Price, D. R., Jewell, W. J. (1975). Energy utilization on beef feedlots and dairy farms. *In* "Energy, Agriculture and Waste Management" (W. J. Jewell, ed.). pp. 29–47. Ann Arbor Sci. Publ., Ann Arbor, Michigan.

Wisconsin Agricultural Reporting Service (Wisc. Dept. Agric. and U. S. Dept. Agric.) (1980). "1980 Wisconsin Agricultural Statistics." Madison.

Wisconsin Statistical Reporting Service (1972). Monthly Report, November, 1972, Madison.

Wisconsin Statistical Reporting Service (1973). Monthly Report, February, 1973, Madison.

Wood-Bradley, R., Flint, D. M., and Wahlqvist, M. L. (1980). Food and nutrition in an independent Papua New Guinea. *Search* **11**, 73–77.

Zavitkovski, J. (1979). Energy production in irrigated, intensively cultured plantations of *Populus* 'Tristis #1' and jack pine. *Forest Sci.* **25**, 383–392.

Chapter XIV

PHASE CHANGES OF WATER IN ECOSYSTEMS:
I. FREEZING AND THAWING

The peculiar nature of the water molecule results in a heat of fusion, 334 kJ kg^{-1}, that is larger than that of other molecules of similar structure, and an anomalously high melting temperature, 273°K. In addition, the large heat of condensation, 2.5 MJ kg^{-1}, high specific heat, 4.18 kJ kg^{-1}, and the radiative absorptivity of water vapor in the long wavelengths make water in an ecosystem a major thermal flywheel, which takes in and releases large quantities of energy with little or no fluctuation in temperature and so reduces the degree to which ecosystem temperature fluctuates to keep its energy budget in balance.

In this chapter we are concerned with changes between the liquid and solid phases as they occur in due season in the annual functioning of an ecosystem. While the quantities of energy involved per kilogram of water are smaller than those involved per kilogram of carbon compounds discussed in Chapters XI–XIII, they are significant because the quantities of water are large, being measured in hundreds of kilograms per square meter. Because radiant-energy intake usually dominates melting, the energy budget can be constructed in large part from radiation considerations in Chapters III–X.

PRESENCE AND CHARACTERISTICS OF SNOW COVER IN ECOSYSTEMS

Formation of Snow Cover

Almost all precipitation in the middle and polar latitudes and at high altitudes everywhere starts as ice or snow particles. The conditions for formation of a snow cover are simply that falling snowflakes not melt on the way down and not melt on the ground after arrival. These regions include maritime mountains, continental mountains, midlatitude lowlands remote from the ocean, and high-latitude lowlands.

Winter accumulation of snow cover in an ecosystem is a contest between deposits made by individual storms and depletion between storms. Models that reproduce additions to the snow cover and depletion of it can simulate either intermittent snow cover or variations in mass and temperature of a winter-long cover.

Ecosystem Characteristics

Important characteristics are slope aspect, slope steepness, and tree cover, which determine incidence of the solar beam and the flux of diffuse solar radiation that reach the snow surface. In the Cooperative Snow Investigations sampling of snow cover was stratified by these terrain characteristics in order to explain differences in melting, which in one spring period varied from 28 mm day^{-1} (24-hr mean energy conversion rate 110 W m^{-2}) on shaded, sheltered slopes to 45 mm day^{-1} (175 W m^{-2}) on exposed slopes (Miller, 1950; U.S. Corps of Engineers, 1956). Later models include even more terrain variables (Meier and Schädler, 1979).

The "caloric equivalent" of variations in terrain in the Sierra crest region was evaluated as 2 W m^{-2} for each 10° change in slope orientation or aspect, 1 W m^{-2} for each 10-m increase in altitude, and 2 W m^{-2} for each unit change in a 5-point scale of vertical curvature (Miller, 1955, p. 192). Hoeck's (1952) estimates of melting on different slopes in the Alps depend on exposure to beam and diffuse solar radiation. Larson et al. (1974) find aspect and steepness data in dissected terrain to be sufficient for a model that calculates the melting of intermittent snow cover, but need forest-cover data in less dissected landscapes. A melting model including elevation, forest cover, and aspect-steepness dimension demonstrates timing of melting at 96 ecosystems in a 108 km^2 drainage basin (Hendrick et al., 1971). Colbeck et al. (1979) feel that forest canopy still presents difficult characterization problems.

Shrub ecosystems exert a similar though smaller effect to that of forest in catching falling snow, shaping its spatial distribution, and exposing it to energy fluxes. Their effect is largest where snow and branches form a single mixture, as in manzanita crushed down under heavy snow cover (Wilken, 1967), or snow in sagebrush (Hutchison, 1965). Metamorphosis is accelerated by increased solar-energy absorption (Sturges, 1977) and melting by increased absorption and roughness (Price and Dunne, 1976). Snow intercepted in forest canopy is also intermingled with plant matter

(Miller, 1977, Chap. VI) that absorbs solar radiation well. The warm stems and needles transfer energy by radiation, conduction, and convection to snow clumps in the canopy, which may be released and fall before melting completely (Satterlund and Haupt, 1970).

A similar mingling of snow with darker material occurs in late spring after rocks and patches of soil are free of snow cover and exposed to solar heating. Complex patterns of microrelief in the Austrian and Swiss Alps produce complex patterns of snow-free areas (Aulitzky, 1965), which accelerate melting of intermingled snow patches. Similar patterns affect melting of thin snow on the Plains (Granger and Male, 1978).

Characteristics of Snow Cover

The thermal and radiative characteristics of a snow layer result from conditions in the storm that delivered it and subsequent metamorphosis. Fresh snow cover, made up of delicate, many-branched flakes that trap 10 to 20 times their volume of air, exhibits the smallest thermal conductivity of any natural substance, but aging increases both density and conductivity. Sources of energy for metamorphosis include kinetic energy (wind packing), radiant energy (vapor translocation and crystal growth), soil warmth (bottom melting), refreezing in ice lenses of meltwater draining down from an irradiated surface, and refreezing at the surface (sun slabs) (Bader et al., 1939).

Temperature of snow cover is usually below 273°K in cold environments, and vapor moves from warmer to colder layers in temperature-gradient metamorphosis (Sommerfeld and LaChapelle, 1970). A dry snow cover builds up a value of "Kältegehalt" or cold content, a term that looks forward to subsequent melting, and that amounts to 10–15 MJ m^{-2} in moderately deep snow cover. It is modeled by simulating the substrate-heat flux (Obled and Rosse, 1977; Colbeck et al., 1979).

In higher-energy environments that hold the bulk of the snow cover at 273°K for long periods, equitemperature metamorphosis occurs and is marked by diffusion of vapor molecules to locations that will minimize the free energy at the surface of the snow crystals, largely by reducing their surface area. Metamorphosis changes the number of ice–air interfaces within the snow cover; refraction and scattering of incoming photons are reduced, and both transmission and absorption increase. Visible-band solar radiation is reflected near the surface of dry snow cover, but in moist snow penetrates deeply in accord with the extinction coefficient ν in Beer's law,

$$I = I_0 e^{-\nu z},$$

where I_0 is initial and I local intensity and z depth into the snow cover. In new snow v can be as large as 50 m^{-1} and decreases by an order of magnitude during metamorphosis (Mellor, 1964). Increased transmission of light in spring aids germination of plants under snow cover and stimulates chlorophyll synthesis (Richardson and Salisbury, 1977).

Density of dry snow in the upper layer of a snow cover affects transmission and absorption of radiation, as does size of snow grains (Bergen, 1975). With large grains density is the major factor; with small grains it is grain size. The large grains and high density produce greater extinction and absorptivity in older snow (Anderson, 1976, p. 81; Arai, 1965), which augments the effect of increasing solar energy in spring.

Emissivity and absorptivity of melting snow for longwave radiation approaches that of a blackbody,* due in part to its highly porous surface, virtually composed of tiny blackbody cavities. Projecting crystals led early investigators to postulate that emissivity was higher than that of a blackbody (Voeikov, 1885); while this is not true, near-blackbody absorptivity does couple snow cover closely with longwave radiators: atmospheric vapor, CO_2 molecules, clouds, or forest canopy.

The turbulent coupling of snow cover and atmosphere depends on ecosystem roughness, e.g., when understory vegetation melts out of the snow (Price and Dunne, 1976). Characteristics of snow and ecosystem—thermal conductivity, absorptivity, emissivity, and roughness—condition the coupling of the snow with the sources of energy—soil, atmosphere, and sun. We will now look at these sources and couplings.

THE ENERGY BUDGET OF MELTING SNOW COVER

Major Sources of Energy

In the source–sink relation the sink is the meltwater that drains out of the melting layer at a cost of 334 kJ kg^{-1}, but sources are diverse.

The Ground. The first snows of autumn are often casualties of warm soil, but this conduction potential diminishes as the bare soil cools by radiative exchange in the lengthening nights or by losing sensible heat to successive invasions of polar air streams. In any case, the quantity of energy is limited to the amount of heat the soil took in during the previous summer and seldom will dispose of more than 200 mm of

---

* Lower values may occur in cold, small-grained snow (Mellor, 1977) in nonmelting situations.

snowfall. Aquatic ecosystems are quite different, of course, due to their enormous heat storage.

The Atmosphere. This source varies from day to day as the synoptic circulation changes. Energy in large quantities is advected poleward from snow-free parts of the world, particularly the Gulf of Mexico for the Midwest, in several forms: potential, kinetic, latent, and sensible. This energy is delivered to the underlying snow cover by radiative and turbulent processes. Sensible-heat flux to snow on the Canadian prairies is indexed by temperature of the air at the 850-mbar pressure level (Granger and Male, 1978). However, warm air, defined by 1000–500 mbar thickness exceeding 5.3 km, does not transgress over snow-covered land more than a few hundred kilometers because it loses energy to "many sorts of cooling processes," including radiative and convective exchanges with the underlying snow (Lamb, 1955). Air streams exceeding 5.4 km thickness do not transgress at all without the aid of subsidence. Ecosystems remote from the equatorward margin of regional snow cover do not have access to this major heat reservoir.

The Sun. Ecosystems at all locations do, however, have access to solar energy in spring, and as spring advances its latitude dependence diminishes. Access is best in subsidence situations, which are frequent over the cordillera of western North America. Intense midday radiation causes snow cover to age and its absorptivity to increase day by day. In spring the increase of radiative flux density, lengthening of days, and greater absorptivity combine to make the sun the dominant energy source.

Soil-Heat Flux

Conduction governs the flow of soil heat into the overlying snow cover, as well as from a warmed upper layer of snow into colder middle layers.* It is small but continues through the winter, the diminishing gradient of temperature being offset by increasing thermal conductivity as the soil gets wetter (U.S. Corps of Engineers, 1956, p. 179). The insulating power of the snow cover traps the upwelling heat from the ground to bring about an equilibrium at 273°K at the snow–soil interface.† The lower layers of snow lie at the melting point, melting only slowly; their condition is like that of an ice and water mixture in a ther-

---

* Conduction can move no heat within isothermal layers of the snow cover, and since the upper layer cannot exceed 273°K, it cannot deliver heat to lower layers also at 273°K.

† Top layers of the soil that had frozen before snow cover forms and insulates them often thaw from the heat that continues to rise from deeper, warmer layers of the soil.

mos flask. In the absence of a large inflow or outflow of energy, impossible under the blanket of snow, neither freezing nor melting gains the upper hand. Flux densities average about 3 W m^{-2} in the Sierra Nevada and produce appreciable melting during the long life of the snow cover. Winter runoff in maritime ranges of Japan is generated at rates of 2 W m^{-2} in Hokkaido and 6–8 W m^{-2} in Honshu (Arai, 1968). In colder climates, even though soil and snow temperatures may not reach 273°K, soil-heat flux occurs; rates in New England average about 3 W m^{-2} [from data in Anderson (1976)]. The lower layers of Arctic snow benefit from trapped heat from the soil and plants, and animals live in an environment that is warm and steady in temperature, humid, silent, and dark (Pruitt, 1957). Electrical heating of bridge decks in Louisville melts snow at its rate of delivery; however, this averages only 4 W m^{-2} and is achieved by conversion of electrical energy at a rate of 185 W m^{-2} (Havens et al., 1979).

Energy Fluxes from the Atmosphere

Rainwater. Rainwater used to be considered to be a large source of heat for melting snow, an idea made plausible by the observed rapid disappearance of snow in a warm rainstorm. However, energy analysis showed this flux to be minor (Horton, 1915). The specific heat of rainwater, 4.2 kJ kg^{-1} deg^{-1}, is small compared with the heat of fusion, 334 kJ kg^{-1}; 1 kg m^{-2} of rainwater at a temperature of +5°C will melt only 5 × 4.2 kg/334 of ice particles or 0.06 kg m^{-2}. The amount of melting ascribed to rainwater in a Vermont basin averaged only 4 mm per season (Anderson, 1976) and the energy-flux equivalent is less than 0.2 W m^{-2}.

Latent Heat of Condensation. The observed rapid snow melting in rainstorms is due to other heat transports from the air, one of which is the latent heat released when vapor condenses on the cold snow surface. Comparing the heat of condensation to the heat of fusion gives a ratio quite different from that in the rainwater case:

$$2500 \text{ kJ}/334 \text{ kJ} = 7.5.$$

This mode of heat transport is episodic, however, and over a season may not bring more heat to the snow cover than the upward flux of vapor by evaporation in interstorm periods removes from it. The net balance depends on frequency of air streams with vapor concentration exceeding that at the snow surface, 3.8 g kg^{-1} (or 6.1 mbar vapor pressure). Snow hanging in tree crowns is fully exposed to wind and sun, two factors we usually associate with strong evaporation; yet because its vapor pressure seldom exceeds atmospheric vapor pressure the direction of the latent-heat flux is usually toward the snow, and melting ensues (Miller, 1967).

Sensible Heat. This turbulent heat flux often parallels the flux of latent heat from air to snow, except under hot, dry air when it has the opposite direction. The two fluxes are carried by the same turbulent eddies and affected by the same conditions of ecosystem roughness and atmospheric kinetic energy and gustiness. Their relative sizes depend on the respective gradients, that is, atmospheric warmth above 273°K and vapor concentration above 3.8 g kg^{-1}. Any storm producing condensation melting produces also a parallel sensible-heat flux, which augments that melting. The sensible-heat flux is important in ordinary spring melting weather when it reaches midday rates of 30–50 W m^{-2}.

Longwave Radiation. This heat flux from air to snow cover is always large (Chapter VI) and reaches a maximum during the massive advection of a major rainstorm, as Hubley (1957) showed for a maritime snowfield in southeastern Alaska. Because rain clouds may be warmer than the snow cover and have equal emissivity, radiative input to the snow outweighs the radiant flux emitted by the snow (315 W m^{-2}) by 20 to 30 W m^{-2} or even more.

Other than in episodes of strong heat advection the downward flux of longwave radiation usually falls in a range from 250 to 300 W m^{-2}, and the net exchange of energy by longwave radiation is a deficit that varies from -15 to -65 W m^{-2}. When night comes, solar energy shuts off but longwave keeps coming and prevents supercooling of the snow. We tend not to see this quiet flow of energy and should look at sites where it is weak, i.e., at high altitudes. From data at 5.4 km in the Andes (Hastenrath, 1978) I estimate the incoming mean longwave flow at 225 W m^{-2} or less, a flux so small that the sun is unable to bring the snow to the melting point.

Absorbed Solar Radiation

This energy flux depends on snow-cover absorptivity, is absent at night, and is small in episodes of strong advection. When it is present, however, it is the central component of the energy budget of the snow cover. In Sierra spring weather absorption of solar energy peaks around 400 W m^{-2} in its daily cycle, averages 200 W m^{-2} over the daylight period and 100 W m^{-2} over the 24-hr cycle; it is a major energy flow on the Great Plains also (Granger and Male, 1978).

The importance of solar absorption in the energy balance of snow cover has often suggested practices to increase it in critical sites or at dates in spring while air temperature is still below freezing. Carbon black, coal dust, cinders, and other substances have been used with expectable results but economic hindrances restrict the practice to airports, deep snow along field edges, or browse areas important in winter survival of game animals. Overenthusiastic application of the darkening agent may increase heat loss by turbulent fluxes or insulate the snow itself (Slaughter, 1969).

The Energy Budget

The energy fluxes described above transport energy from three reservoirs, ground, atmosphere, and sun. Their flux densities depend on the strength or energy density of these sources, but also upon the special characteristics of melting snow, especially its restricted temperature range.

Surface Temperature. Snow contrasts with other ecosystem surfaces, which warm up in response to radiant-energy intakes to a temperature that comes to exceed air temperature and cuts off gain of sensible heat from the air (Chapter VIII). By staying cold, snow may retain a downward temperature gradient regardless of its heat intake. Its correspondingly low vapor pressure similarly retains a downward humidity gradient.

The energy-budget approach, in which surface temperature mediates between the absorption of radiant energy and the disposition of the same quantity of energy, is facilitated by the fixed temperature of melting snow and ice. The flux of emitted longwave radiation remains constant, as do surface temperature and vapor pressure that determine the gradients along which sensible and latent heat flow. These gradients therefore can be determined by air temperature and vapor pressure as measured conventionally.

The Energy-Budget Concept. The connection of glacier changes to climatic change led to the study of snowfields in Scandinavia in the 1920s and 1930s, and the investigators adopted the energy-budget approach (e.g., Sverdrup, 1935; Wallén, 1949), based on earlier studies of radiation. The concept was later applied to snow cover in the Alps, Russia, and North America. In the western United States it replaced degree-day or air-temperature methods, inadequate for the hydrologic design of reservoirs and dam spillways being built after 1946, and was modified for topographic and forest environments of the snow fields that yield the water supporting Western agriculture and cities (U.S. Corps of Engineers, 1956; Miller, 1953, 1955). Forest effects were specifically examined in later work by the Forest Service (Anderson et al., 1958), and the methods were extended to large drainage basins by the U.S. Corps of Engineers (1960) and U.S. Geological Survey (Rantz, 1964).

Subsequently, the energy-budget model, usually as adapted from its detailed explication by the Snow Investigations (U.S. Corps of Engineers, 1956), was expanded by computers to allow rapid calculation of melting rates through a whole season. These included submodels of snowfall, so that snow-cover mass and meltwater are determined day by day but remain in need of field measurements to validate them (McKay and Thurtell, 1978). Applications of the Snow Investigations model have been made in the former snow laboratories (Riley et al., 1973), the Rockies (Leaf and Brink, 1973), the Southwestern plateaus (Solomon et al., 1975), New England (Hendrick et al., 1971), the boreal forest (Price and Dunne, 1976), the Arctic (Outcalt et al., 1975), and the Alps (Obled and Rosse, 1977), as well as in many other environments. Some models generate a surface temperature that is converted to melting if it exceeds 273°K (Outcalt et al., 1975) and shows the effects of cold spells (Obled and Rosse, 1977). Others improve on the formulation of turbulent heat fluxes (Price and Dunne, 1976); some incorporate submodels for estimating solar radiation from more commonly observed parameters (Solomon et al., 1976).

Some melting models are incorporated in a program of river forecasting (Rockwood, 1972). In the Soviet Union, the energy balance method has been extended in snow hydrology as well as in other geophysical processes (e.g., Kuz'min, 1960). Even with its uncertainties, it is far superior to methods based on air temperature (McKay and Thurtell, 1978), which of course is not a measure of energy or energy flux.

EFFECTS OF TERRAIN AND WEATHER ON THE ENERGY BUDGET

Transformations of Solar Energy in Anticyclonic Weather

Forest canopy separates the snow cover from sun and atmosphere visually, but in other ways is a thermodynamic intermediary. Its role of shelter from the wind has minor significance in settled anticyclonic weather, but its role in transforming solar radiation into other forms of energy is significant. The irradiated foliage is warmed, vapor pressure at leaf surfaces increases, and the canopy generates sensible and latent heat and longwave radiation. Its lower levels radiate to the snow on the forest floor, often at a rate higher than the 315 W m^{-2} that the snow is emitting. Small fluxes of sensible heat and vapor make their way from crowns to snow, though they are difficult to measure (Miller, 1955; Colbeck et al., 1979).

Where forest is open, or where geological contrasts, shallow soils, or fire have made it discontinuous, snow cover on meadows or shrub ecosystems is interspersed with snow-free canopy of tree ecosystems. Sensible and latent heat generated in the tree stands finds its way to the snow in the openings, carried by turbulence caused by the rough mixture of stands and openings. The earmarks of this mesoscale transport in the local air are air temperatures exceeding $+10°C$, even in the presence of snow 2 to 3 m deep, and vapor pressures exceeding 10 mbar (Miller, 1950), which can have no other source than the cell walls of the irradiated needles.

Unfrozen soil under the snow does not limit water uptake, and transpiration powers a local cycling of water from soil to needles, into the local air, and in part downward to the snow cover, a mass flow that transports energy from tree crowns to snow and augments midday melting.

Mesoscale transports late in the melting season move energy from bare ground to remaining patches of snow, both in mountains and on plains; turbulent heat fluxes are strong at the edge of a snow patch and diminish inward (Khodakov, 1965), just as in an oasis. On an average day in spring 40 km^2 of bare patches are formed in the basin of the American River (5600 km^2) (Schneider et al., 1976) and become new heat sources. Kinetic energy for this transport derives from unequal fluxes of sensible heat from forest and meadows, and in dissected topography from the mountain–valley breeze circulation, both generated by solar energy.

Anticyclonic subsidence of air into mountain ranges in radiation weather brings heat to ridge ecosystems, especially at night, in quantities as large as 10–20 W m^{-2} (Miller, 1955, p. 164). Beam radiation dominates in anticyclonic weather, and the shade pattern near a forest stand depends on its height and on sun altitude and azimuth, as well as ground slope, and can be delineated at each hour of the day (Satterlund, 1977). Longwave emission by tree borders replaces a smaller flux density from the part of the sky it blocks out. The net gain is usually greatest from trees on the north side of an opening, and meadow or strip-cut forest is a beneficiary (Anderson et al., 1958).

Diurnal March. Measurements of the energy fluxes at the surface of a snow cover 1.2 m deep in a mountain meadow on a clear spring day (22 April 1954, solar declination $+11°$ 35') from Snow Investigations sources (Hildebrand and Pagenhart, 1955; U.S. Corps of Engineers, 1956, Plate 5-8) show clear diurnal variations. Decreasing longwave radiation from the cooling local air in the valley during the night allowed the snow surface to chill to 267°K by dawn, at which time its emission, 290 W m^{-2}, was met from incoming longwave radiation (245 W m^{-2}), from the freezing of meltwater into a crust (25 W m^{-2}), from heat released as the snow cooled (substrate heat flux) (10 W m^{-2}), and from condensation of vapor (10 W m^{-2}). Local heat sources maintained its energy budget in balance.

This low-energy condition altered rapidly when the sun rose. By noon absorption of solar energy by the snow reached 385 W m^{-2}, and the impulse of solar energy absorbed by the lodgepole pine ecosystems that bordered the meadow set other transformations into operation. Their irradiated foliage warmed the local air to 15°C and the air on higher slopes still more, bringing into existence an up-valley wind, which moved air from tree stands over the meadow with kinetic energy sufficient to generate mechanical turbulent exchange with the snow surface. The turbulent flux of sensible heat brought 45 W m^{-2} to the melting snow at noon and was accompanied by a smaller turbulent flux of latent heat. In addition, the warm air had increased its radiation downward to the snow. These auxiliary fluxes, which augmented the solar regime already manifested in the absorption of shortwave radiation by the snow, are forms of local heat advection that have been noted also in sagebrush ecosystems (Meiman, 1969, p. 62).

The midday radiant-energy intake, 755 W m^{-2}, is diminished by the 315 W m^{-2} emitted by the snow but augmented by 70 W m^{-2} in nonradiative but solar-generated fluxes. The energy sink had to dispose of 510 W m^{-2}, which went to melt snow. The meltwater equivalent, found by dividing by 335 kJ kg^{-1}, is equal to 1.52×10^{-3} kg m^{-2} sec^{-1} = 5.47 kg m^{-2} hr^{-1}, i.e., 5.5 mm hr^{-1}, confirmed by measurements of water draining from the snow (Miller, 1977, p. 177).

The daily waves of water yielded by the meadow evidence solar control (Fig. 1) during clear weather. The spring climate of this snow cover ecosystem is a series of strong, regular pulses, the impacts of the intermittent flux of solar energy, and can be accurately modeled (Miller, 1955; Anderson, 1968). In the Alps also the time and space patterns of melting out (Ausaperung) of ecosystems can be modeled from net shortwave radiation (Meier and Schädler, 1979). Radiative melting is not confined to mountains or to middle latitudes; open spruce forest in Ungava at 55°N displayed midday rates of 165 W m^{-2} in a plot facing northeast, and 200 W m^{-2} in one facing southwest (Price and Dunne, 1976), with a well-marked diurnal regime.

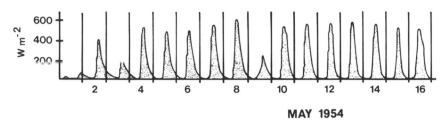

MAY 1954

Fig. XIV.1. Diurnal pulses of meltwater from lysimeter in the Lower Meadow, Central Sierra Snow Laboratory, in the spring of 1954, expressed in terms of energy equivalent (watts per square meter). Clouds on 3 May and 2 mm rain on 9 May were the only interruptions to a regular series of daytime energy pulses that started following a cold, dry outbreak after a snowstorm on 30 April. Depth of snow on the lysimeter decreased steadily through this period, and bare spots began to show by 12 May. (Hildebrand and Pagenhart, 1955.)

Transformation of Advected Atmospheric Energy

In advective melting a warm, humid, cloudy air stream delivers energy to the snow in three modes:

(1)   The most important is longwave radiation, which is large (300–350 W m^{-2}) by reason of the warmth of the radiators and the high emissivity of the vapor concentration and clouds.

(2)   Turbulent transport of sensible heat is large since the kinetic energy of the storm easily overcomes the negative buoyancy of the stable air overlying the snow.

(3)   At moderate to high vapor concentrations, the latent-heat flux from air to snow is large.

The turbulent fluxes are larger at low altitudes in a drainage basin than at high and are reduced by continuous forest canopy. Episodes of advective thawing of winter snow cover occur in midlatitude and continental interiors and lower-altitude* maritime mountains. Spring melting in the northern Rockies in years of frequent advection may average as much as 4 mm day^{-1} more than in other years (Packer, 1971). High-altitude melting in midlatitude mountains like the Sierra occurs only in air streams from low latitudes, as in December 1937 and November 1950, and creates catastrophic flooding.

Hubley (1957) measured high 24-hr melting rates in summer rainstorms in southeastern Alaska, which were due to 350 W m^{-2} of long-

---

* The effect of Pacific air entering the northern Rockies is shown by a snowmelt lysimeter (Haupt, 1969), which suggests that much winter meltwater originates in the melting of intercepted snow in the forest canopy (Haupt, 1972).

wave radiation from warm, low clouds and 200 W m^{-2} in the turbulent fluxes. A design-flood sequence in the Oregon Cascades postulates melting rates in the maximum 6-hr period to be supported by turbulent fluxes equal to 50 W m^{-2} sensible heat and 130 W m^{-2} latent heat, averaged over all elevations of the drainage basin (U.S. Corps of Engineers, 1960, Plate 6). Design rates of melting in central Russia (Kuz'min, 1960) include short-term turbulent energy fluxes of 300 W m^{-2}. Such rates of melting, often accompanied by heavy rain, create a layer of slush at the snow–soil interface (Horton, 1936) that is hydraulically unstable.

Structure of Energy Budgets

Advection of warm, humid air can sustain a rate of delivery of energy to the underlying snow of 500 W m^{-2} or more by three modes of heat transfer, a figure comparable to midday rates of energy delivery in radiation weather. However, the structure of the energy budget is different in the two cases. The structure is outlined as follows:

$$K_{abs} + L_{abs} + H + 2.5 \times 10^6 E = 315 + M,$$

in which the gain by absorption of radiant energy by the snow cover is $K_{abs} + L_{abs}$ (where K and L stand for shortwave and longwave radiation, respectively), the gain by sensible heat is H, and latent heat is $2.5 \times 10^6 E$, where E is condensation in millimeters per second. The sum of radiative and turbulent heat intake is equal to emitted longwave radiation, constant at 315 W m^{-2}, and a residual variable, the latent heat of fusion M. All fluxes are in watts per square meter.

In radiative weather in the Sierra at midday, this equation becomes

$$385 + 370 + 45 + 25 = 825 = 315 + 510 \quad W\,m^{-2},$$

in which 510 W m^{-2} is the rate of energy conversion in melting. Radiation absorbed by the snow cover is 755 W m^{-2}, half solar and half longwave, and contributes 0.91 of the total intake of energy. The solar contribution itself is 0.46 and beam radiation approximately 0.40.

The structure of the energy budget over the 24-hr day is less dominated by solar energy:

$$120 + 284 + 18 + 8 = 430 = 305 + 125 \quad W\,m^{-2}.$$

Radiant-energy intake (404 W m^{-2}) is 0.94 of the total energy intake of the snow cover, and the purely solar contribution is 0.38. After longwave emission is subtracted from the total intake, 125 W m^{-2} remains for the heat of fusion, typical for high-altitude spring at a middle latitude.

Summer in Lappland, at a much higher latitude (Wallén, 1949), is characterized by energy advection and the mean equation is

$$90 + 280 + 30 + 20 = 420 = 315 + 105.$$

Absorbed radiation contributes 370 W m^{-2}, mostly in the long wavelengths. The atmosphere contributes this longwave flux, sensible-heat flux, and condensation, or $(280 + 30 + 20) = 330$, for 0.79 of the total. Midlatitude cyclonic storms in the polar Urals bring a total heat intake up to 590 W m^{-2} (Kotliakov, 1968, data in Table 26), 0.37 of this via the turbulent fluxes; high-latitude cyclones brought less heat, only 0.17 of which was delivered by turbulence. Solar energy provides 0.3 of the total in anticyclonic weather and 0.1–0.15 in cyclonic.

The structure of the energy budget of ecosystems during the melting-out period varies with synoptic conditions. Solar energy is dominant in anticyclonic periods, providing 0.4 of the total intake of energy by the snow and evoking diurnal regimes in the other energy fluxes. In advective weather it provides 0.2, or less in large rainstorms, and exerts little influence on the other fluxes.

Removal of all the snow in an ecosystem at the end of winter takes energy flow over time, but delay in spring warming of the ecosystem and its soil is offset in part by the fact that the snow cover has protected the soil from freezing deeply, or even freezing at all. The size of this benefit depends on whether or not it was established early enough in fall to have forestalled deep chilling of the soil.

Spring snow cover in the mountains of western North America holds the water needed in summer by crop and urban ecosystems in the valleys below, and the later the snow melts the better. Managers of mountain ecosystems seek to delay and prolong the melting season. It appears attractive to manipulate vegetation for this purpose (Meiman, 1969, p. 67), but requires an understanding of forest effects on radiative and turbulent fluxes (Colbeck et al., 1979; Reifsnyder, 1979).

Spatial Distributions in Radiative and Advective Melting

Radiative melting displays large place-to-place differences in dissected terrain because the directional solar beam is dominant. As a result, there is a spatial sequence of melting—shaded slopes after sunny slopes— that prolongs the period of meltwater yield from a drainage basin. Late snow in a forest opening also melts rapidly by receiving sensible heat transferred from trees and bare sun-heated forest floor (DeWalle and Meiman, 1971), to an amount equal to 85 W m^{-2}, according to a Colorado study.

Advective melting displays spatial differences associated with altitude
and aerodynamic shelter by topography and forest. These factors produce a different spatial pattern than in radiative melting, but similarly
prolong the melting season of a drainage basin. Altitude effects in the
northern Rockies produce a delay of 4 days per 100-m increase (Packer,
1971) and interact with spatial patterns of radiation melting. These interactions have been evaluated in New England (Hendrick et al., 1971)
and the Alps, where the snow cover is particularly patchy over a period
of more than 2 months (Aulitzky, 1958, 1965; Meier and Schädler, 1979).
Successive measurements of snow-covered area in a drainage basin by
air photography, aerial reconnaissance, and satellite-borne sensors are
employed to determine shrinkage of the cover and provide data from
which melting rates and energy supply can be assessed (U.S. Corps of
Engineers, 1956; Chernogorov, 1966; Leaf, 1975).

THAWING OF SOIL AND ICE

Different physical characteristics of snow cover and frozen soil govern
the respective couplings with their environments.

(a)   Soil, wet or frozen, has absorptivity of the order of 0.85–0.90 for
solar radiation, about four times that of snow cover.

(b)   Although soil–ice temperature remains at 273°K or below, the
ice is intimately mixed with mineral soil particles that have no such
restraint and can transfer heat to adjacent ice. The mixture of ice and
warmer soil particles can emit more radiation than snow cover (except
in low-emissivity high-silica soils).

(c)   Thin layers of frozen soil that form on autumn nights are subject
to melting from below, like an early snowfall.

Several centimeters of the top of a layer of frozen soil can melt on a
sunny day in spring, although meltwater cannot easily drain away. Mud
is a constant factor in military or any other off-road operations in spring,
and in summer is a problem in permafrost lands, confining heavy transport to winter.

The high absorptivity of ice cover of aquatic ecosystems in comparison
with snow cover increases the importance of solar radiation. High absorptivity also accentuates spatial differentiation in melting rates caused
by shading of a glacier surface by valley walls or by the orientation of
sectors of the glacier surface relative to the angle of the sun.

FREEZING

Water in leaves of ecosystems freezes under the control of biophysical forces, but nevertheless in response to leaf temperatures of 273°K or lower which are governed by the nocturnal energy budget, in particular the exchange of longwave radiation and sensible heat. Little water is involved, and so the quantity of latent heat of fusion is small.

Freezing of Wet Snow or Soil

Meltwater, held on the surfaces of snow grains at and near the upper surface of the snow cover, can be measured by calorimetry or remotely (Stiles and Ulaby, 1980). It begins to freeze if incoming longwave radiation and turbulent heat fluxes diminish to a sum less than 315 W m^{-2}. As long as liquid water is present, its freezing will supply energy at the rate of 335 kJ kg^{-1} and the snow will not drop in temperature. The fact that surface temperature did in fact decrease to 267°K through the night in the Sierra example discussed earlier in this chapter indicated that refreezing of meltwater occurred. The refrozen water cemented snow grains together to a depth of several centimeters, and this crust cooled substantially below the freezing point, releasing additional heat. The amount of water frozen in a crust is limited by several factors:

(a) The nocturnal intake of longwave radiation is usually not less than 200–250 W m^{-2}, so that the deficit below the 315 W m^{-2} threshold never becomes very large.

(b) The duration of the night is limited.

(c) As water drains downward, its initial concentration (0.05–0.10 of the snow mass) decreases, less is available to freeze, and more has escaped deeper into the snow cover and farther from the radiating surface.

The total energy turned over by refreezing in the situation discussed was 1.0 MJ m^{-2}, a typical quantity.

This carryover of energy from day to night makes an important contribution to the nighttime energy budget and forestalls strong cooling of the snow. Surface and air temperatures in the Sierra seldom fall lower than −5 to −10°C at night, partly as a result of the buffering action of this stored heat (Miller, 1955, p. 153). A similar carryover occurs when a cold dry air stream overspreads a snow cover following warm humid air, as was shown by careful measurements at Guelph, Ontario (McKay and Thurtell, 1978).

Storage, transfer, and release of energy also take place at the surface of wet soil in fall on a cloudless night with strong cooling by exchange of longwave radiation. Needle ice as a special case under specific moisture conditions is amenable to energy modeling (Outcalt, 1971).

The freezing of wet soil brings about the same diurnal cycle in bearing strength and off-road trafficability as freezing does in mountain snow. Movement may be possible only as long as the frozen crust remains and bogs down after it thaws in midmorning.

Anticyclonic weather in Indian summer produces a sequence of nights with steadily greater depths of soil cooling and freezing, which comes to involve soil-heat flux (Chapter XVII). Models of soil freezing may combine the surface energy budget and soil-heat flux (Cary *et al.*, 1978), the latter being perhaps more important.

Freezing of Bulk Water

Latent heat released when water freezes is utilized in the frost-protection practice of sprinkling orchard foliage, usually at a rate of 2 mm hr^{-1} (Bagdonas *et al.*, 1978, p. 57). Ice that forms on the leaves releases heat at a rate of 100–200 W m^{-2} that holds leaf temperatures at 0°C so long as the night is not too long, nocturnal chilling not too vigorous, or the ice too heavy. The objective is to form "icicles with clear ice encasing all parts of the plant" (Bagdonas *et al.*, 1978, p. 59); this practice walks a tightrope at 273°K in the hope that the sun will rise before the ice load breaks the trees.

Formation of an ice cover on shallow water bodies displays a diurnal freeze-and-thaw cycle similar to that at the surface of snow cover or wet soil, and similarly carries energy over from sunny days to clear cool nights. The establishment and thickening of ice cover on aquatic ecosystems involves substrate heat fluxes and will be discussed in Chapter XVIII.

REFERENCES

Anderson, E. A. (1968). Developing and testing of snow pack energy balance equation. *Water Resour. Res.* **4,** 19–37.
Anderson, E. A. (1976). A point energy and mass balance model of a snow cover. *NOAA (Natl. Oceanic Atmos. Adm.) Tech. Rep.* **NWS-19.**
Anderson, H. W., Rice, R. M., and West, A. J. (1958). Forest shade related to snow accumulation. *Proc. West. Snow Conf., 26th* pp. 21–31.
Arai, T. (1965). On the relationship between albedo and the properties of snow cover. *Shigen Kagaku Kenkyusho Iho* No. 64.

Arai, T. (1968). Hydro-climatological study on the mid-winter runoff from the snowy regions in Japan. *Geogr. Rev. Jpn.* **41**, 615–622. (Engl. summ.)

Aulitzky, H. (1958). Waldbaulich-ökologische Fragen an der Waldgrenzen. *Centralbl. Gesamte Forstwes.* **75**, 18–33.

Aulitzky, H. (1965). Waldbau auf bioklimatischer Grundlage in der subalpinen Stufe der Innenalpen. *Centralbl. Gesamte Forstwes.* **82**, 217–245.

Bader, H., Haefeli, R., Bucher, E., Neher, J., Eckel, O., and Thams, C. (1939). Der Schnee und seine Metamorphose. *Beitr. Geol. Schweiz, Geotech. Ser., Hydrol.* No. 3.

Bagdonas, A., Georg, J. C., and Gerber, J. F. (1978). Techniques of frost prediction and methods of frost and cold protection. *WMO Tech. Note* No. 157.

Bergen, J. D. (1975). A possible relation of albedo to the density and grain size of natural snow cover. *Water Resour. Res.* **11**, 745–746.

Cary, J. W., Campbell, G. S., and Papendick, R. I. (1978). Is the soil frozen or not? An algorithm using weather records. *Water Resour. Res.* **14**, 1117–1122.

Chernogorov, V. P. (1966). "Aerofotos"emka Snezhnogo Pokrova v Verkhov'iakh R. Angren dlia Gidrologicheskikh Tselei." Gidrometeoizd., Leningrad.

Colbeck, S. C., Anderson, E. A., Bissel, V. C., Crook, A. G., Male, D. H., Slaughter, C. W., and Wiesnet, D. R. (1979). Snow accumulation, distribution, melt, and runoff. *EOS, Trans. Am. Geophys. Union* **60**, 465–468.

DeWalle, D. R., and Meiman, J. R. (1971). Energy exchange and late season snowmelt in a small opening in Colorado subalpine forest. *Water Resour. Res.* **7**, 184–188.

Granger, R. J., and Male, D. H. (1978). Melting of a prairie snowpack. *J. Appl. Meteorol.* **17**, 1833–1842.

Hastenrath, S. (1978). Heat-budget measurements on the Quelccaya Ice Cap, Peruvian Andes. *J. Glaciol.* **20**(82), 85–97.

Haupt, H. F. (1969). A simple snowmelt lysimeter. *Water Resour. Res.* **5**, 714–718.

Haupt, H. F. (1972). The release of water from forest snowpacks during winter. *U.S. For. Serv., Res. Pap.* **INT-114**.

Havens, J. H., Azevedo, W. V., Rahal, A. S., and Deen, R. C. (1979). Heating bridge decks by electrical resistance. *In* "Snow Removal and Ice Control Research." Nat. Res. Counc., Transp. Res. Bd., *Spec. Rep.* **185**. 159–168, Nat. Acad. Science, Washington, D.C.

Hendrick, R. L., Filgate, B. D., and Adams, W. M. (1971). Application of environmental analysis to watershed snowmelt. *J. Appl. Meteorol.* **10**, 418–419.

Hildebrand, C. E., and Pagenhart, T. H. (1955). Lysimeter studies of snow melt. *U.S. Army Corps Eng. Snow Invest. No. Pac. Div., Portland, Ore., Res. Note* No. 25.

Hoeck, E. (1952). Der Einfluss der Strahlung und der Temperatur auf den Schmelzprozess der Schneedecke. *Beitr. Geol. Schweiz, Geotech. Ser., Hydrol.* No. 8. (SIPRE Transl. No. 49, 1958.)

Horton, R. E. (1915). The melting of snow. *Mon. Weather Rev.* **43**, 599–605.

Horton, R. E. (1936). Phenomena of the contact zone between the ground surface and a layer of melting snow. Int. Union Geod. Geophys., *Bull. Int. Assoc. Hydrol.* **23**, 545–561.

Hubley, R. C. (1957). An analysis of surface energy during the ablation season on Lemon Creek Glacier, Alaska. *Trans. Am. Geophys. Union* **38**, 68–85.

Hutchison, B. A. (1965). Snow accumulation and disappearance influenced by big sagebrush. *U.S. For. Serv., Res. Note* **RM-46**.

Khodakov, V. G. (1965). Nekotorye osobennosti taianiia bol'shikh lednikov i snezhnikov. *In* "Teplovoi i Vodnyi Rezhim Snezhno-Lednikovykh Tolshch" (G. A. Avsiuk, ed.), pp. 81–86. Izd. Nauka, Moscow.

Kotliakov, V. M. (1968). "Snezhnyi Pokrov Zemli i Ledniki." Gidrometeoizd., Leningrad. (Engl. summ.)

Kuz'min, P. P. (1960). Itogi issledovanii teplovogo i vodnogo balansov snegotaianiia i zadachi dal'neishikh rabot v etoi oblasti. In "Geografiia Snezhnogo Pokrova" (G. D. Rikhter, ed.), pp. 138–150. Izd. Akad. Nauk SSSR, Moscow.

Lamb, H. H. (1955). Two-way relationship between the snow or ice limit and 1,000–500 mb thickness in the overlying atmosphere. Q. J. R. Meteorol. Soc. 81, 172–189.

Larson, F. R., Ffolliott, P. F., and Moessner, K. E. (1974). Using aerial measurements of forest overstory and topography to estimate peak snowpack. U.S. For. Serv., Res. Note RM-267.

Leaf, C. F. (1975). Application of satellite snow cover in computerized short-term stream-flow forecasting. In "Operational Applications of Satellite Snowcover Observations" (A. Rango, ed.), pp. 175–186. Natl. Aeronaut. Space Adm., Washington, D.C.

Leaf, C. F., and Brink, G. E. (1973). Computer simulation of snowmelt within a Colorado subalpine watershed. U.S. For. Serv., Res. Pap. RM-99.

McKay, D. C., and Thurtell, G. W. (1978). Measurements of the energy fluxes involved in the energy budget of snow cover. J. Appl. Meteorol. 17, 339–349.

Meier, R., and Schädler, B. (1979). Die Ausaperung der Schneedecke in Abhängigkeit von Strahlung und Relief. Arch. Meteorol., Geophys. Bioklimatol., Ser. B 27, 151–158.

Meiman, J. R., ed. (1969). Proc. Workshop Snow Ice Hydrol., Colo. State Univ.

Mellor, M. (1964). Snow and ice on the earth's surface. In "Cold Regions Science and Engineering" (F. J. Sanger, ed.), Part II, Section C1. U.S. Army Corps of Engineers Cold Regions Research and Engineering Laboratory, Hanover, New Hampshire.

Mellor, M. (1977). Engineering properties of snow. J. Glaciol. 19(81), 15–66.

Miller, D. H. (1950). Insolation and snow melt in the Sierra Nevada. Bull. Am. Meteorol. Soc. 31, 295–299.

Miller, D. H. (1953). Snow cover and climate in the Sierra Nevada, California. Ph.D. Thesis, Univ. California, Berkeley.

Miller, D. H. (1955). Snow cover and climate in the Sierra Nevada California. Univ. Calif., Publ. Geogr. 11.

Miller, D. H. (1967). Sources of energy for thermodynamically-caused transport of inter-cepted snow from forest crowns. In "Forest Hydrology" (W. E. Sopper and H. W. Lull, eds.), pp. 201–211. Pergamon, Oxford.

Miller, D. H. (1977). "Water at the Surface of the Earth." Academic Press, New York.

Obled, C., and Rosse, B. (1977). Mathematical models of a melting snowpack at an index plot. J. Hydrol. (Amsterdam) 32, 139–163.

Outcalt, S. I. (1971). The climatonomy of a needle ice event: An experiment in simulation climatology. Arch. Meteorol., Geophys. Bioklimatol., Ser. B 19, 325–338.

Outcalt, S. I., Goodwin, C., Weller, G., and Brown, J. (1975). Computer simulation of the snowmelt and soil thermal regime at Barrow, Alaska. Water Resour. Res. 11, 709–715.

Packer, P. E. (1971). Terrain and cover effects on snowmelt in a western white pine forest. For. Sci. 17, 125–134.

Price, A. G., and Dunne, T. (1976). Energy balance computations of snowmelt in a subarctic area. Water Resour. Res. 12, 686–694.

Pruitt, W. O. (1957). Observations on the bioclimate of some taiga mammals. Arctic 10, 130–138.

Rantz, S. E. (1964). Snowmelt hydrology of a Sierra Nevada stream. U.S. Geol. Surv., Water-Supply Pap. 1779-R.

Reifsnyder, W. E. (1979). In "Symposium on Forest Hydrology" (W. E. Reifsnyder, ed.), pp. xii–xiii. World Meteorolical Organization (WMO No. 527) Geneva.

Richardson, S. G., and Salisbury, F. B. (1977). Plant responses to the light penetrating snow. Ecology 58, 1152–1158.

Riley, J. P., Israelson, E. K., and Eggleston, K. O. (1973). Some approaches to snowmelt prediction. *Int. Assoc. Hydrol. Sci., Publ.* **107,** 956–971.

Rockwood, D. M. (1972). New techniques in forecasting runoff from snow. *Int. Assoc. Hydrol. Sci. Publ.* **107:** 1058–1061.

Satterlund, D. R. (1977). Shadow patterns located with a programmable calculator. *J. For.* **70,** 262–263.

Satterlund, D. R., and Haupt, H. F. (1970). The disposition of snow caught by conifer crowns. *Water Resour. Res.* **6,** 649–652.

Schneider, S. R., Wiesnet, D. R., and McMillan, M. C. (1976). River basin snow mapping at the National Environmental Satellite Service. *NOAA (Natl. Oceanic Atmos. Adm.) Tech. Memo NESS* No. 83.

Slaughter, C. W. (1969). Snow albedo modification/A review of the literature. *U.S. Army Corps Eng., Cold Reg. Res. Eng. Lab., Tech. Rep.* No. 217.

Solomon, R. M., Ffolliott, P. F., Baker, M. B., Jr., Gottfried, G. J., and Thompson, J. R. (1975). Snowmelt runoff efficiencies on Arizona watersheds. *Univ. Ariz. Agric. Exp. Stn., Res. Rep.* No. 274.

Solomon, R. M., Ffolliott, P. F., Baker, M. B., Jr., and Thompson, J. R. (1976). Computer simulation of snowmelt. *U.S. For. Serv., Res. Pap.* **RM-174.**

Sommerfeld, R. A., and LaChapelle, E. (1970). The classification of snow metamorphism. *J. Glaciol.* **9**(55), 3 17.

Stiles, W. H., and Ulaby, F. T. (1980). The active and passive microwave response to snow parameters I. Wetness. *J. Geophys. Res.* **85,** 1037–1044.

Sturges, D. L. (1977). Snow accumulation and melt in sprayed and undisturbed big sagebrush vegetation. *U.S. For. Serv., Res. Note* **RM-348.**

Sverdrup, H. U. (1935). Scientific Results of the Norwegian–Swedish Spitzbergen Expedition in 1934. Part IV. The Ablation on Isachsen's Plateau and on the 14th of July Glacier. *Geogr. Ann.* **17,** 53–166.

U.S. Corps of Engineers (1956). "Snow Hydrology." Portland, Oregon. (Reprinted by U.S. Gov. Print. Off., Washington, D.C., 1958.)

U.S. Corps of Engineers (1960). "Runoff from Snowmelt," Eng. Design Man. EM-1110-2-1406. U.S. Gov. Print. Off., Washington, D.C.

Voeikov, A. (1885). On the influence of accumulations of snow on climate. *Q. J. R. Meteorol. Soc.* **11,** 299–309.

Wallén, C. C. (1949). Glacial–meteorological investigations on the Kårsa Glacier in Swedish Lappland 1942–1948. *Geogr. Ann.* **30,** 451–672.

Wilken, G. C. (1967). Snow accumulation in a manzanita brush field in the Sierra Nevada. *Water Resour. Res.* **3,** 409–422.

Chapter XV

PHASE CHANGES OF WATER IN ECOSYSTEMS: II. VAPORIZATION

Ecosystems leak water at the same time they take in the CO_2 that feeds their photosynthetic activity. The water leaking through cell walls in the leaves is converted from the liquid state, in which it has played physiological roles in the plant, to the mobile vapor state at an energy cost that is much greater than the cost of melting ice to liquid. This important conversion of energy into a latent, or hidden, form removes energy from the thermal arena and prevents leaves from overheating at the times when photosynthesis is in need of a stable temperature regime. Vaporization also removes excess water, an obvious benefit in spring when waterlogged soil becomes low in oxygen for ecosystem roots.

If we think of these occasions as simply due to unseasonable rain, it is because rain is basically unpredictable and we do not recognize how much we rely on the steadiness of the conversion of energy in vaporizing water. Yet the predictability of evapotranspiration is one of the values of this energy conversion in the life of ecosystems. The energy conversions associated with carbon (Chapters XI and XII) and water overlap in this process.

ENERGY CONVERSION BY VAPORIZATION

Surface Temperature

Surface temperature T_0, the key to all the conversions and fluxes of energy in ecosystems, is particularly important for the latent-heat conversion, and maintaining surface temperature in the face of heat-removing conversions means that energy has to be available. The energy demands of evaporation are large, a fact of molecular interaction expressed in the anomalously large latent heat of vaporization of water: 2.5 MJ kg^{-1}, varying slightly with temperature.

Unlike melting, vaporization is not restricted to a specific point on the temperature scale, but quietly proceeds in the cold as well as in the heat. At whatever temperature it occurs, however, it sequesters energy that the ecosystem must take in from one source or another.

Energy Sources

In the most common case most of the energy for vaporization comes from radiant-energy intake: the absorption, independent of surface temperature, of incoming longwave and solar radiation. Other sources are important only in such special sites as aquatic ecosystems and oases.

Radiant-Energy Intake. When leaf stomata open during the hours when photosynthetically active radiation is being received by the ecosystem to let CO_2 molecules enter they let H_2O molecules escape. Transpiration is powered in large part by absorption of radiation not used in photosynthesis because it is of the wrong wavelength or is delivered in excess of the assimilative capacity of the photosynthesizing apparatus.

Ecosystem albedo is important in modeling how evapotranspiration will be changed if, for instance, New Zealand tussock is converted to improved pasture (Fitzharris, 1974). Regional differences in longwave radiation also enter models of soil evaporation (Kalma *et al.*, 1977).

Densities of the absorbed fluxes approximate 500–600 W m^{-2} in midday hours during the growing season, of which photosynthesis claims only 10–20 W m^{-2}. Another 315 W m^{-2} or more must go as longwave radiation emitted at leaf temperatures necessary for photosynthesis, but much energy remains for other conversions, including latent heat, a major "dissipative" term in the budget (Lowry, 1969, p. 130).

This partitioning of radiant-energy intake into emitted radiation, sensible heat, and latent heat provides one basis for estimating the latent-heat conversion. Typically, the net surplus of all-wave radiation, soil-heat flux, and wind and temperature profiles are measured to determine sensible heat, from which latent heat is found. This was the best method over grassland in Canada (Ripley and Saugier, 1978), barley in England (Grant, 1975), and from a Netherlands catchment (Stricker and Brutsaert, 1978), and it was used in a model (Saltzman and Ashe, 1976; Saltzman, 1980).

The geometrical pattern of incoming radiant energy is important. An increase in the fraction of diffuse solar radiation, which penetrates ecosystems deeply, results in the opening of more leaf stomata and a reduction in the resistance to vapor diffusion, hence a potential increase in the latent-heat conversion (Webb, 1975).

Special Energy Sources. The salient thermal characteristics of aquatic ecosystems are their enormous capacity to store energy internally and their free-water surface. Since water and energy are both at hand during the night, latent-heat conversion continues, as it also does in the autumn after terrestrial ecosystems have gone into winter standby operation. The distinction is enhanced when fossil energy is added to the substrate heat supply. Cooling ponds steam all through the winter (Fig. 1).

Another energy source comes into action when moist and dry ecosystems lie side by side in a landscape mosaic (irrigation districts, lands with nonuniform soil-moisture availability, riparian ecosystems amid drier uplands) because air warmed over the drier systems drifts over the moister, cooler ones and supplies a downward flux of sensible heat. This source provides a supplement that may come to rival the radiation source in small ecosystems and at leading edges, as, for example, in

Fig. XV.1. Steam fog rising from near-shore water that is warmed by dump heat from Oak Creek power plant near Milwaukee makes visible the high rate of latent-heat flux into the dry cold air of January. Also visible is the effect of the buoyancy imparted by the sensible-heat flux from the stacks.

Tamarix in the Rio Grande floodplain (Gay and Fritschen, 1979). (Mesoscale transports among ecosystems will be discussed in Chapter XXI.) Heat advection on a regional scale occurs in the dry Great Plains, being an almost constant source to individual irrigated fields (Verma *et al.*, 1978).

ECOSYSTEM CHARACTERISTICS

Water Availability

A requisite for latent-heat conversion is an input of liquid-state water into the transpiring ecosystem, the plant-moisture status of which is often approximated by soil-moisture status. [For specific plant–water relations the reader is referred to Slatyer (1967) and Kramer (1969).] Amounts of energy needed in taking up water from small pores in the soil, moving it through the roots and stem, lifting it to the canopy level, and moving it out to the leaves are minor compared to the energy cost of vaporizing it. The transpiration stream carries nutrients in dissolved form, and in the sense that it is powered by the evaporating process at the end of the wick, so to speak, the radiant energy converted into latent heat represents a subsidy to the ecosystem from an outside source (Odum, 1971, pp. 47, 376). Nutrients influence the vigor of the ecosystem and hence its generation of latent heat; nitrogen fertilization of irrigated prairie increased latent-heat flux (Lauenroth and Sims, 1976) over 125 days by 17 W m^{-2}.

The rate of transpiration begins to decrease as water becomes less available in a drying ecosystem (Miller, 1977, Chaps. XI and XII; Black, 1979), and, if energy intake remains the same, a shift has to take place in the way it is partitioned. This shift usually requires a rise in leaf temperature, which steepens the temperature gradient along which sensible heat moves and increases the emission of longwave radiation, and surface temperature excess over air temperature can be used as a stress index (Jackson *et al.*, 1977). While the higher leaf temperature T_0 means a higher vapor concentration q_0, the effect of this in the vapor gradient is countered by narrowing of the stomata, which increases the internal resistance r_l in the diffusion pathway. This brings about a decrease in the inward diffusion of CO_2 and in the energy conversion in photosynthesis. All the energy fluxes shift as soil moisture is drawn down.

Ecosystem Structure

The soil component of an ecosystem contains moisture, but is not accessible to the large energy inputs that would support rapid evaporation. The more active component of terrestrial ecosystems, where "spatial and temporal confluence of water and thermal energy" occurs (Goodell, 1967), lies in the atmosphere where leaf surfaces are deployed to facilitate the exchange of gases between their organs and the air.

An index of the openness of an ecosystem canopy is its leaf-area density, expressed, for example, as 0.5 m^2 of leaf area per cubic meter of plant–air volume, which would correspond to total leaf area of 5 m^2 m^{-2} in a depth of 10 m. Considering how thin most leaves are, it is evident that the leaf volume in a cubic meter of plant–air volume is only 1–2×10^{-3}, a fraction of 1%. Ventilation of individual leaves is good in such a dilute mixture of leaves in air,* as is shown when transfer coefficients are measured (Murphy and Knoerr, 1977). Ventilation is also increased when a water surface is disturbed (Hicks, 1980).

Unfolding of leaves of deciduous ecosystems in spring increases the transpiring area, but in some species, such as *Quercus rubra* in New Hampshire, this increase is countered by a temporary rise in the internal resistance to vapor diffusion, alleviating vapor loss while the water conduction mechanism is still being developed (Federer, 1976). Regrowth of leaves following hay cutting increases evapotranspiration (Mustonen and McGuinness, 1967) at a daily rate equal to about 2.5% of the full-leaf rate.

Surface Vapor Concentration.   The gradient along which vapor molecules leave an ecosystem is that of vapor concentration. This parameter, at saturation in plant leaves, is a function of leaf temperature—in other words, ecosystem surface temperature. Because surface temperature plays an analogous role in forming the gradient along which sensible heat moves from an ecosystem, a useful parallelism exists between the fluxes of latent and sensible heat, which will be discussed in Chapters XVI and XXIII. High surface temperature generates free convection, which amplifies "the forced convective eddies generated by vertical wind shear alone and vertical transfer processes are substantially enhanced" (Thom, 1975), thereby accelerating the removal of vapor from an ecosystem.

The relation of the temperature of an evaporating surface, such as the wall of a leaf cell, to the saturation vapor concentration q or vapor pressure e in the overlying space is called the Clausius–Clapeyron relation (Table I). The slope of the vapor–temperature relation S increases with temperature, and a degree rise in surface temperature near freezing

---

* Self-sheltering of needles in a cluster or shoot (Jarvis *et al.*, 1976) reduces the almost excessive ventilation of these narrow leaves, which have a small aerodynamic resistance to vapor diffusion.

TABLE I

Clausius–Clapeyron Relation[a]

Surface temperature T_0 (°C)	Vapor concentration[b]	
	Specific humidity q (parts per thousand)	Vapor pressure e (mbar)
−20	0.8	1.2
0	3.8	6.1
+20	14.9	23.4
+40	49.7	73.8

[a] From Smithsonian Institution (1966).
[b] Note: Specific humidity q and vapor pressure e are related through the equation $q = 0.622\, e/(p - 0.378e)$, where p is atmospheric pressure. Both pressures are measured in millibars (mbar) or kilopascals (kPa); 1 mbar = 0.1 kPa.

creates only a small rise in vapor concentration. The same degree temperature rise under midday conditions in the growing season creates a large increase (1.2 parts per thousand at $T_0 = 25°C$) in specific humidity, which steepens the outward vapor gradient.

Diffusion of Vapor

Transpiration and photosynthesis operate in parallel in C_3 and C_4 plants, although productivity for a given water transport may be higher in C_4 ones (Milthorpe and Moorby, 1979, p. 111).Molecules of CO_2 diffuse inward from the local air through the laminar film coating the leaf, through the stomata into the mesophyll spaces and to the wet cell walls, where they are dissolved and carried into the cells that fix them into photosynthates. At the same time, vapor molecules generated at the cell walls diffuse outward through the mesophyll spaces and stomata, through the laminar film into the local air, and on into the free atmosphere. This well-known association of photosynthate and vapor production (described in Chapter XI) displays variations that permit distinguishing of local exposures of ecosystems (Rauner, 1973). Forest stands studied in the International Biological Programme produce 1–2 g of aboveground net primary production per kilogram of water vaporized, and hot-desert ecosystems produce 0.1–0.3 g (Webb et al., 1978).

The parallel paths of CO_2 and H_2O molecules encounter resistances to diffusion within the leaf and across the laminar film which qualify the simple vapor-concentration difference between q_0 in the leaf to q_a in the air:

$$E = \rho \, (q_0 - q_a)/(r_l + r_a),$$

in which E is the flux of vapor molecules (kg m^{-2} sec^{-1}); ρ is air density (kg m^{-3}); q_0 and q_a are dimensionless ratios of mass of vapor to mass of air at the evaporating cell walls and in the air, respectively; and r_l and r_a denote resistances within the leaf and air film, respectively. Dimensions of r (sec m^{-1}) are the inverse of velocity. The utility of this form rather than its reciprocal, the diffusion coefficient, is that resistances in each locus traversed by the diffusing vapor molecule can be added.

When leaf stomata are partly closed r_l increases, an effect that is evaluated from independent determinations of latent-heat flux under varying conditions of soil or plant moisture (Davies, 1972). Field measurement of r_l is now also feasible (Fetcher, 1976), and porometer data in Douglas fir were adequate, with leaf-area index, to calculate vapor diffusion (Tan et al., 1978). In some plant stands r_l has been found related to specific-humidity deficit; as measured in cut trees that were recut under water and had their uptake measured, r_s was 200 sec m^{-1} at δq $= 8 \times 10^{-3}$ (Roberts, 1978). The related, "bulked-up" canopy resistance, which is inversely proportional to $1 - 0.27 \times 10^2 \, \delta q$ and a sinusoidal seasonal term, fits into a transpirational model that reconstitutes the annual cycle of latent-heat flux from a stand of spruce (Calder, 1978).

The physiological control of water loss from an ecosystem is a survival mechanism in which the stomata play a major role (Waggoner and Zelitch, 1965) though at times their effect on vapor diffusion has been poorly modeled (Lee, 1967). Stomatal control prevents excessive loss of water from leaves and prevents water potential in the xylem from diminishing (Federer and Gee, 1976). Chemical antitranspirants that close stomata reduced transpiration enough to result in an increase in streamflow in an experiment in Idaho (Belt et al., 1977), reducing mean latent-heat flux over a 63-day period in summer by 15 W m^{-2}. Dunin and Reyenga (1978) find about 0.2 of total latent-heat conversion in native Australian grassland occurs under meteorological influences, but 0.8 to be "potentially subject to stomatal and other control mechanisms;" streamflow is thereby affected by species changes (Bell, 1980).

Diffusion outside the leaf, r_a, depends on air motion and leaf size. Data on evaporation of intercepted rain from a canopy, which depends only on r_a and can be very large (Singh and Szeicz, 1979; Massman, 1980) can be compared with evapotranspiration from an externally dry canopy to evaluate r_l, which typically has a numerical value around 60 sec m^{-1} (Webb, 1975). Comparison of wet pines and wet grass show the effects of superior ventilation of the taller ecosystem (Pearce *et al.*, 1980). Evaluation of $r_a + r_l$ can be also done by analogy with the inward diffusion of CO_2 from the air if an additional resistance for CO_2 at the cell wall is added and an adjustment is made for the heavier CO_2 molecule. Ecosystems in the field generally experience enough air motion that r_a seldom dominates diffusion to the exclusion of stomatal control (Lee, 1967): "The purely mechanical resistances to gaseous diffusion set up by a system of small pores is effective even in still air; its effectiveness increases in moving or turbulent air."

Diffusion through the air of the ecosystem and into the free air above can be conceptualized in a similar way, but now including the buoyant forces associated with the concomitant flux of sensible heat. The structure of ecosystem canopies, evolved to facilitate the uptake of CO_2, which is hardly more abundant than a trace substance in the atmosphere, and to optimize the absorption of solar photons, is well suited to disperse H_2O molecules from an extensive evaporating surface, i.e., the hundreds of square meters of cell walls in a square meter of ecosystem area. Even though the initial steps of the journey of each molecule are taken in the relatively slow mode of molecular diffusion, the net result is one of fairly rapid flux of vapor and the latent heat it carries, which attain densities of 600–700 W m^{-2}.

ATMOSPHERIC REMOVAL OF VAPOR FROM ECOSYSTEMS

Terrestrial ecosystems are arenas for interaction of converging streams of liquid water and radiant energy and diffuse the resulting vapor molecules into the plant-air space. These molecules must be removed from the ecosystem as a whole to keep vaporization going.

Exchanges of vapor, CO_2, and air between the plant-air volume in an ecosystem and the air above it depend on the openness of the canopy and its roughness, which augments mechanical turbulence in the air stream, as well as on the thermal stability determined by relative temperatures of canopy and local air. The first two characteristics pertain to mechanical turbulence as a mode of vertical eddy diffusion, and the third to free or thermal convection.

Because the destabilizing sensible-heat flux is usually in operation during the sunny midday periods when latent heat is being generated, thermal convection is important in removing vapor from an ecosystem. The destabilizing effect is augmented by the low density of vapor (molecular weight 18) compared with air (molecular weight ~29) (Stricker and Brutsaert, 1978). Because the L-scale shrinks close to the earth in midday conditions, with L (Monin–Obukhov length) being only a few meters, thermal convection takes over close to the transpiring ecosystem and can rapidly move vapor to a distance of a kilometer or so above the system. Moreover, midday destabilization brings high speeds down near the ground to increase mechanical convection by the additional turbulent kinetic energy, which helps move vapor through the air in and just above the ecosystem.

Expressions for Flux Density

Equations for latent-heat flux density involve a gradient of atmospheric specific humidity through the layer overlying an ecosystem or a difference between surface and atmospheric humidity above it. Both can be interpreted in terms of ecosystem temperature and energy budget.

Gradient Form. The gradient form is

$$\lambda E = \lambda \rho K_W \, \partial q / \partial z,$$

in which λ is the specific heat of vaporization (2.5 MJ kg^{-1}), ρ is air density, q specific humidity. K_W, diffusivity for water vapor ($m^2 \ sec^{-1}$), is an analog of K_H for sensible heat, but they are not necessarily equal (Munn, 1966, p. 95); K_W is perhaps systematically smaller than K_H in ecosystems receiving substantial supplements of advected sensible heat (Verma *et al.*, 1978). In situations of approximate equality this circumstance is expressed in the Bowen (1926) ratio, which is utilized in a common method (see Chapter XXIII) of partitioning the total flux of latent and sensible heat from an ecosystem. The aerodynamic equation of Thornthwaite and Holzman (1939) evaluates K_W from wind shear in conditions of neutral stability, but in other conditions cannot be applied directly.

The gradient of specific humidity $\partial q / \partial z$ varies with height (Lowry, 1969, p. 76) and varies inversely with z^{-2} and wind shear; under unstable conditions (Webb, 1975), the effect of instability is shown by a profile shape factor that is related to the relative height z/L and other parameters identified in painstaking field studies. The vertical gradient should not be regarded as the *cause* of vapor removal but rather as "a function of the evaporation rate," i.e., the rate of flux generation (Thornwaite and Hare, 1965).

Surface versus Air. The humidity-difference approach uses q_0 at the wet surface, a function of leaf temperature, and q_z at some height above the ecosystem:

$$\lambda E = \lambda \rho (q_0 - q_z) f(u_z),$$

in which $f(u_z)$ is a function of wind speed at height z that "represents the efficiency of transfer through the atmosphere" (Webb, 1975) and can be varied to take account of surface roughness and stability or instability. It can be replaced by r_a^{-1}, the reciprocal of the aerodynamic resistance to vapor diffusion from leaf surfaces to the free air.

Combination Form. A combination of this equation with the analogous sensible-heat flux equation gives a sum that is equal to the radiant-energy intake of an ecosystem less emitted longwave radiation and soil-heat flux and that we will call $R_n - G$. The combination of the turbulent fluxes at wet surfaces employs the relation of saturation specific humidity to surface temperature, the slope S of the Clausius–Clapeyron relation mentioned earlier in this chapter. Combining the two surface-based expressions eliminates surface temperature and surface humidity and arrives at an expression employing only shelter-height measures of temperature and humidity, transformed into saturation and actual specific humidity (q_z' and q_z), which is

$$\lambda E = \frac{S}{S + \gamma} (R_n - G) + \frac{\gamma}{S + \gamma} (q_z' - q_z) f(u) \lambda \rho,$$

in which γ is the ratio of specific heat of air to the latent heat of vaporization $\approx 0.4 \times 10^{-3}$ deg^{-1} (Webb, 1975).

The first term on the right is called a radiation term, since it includes net all-wave radiation R_n (Chapter X) and has as a multiplier

$$\frac{S}{S + \gamma} = \frac{1 \times 10^{-3}\,\mathrm{deg}^{-1}}{(1 \times 10^{-3}\,\mathrm{deg}^{-1}) + (0.4 \times 10^{-3}\,\mathrm{deg}^{-1})} = 0.7,$$

at temperatures around 25°C. Thus the latent-heat flux claims from this term about 0.7 of the total radiative surplus less soil-heat flux, or 0.7 of the radiant-energy intake (Chapter VII) less nonatmospheric energy removals, namely, emitted radiation (Chapter IX) and the substrate-heat flux (Chapter XVII).

How much the latent-heat flux claims through the second term depends on its four parts:

(1) The ratio $\gamma/(S + \gamma) = 0.4/1.4 = 0.3$ at temperatures of about 25°C and is larger in colder conditions (see Table II).

(2) The expression $(q_z' - q_a)$ indicates the dryness of the air above the ecosystem; its smallness in cold conditions tends to compensate for the larger ratio $\gamma/(S + \gamma)$.

TABLE II

Variation in S and γ Relation with Temperature[a]

	T		
	10°C	20°C	30°C
$S/(S + \gamma)$	0.55	0.68	0.78
$\gamma/(S + \gamma)$	0.45	0.32	0.22

[a] Data from Slatyer and McIlroy (1961, p. V-7).

(3)   The function of wind speed $f(u)$ includes adjustments for stability and ecosystem roughness that are hard to quantify.*

(4)   The expression $\lambda\rho$ (in J m^{-3}) makes the conversion into energy units.

The entire second term on the right-hand side of the equation is called aerodynamic and incorporates atmospheric dryness and turbulence. The second part of this term has special interest: the drier the air the more it will give up internal energy to power ecosystem evaporation. Moreover, in dry climates $q'_z - q_z$ interacts with soil moisture to influence the leaf resistance to vapor diffusion (r_l) in pine needles (Fetcher, 1976). Conversely, where humidity deficit is small, transpiration is reduced even when stomata are open for full access of CO_2. A given quantity of irrigation water "will result in more plant growth in an area of high atmospheric humidity than in one of low humidity" (Arkley, 1963). When this second term becomes zero in saturated air the generation of latent heat does not cease but comes to depend solely on the radiation term in the basic equation. This accounts in large part for the fact that latent-heat flux expressions are controlled by the underlying surface (Davies, 1972). In addition, the wind-speed function $f(u)$ (or r_a) is also surface dependent.

The slope S of the Clausius–Clapeyron relation is large in warm conditions and amplifies the radiation term. The diurnal period in vaporization can be interpreted in two ways:

(1)   Several terms of the combination equation become small or even negative at night, specifically $R_n - G$ and $q'_z - q_z$, and wind speed is low.

---

* In fact, ecosystems of large roughness may or may not be appropriate for application of aerodynamic methods (Thom et al., 1975; Hicks et al., 1979).

(2)   Alternatively, there is no solar energy to activate photosynthesis and cause stomata to open, provide energy for vaporization of water within the leaves, and destabilize the local air.

Downward Vapor Flux

Reversal of the latent-heat flux usually occurs only at night. This flux is never large, for reasons of energy balance. For example, formation of dew at a latent-heat flux density of 7 W m^{-2} under a clear sky halted with the arrival of clouds that increased incoming longwave radiation by 50 W m^{-2}; in fact, the deposit that had already formed evaporated (Hofmann, 1956). A typical flux density on dewfall nights is 5–10 W m^{-2} (Monteith, 1957), and the maximum rate seldom exceeds 40 W m^{-2} in central Europe (Hofmann, 1956).

The downward flux is small because of weak turbulence in stable air, the smallness of the $R_n - G$ term, and competition by the associated downward flux of sensible heat to the ecosystem (Miller, 1977, pp. 362–366). As noted in Chapter X, the net exchange of energy by long-wave radiation at night R_n, under the most favorable conditions of warm soil (large upward flux) and clear, dry air (small downward flux), lies in the 70–100 W m^{-2} range. Yet, as Wells showed in 1814, it is the principal factor in dew formation (Middleton, 1966, p. 188). Soil conditions at times of high atmospheric humidity usually favor a large soil-heat flux G, further reducing the $R_n - G$ term.

The G term is very small in dry snow, and condensation is frequent in the Alps (Lauscher, 1977) and Sweden (Lauscher, 1978), usually compensating evaporation from winter snow cover.

Heat sinks that maintain low values of T_0 and q_0 under advection of humid air experience large downward latent-heat fluxes. Cold lakes can take in any quantity of heat that condensation could release at their surfaces, and melting snow, as described in Chapter XIV, can also dispose of any quantity of the heat of condensation. Such augmentation of the downward flux occurs to a lesser degree on cold streets when moist air arrives in a city, and in winter may produce an ice layer (Kraus, 1964).

Two cases of extraction of vapor from unsaturated air may be noted because of their interesting energetics:

(a)   Hygroscopic salts take up vapor and condense it on the leaves of a plant, *Nolana mollis*, studied by Mooney et al. (1980), in a desert on the coast of Chile where "high fog" (i.e., stratus) occurs but the air near the ground is usually not at saturation. At 85% relative humidity, salt exudates on the leaves of *Nolana* take up atmospheric water at a rate of $30 \text{ mg m}^{-2} \text{hr}^{-1}$. This salty water drips to the ground and forms a saline soil solution from which the plant roots extract the small quantity of water needed at the expenditure of about 0.1 W m^{-2}, which is a large fraction of its respiratory energy conversion.

(b)   Humans in need of the cooling effect of skin evaporation can reduce the vapor concentration in their environment by mechanical means that convert latent heat to sensible. Part of the flux of sensible heat from this air conditioner comes from its cooling of room air and its own energetic operating costs, but about half represents heat released when vapor is condensed and water drips out of the machine.

Other Methods of Evaluating Vaporization

Observations of loss of weight of ecosystem moisture or of vapor carried upward are also used to evaluate latent-heat conversion. Energy-budget models, first used by Ångström (1920) and Bowen (1926), are in frequent use (Fitzharris, 1974; Black, 1979; Stricker and Brutsaert, 1978) and, indeed, are responsible for much interest in the energy budget per se.

The water-budget method requires considerable periods of time unless the scales of the lysimeter are unusually precise. As applied in the field, it encounters sampling difficulties in determining changes in soil moisture. For example, in rainless periods with strong evapotranspiration, with six or more sampling sites for soil moisture, intervals of at least 4 days are still required for "reasonable accuracy" (Rouse and Wilson, 1972). Longer periods are required when transpiration is weak or fewer soil–moisture sampling sites are used and only time-averaged rates of energy conversion can be found. On a larger spatial scale precipitation and runoff measurements are used to estimate evapotranspiration from drainage basins, apparently first by Ångström (1925), who recognized the unrepresentative nature of evaporation pans.

The direct aerodynamic method (Thornthwaite and Holzman, 1939; Dyer, 1974) makes use of similarities between the vertical profile of wind speed and that of water vapor concentration. The effects of stability are included in various empirical ways, most successfully over low vegetation such as grassland in Saskatchewan (Saugier and Ripley, 1978), and New South Wales (Dunin et al., 1978).

The eddy correlation method meets problems in finding uniform surface conditions; in addition, humidity sensors are not quick to respond and levels where gusts are small near ecosystem surfaces cannot be sampled. Under the right conditions, however, it provides a direct measure of vapor flux (Dyer and Maher, 1965), which is to be preferred over estimates based on aerodynamic profiles. It can now be measured in the field "with the same ease as profiles" (Dyer, 1974) and converted to energy terms.

VARIATIONS IN LATENT-HEAT CONVERSIONS

Variations over Time

Short-term variations in the latent-heat flux occur as responses to momentary fluctuations in the solar beam, e.g., its interruption by a passing cloud, because leaves have little heat storage to tide them over interruptions in energy input.

Diurnal Cycle. The diurnal variation in the flux stems basically from the daytime-only intake of CO_2 in most C_3 and C_4 terrestrial ecosystems. Exceptions are those plants that can take in CO_2 at night and open their stomata without incurring much leakage of water, because energy input is small. Pineapple, one of the species using the crassulacean-acid metabolism route in photosynthesis, suspends transpiration after midday in spite of extensive leaf area and high leaf temperatures (caused in part by their thickness) (Ekern, 1965). Its 24-hr mean rate of latent-heat conversion, 30 W m^{-2}, is a fraction of that of Bermuda grass at the same site.

Energetic conditions do not favor evaporation at night. Much of the deficit in net all-wave radiation is met by the heat flux from the soil; so the first or radiative term $R_n - G$ in the combination equation indicates zero latent-heat conversion, or negative in the periods of dew formation. If the air remains warm, the second term in the equation can provide energy but it is weakened by low wind speeds and the stability correction. The $q_z^* - q_z$ expression also diminishes at night.

Evaporation seems more likely to linger in the dusk than to make a fast start at sunrise. For example, the conversion rate in June at Hamburg at 1600 is double that at 0800, and that at 1800 triple that at 0600 (Frankenberger, 1960), and sap flow in trees (Swanson, 1967) suggest a small flux continuing after sunset. In general, however, as Kozlov shows [see diagrams in Miller (1977, p. 279)], the daily course of transpiration is closer to that of solar radiation than to that of air temperature.

A midday reduction of latent-heat generation occurs in some ecosystems, which may (Ittner, 1968) or may not be reversed in later hours. Field measurements of r_l in lodgepole pine in Wyoming showed increases from 300–500 to 1000–1400 sec m^{-1} at mesic sites and 5700 sec

m^{-1} on the driest site (Fetcher, 1976); sunflowers, on the other hand, generate more latent heat in the afternoon than in the morning (Saugier, 1976).

A small contribution of vapor from the soil, still warm from the day's heating (Thornthwaite and Hare, 1965), may extend the pulse. After the first phase of soil drying the locus of evaporation goes deeper, and latent-heat flux does not begin until several hours after sunrise (Zito *et al.*, 1978).

Exceptions to the nocturnal shut-off are found in such species as pineapple, in terrestrial ecosystems still wet from rain, and in wet soil and aquatic ecosystems. If advected sensible heat is available, intercepted rain on leaves evaporates rapidly. Aquatic ecosystems have a free water surface plus availability of substrate heat; a shallow lake in Canada converted most of the solar heat it took in during the long days into nocturnal evaporation (Stewart and Rouse, 1976), and this is typical behavior.

Synoptic-Scale Variations. Day-to-day variations in the conversion of latent heat in terrestrial ecosystems are most likely to be caused by depletion of underground water reserves and so reflect root depth and the stochastic nature of rainfall. For example, the careful measurements at O'Neill, Nebraska, showed daytime flux density declining from 225 to 45 W m^{-2} as the soil dried out, and r_l increased manyfold.

Another source of variance lies in day-to-day changes in atmospheric dryness and the radiant-energy intake. These can occur at different time scales than rain and are superimposed on the steady depletion of soil moisture during an interstorm period, but not in a simple additive way. The latent-heat conversion on a high-radiation day may be reduced even when soil moisture is high enough to support full transpiration on a day of lesser energy intake (Denmead and Shaw, 1962); the soil-moisture effect is not constant, but varies with daily radiant-energy intake.

An idea of interdiurnal variability in the latent-heat flux is given by a frequency distribution of mean daily flux densities from grass at Hamburg-Quickborn (Table III). For example, the daily mean flux in June was upward on every day and exceeded 25 W m^{-2} on 0.97 of the days and 100 W m^{-2} on a third of them. The standard deviation among days, 35 W m^{-2}, is about 0.3 of the overall mean. It reflects, in addition to declining interstorm soil moisture and variable radiant-energy intake due to cloudiness, the effects of vapor concentration and temperature in different air streams on the aerodynamic term of the combination equation; dry air increases the latent-heat conversion, and so does warm air. However, advection has a different, usually lesser, significance for vaporization than it does for melting.

TABLE III

Latent-Heat Conversion by a Meadow near Hamburg[a]

	J	F	M	A	M	J	J	A	S	O	N	D
Percent of days (in 1958) when upward flux averaged												
>0 W m⁻²	97	97	100	100	100	100	100	100	100	100	100	94
>25 W m⁻²	0	7	42	94	97	97	100	100	97	42		0
>50 W m⁻²	0	0	3	33	89	70	77	89	53	3		0
>100 W m⁻²	0	0	0	0	22	33	6	16	0	0		0
Midday intensity (1100–1200)[b]												
Mean of all days	−45	−35	−45	−128	−240	−220	−197	−197	−140	−60	−35	−25
Mean of clear days	−45	−25	−25	−130	−230	−370	−310	−280	−175	−100	−60	—
Mean monthly flux average	−8	−10	−21	−42	−85	−88	−82	−78	−50	−22	−12	−6

[a] Data from Frankenberger (1955, 1960).
[b] Data for period 1953–1954. Data given in watts per square meter.

The Annual Cycle. Table III presents frequency distributions of daily values of the latent-heat conversion in each month of 1953–1954 and shows the rather short season in which a substantial fraction of days average more than 100 W m^{-2}. On the other hand, days of moderate-sized mean rates (exceeding 25 W m^{-2}) continued to occur through October. This hysteresis lag appeared in observations in Melbourne by McIlroy and Angus (1964), who found that at a given level of the net all-wave radiation surplus the latent-heat flux on a fall day averaged about 30 W m^{-2} more than on a spring day. They ascribe part of this effect to warmer soil in fall, part to large-scale heat advection.

In contrast, other ecosystems reduce transpiration in late summer; "stomatal control of transpiration is more important toward the end of the growing season" as shown by the onset of increased afternoon values of r_l (Fetcher, 1976). Soil-moisture shortages often suppress evaporation in late summer in the Midwest after spring stores of soil moisture have been expended by summer evaporation in excess of summer rainfall. Grassland in Saskatchewan converts a fraction of its radiation surplus into latent heat in dependence on green leaf area in spring and early summer, but on soil moisture in late summer (Ripley and Saugier, 1978). In other climates, however, there is a late-summer minimum in canopy resistance to vapor diffusion (Calder, 1978).

Evaporation continues in winter if ecosystems are not dormant, but even in mild-winter climates its rate is an order of magnitude less than in summer. The annual cycle of latent-heat conversion reflects the cycles of water availability, ecosystem surface development, and radiant-energy intake. The fourth factor, atmospheric removal of vapor, does not appear to be important in the yearly regime, but there is a seasonal change of $S/(S + \gamma)$ since S is temperature dependent.

Phenological stage obviously affects a process so intimately associated with biological production in an ecosystem and may amplify seasonal changes in the energy regime (Tables IV and V). Both examples are from high latitudes, where the regime of radiation is markedly seasonal, to the extent that liquid water is absent in the low-sun season.

Annual ecosystems display a marked change in latent-heat generation as they develop their leaf canopies. Maize in the emergent state converts into latent heat as much energy as meadow, but in the tasseling–silking period 0.1 more at 10°C and 0.5 more at 20°C (Konstantinov, 1968, p. 386). In these vulnerable stages it increases its latent-heat conversion by 1.2 W m^{-2} for each 1-mm increase in available soil moisture, whereas in the fully mature state its energy response is only half as great (Konstantinov, 1968, p. 412).

A patternless variation in latent-heat flux over time occurs in warm

TABLE IV

Mean Daily Flux of Latent Heat from Tundra Meadow Ecosystems[a]

Phase		
I	Midwinter	~5 downward
II	Premelting	~10 downward
III	Melting (June)	35
IV	Postmelting	125 (evaporation of meltwater pools)
V	Summer	80
VI	Freeze-up (early September)	35

[a] Data from Lewis and Callaghan (1976, pp. 410–414). Data given in watts per square meter.

regions where rainfall is patternless, as is true in semideserts and deserts such as central Australia and southern Arizona. If the random rainy periods come in the low-energy season, ecosystems remain active several months and generate latent-heat fluxes of 10–15 W m^{-2}; if in the high-energy season, their activity is briefer and more intense, drawing down the small store of soil moisture in a few weeks (Cable, 1977) at flux rates around 30 W m^{-2}.

Variations over Space

Ecosystem Scale. Plant density within ecosystems affects rates of both transpiration and soil evaporation, not necessarily in compensating degrees. In saltbush in Western Australia increased density raised the rate of transpiration but had no effect on soil evaporation, suggesting that closer planting might be beneficial in lowering a saline water table (Greenwood and Beresford, 1980).

TABLE V

Mean Daily Flux of Latent Heat from a Deciduous Forest Ecosystem near Moscow[a]

Period		
I	Snow melting	35
II	Postmelting	70
III	Leafing-out	125
IV	Full foliage	150
V	Yellowing and leaf fall	90
VI	After leaf fall	15

[a] Data from Rauner (1976). Data given in watts per square meter.

Early-season evaporation from soil in row-crop ecosystems is a third to half the total latent-heat conversion (Thornthwaite and Hare, 1965), but by full leafing out it is nearly an order of magnitude smaller than transpiration. It has only the minor effect on the diurnal regime that was noted earlier in this chapter. The ratio of transpiration to total evapotranspiration is

$$1 - \exp(-\alpha \, LAI),$$

in which α is an extinction coefficient similar to that for radiation entering the ecosystem and LAI is leaf-area index (Uchijima, 1976). In typical foliage (LAI = 4 and extinction coefficient 0.5) this ratio is 0.87; i.e., transpiration makes up 0.87 of the total latent-heat conversion and soil evaporation 0.13. Other aspects of flux generation with depth will be discussed in Chapter XX, but one unusual case can be noted: the generation of a latent-heat flux of 4 W m^{-2} at the summit of Mauna Kea from permafrost deep in the volcanic ash (Woodcock and Friedman, 1980).

Mosaic Scale. Differences among ecosystems reflect contrasts in the basic factors of water availability, transpiring surfaces, radiant-energy input, and vapor removal, which can be expressed in the terms in the combination equation.

(1)   Moisture status of ecosystem and soil affects canopy resistance (Webb, 1975);

(2)   The net surplus of all-wave radiation R_n shows contrasts in beam angle, ecosystem absorptivity and surface temperature;

(3)   contrasts in storage of energy in the soil affect soil-heat flux G;

(4)   differences in air temperature affect S and q_z' and dryness;

(5)   Contrasts in wind speed, stability and roughness modify the wind term.

Differences in radiation input to ecosystems in dissected terrain result from differences in beam-radiation absorption and in terrain-bound clouds (Webb, 1975). In conditions of uniform moistness and in quiet anticyclonic weather only minor differences in latent-heat generation in grass ecosystems in southern England occurred as long as cloud streets that would cause spatial differences in radiation intake did not form (Wood, 1977). Different ecosystems that contrast in albedo and roughness can be compared by energy budgets (Fitzharris, 1974).

Opposing effects of radiation intake and soil moisture were found on the north and south slopes of one of the volcanic hills near Montréal (Rouse, 1970). The deciduous forest on the south slope enjoyed a larger surplus of all-wave radiation (10–15 W m^{-2} more), but experienced a shortage of soil moisture caused by too-early snow melting. The moisture factor outweighed the radiant-energy factor, and the south-slope stands generated less latent heat than those on the north slope. We see the integrated effects of these factors in such contrasts (to be discussed more comprehensively in Chapter XXI) as those between forest stands and low vegetation (Baumgartner, 1967), between ecosystems on thin and deep soils, ecosystems at different phenological stages, and ecosystems of different species compositions (Bell, 1980). In comparing oasis versus desert, or riparian versus upland ecosystems in dry country, or terrestrial versus aquatic ecosystems, the contrast is sharpened because the small latent-heat conversion in the drier systems shunts energy into sensible heat that is in part carried over to the moister systems to augment generation of latent heat in them.

Synoptic and Regional Scales. On larger spatial scales the patterns of latent-heat conversion follow those of energy and water inputs. Replenished ecosystems behind an isolated convective rainstorm resume photosynthesis and transpiration while systems outside the wetted swath keep on waiting. The wetted systems are affected by the dryness and warmth of the regional atmosphere; region-wide moisture conditions influence local latent-heat conversions (Tweedie, 1956; Bouchet, 1963; Seguin, 1975).

An equilibrium tends to develop between a regional-scale expanse of well-watered ecosystems and the air in steady nonadvective weather, and the second term of the combination equation tends to become constant,* in some cases at about one-quarter of the first term (Priestley and Taylor, 1972). The effect of wind has vanished, but the temperature factor continues through its effect on $S/(S + \gamma)$.

In such a large equilibrium system surface evapotranspiration provides the raw material for middle-atmosphere condensation into clouds and rain, clearly seen day after day in satellite imagery of central Africa (Weischet, 1980). This circulation of water substance is paralleled by a circulation of energy, as heat made available in the surface energy budget is transferred and realized at cloud level in another energy budget. This budget is, however, balanced at a lower temperature than the surface budget. The rain that falls to earth to complete the mass circulation brings with it the coldness of the middle atmosphere in such quantities as to require separate entry in the surface energy budget, as Albrecht (1940) showed in a budget for Indonesia. While the mass circulation is indeed closed, the energy circulation is not; rather, daily

* or to vanish, as in midocean conditions of saturated air (Davies, 1972), in which the control of the latent-heat equation by the underlying surface has become complete.

pulses of solar energy are needed to keep these upward and downward transports in operation.

Latent heat is generated at the surface of the earth by the convergence of water and energy, and in terrestrial ecosystems energy usually means radiation. Net all-wave radiation surplus should be mapped on an operational basis to support "the specification of heat flux and evaporation over the land surface of the globe" (Priestley and Taylor, 1972) for an understanding of global weather, and such data would also be useful in comparative ecosystem-scale energy analysis.

The conversion of radiative and other forms of energy into latent heat plays a central role in the energy budget of all ecosystems, being intimately associated with their photosynthetic production. It depends primarily on the inputs of energy and water, secondarily on vapor removal into the atmosphere. It is more prevalent than the energy converted in the other change of physical state of water (that of melting) and while it permits some variation to occur in ecosystem temperature, these changes are usually small because this conversion sequesters energy in latent form, outside the thermal arena. It is a stabilizing factor in ecosystem functioning. Its removal runs parallel to the removal of sensible heat from ecosystems, the subject of Chapter XVI.

REFERENCES

Albrecht, F. (1940). Untersuchungen über den Wärmehaushalt der Erdoberfläche in verschieden Klimagebieten. Germany. Reichsamt. Wetterdienst, *Wiss. Abhandl.* **8**, No. 2.

Ångström, A. (1920). Applications of heat radiation measurements to the problems of the evaporation from lakes and the heat convection at their surfaces, *Geog. Annaler* **2**, 237–252.

Ångström, A. (1925). On radiation and climate. *Geog. Annaler* **7**, 122–142.

Arkley, R. J. (1963). Relationships between plant growth and transpiration. *Hilgardia* **34**, 559–584.

Baumgartner, A. (1967). Energetic bases for differential vaporization from forest and agricultural lands. *In* "Forest Hydrology" (W. E. Sopper and H. W. Lull, eds.), pp. 381–389. Pergamon, Oxford.

Bell, A. (1980). Sydney's water, and land use on the Shoalhaven. *ECOS (Melbourne)* **23**, 8–12.

Belt, G. H., King, J. G., and Haupt, H. F. (1977). Augmenting summer streamflow by use of a silicone antitranspirant. *Water Resour. Res.* **31**, 267–272.

Black, T. A. (1979). Evapotranspiration from Douglas fir stands exposed to soil water deficits. *Water Resour. Res.* **15**, 164–170.

Bouchet, R. J. (1963). Évapotranspiration réele et potentielle; signification climatique. *Assoc. Int. Hydrol. Sci., Publ.* **62**, 134–142.

Bowen, I. S. (1926). The ratio of heat losses by conduction and by evaporation from any water surface. *Phys. Rev.* **27**, 779–787.

Cable, D. R. (1977). Soil water changes in creostebush and bur-sage during a dry period in southern Arizona. *J. Ariz. Acad. Sci.* **12**(1), 15–20.

Calder, I. R. (1978). Transpiration observations from a spruce forest and comparisons with predictions from an evaporation model. *J. Hydrol.* (*Amsterdam*) **38**, 33–47.

Davies, J. A. (1972). Actual, potential and equilibrium evaporation for a beanfield in southern Ontario. *Agric. Meteorol.* **10**, 331–348.

Denmead, O. T., and Shaw, R. H. (1962). Availability of soil water to plants as affected by soil moisture content and meteorological conditions. *Agron. J.* **45**, 385–390.

Dunin, F. X., and Reyenga, W. (1978). Evaporation from a *Themeda* grassland. I. Controls imposed on the process in a sub-humid environment. *J. Appl. Ecol.* **15**, 317–325.

Dunin, F. X., Aston, A. R., and Reyenga, W. (1978). Evaporation from a *Themeda* grassland. II. Resistance model of plant evaporation. *J. Appl. Ecol.* **15**, 847–858.

Dyer, A. J. (1974). A review of flux-profile relationships. *Boundary-Layer Meteorol.* **7**, 363–372.

Dyer, A. J., and Maher, F. J. (1965). Automatic eddy-flux measurement with the evapotron. *J. Appl. Meteorol.* **4**, 622–625.

Ekern, P. C. (1965). Evapotranspiration of pineapple in Hawaii. *Plant Physiol.* **40**, 736–739.

Federer, C. A. (1976). Differing diffusive resistance and leaf development may cause differing transpiration among hardwoods in spring. *For. Sci.* **22**, 359–364.

Federer, C. A., and Gee, G. W. (1976). Diffusion resistance and xylem potential in stressed and unstressed northern hardwood trees. *Ecology* **57**, 975–984.

Fetcher, N. (1976). Patterns of leaf resistance to lodgepole pine transpiration in Wyoming. *Ecology* **57**, 339–345.

Fitzharris, B. B. (1974). Land-use change and the water balance—an example of an evapotranspiration simulation model. *J. Hydrol. (N.Z.)* **13**, 98–114.

Frankenberger, E. (1955). Über vertikale Temperatur-, Feuchte- und Wind-gradienten der Atmosphäre, den Vertikalaustausch und den Wärmehaushalt an Wiesboden bei Quickborn/Holstein 1953/1954. *Ber. Dtsch. Wetterdienstes* **3**, No. 20.

Frankenberger, E. (1960). Beiträge zum Berechnungen zum Wärmehaushalt der Erdoberfläche. *Ber. Dtsch. Wetterdienstes* **10**, No. 73.

Gay, L. W., and Fritschen, L. J. (1979). An energy budget analysis of water use by saltcedar. *Water Resour. Res.* **15**, 1589–1592.

Goodell, B. C. (1967). Watershed treatment effects on evapotranspiration. *In* "Forest Hydrology" (W. E. Sopper and H. W. Lull, eds.), pp. 477–482. Pergamon, Oxford.

Grant, D. R. (1975). Comparison of evaporation measurements using different methods. *Q. J. R. Meteorol. Soc.* **101**, 543–550.

Greenwood, E. A. N., and Beresford, J. D. (1980). Evaporation from vegetation in landscapes developing secondary salinity using the ventilated-chamber technique. II: Evaporation from *Atriplex* plantations over a shallow saline water table. *J. Hydrol. (Amsterdam)* **45**, 313–319.

Hicks, B. B. (1980). Investigations of sparging as a method for promoting cooling pond heat transfer. *J. Appl. Meteorol.* **19**, 193–198.

Hicks, B. B., Hess, G. D., and Wesely, M. L. (1979). Analysis of flux-profile relationships above tall vegetation—an alternative view. *Q. J. R. Meteorol. Soc.* **105**, 1074–1077.

Hofmann, G. (1956). Verdunstung und Tau als Glieder des Wärmehaushalts. *Planta* **47**, 303–322.

Ittner, E. (1968). Der Tagesgang der Geschwindigkeit des Transpirationsstromes im Stamm einer 75-jährigen Fichte. *Oecol. Plant.* **3**, 177–183.

Jackson, R. D., Reginato, R. J., and Idso, S. B. (1977). Wheat canopy temperature: A practical tool for evaluating water requirements. *Water Resour. Res.* **13**, 651–656.

Jarvis, P. G., James, G. B., and Landsberg, J. J. (1976). Coniferous forest. *In* "Vegetation and the Atmosphere" (J. L. Monteith, ed.), Vol. 2, pp. 171–240. Academic Press, New York.

Kalma, J. D., Fleming, P. M., and Byrne, G. F. (1977). Estimating evaporation: Difficulties of applicability in different environments. *Science* **196**, 1354–1355.

Konstantinov, A. R. (1968). "Isparenie v Prirode," 2d ed. Gidrometeorol. Izd., Leningrad.

Kramer, P. J. (1969). "Plant and Soil Water Relationships. A Modern Synthesis," McGraw-Hill, New York.

Kraus, H. (1964). Die Entstehung von Strassenglätte durch gefroren Tau, Reif und Beschlag. Univ. München, Meteorol. Inst., *Wiss. Mitt.* **9**, 107–128.

Lauenroth, W. K., and Sims, P. L. (1976). Evapotranspiration from a shortgrass prairie subjected to water and nitrogen treatments. *Water Resour. Res.* **12**, 437–444.

Lauscher, F. (1977). Reif und Kondensation auf Schnee und die wahre Zahl der Tage mit Reif. *Wetter Leben* **29**, 175–180.

Lauscher, F. (1978). Eine neue Analyse von Hilding Köhlers Messungen der Schneeverdunstung auf dem Haldde-Observatorium aus dem Winter 1920/1921. *Arch. Meteorol., Geophys. Bioklimatol., Ser. B* **26**, 193–198.

Lee, R. (1967). The hydrologic importance of transpiration control by stomata. *Water Resour. Res.* **3**, 737–752.

Lewis, M. C., and Callaghan, T. V. (1976). Tundra. *In* "Vegetation and the Atmosphere" (J. L. Monteith, ed.), Vol. 2, pp. 399–430. Academic Press, New York.

Lowry, W. P. (1969). "Weather and Life. An Introduction to Biometeorology." Academic Press, New York.

McIlroy, I. C., and Angus, D. E. (1964). Grass, water and soil evaporation at Aspendale. *Agric. Meteorol.* **1**, 201–224.

Massman, W. J. (1980). Water storage on forest foliage: a general model. *Water Resour. Res.* **16**, 210–216.

Middleton, W. E. K. (1966). "A History of the Theories of Rain and Other Forms of Precipitation." Franklin Watts, New York.

Miller, D. H. (1977). "Water at the Surface of the Earth." Academic Press, New York.

Milthorpe, F. L., and Moorby, J. (1979). "An Introduction to Crop Physiology" (2d ed.). Cambridge Univ. Press, London and New York.

Monteith, J. L. (1957). Dew. *Q. J. R. Meteorol. Soc.* **83**, 322–341.

Mooney, H. A., Gulmon, S. L., Ehleringer, J., and Rundel, P. W. (1980). Atmospheric water uptake by an Atacama Desert shrub. *Science* **209**, 693–694.

Munn, R. E. (1966). "Descriptive Micrometeorology." Academic Press, New York.

Murphy, C. E., Jr., and Knoerr, K. R. (1977). Simultaneous determination of the sensible and latent heat transfer coefficients for tree leaves. *Boundary-Layer Meteorol.* **11**, 223–241.

Mustonen, S. E., and McGuinness, J. L. (1967). Lysimeter and watershed evapotranspiration. *Water Resour. Res.* **3**, 989–996.

Odum, E. P. (1971). "Fundamentals of Ecology," 3rd ed. Saunders, Philadelphia, Pennsylvania.

Pearce, A. J., Gash, J. H. C., and Stewart, J. B. (1980). Rainfall interception in a forest stand estimated from grassland meteorological data. *J. Hydrol. (Amsterdam)* **46**, 147–163.

Priestley, C. H. B., and Taylor, R. J. (1972). On the assessment of surface heat flux and evaporation using large-scale parameters. *Mon. Weather Rev.* **100**, 81–92.

Rauner, Iu. L. (1973). Energeticheskaia effektivnost' produktsionnogo protessa rastitel'nykh soobshchestv. *Izv. Akad. Nauk SSSR, Ser. Geogr.* No. 6, 17–28.

Rauner, Iu. L. (1976). Deciduous forest. *In* "Vegetation and the Atmosphere" (J. L. Monteith, ed.), Vol. 2, pp. 241–262. Academic Press, New York.

Ripley, E. A., and Saugier, B. (1978). Biophysics of a natural grassland: evaporation. *J. Appl. Ecol.* **15**, 459–479.

Roberts, J. (1978). The use of the "tree cutting" technique in the study of the water relations of Norway spruce, *Picea abies* (L.) Karst. *J. Exp. Bot.* **29**, 465–471.

Rouse, W. R. (1970). Relations between radiant energy supply and evapotranspiration from sloping terrain: an example. *Can. Geogr.* **14**(1), 27–37.

Rouse, W. R., and Wilson, R. G. (1972). A test of the potential accuracy of the water-budget approach to estimating evapotranspiration. *Agric. Meteorol.* **9**, 421–446.

Saltzman, B. (1980). Parameterization of the vertical flux of latent heat at the earth's surface for use in statistical-dynamical climate models. *Arch. Meteorol. Geophys. Bioklimatol. Ser. A* **29**, 41–53.

Saltzman, B., and Ashe, S. (1976). The variance of surface temperature due to diurnal and cyclone-scale forcing. *Tellus* **28**, 307–322.

Saugier, B. (1976). Sunflower. *In* "Vegetation and the Atmosphere" (J. L. Monteith, ed.), Vol. 2, pp. 87–119. Academic Press, New York.

Saugier, B., and Ripley, E. A. (1978). Evaluation of the aerodynamic method of determining fluxes over natural grassland. *Q. J. R. Meteorol. Soc.* **104**, 257–270.

Seguin, B. (1975). Influence de l'évapotranspiration régionale sur la mesure locale d'évapotranspiration potentielle. *Agric. Meteorol.* **15**, 355–370.

Singh, B. and Szeicz, G. (1979). The effect of intercepted rainfall on the water balance of a hardwood forest. *Water Resour. Res.* **15**, 131–138.

Slatyer, R. O. (1967). "Plant-Water Relationships." Academic Press, New York.

Slatyer, R. O., and McIlroy, I. C. (1961). "Practical Microclimatology, with Special Reference to the Water Factor in Soil–Plant–Atmosphere Relationships." UNESCO. Austral. CSIRO, East Melbourne.

Smithsonian Institution (1966). "Smithsonian Meteorological Tables" (R. J. List, ed.), 6th ed., Publ. No. 4014. Washington, D.C.

Stewart, R. B., and Rouse, W. R. (1976). A simple method for determining the evaporation from shallow lakes and ponds. *Water Resour. Res.* **12**, 623–628.

Stricker, H., and Brutsaert, W. (1978). Actual evapotranspiration over a summer period in the "Hupsel Catchment." *J. Hydrol. (Amsterdam)* **39**, 139–157.

Swanson, R. H. (1967). Seasonal course of transpiration of lodgepole pine and Engelmann spruce. *In* "Forest Hydrology" (W. E. Sopper and H. W. Lull, eds.), pp. 419–434. Pergamon, Oxford.

Tan, C. S., Black, T. A., and Nnyamah, J. U. (1978). A simple diffusion model of transpiration applied to a thinned Douglas-fir stand. *Ecology* **59**, 1221–1229.

Thom, A. S. (1975). Momentum, mass and heat exchange of plant communities. *In* "Vegetation and the Atmosphere" (J. L. Monteith, ed.), Vol. 1, pp. 57–109. Academic Press, New York.

Thom, A. S., Stewart, J. B., Oliver, H. R., and Cash, J. H. C. (1975). Comparison of aerodynamic and energy budget estimates of fluxes over a pine forest. *Q. J. R. Meteorol. Soc.* **101**, 93–105.

Thornthwaite, C. W., and Hare, F. K. (1965). The loss of water to the air. *Meteorol. Monogr.* **6**(28), 163–180.

Thornthwaite, C. W., and Holzman, B. (1939). The determination of evaporation from land and water surfaces. *Mon. Weather Rev.* **67**, 4–11.

Tweedie, A. D. (1956). The measurement of water need in Queensland. *Austral. Geogr.* **6**, 34–39.

Uchijima, Z. (1976). Maize and rice. *In* "Vegetation and the Atmosphere" (J. L. Monteith, ed.), Vol. 2, pp. 33–64. Academic Press, New York.

Verma, S. B., Rosenberg, N. J., and Blad, B. L. (1978). Turbulent exchange coefficients for sensible heat and water vapor under advective conditions. *J. Appl. Meteorol.* **17,** 330–338. Comments by Hicks, B. B. and Everett, R. G. (1979). **18,** 381–382.

Waggoner, P. E., and Zelitch, I. (1965). Transpiration and the stomata of leaves. *Science* **150,** 1413–1420.

Webb, E. K. (1975). Evaporation from catchments. *In* "Prediction in Catchment Hydrology" (T. G. Chapman and F. X. Dunin, eds.), pp. 203–236. Austral. Acad. Sci., Canberra.

Webb, W., Szarek, S., Lauenroth, W., Kinderson, R., and Smith, M. (1978). Primary productivity and water use in native forest, grassland, and desert ecosystems. *Ecology* **59,** 1239–1247.

Weischet, W. (1980). Klimatologische Interpretation von METEOSAT-Aufnahmen. *Geog. Rundschau* **32**(2), 80–81, 84.

Wood, N. L. H. (1977). A field study on the representativeness of turbulent fluxes of heat and water vapor at various sites in southern England. *Q. J. R. Meteorol. Soc.* **103,** 617–624.

Woodcock, A. H., and Friedman, I. (1980). Mountain breathing—preliminary studies of air-land interaction on Mauna Kea, Hawaii. *U.S. Geol. Surv. Prof. Pap.* 1123A.

Zito, G., Mongelli, F., and Loddo, M. (1978). Influence of accurate heat flux measurements on the evaluation of evaporation from a bare soil by energy balance. *Arch. Meteorol., Geophys. Bioklimatol., Ser. A* **27,** 155–169.

Chapter XVI

THE FLUX OF SENSIBLE HEAT FROM ECOSYSTEMS

An ecosystem under strong radiation can respond by using the local air as a heat sink. The question is how much heat the air can quickly accept or admit as the sensible-heat flux H from the ecosystem. A quick response to a sudden loading of an ecosystem by absorbed radiation forestalls brief but damaging increases above the lethal point in cell temperature, which is 45–50°C in many species. Other modes of heat removal also operate in this situation, but without the sensible-heat flux would often be inadequate to prevent overheating, especially when moisture conditions are unfavorable.

The less conspicuous delivery of atmospheric heat to an ecosystem threatened at night by a net radiation deficit that might bring its leaf cells to the freezing point is also important. This downward flux is seldom as large as the upward flux at midday, but the net radiation deficit is not large either.

Sensible heat is also translocated within an ecosystem, from soil to leaves, from sun leaves to shade leaves, from the sunny side of a tree to its shaded side, from upper levels to lower. These tiny fluxes help maintain optimum operating temperatures throughout the entire system.

Both air and soil receive thermal energy by day from the irradiated ecosystem surface, and in an early energy-budget analysis Ångström (1925) computed the sum of the fluxes of sensible heat into these adjacent media. He called it "temperature effective energy" because its flow is manifested in a change of temperature, and found its maximum in the area around Stockholm to occur in May, when it reached 75 W m^{-2}.

Here we take the point of view of the ecosystem, rather than the soil or atmosphere, and consider the large morning flux of sensible heat into the local air as relieving possible heat stress on the ecosystem rather than as a heat source that destabilizes the lowest layers of the atmosphere, the latter being a topic in micrometeorology. Similarly, we view the flux at night as restraining ecosystem cooling, rather than for its effect on atmospheric structure, although temperature changes in the local air (and soil) naturally interact with the fluxes themselves, as will be seen below.

ECOSYSTEM CHARACTERISTICS

As foreshadowed in Chapter VIII, the temperature of the outer active surface of an ecosystem influences the heat-removing energy fluxes. It is particularly important in determining the density of the flux of sensible heat, and while sensible heat seldom removes as much energy as is emitted as longwave radiation, its response is more acute. A 10°C rise in surface temperature, which increases the emission of longwave radiation by a mere 0.10–0.15, may well double or triple the sensible-heat flux.

Ecosystem structure is open to ventilation by even light air movement, as we saw in discussing the intake of CO_2 in Chapter XI and the removal of water vapor in Chapter XV. Thin small leaves have favorable ratios of surface to volume and enjoy good ventilation, such that a steep gradient of temperature into the air is seldom necessary to remove excess heat. As leaf width decreases, convective heat removal increases by the square root of the width change (Gates, 1968), and laminar flow dominates (Murphy and Knoerr, 1977). Removal of heat is especially rapid from needle leaves, depending on their clustering habit and orientation on the shoots (Jarvis et al., 1976). Large, fleshy leaves and stems tend to overheat; cactus often develops high temperatures that can be modeled (Nobel, 1980). A porous zone of heat-exchanging surfaces is less likely to become overheated than a thin zone; forest canopy is less likely to overheat than a dense sward of grass. Surface roughness, which affects the wind structure, is another ecosystem property significant in moving sensible heat into the atmospheric heat sink, again favoring forest ecosystems.

THE ATMOSPHERIC ENVIRONMENT AS A SINK OR SOURCE FOR SENSIBLE HEAT

Structure

The local air above an ecosystem is not a uniform, homogeneous fluid, but rather is structured by its lower boundary and displays distinctive vertical profiles of wind speed, temperature, concentrations of vapor and CO_2, and so on. Some of these elements of structure interact with the daytime upward flux of sensible heat and give a means of evaluating it as well. One means of characterizing structure of the local air is the Monin–Obukhov length L, which expresses the ratio of shearing stress τ to the flux density of sensible heat H, multiplied by a quasi-constant term:

$$L = (\tau^{3/2}/H)(\rho c_p T/kg\rho^{3/2}),$$

in which ρc_p is volume heat capacity of the air (J m^{-3} deg^{-1}), T absolute temperature, k the von Karman number (0.35–0.4), and g the acceleration of gravity. In calm, sunny conditions L may be as small as a meter, and in windy conditions up to 50 m (Webb, 1964). Like the Richardson number it is a parameter of stability (Priestley, 1959, p. 23), and can be used to define structure:

Layer

(a)	Surface to 0.03L:	Temperature gradient is steepest near the surface and decreases with height; turbulence generated by wind is dominant, and "heat is carried through as an inert passenger" (Webb, 1964)
(b)	0.03–1.0L:	Both forced and thermal convection are present
(c)	>L:	Buoyancy forces in rising bubbles or plumes carry heat; there is "no appreciable effect of wind turbulence on the convective currents" (Webb, 1964); the gradient is small or zero, so eddies transport no heat. Free convection becomes the principal mechanism (Thom, 1975)
(d)	>10L:	Plumes seem to give way to thermals (Deacon, 1969)

Sensible heat feeding into these thermals provides energy that maintains soaring scavengers (Cone, 1962) in surveillance of the fauna of ecosystems below and powers their long cross-country trips to food sources (Pennycuick, 1973). Thermals can be looked at as upward extensions of the power of the underlying ecosystems to support essential functions at a distance.

The Gate Function

Sensible-heat flux acts as a gate function of the regime of wind speed near the surface of ecosystems because its destabilizing effect brings the lower air into connection with more rapidly moving upper air, which produces the usual daytime maximum of wind speed near the ground and consequent increased conversion of kinetic energy into mechanically generated turbulence. One form of energy flux is a gate or control over the conversion of another kind.

Mechanical turbulence is in most cases capable of carrying away most if not all of the available sensible heat, which flux is in general "determined by the gross parameters (insolation, wetness and general nature of ground, etc.) operating through the principle of energy balance" (Priestley, 1959, p. 49). Mechanical turbulence is dominant to a greater height when the wind is strong and sensible-heat flux small than on days of weak wind and strong sunshine when buoyancy takes over near to the ecosystem surface, as the L concept indicates.

The combination of horizontal and vertical motion in the air supports ecologically significant transports of dust from desert or eroding fields and wide dispersion of pollen and spores. Many plants release these viable particles only at times of day when this vertical transport is likely to occur, i.e., when the sensible-heat flux is large.

The Valve Effect

Positive buoyancy due to the input of sensible heat into the local air favors upward fluxes. Negative buoyancy, conversely, restricts vertical movement so that the downward heat flux that ecosystems can receive at night is smaller than the upward daytime flux by an order of magnitude. The atmosphere is a weak source of heat for an ecosystem.

Weak Diffusion. Low wind speed near the surface at night provides little kinetic energy to maintain turbulence against negative buoyancy, i.e., the downward decrease in temperature and increase in air density (Eskinazi, 1975, pp. 213, 297). Little eddy diffusion of sensible heat occurs, as Oke (1970) found in measurements under these conditions. The association of light nocturnal winds with small quantities of heat that a grassland ecosystem can extract from the air is shown in Table I. The soil of this ecosystem had normal thermal conductivity and could make some contribution to the deficit in the radiation budget. Drier soil, as at O'Neill, Nebraska, in late summer yields less heat, and the atmosphere at night supplied 20–25 W m^{-2} (Lettau and Davidson, 1957); the soil and air fluxes were inversely correlated. Dry snow yields still less heat, and measurements in and above it in the Arctic night were one of the first means of evaluating the "convectivity" (or eddy conductivity) of the air (Ångström, 1918).

Strong Diffusion. Downward diffusion of heat over snow or an aquatic ecosystem can be large, given enough kinetic energy to force air

TABLE I

Downward Fluxes of Sensible Heat to a Grassland Ecosystem in the
North Sea Coastal Zone[a]

Speed (m sec^{-1})	Sensible-heat flux (W m^{-2})	Speed (m sec^{-1})	Sensible-heat flux (W m^{-2})
0–1	+8	3–4	+16
1–2	+16	4–5	+33
2–3	+26	5–6	+70

[a] Data from Frankenberger (1955). Measurements at Hamburg-Quickborn.

parcels down into the denser, lower air, because in these two systems surface temperature T_0 rises only a little or not at all, and the temperature difference between ecosystem and air remains large. Flux rates reach 150–200 W m^{-2} in strong winds (Price and Dunne, 1976).

Nocturnal stability nevertheless diminishes the flux rate $H = \rho c_p D(T_0 - T_a)$ by decreasing the usual exchange coefficient D to the stable case D_s, where

$$D_s = D/(1 + 10\mathrm{Ri}),$$

in which Ri is the Richardson number.

$$\mathrm{Ri} = (g\,\Delta T_a)/(T_a(\Delta u)^2)$$

(Price and Dunne, 1976), with T_a and u representing air temperature and wind speed at height z and ΔT representing warmth of advected air relative to the melting snow cover.

The sensible-heat flux over snow displays no diurnal periodicity because it does not draw upon the radiation surplus, as the upward flux does over terrestrial ecosystems in the growing season (McKay and Thurtell, 1978). Flux densities as large as 100 W m^{-2} were measured with a pressure-sphere anemometer in advective weather. Similar fluxes occur to evaporating alfalfa in advective weather on the Great Plains. (Motha et al., 1979).

In strong wind at night, powered by such synoptic-scale energy sources as foehn, the downward heat flux to terrestrial ecosystems may still be limited when air and surface can approach thermal equilibrium. Nocturnal fluxes in foehn in central Asia have been measured as 50–60 W m^{-2} (Zuev and Pavlov, 1959) through the night, smaller flux densities than those to surfaces kept cool by phase changes of water. Here we see one of the many interactions among energy fluxes.

ECOSYSTEM–ATMOSPHERE RELATIONSHIPS

The temperature difference between an ecosystem and the air and the atmospheric ventilation of the ecosystem are not independent factors, but are mutually involved in a complex interaction. This involvement is close when strong radiation absorption forces the ecosystem to rid itself of a large quantity of heat, which the foliage–air system does by increasing the temperature difference or the ventilation or both.

The flow of energy H expresses the relationship between these factors in two ways, depending on how the temperature function is stated,

$$H = \rho c_p K_H \, \partial T / \partial z \tag{1}$$

$$H = \rho c_p D (T_0 - T_a) \tag{2}$$

in which ρc_p, the volumetric heat capacity of the air, with dimensions J m^{-3} deg^{-1}, takes care of the conversion from temperature to energy. Its numerical value is approximately 1300. K_H is diffusivity (m^2 sec^{-1}) for heat, and D (m sec^{-1}) is a height-integrated diffusion coefficient (Budyko, 1974, p. 71). Both terms are complex expressions of micrometeorological processes, interacting with the thermal structure of the air, extraction of its momentum by the underlying ecosystems, and other factors. $\partial T / \partial z$ is the gradient of temperature in the air, T_0 is surface temperature of the ecosystem, and T_a air temperature at a meter or so distance above it. Thermal admittance is $\rho c_p \sqrt{K_H}$ and can be compared with admittance into the soil, to be discussed in Chapter XVII.

Temperature Gradient or Difference

Leaf temperature responds quickly to changes in energy loading or removal, and the temperature difference between leaf and air fluctuates correspondingly. An increase in radiant-energy loading or the cut-off of another heat removal process (as when leaf transpiration is stopped) causes an increase in the temperature gradient. Network measurements of temperature gradient between 0.2 and 1.5 meters over short grass at weather observing stations in the Soviet Union exhibit a close relation to solar radiation.

Different temperature profiles characterize each zone: the laminar layer on the surface of each leaf, through which heat moves by molecular diffusion; the layer of air held within the foliage of the ecosystem; and the part of the overlying atmosphere that comes under the influence of its lower boundary, in which gradient is approximately inversely proportional to height, in forced convection (Priestley, 1959, p. 39). A gradient through the planetary boundary layer is used in one model (Saltz-man and Ashe, 1976a,b). The surface is usually colder than the air at night, being a better radiator, and the temperature gradient is directed downward. On a calm clear night it may exceed 10°C in a meter. Such a gradient represents high stability, which by reducing diffusion isolates the ecosystem from the reservoir of heat in the atmosphere. Laikhtman (1964, p. 101) found nocturnal diffusivity K to be 0.03 m^2 sec^{-1} when heat flux was less than 15 W m^{-2}. Thermal admittance of the air, $\rho c \sqrt{K}$, 200 J m^{-2} deg^{-1} $sec^{-1/?}$, was only one-fifth of the thermal admittance of the soil, and little of the nocturnal heat demand of the surface was met by extracting heat from the air.

Maintaining Temperature Gradient or Difference

Maintenance of a temperature gradient in the local air or a temperature difference between surface and air implies that heat is steadily being made available to the warm end and/or being removed from the cold end. Ecosystems, in the general daytime case, are warmed by absorption of radiant energy, variations in which are due mostly to the changing altitude of the sun and are the source in the source–sink system. Ecosystems with large thermal inertia, e.g., aquatic systems, are warm relative to the air at night and in winter and the source of heat at the ecosystem surface is the flow from the substrate, to be described in Chapter XVIII. The source–sink relation is also maintained by changes in physical state of water: condensation in storm air keeps it warm, and melting at the earth's surface maintains a heat sink.

Expressions of Heat Flow

The familiar starting place in considering the diffusivity term K_H in Eq. (1) for the sensible-heat flux is the molecular conduction of heat: the traditional iron rod, hot at one end and cold at the other, or the laminar film of air that coats the leaves of an ecosystem. The term $\rho c_p K_H$ in Eq. (1) has these units: kg m^{-3} J kg^{-1} deg^{-1} m^2 sec^{-1} or J m^{-1} sec^{-1} deg^{-1}, that is, energy moving per second and degree difference in temperature over a meter length of rod. This expresses conductivity and can be written

$$J\ m^{-2}\ sec^{-1}/deg\ m^{-1} = W\ m^{-2}/deg\ m^{-1},$$

in which the denominators express temperature gradient and the numerators express energy flux density.

In molecular conduction the numerical value of $\rho c_p K_H$ depends upon properties of the material; it is large in metals, small in nonmetals, still smaller in such porous substances as soil and snow. The transport of heat by convection in fluids is by analogy called eddy conductivity, and K now refers to the turbulent diffusion of some property within a fluid. It can inversely be regarded, from dimensional analysis (Thom, 1975), as an aerodynamic resistance per unit distance, or sec m^{-1} per meter, or sec m^{-2}.

The terms designated K for sensible heat, water vapor, CO_2, momentum, and other constituents of the atmosphere have similarities that are utilized in aerodynamic methods of examining the relationships between ecosystems and the air, qualified by the spectrum of turbulent elements at different heights and by inhomogeneities in the air stream due to height, openness, and roughness of upwind ecosystems that have injected heat into the airstream and extracted momentum from it. Turbulence scale and intensity have ecological implications (Moen, 1974), such as the penetration of ecosystems or animal pelage by wind. Moreover, heat is not necessarily exchanged at the same locus in an ecosystem as other atmospheric attributes like CO_2 or vapor. Momentum may be extracted from the air stream by sparse projecting stalks, whereas heat and vapor are transferred at leaves lower in the system. Latent and sensible heat often are generated at quite different levels within a forest system.

The integrated coefficient of diffusion D (Budyko, 1974, pp. 71–74; Thom, 1975) changes little with the height above the ecosystem surface to which integration is taken, and in daytime is equal to about 0.01 m sec^{-1}. It is smaller in inversions, and 24-hr means are of the order of 0.006–0.007 m sec^{-1} in humid regions, larger in arid ones. It is relatively insensitive to wind speed over terrestrial ecosystems; if wind decreases and very steep temperature gradients develop, thermal convection replaces mechanical in transporting sensible heat out of an ecosystem.

Thermal Admittance. Another way to look at the removal of sensible heat from an ecosystem is by the concept of thermal admittance, the product of the square root of diffusivity $\sqrt{K_H}$ and heat capacity ρc_p, with units J m^{-2} deg^{-1} sec$^{-1/2}$ (Brooks and Rhoades, 1954; Kuo, 1968). This term is variable in the lower air in response to the fluctuating state of turbulence as indexed by K_H, and in stirred air lies numerically between the values in soil and those in stirred water. Stirred daytime air accepts sensible heat from an irradiated ecosystem more avidly than the soil does; the atmospheric heat flux outruns the soil heat flux. Conversely, when air is paired against a stirred water body with a still greater value of thermal admittance, the atmospheric heat flux takes second place.

The concept of thermal admittance illustrates the relative celerity with which an ecosystem surface can rid itself of sensible heat, and in what direction it moves, to be developed in Chapters XVII and XVIII.

VARIATIONS IN THE REMOVAL OF SENSIBLE HEAT FROM AN ECOSYSTEM

The movement of sensible heat from terrestrial ecosystems is small most of the time. Periods of net deficit in the radiation budget, which prevail more than half of the hours of the year, are periods when non-radiative energy fluxes are also small and the stable structure of the air hampers vertical motion. The downward flux attains moderate size only when the ecosystem is experiencing a phase change of water in the presence of heat advection. The upward sensible-heat flux reaches high intensities only in the midday hours of large radiation surplus in relatively dry conditions, when in models it can be approximated as a quarter of the incoming flux of solar energy (Greenland, 1980).

Sensible-heat flux density is directly measured by eddy correlation methods in which the temperature (and ρc_p) of ascending and descending parcels of air and the respective upward and downward speeds are sensed, multiplied to form the individual pulses of sensible heat, and put into storage (Swinbank, 1951; Dyer et al., 1967; Frankenberger, 1958). After half an hour the ascending pulses are retrieved from storage and added up, as are the descending fluxes, to yield the net upward or downward flux of heat during that period. The pulses are shorter than about 10 sec, even when measured 4 m above the surface, at a height where turbulence elements are fairly large. Elements are smaller closer to the surface and the fluctuations in vertical velocity more rapid, to the degree that the response of the wind and temperature sensors may be taxed (Munn, 1966, p. 88). Siting the sensors higher, however, increases the possibility of coming under the influence of upwind ecosystems differing in roughness or T_0 from the ecosystem under study. The requirement that fetch should be 100–200 times the height of the sensors tends to be restrictive for ecosystems of usual sizes and especially forest, for which a required fetch of 0.8 km was determined in a South Australian pine stand (Hicks et al., 1975).

Sensible-heat flux density is calculated along the lines given in Eqs. (1) and (2), with various means of adjusting K or D for height, stability, and other parameters of turbulence. Varying K with height yielded good reconstitutions of diurnal temperature regimes up to 500 m (Kuo, 1968); in neutral conditions the temperature profile and wind speed can serve

(Stricker and Brutsaert, 1978). The flux is also obtained as the residual quantity where other fluxes in the energy budget are measured.

The shortest variations in sensible-heat flux are associated with the individual eddies. Over longer periods the net flux is steadier (Desjardins, 1977), showing, for example, only one or two dips in the afternoon decline.

The Daily March

The persistence of stable stratification of the lower air in the morning tends to delay the rise of the sensible-heat flux, but its peak is still reached by midday from ecosystems in which the soil is isolated from the irradiated surfaces. Midday peaks from grass at Hamburg came at 1200, 1100, 1230, 1130, and 1200 in the months April–September, respectively, in 1958 (Frankenberger, 1960) (Fig. 1); Yap and Oke (1974a) also found no phase lag after the net radiation regime over grass.

The morning pulse of sensible heat is correlated with the rise in air temperature at 2 m (Fig. 2). For example, in February the total heat transferred upward from the morning crossover until 1400 was 0.25 MJ m^{-2}, and air temperature rose 3.1°C; in April the pulse was 2.75 MJ m^{-2}, and temperature rose 5.6°C. Ångström (1925) notes the close relation of morning radiation to heating of the air, and Neiburger (1941) shows how forecasting the morning heat pulse can be employed to predict the daily maximum air temperature in nonfrontal weather and to assess the possibility of dissipating an inversion limiting upward dilution of pollutants (Neiburger, 1957). Where this heat pulse burns off morning stratus or, conversely, initiates afternoon convective cumulus,

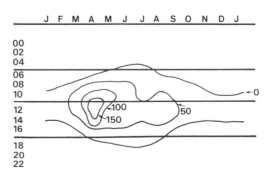

Fig. XVI.1.   Isopleth diagram of sensible-heat fluxes at Hamburg-Quickborn in 1958. Values are means in the hour beginning at time indicated (W m^{-2}). [Data from Frankenberger (1960).] The maximum in the midday hours appears in April, but the length of the daytime period with upward sensible-heat flux increases until early July.

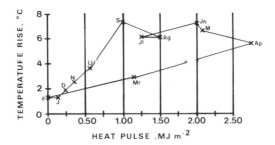

Fig. XVI.2. Rise in 2-m air temperature above grass at Hamburg-Quickborn from the hour when the sensible-heat flux became directed upward until 1400, as a function of the pulse of sensible heat during the same period (MJ m^{-2}). [Data from Frankenberger (1960).]

the effect of cloudiness on the radiant-energy intake feeds back to alter the diurnal regime of the sensible-heat flux.

The upward flux declines through the afternoon and ends at Hamburg about 1830 in summer, $1\frac{1}{2}$–2 hr before sunset, as was true also in Wisconsin (1700 in September) (Tanner and Thurtell, 1970). It usually ends earlier than the flux of latent heat.

Rates exceeding 60 W m^{-2} occur for about 8 hr in summer days at Hamburg and for about 10 hr above dry surfaces. Peak densities at Hamburg reached 160 W m^{-2}, but from dry surfaces may reach 400 W m^{-2} on a desert in central Asia (Aizenshtat, 1960), or even 500 W m^{-2} on a south slope at 3.2 km altitude in the Turkestan range (Aizenshtat, 1966) or a California dry lake (Brooks and Rhoades, 1954).

Downward fluxes at night, as noted earlier in the chapter, are small because of the valve effect, and at Hamburg (Fig. 1) they are of the order of 10–20 W m^{-2}. As a result, the net 24-hr transport of sensible heat in summer is ~20 W m^{-2} upward. The 24-hr mean is small on most winter days, or even net downward in marine climates.

Synoptic-Scale Variations

Day-to-day variations in the heat flux from an ecosystem reflect variations in its absorption of radiation, in dryness, or in air-stream warmth. On winter days with a small flux of downward longwave radiation at Hamburg, the sensible-heat flux brought in +15 W m^{-2}. In the spring transition, from February days of generally downward heat flux to April days of generally upward flux (Table II), the day-to-day variation increased. April displays a bimodal distribution.

The heat flux increases as a field dries out after rain, as illustrated by midday flux rates at a steppe site (Table III). Drying allowed higher surface temperatures to develop, and the steeper gradient supported a larger flux of sensible heat; latent-heat flux diminished.

Seasonal Regime

The seasonal variation in absorption of radiant energy by ecosystems in middle latitudes, modified by changes in ecosystem dryness and roughness in the yearly cycle, produces a variation in the flux of sensible heat. These effects are illustrated by eddy-flux measurements over short grass in interior southern Australia (Table IV). Tall, green, actively transpiring grass of September, under a net all-wave radiation surplus averaging 240 W m^{-2} in the hours from 1000 to 1600, puts only 0.2 of this energy into sensible heat. Somewhat shorter grass on warm days in late October converted half its radiation surplus into sensible heat. Dry short grass in late summer converted 0.75 of its radiation surplus into sensible heat. The sensible-heat flux densities over 6-hr periods increased from 45, 145, 240, to 290 W m^{-2} and represented the succession of phenological changes.

TABLE II

Frequency Distributions of Net Daily Fluxes of Sensible Heat at Hamburg in Spring[a]

	Net 24-hr flux of sensible heat (MJ m^{-2})	February[b]	March[b]	April[b]
Downward	>1.7	11	0	0
	0.9–1.7	29	0	0
	0–0.9	43	16	3
Upward	0–0.9	17	36	7
	0.9–1.7	0	19	10
	1.7–2.6	0	23	20
	2.6–3.4	0	3	17
	3.4–4.2	0	0	10
	4.2–5.1	0	0	7
	5.1–5.8	0	3	20
	5.8–6.7	0	0	3
	6.7–7.5	0	0	3
Median value		+0.6	−0.8	−3.0

[a] Data from Frankenberger (1960) Appendix 3.

[b] Data given in percent.

TABLE III

Midday Sensible-Heat Flux (1300) at a Steppe Site before and after Rain[a]

Sept	T_0	T_{2m}	$(T_0 - T_{2m})$	Heat Flux (W m^{-2})
10	42	27	15	−220
11[b]	24	17	7	−80
12	37	21	16	−170
13	40	25	15	−180
14	42	27	15	−220
15	47	28	19	−220

[a] Data from Rauner (1960), pp. 135, 146, 154.
[b] Rain of 7 mm in morning.

The regime over a year can be illustrated from Hamburg data (Table V). The early peak and the decline after May as transpiration began to claim more heat is noteworthy. A spring peak is also found in the troposphere (Oerlemans, 1980). The mean downward flux during winter at Hamburg shows how much energy this coastal ecosystem extracts from marine air streams.*

Dry regions or those with stronger solar radiation display larger 24-hr mean fluxes in summer than the 25–30 W m^{-2} of Table V. While Nigerian rainforest generates 10 W m^{-2}, steppe ecosystems generate 80 (Oladipo, 1980). Estimates by Budyko (1974) that show 50 W m^{-2} as mean flux density from Midwest ecosystems show 100–110 W m^{-2} from desert ecosystems.

SPATIAL CONTRASTS

Small-Scale Spatial Variations

Variations in sensible-heat flux from shade leaves and sun leaves are obvious and may even differ in sign when sun leaves generate heat while shade leaves extract it from the air. As will be seen in Chapter XX, differences in the sensible-heat exchange between leaves and air often exist at different levels in an ecosystem. Horizontal differences in the same ecosystem are, however, small (Yap and Oke, 1974a).

* It also benefits from a high rate of incoming longwave radiation from the moist, cloudy air. The two modes of delivery of advected oceanic heat make possible a high ecosystem temperature in the dark days of winter at this latitude.

TABLE IV

Eddy-Correlation Measurements of Sensible Heat over Grass at Kerang, Australia[a]

Condition	Temperature difference 1–4 m (°C)	Sensible-heat flux density (W m^{-2})				Sensible heat as fraction of all nonradiative heat fluxes			
		10–12	12–14	14–16	Sum (MJ m^{-2})	10–12	12–14	14–16	Average
10 cm, green	0.45	−80	−60	−10	1.0	0.23	0.22	0.07	0.20
7 cm, green	0.93	−90	−230	−115	3.2	0.38	0.45	0.34	0.49
5 cm, green	1.08	−260	−315	−150	5.2	0.48	0.53	0.41	0.62
5 cm, dry	1.36	−315	−315	−270	6.4	0.75	0.75	0.76	0.75

[a] Data from Swinbank and Dyer (1968).

TABLE V

Mean Monthly Fluxes of Sensible Heat from Grassland at Hamburg-Quickborn[a] in 1957[b]

D	J	F	M	A	M	J	J	A	S	O	N	Mean
+6	+8	+7	−12	−37	−25	−23	−19	−19	−12	−1	+1	−11

[a] Latitude 54°N.
[b] Data from Frankenberger (1960). Data given in watts per square meter.

Ecosystems on north and south slopes in dissected terrain discharge different quantities of sensible heat as a result of different exposures to beam radiation. A summer example in a dry mountain region is seen in Table VI (Aizenshtat, 1966), and one from a polar range in Table VII (Brazel and Outcalt, 1973).

The differences in heat injected into the local air produce differential expansion, which generates potential energy that powers the typical mountain wind system. The tertiary circulations of the lower atmosphere that have a diurnal period manifest the daytime injection of pulses of sensible heat from the underlying ecosystems. Even in such humid terrain as the Paris Basin, mesoscale differences in sensible-heat flux are located and used by soaring birds and sailplane pilots searching for thermals (Fig. 3) (Pédelaborde, 1957, p. 463).

Contrasts between the large upward pulses of sensible heat from dry ecosystems and the small downward fluxes of heat extracted from the air by irrigated crops become large in midday hours. At Davis, California, the downward flux of sensible heat was 90 W m^{-2} at 170 m from the edge of an irrigated field and 160 W m^{-2} at the edge itself (Dyer and Crawford, 1965).

Miess (1968, p. 164) found June heat flows from high moor in northwestern Germany to be 0.6 greater than those from low moor. Another kind of water-related contrast is seen between aquatic and adjacent terrestrial ecosystems. The large sensible-heat fluxes from terrestrial ecosystems by day contrast with the small fluxes from aquatic surfaces; the situation is reversed at night.

TABLE VI

Midday Flux Densities of Sensible Heat from Slopes in the Turkestan Range[a]

Flat surface	North	East	South	West
−440	−135	−290	−520	−345

[a] Altitude 3.2 km, latitude 40°N. Data given in watts per square meter.

TABLE VII

Daytime Mean Flux Densities of Sensible Heat from Slopes in Alaska[a]

Flat surface	Northwest slope	Southeast slope
−65	−55	−110

[a] Altitude 1.8 km, latitude 61°N. Data given in watts per square meter.

Rock ledges in the Sahara (de Félice, 1968), where surface temperature did not go lower than 13°C, generated sensible-heat flux all night long into cold (10°C) air. This air had been chilled over cold dune sands, which had extracted heat from it at a rate of +70 W m^{-2} early in the night decreasing to +15 at dawn.

Heat stored in concrete in cities adds to the winter leakage of heat from buildings to provide an all-day source of sensible-heat flux. Midday

Fig. XVI.3.   Hang-glider seeking column of rising air on east slope of Mount Mansfield, Vermont, in June. Ski runs below are less likely than the paved parking lot beyond to generate large sensible-heat fluxes.

fluxes of sensible heat from a building roof in Vancouver reached 300 W m^{-2} (Yap and Oke, 1974b), comparable with the heat fluxes from deserts. This flux includes the heat given off by thousands of air-conditioners that remove solar heat and latent heat from buildings to dump it into the local air. The quantities are particularly large in high-density cities like Hong Kong (Newcombe, 1979) and New York.

Injection of urban heat from Washington, D. C. into an atmosphere approaching instability produced showers only downwind from the city (Harnack and Landsberg, 1975). This sensible heat was shown by parcel theory analysis to be "the likely trigger force for shower development."

Larger-Scale Spatial Variations

Cloud-free areas in flows of unstable air in summer also can localize the final destabilization that triggers severe storms. This appears to have happened in the second flash flood at Kansas City in 1977 (Hales, 1978).

Marine or lake air in a sea-breeze circulation or synoptic-scale advection in summer crosses coastal lands as a cold flood that extracts a disproportionate fraction of the energy available in ecosystems and reduces their transpiration. Cold advection characterizes, for example, ecosystems of the Maine coast, some parts of the coastal zone of the Great Lakes, the Arctic Coast, and the coastal Ukraine, and reduces their evapotranspiration correspondingly.

The warm-season flux of sensible heat in the U.S.S.R. takes away 0.8 of the net surplus of all-wave radiation from desert ecosystems in central Asia and progressively smaller fractions as one goes toward the moist forest zone in the north and northwest. North of Moscow to the White Sea it claims only 0.2 of the radiation surplus (Berliand et al., 1972). Extreme air temperatures along the coast of Queensland (mean flux 40 W m^{-2}) run about 10°C lower than extremes observed inland (Dury, 1972), where the mean flux is 60–65 W m^{-2} (Budyko, 1963) from dry ecosystems.

The high, dry mountains and plateaus of interior Asia generate large sensible-heat fluxes in summer (Table VI), and daytime means of 200 W m^{-2} occur at 4.0 km in the Pamirs (Muminov, 1959) and even larger pulses in the Kara-Kul' valley at the same high altitude (Zuev, 1960). These suggest to Flohn (1968) important dynamic implications because they are injected into the middle levels of the planetary atmosphere. Calculations of the sensible-heat flux on a large scale figure in models of the circulation of the atmosphere (e.g., Gates and Schlesinger, 1977; Saltzman and Ashe, 1976b) and define the main patterns of the global energy budget.

Dry low-latitude continents generate mean annual flux densities of about 65 W m^{-2} (Budyko, 1974), and the consequent thermal expansion of the atmosphere over Australia in summer draws in a large volume of lower-tropospheric air, shown by computations of mass convergence (Troup, 1974). This volume is equivalent to lifting the continental atmosphere 20 m, which represents a conversion of 25 W m^{-2} of sensible heat into potential energy, or one-third of the seasonal sensible-heat flux into the atmosphere from the underlying ecosystems.

Removal of energy from ecosystems by way of the sensible-heat flux is negligible much of the time, small in daytime from well-watered ecosystems, but can become large during dry periods in the growing season. These periods, often unpredictable, threaten ecosystems with not simply the loss of potential biomass production but, more seriously, the loss of life. The rapid response of the sensible-heat flux to remove heat in these crises makes it a vital part of the ecosystem energy budget.

REFERENCES

Aizenshtat, B. A. (1960). "The Heat Balance and Microclimate of Certain Landscapes in a Sandy Desert" (transl. by G. S. Mitchell). U.S. Weather Bur., Washington, D.C.

Aizenshtat, B. A. (1966). Issledovaniia teplovogo balansa Srednei Azii. In "Sovremennye Problemy Klimatologii" (M. I. Budyko, ed.) pp. 94–129. Gidrometeorol. Izd., Leningrad.

Ångström, A. (1918). On the radiation and temperature of snow and the convection of the air at its surface. Observations at Abisko (68°21'N, 18°47'E), in January 1916. Ark. Matem. Astron. Fysik 13, No. 21.

Ångström, A. (1925). On radiation and climate. Geogr. Ann. 7, 122–142.

Berliand, T. G., Rusin, N. P., Efimova, N. A., Zubenok, L. I., Mukhenberg, V. V., Ogneva, T. A., Pivovarova, S. I., and Strokina, L. A. (1972). Issledovanie radiatsionnogo rezhima i teplovogo balansa zemnogo shara. Vses. Meteorol. S"ezda, V, Tr. 3, 57–77.

Brazel, A. J., and Outcalt, S. J. (1973). The observation and simulation of diurnal evaporation contrast in an Alaskan alpine pass. J. Appl. Meteorol. 12, 1134–1143.

Brooks, F. A., and Rhoades, D. G. (1954). Daytime partition of irradiation and the evaporation chilling of the ground. Trans. Am. Geophys. Union 35, 145–152.

Budyko, M. I., ed. (1963). "Atlas Teplovogo Balansa Zemnogo Shara." Akad. Nauk SSSR, Mezhduved. Geofiz. Kom., Moscow.

Budyko, M. I. (1974). "Climate and Life" (transl. ed. by D. H. Miller). Academic Press, New York.

Cone, C. D., Jr. (1962). Thermal soaring of birds. Am. Sci. 50, 180–209.

Deacon, E. L. (1969). Physical processes near the surface of the earth. In "General Climatology" (H. Flohn, ed.), World Survey of Climatology, Vol. 2, pp. 39–104. Elsevier, Amsterdam.

de Félice, P. (1968). Étude des échanges de chaleur entre l'air et le sol sur deux sols de nature differente. Arch. Meteorol., Geophys. Bioklimatol., Ser. B 16, 70–80.

Desjardins, R. L. (1977). Energy budget by an eddy correlation method. J. Appl. Meteorol. 16, 248–250.

Dury, G. H. (1972). High temperature extremes in Australia. *Ann. Assoc. Am. Geogr.* **62,** 388–400.

Dyer, A. J., and Crawford, T. V. (1965). Observations of the modification of the microclimate at a leading edge. *Q. J. R. Meteorol. Soc.* **91,** 345–348.

Dyer, A. J., Hicks, B. B., and King, K. M. (1967). The fluxatron—a revised approach to the measurement of eddy fluxes in the lower atmosphere. *J. Appl. Meteorol.* **6,** 408–413.

Eskinazi, S. (1975). "Fluid Mechanics and Thermodynamics of Our Environment." Academic Press, New York.

Flohn, H. (1968). Contributions to a meteorology of the Tibetan Highlands. *Colo. State Univ., Dep. Atmos. Sci., Pap.* No. 130.

Frankenberger, E. (1955). Über vertikale Temperatur-, Feuchte- und Windgradienten der Atmosphäre, den Vertikalaustausch und den Wärmehaushalt an Wiesboden bei Quickborn/Holstein 1953/1954. *Ber. Deutsch. Wetterdienstes* **3,** No. 20.

Frankenberger, E. (1958). Ein Messgerät für vertikalgerichtete atmosphärische Wärmeströme. Germany. Wetterdienst, *Tech. Mitt. Instrumentw.,* N. F., **4,** 21–28.

Frankenberger, E. (1960). Beiträge zum Berechnungen zum Wärmehaushalt der Erdoberfläche. *Ber. Dtsch. Wetterdienstes* **10,** No. 73.

Gates, D. M. (1968). Transpiration and leaf temperature. *Ann. Rev. Plant Physiol.* **19,** 211–238.

Gates, W. L., and Schlesinger, M. E. (1977). Numerical simulation of the January and July global climate with a two-level atmospheric model. *J. Atmos. Sci.* **34,** 36–76.

Greenland, D. (1980). Atmospheric dispersion in a mountain valley. *Ann. Assoc. Am. Geogr.* **70,** 199–206.

Hales, J. E., Jr. (1978). The Kansas City flash flood of 12 September 1977. *Bull. Am. Meteorol. Soc.* **59,** 706–710.

Harnack, R. P., and Landsberg, H. E. (1975). Selected cases of convective precipitation caused by the metropolitan area of Washington, D. C. *J. Appl. Meteorol.* **14,** 1050–1060.

Hicks, B. B., Hyson, P., and Moore, C. J. (1975). A study of eddy fluxes over a forest. *J. Appl. Meteorol.* **14,** 58–66.

Jarvis, P. G., James, G. B., and Landsberg, J. J. (1976). Coniferous forest. *In* "Vegetation and the Atmosphere" (J. L. Monteith, ed.), Vol. 2, pp. 171–240. Academic Press, New York.

Kuo, H. L. (1968). The thermal interaction between the atmosphere and the earth and propagation of diurnal temperature waves. *J. Atmos. Sci.* **25,** 682–706.

Laikhtman, D. L. (1964). "Physics of the Boundary Layer of the Atmosphere." Isr. Program Sci. Transl., Jerusalem. (Transl. by I. Shectman from "Fizika Pogranichnogo Sloia Atmosfery." Gidrometeorol. Izd., Leningrad, 1961).

Lettau, H. H., and Davidson, B. (1957). "Exploring the Atmosphere's First Mile," 2 vols. Pergamon, Oxford.

McKay, D. C., and Thurtell, G. W. (1978). Measurements of the energy fluxes involved in the energy budget of a snow cover. *J. Appl. Meteorol.* **17,** 339–349.

Miess, M. (1968). Vergleichende Darstellung von meteorologischen Messergebnisse und Wärmehaushaltsuntersuchungen an drei unterschiedlichen Standorten in Norddeutschland. *Tech. Univ. Hannover, Inst. Meteorol. Klimatol., Ber.* **2.**

Moen, A. N. (1974). Turbulence and the visualization of wind flow. *Ecology* **55,** 1420–1424.

Motha, R. P., Verma, S. B., and Rosenberg, N. J. (1979). Turbulence under conditions of sensible heat advection. *J. Appl. Meteorol.* **18,** 467–473.

Muminov, F. A. (1959). Radiatsionnyi i teplovoi balansy Alaiskoi Doliny v raione Sarytasha. *Tr. Sredneaziat. Nauchno-Issled. Gidrometeorol. Inst.* No. 2, 165–174.

Munn, R. E. (1966). "Descriptive Micrometeorology." Academic Press, New York.
Murphy, C. E., Jr., and Knoerr, K. R. (1977). Simultaneous determination of the sensible and latent heat transfer coefficients for tree leaves. *Boundary-Layer Meteorol.* **11**, 223–241.
Neiburger, M. (1941). Insolation and the prediction of maximum temperatures. *Bull. Am. Meteorol. Soc.* **22**, 95–102.
Neiburger, M. (1957). Weather modification and smog. *Science* **126**, 637–643.
Newcombe, K. (1979). Energy use in Hong Kong: Part IV. Socioeconomic distribution, patterns of personal energy use, and the energy slave syndrome. *Urban Ecol.* **4**, 179–205.
Nobel, P. S. (1980). Morphology, surface temperatures, and northern limits of columnar cacti in the Sonoran Desert. *Ecology* **61**, 1–7.
Oerlemans, J. (1980). An observational study of the upward sensible heat flux by synoptic-scale transients. *Tellus* **32**, 6–14.
Oke, T. R. (1970). Turbulent transport near the ground in stable conditions. *J. Appl. Meteorol.* **9**, 778–786.
Oladipo, E. O. (1980). An analysis of heat and water balances in West Africa. *Geogr. Rev.* **70**, 194–208.
Pédelaborde, P. (1957). "Le Climat du Bassin Parisien. Essai d'une Méthode rationelle de Climatologie physique," 2 vols. Libraire de Médicis, Paris.
Pennycuick, C. J. (1973). The soaring flight of vultures. *Sci. Am.* **229**(6), 102–109.
Price, A. G., and Dunne, T. (1976). Energy balance computations of snowmelt in a subarctic area. *Water Resour. Res.* **12**, 686–694.
Priestley, C. H. B. (1959). "Turbulent Transfer in the Lower Atmosphere." Univ. of Chicago Press, Chicago, Illinois.
Rauner, Iu. L. (1960). "Zakonomernosti Formirovaniia Teplovogo Balansa i Mikroklimata v Zasushlivyky Usloviiakh." Izd. Akad. Nauk SSSR, Moscow.
Saltzman, B. and Ashe, S. (1976a). The variance of surface temperatures due to diurnal and cyclone-scale forcing. *Tellus* **28**, 307–322.
Saltzman, B., and Ashe, S. (1976b). Parameterization of the monthly mean vertical heat transfer at the earth's surface. *Tellus* **28**, 323–332.
Stricker, H., and Brutsaert, W. (1978). Actual evapotranspiration over a summer period in the "Hupsel Catchment." *J. Hydrol. (Amsterdam)* **39**, 139–157.
Swinbank, W. C. (1951). The measurement of vertical transfer of heat and water vapor by eddies in the lower atmosphere. *J. Meteorol.* **8**, 135–145.
Swinbank, W. C., and Dyer, A. J. (1968). Micrometeorological expeditions 1962–1964. *CSIRO, Div. Meteorol. Phys., Tech. Pap.* No. 17.
Tag, P. M. (1979). A numerical simulation of fog dissipation using passive burner lines. Part II: Sensitivity experiments. *J. Appl. Meteorol.* **18**, 1455–1471.
Tanner, C. B., and Thurtell, G. W. (1970). Sensible heat flux measurements with a yaw sphere and thermometer. *Boundary-Layer Meteorol.* **1**, 195–200.
Thom, A. S. (1975). Momentum, mass and heat exchange of plant communities. *In* "Vegetation and the Atmosphere" (J. L. Monteith, ed.), Vol. 1, pp. 57–109. Academic Press, New York.
Troup, A. J. (1974). Mean flow at the boundary of the Australian continent. *Austral. Meteorol. Mag.* **22**, 61–66.
Webb, E. K. (1964). Daytime thermal fluctuations in the lower atmosphere. *Appl. Opt.* **3**, 1329–1336.
Yap, D., and Oke, T. R. (1974a). Eddy-correlation measurements of sensible heat fluxes over a grass surface. *Boundary-Layer Meteorol.* **7**, 151–163.
Yap, D., and Oke, T. R. (1974b). Sensible heat fluxes over an urban area—Vancouver, B.C. *J. Appl. Meteorol.* **13**, 880–890.

Zuev, M. V. (1960). "On the Heat Balance of the Kara-Kul' Lake (Ozero) Valley in the East Pamirs." U. S. Weather Bur., Washington, D. C. (Transl. by N. A. Stepanova of O teplovom balanse doliny ozera Kara-Kul'. *In* "Sovremmenye Problemy Meteorologii Prizemnogo Sloia Vozdukha," pp. 61–66. Gidrometeorol. Izd., Leningrad, 1958.

Zuev, M. V., and Pavlov, D. F. (1959). Osobennosti teplovogo balansa Golodnoi Stepi pri ursat'evskom vetre. *Tr. Sredneaziat. Nauchno-Issled. Gidrometeorol. Inst.* No. 2, 41–53.

Chapter XVII

SUBSTRATE HEAT FLUX IN TERRESTRIAL ECOSYSTEMS

The subterranean part of a terrestrial ecosystem has a substantial capacity to take in, store, and pay out heat, which buffers the fluctuations in ecosystem temperature that result from fluctuations in absorbed radiant energy. The substrate flows of heat determine the warming and cooling of ecosystem soil and its overall temperature, which affects water uptake, decomposition of carbon compounds, fixing of atmospheric nitrogen and other vital processes.

ROLE OF THE SOIL IN THE ECOSYSTEM

The rate of nitrogen fixation declines at temperatures below 15°C (1.0) to 0.6 at 10° and only 0.1 at 6° (Crofts, 1969), and the food value of pasture decreases in winter. All processes involved in nitrogen flow, including transformations of soil ammonium and nitrates and the role of live and dead roots and litter, can be modeled from two driving variables: soil moisture and soil temperature (Reuss and Innis, 1977). When run at the Pawnee grassland in Colorado, the model reads out yields, top biomass and other ecosystem variables. Alaskan soil, insulated by an accumulation of sphagnum moss, remains so cold in summer that little soil nitrogen is available for tree growth; where the moss has been removed by fire, the distribution of nitrogen in the soil is altered and becomes highest near the surface (Heilman, 1966), where soil is the warmest. An increase in soil-heat flux accelerates its circulation in the ecosystem and biological productivity.

The heat reservoir of the soil forms the environment of roots and the renewal buds of cryptophytes, hemicryptophytes, and geophytes and is more or less closely coupled with above-ground biomass. In midlatitude mountains this thermal coupling with small plants is very close; they find a favorable environment in and near the relatively warm soil.

Temperatures in ecosystem soil are controlled by and at the same time generate soil-heat fluxes in the flux-gradient relation expressed by thermal conductivity λ:

$$\lambda = \text{heat flux/temperature gradient.}$$

Dimensions are $J \ sec^{-1} \ m^{-1} \ deg^{-1} = (W \ m^{-2})/(deg \ m^{-1})$. Heat fluxes can be calculated as the product of measured gradients and conductivity, but changes in gradients with depth and changes in conductivity with soil moisture often make a more indirect method preferable (e.g., Kimball and Jackson, 1975). Bridging of pores by water up to 25% increases λ rapidly, but after 25% the increase is only that due to water replacing air (Marshall and Holmes, 1979, p. 279)

Thermal Capacity

Ecosystem soil forms a closed system, and in analysis heat flux data are combined with thermal capacity. The thermal capacity of a substrate is one factor governing how much heat can be removed from the surface and is the product of specific heat per kilogram of the material and its density, i.e., $c\rho$. Specific heat c is the number of joules required to raise the temperature of 1 kg of a substance 1°. For water c is 4185 J kg^{-1} deg^{-1} and less for most other substances because their molecules respond more actively to an input of energy than do those of water, linked by their polar nature. The specific heat of most rock and soil minerals is approximately 10^3 J kg^{-1} deg^{-1}.*

In energy-budget studies we are interested in heat capacity in each layer and use the conversion from kilogram to cubic meters via density ρ. For water $\rho = 10^3$ kg m^{-3}; for most kinds of rock ρ lies between 2 and 3×10^3 kg m^{-3}. Soil, as a mixture of rock particles, air, and organic matter with varying amounts of water replacing air in some of the pores, has a smaller bulk density than solid rock, and ρ ranges from 1.3 to 1.6 $\times 10^3$ kg m^{-3}.

Combining ρ and c to obtain thermal capacity per unit volume gives units of (kg m^{-3})(J kg^{-1} deg^{-1}) = J m^{-3} deg^{-1}, the energy required for a 1° warming of 1 m^3 of substrate. For water $\rho c = 4.185$ MJ m^{-3} deg^{-1}; for dry soil ρc is usually between 0.8 and 1.5 MJ m^{-3} deg^{-1}. The high density of the rock particles compensates somewhat for their small specific heat.

* Artificial energy storage is expressed in kilojoules per kilogram; e.g., a lead–acid battery stores 140 kJ kg^{-1} (Kalhammer, 1979).

Effects of Soil Moisture.   Water increases both ρ and c. It fills the empty pores, increasing the bulk density, and increases the specific heat of the soil. For wet soils ρc is of the order of 2 to 2.5 MJ m^{-3} deg^{-1}.

Considering how much the moisture of ecosystem soil layers changes with time, it can be seen that its thermal properties are neither uniform nor constant. For this and other reasons the laws of heat conduction in uniform solids are difficult to apply precisely. Although often used in atmospheric modeling and helpful toward a qualitative understanding of how heat moves into and out of substrates, they should be accepted with caution in ecosystem analysis.

Freezing and thawing increases the capacity of a given layer to store heat by adding the latent heat of fusion, often accompanied by an upward movement of liquid water (Outcalt, 1971) and associated with heaving of overwintering crops like alfalfa (Rohweder and Smith, 1978). In the annual cycle the active layer in permafrost regions is only 1–2 m thick, but the quantity of heat stored in summer and released in winter may be as great as in a midlatitude soil. For instance, on the Kara Sea coast tundra took in 130 MJ m^{-2} of heat in summer and thawed to 0.8 m; three-quarters of the heat intake went to thawing (Bakalov *et al.,* 1961). Different kinds of permafrost seem to be associated with different upper limits in soil temperature (Washburn, 1980).

Different Substrates.   Thermal capacity differs in different substrates (Table I). Ice as a bulk solid has different thermal properties than water, and its finely divided state as snow is still different. Granite is a typical bulk rock with high density and low specific heat; mineral soil as a divided state of rock has values of ρc of about 1 MJ m^{-3} deg^{-1}. Organic soil has low density and resulting small heat capacity.

Depth of Zone in Which Heat Is Stored

This dimension of an ecosystem is important in its energy budget because it determines the mass that can store energy. It must therefore be considered along with heat capacity ρc per unit volume.

The day-and-night cycle is not long enough for heat to move to great depths because conductivity (J sec^{-1} m^{-1} deg^{-1}) has a time dimension and the duration of daytime radiant-energy availability is limited. Time is less restrictive in the summer-and-winter cycle, and a greater depth of soil can take part in the heat-storing process. The depths at which a cyclic temperature variation is reduced to 0.05 of its amplitude at the surface are 0.2–0.3 m in the daily cycle and a few meters in the annual cycle in poorly conducting substrates. In denser, moister soils of agricultural ecosystems depths of daytime heat storage are a meter or so, and of annual storage 10–15 m.*

The downward heat flux does not continue indefinitely downward year after year because it meets the slow upward flux of geothermal heat from the crust, which is about 0.05 W m^{-2}, depending on age of the crust, diminishing to 0.046 W m^{-2} after 800 million years (Sclater et al., 1980).† This tiny flux, because it is always directed upward, is able "to check the many-times-faster inflow of the vadose energy at such a short distance from the point of entrance" (Nikiforoff, 1959). Near-surface heat storage in both the daily and annual cycles resembles soil-moisture storage in that turnover is localized in the layers nearest the soil surface and is more or less captive in the ecosystem.

Biomass Storage of Heat

Vegetation tissue has low mass, and the air in the plant volume has very small mass and heat capacity; the aboveground part of an ecosystem, except the largest branches and trunks, heats and cools easily. A formula for heat storage in biomass (Thom, 1975) is based on specific heat c being about 0.7 that of water and mass of material involved in heat turnover varying from 10 g m^{-2} for short grass to more than 10 kg m^{-2} in forest. If temperature in forest increases 3° hr^{-1}, the uptake of heat is

$$\frac{3 \times 10 \times 3000 \, \text{J m}^{-2}}{3600 \, \text{sec}} = 25 \, \text{W m}^{-2}.$$

Dense forest may take in 0.1–0.2 MJ m^{-2} per degree of temperature rise (Hicks et al., 1975). On frost-danger nights this stored heat, less than 1 MJ m^{-2}, contributes 0.1 or less to the supply of heat needed to meet the deficit in the radiation budget (Brooks et al., 1952).

---

* The depth to which heat penetrates is immensely increased where it is carried downward in water injected into aquifers (Molz et al., 1979), for retrieval the next winter. Underground space also becomes a reservoir for energy storage in compressed-air systems (Chiu et al., 1979).

† In geothermal areas this heat flux is much greater. For example, in the Wairakei area of New Zealand it is about 100 W m^{-2} (White, 1965), and in the more dispersed thermal basins of Yellowstone Park it is about 12 W m^{-2}. In the developed Geysers dry-steam field in California it is calculated as 130 W m^{-2} at the well head, of which 19 W m^{-2} is delivered as electrical energy to the consumers at an energy cost of construction and materials of 1.5 W m^{-2} (Gilliland, 1975).

TABLE I

Substrate Properties Affecting Substrate-Heat Flux with Special Reference to the Coefficient of Thermal Admittance[a]

	Thermal conductivity λ (J m^{-1} deg^{-1} sec^{-1})	Density ρ (kg m^{-3})	Specific heat c (J kg^{-1} deg^{-1})	Specific heat per unit volume (heat capacity) ρc (J m^{-3} deg^{-1})	Thermal diffusivity $K = \lambda/\rho c$ (m^2 sec^{-1})	Thermal admittance coefficient $\sqrt{\rho c \lambda}$ (J m^2 deg^{-1} sec$^{-1/2}$)
Still air	0.02	1.3	1000	1300	16×10^{-6} (stirred: 10)	5.2 (stirred: 5×10^2– 5×10^4)
Granite	2.5	2700	790	2.13×10^6	1.17×10^{-6}	2300
Dry sand	0.2	1600	825	1.25×10^6	0.17×10^{-6}	510
Moor soil (dry)	0.08	400	1650	0.66×10^6	0.17×10^{-6}	230
Still water	0.6	1000	4185	4.185×10^6	0.15×10^{-6} (stirred: 0.009)	1600 (stirred: 5×10^4– 5×10^5)
Wet sand	1.7	1800	1250	2.25×10^6	0.7×10^{-6}	1900
Wet clay	1.7	1900	1250	2.4×10^6	0.7×10^{-6}	2000
Marsh soil, wet	0.8	—	—	2.9×10^6	0.3×10^{-6}	1550

Ice	2.1	910	2120	1.93×10^6	1.09×10^{-6}	2000
New snow	0.08	100	2120	0.21×10^6	0.4×10^{-6}	130
Old snow	0.4	400	2120	0.85×10^6	0.47×10^{-6}	580
Silver	418	10,500	234	2.46×10^6	170×10^{-6}	32,000
Experimental sites						
Shafrikan, sandy soil	0.23	1350	825	1.11×10^6	0.21×10^{-6}	510
El Mirage, lake sediments	0.86	—	—	1.00×10^6	0.87×10^{-6}	930
Copenhagen, grassland soil (0–17 cm)	2.13	—	—	1.68×10^6	1.27×10^{-6}	1900
Finland, granite ledge	3.8	—	—	2.10×10^6	1.8×10^{-6}	2700
Argonne, soil depth 0.1–0.5 m	1.2	—	—	2.75×10^6	0.35×10^{-6}	1700
1–3 m	1.7	—	—	2.75×10^6	0.7×10^{-6}	2100

[a] Data for first 12 rows from Geiger (1961), Sutton (1953), Haltiner and Martin (1957), and Sverdrup (1942).

THERMAL DIFFUSIVITY AND ADMITTANCE

We have discussed heat-storage capacity in the soil, ρc, and noted that conductivity λ and hence depth of storage have a time dimension in the heat-storage cycles of day and year. These quantities can be combined in two ways.

Thermal Diffusivity

Their quotient $\lambda/\rho c$ indicates how temperature rises when heat diffuses through a substance, e.g., downward from the surface into the substrate. A quantity of heat moving into a substance with small heat capacity will warm a large volume; if it diffuses into a substrate with a large heat capacity, much of the heat is taken up near the admitting surface. The same quantity of heat diffusing into a substrate with high thermal conductivity will warm a greater volume than in one with small thermal conductivity. This quotient is thermal diffusivity K, and its dimensions are found from those of λ and ρc:

$$\frac{J \sec^{-1} m^{-1} \deg^{-1}}{J m^{-3} \deg^{-1}} = m^2 \sec^{-1}.$$

The coefficients of diffusivity of sensible heat or water vapor from an ecosystem into the atmosphere (Chapters XV and XVI) have the same unusual dimensions, square meters per second, which can perhaps be visualized as the spreading of fire (Chapter XII) or of such a disturbance as ripples on a smooth pond: each second an additional square meter area is involved. Thermal diffusivity of the soil focuses on the substrate to indicate how a heat flux affects it, rather than on the concern of this book, the interface.

Thermal Admittance

A different combination of thermal capacity and conductivity, thermal admittance, emphasizes the removal of heat from the active surface of an ecosystem and is a more fundamental parameter (Carlson and Boland, 1978) than diffusivity for ecosystem analysis. The quantity of energy stored in the "active layer" of the substrate (i.e., between the surface and the depth to which the daily wave penetrates) is proportional to its depth or thickness and its thermal capacity. Effective thickness is proportional to the square root of the thermal diffusivity $\sqrt{K}$ (Petterssen, 1969, p. 61). The thermal capacity per unit depth is ρc, and the product of these factors indexes the capacity of the substrate to take in and store energy in a given length of time.

This product has no generally accepted name: "conductive capacity," "contact coefficient," "thermal inertia," and "thermal admittance coefficient" have all been applied to it, each with reason. It indexes the capacity of the substrate to receive conducted heat; it represents heat abstracted by contact between the surface and the substrate*; it expresses the inertia of the substrate in reacting to fluctuations in surface temperature, and it indicates admission of sensible heat into the soil, especially in competition at the soil-air interface with sensible-heat flow upward into the air (Lettau, 1952).

Conversely, it also indicates the readiness of a substrate to give up heat. For example, when a cold ($-17°C$) air stream invaded an experimental site near Leningrad, a plot of frozen soil (admittance $= 1000$ J m^{-2} deg^{-1} $sec^{-1/2}$) gave up a flux of 45 W m^{-2}, while the snow on an adjacent plot (admittance about 330 J m^{-2} deg^{-1} $sec^{-1/2}$) gave up only 15 W m^{-2} [Data from Serova (1959)].

Admittance can be calculated directly from thermal conductivity λ and from thermal capacity ρc, as the square root of their product $\sqrt{\rho c \lambda}$. It also follows from the preceding definition as $\rho c \sqrt{K}$, because $K = \lambda/\rho c$.

$$\rho c \sqrt{K} = \rho c \sqrt{\lambda/\rho c} = \sqrt{\rho c}\ \sqrt{\lambda} = \sqrt{\rho c \lambda}.$$

Conductivity and heat capacity are not working *against* each other, as they are in the quotient $\lambda/\rho c$ that represents thermal diffusivity but, rather, enhance each other (Brooks, 1959, p. 85). The faster heat moves in the soil (a function of λ), the more is admitted from the surface. The less the soil is heated by a given intake of energy (a function of ρc), the steeper its temperature gradient will remain, allowing heat to continue to enter freely. Each factor is important, and changes in either affect heat intake. Artificial modification of admittance, intake and quantity of heat stored is easily brought about by increasing λ, an example being the heat pump, which brings a deeper mass of soil and regolith into thermal communication with the overlying surface. Admittances that result from computations that begin with density, specific heat, and thermal conductivity are shown in Table I for a variety of ecosystem

---

* The term "contact coefficient" (Businger and Buettner, 1961) suggests an everyday illustration. The temperatures of a carpeted and a bare wooden bedroom floor are about the same, but their effect on bare feet is entirely different. With low conductivity, density, and specific heat the carpet takes little heat away from the feet and is comfortable to walk on. Values of $\sqrt{\rho c \lambda}$ (van Straaten, 1967, p. 212) help us assess the "thermal shock" of standing on the floor: carpet, 120 J m^{-2} deg^{-1} $sec^{-1/2}$; wood floor, 500; clay tiles, 1000–1250; and iron, 12,500.

substrates and some sample ecosystems where energy fluxes have been measured.

Although difficult to visualize in fluids, this coefficient is basic (Lettau, 1952) for such problems as partitioning the energy made available at an interface into shares that move upward and downward. For example, it helps determine how much heat flows into a sunlit house wall during half the daily cycle (Meinel and Meinel, 1976, p. 480). Intuitively, it may be regarded as a kind of absorbency, like the infiltration index. Water is sucked or pulled into soil at a rate dependent on the dryness of each soil layer; the depth to the wetting front is analogous to the depth of soil being heated. Similarly, energy from absorption of radiation at the surface is pulled both ways at rates and in total quantities that depend on the relative admittances of soil and atmosphere. The flows are reversed at night when the incoming and emitted radiation fluxes would balance at a surface temperature much lower than either air or soil temperature; the heat flows drawn from air and soil then depend on their relative admittances and bring the surface to a temperature at which their sum equals the net deficit of radiation.

Most of the nonradiative energy in a desert energy budget at night comes out of the bare soil, rather than the air, except in windy weather. The decline of surface temperature during the night depends mostly on the income of longwave radiation and soil admittance [see Haltiner and Martin (1957, p. 132)].* An idealized expression that assumes little or no heat supply from the air describes the drop of surface temperature: Temperature drop = $(2)(-$net radiation deficit$)(\sqrt{t/\pi})(1/\sqrt{\rho c\lambda})$. This formula gives a drop on a freeze night at Riverside, California (24 February 1939) (Brooks, 1959; Brooks et al., 1952) of 12° for soil admittance of 1300 J m^{-2} deg^{-1} sec$^{-1/2}$. The observed drop was 10°, indicating that the air supplied some sensible heat. In July a change of 10% in admittance in the Sahel, the Sinai, and South Dakota is calculated to change the daily range of surface temperature by 6 to 7% (Saltzman and Pollack, 1977).

Variations in Admittance. Increasing soil moisture raises both components of thermal admittance. The upper 30-cm layer of Yolo soil (Brooks, 1959, p. 76) at the end of the rainy season has an admittance coefficient of 1750, but in summer only 1250 J m^{-2} deg^{-1} sec$^{-1/2}$. Shal-

---

* This relation can be reversed to infer "thermophysical properties of the upper few centimeters of Mercurian soil" (Chase et al., 1974), as expressed in admittance values at different places on the planet's surface. It has been applied to similarly remote surfaces of the airless moon.

low aquifers have been detected by admittance surveys (Huntley, 1978) by their effect on soil moisture.

An experiment with orchard irrigation for frost protection in California increased thermal admittance of the soil to the extent that daytime heat intake increased from 1.9 to 5.0 MJ m^{-2} (Brooks and Rhoades, 1954). Peak rates of heat flux into the soil increased from 115 to 245 W m^{-2}, and nocturnal outflow also increased. After a period of evaporative chilling, the plot warmed up, "and two weeks after flooding was definitely less liable to frost than the unwatered plots." It is recommended that such irrigation be done several days before a frost to avoid latent-heat loss at a critical time; benefits are evaluated as 0.9–1.3°K increase in nocturnal temperature (Bagdonas et al., 1978, p. 102).

Rolling the loose top soil of vineyards in Australia was found effective in increasing nocturnal temperatures as high as 90 cm above the surface, the level of the vulnerable new shoots (Bridley et al., 1965). More heat supplied to the surface from the compacted soil meant less heat extracted by the surface from the air (Fig. 1). Rolling and irrigation together raised soil moisture from 0.12 to 0.20, bulk density to 1500, increased thermal admittance by at least 400 J m^{-2} deg^{-1} sec$^{-1/2}$ or half its initial value, and resulted in the largest improvement of nocturnal air temperature at the height of the vines, 0.6°.* In marginal circumstances this extra warmth from the soil can be crucial.

Moisture, conductivity, and thermal capacity in stratified soil are complex functions of depth and time. Temperature data alone are inadequate for determining thermal admittance or heat flux; rather, direct measurements of heat flux are necessary, and those of conductivity and diffusity (i.e., admittance) are desirable (Lettau, 1954). Data on admittance in different layers of the soil can be applied to analyze the relation of heat flux to surface temperature (Byrne and Davis, 1980). The horizontal pattern of admittance is also complicated, depending on soil density, composition, moisture, and geological formations.

HEAT STORAGE IN THE DAILY CYCLE

What effect does thermal admittance into a substrate have on the way that surplus energy made available at the interface during daytime is partitioned? What effect does it have on the way that the nocturnal radiation deficit is met? These partitions, which will be discussed for the

---

* Larger on the coldest nights.

general energy budget in Chapter XXIII, are introduced here as they occur in dry ecosystems, where the soil surface is the active surface of the ecosystem.

Two Desert Ecosystems

Early in the morning (Table II), the soil-heat flux in sandy soil at Shafrikan takes 0.6 of the total heat moving from the interface, but its share diminishes in competition with upward heat flow as turbulent mixing of the air increases its eddy conductivity, hence its thermal admittance. During the 10.5 hr (from 0500 to 1530) that heat was moving into the soil, total intake was 3.2 MJ m^{-2}, about a quarter of the net surplus of all-wave radiation and 0.1 of the radiant-energy intake. This 3 MJ of energy is manifest as a warming of each soil layer, especially those near the surface, and is retrieved during the following night when it meets most of the net deficit in all-wave radiation.

A similar set of measurements at a dry lake bed in California with

Fig. XVII.1.   Raised vine trellises elevate susceptible shoots above the cold air formed on calm nights when soil-heat flux receives no help from a downward sensible-heat flux in the quiet air and is unable by itself to meet the longwave radiation deficit at the soil surface. Barossa Valley wine district, South Australia.

TABLE II

Partition of Sensible Heat at the Sand–Air Interface during the Heating Period of an Average July Day[a] at Shafrikan, Uzbek SSR[b]

Energy flux	Hour			
	06	08	10	12
From external sources (absorbed radiation)	+440	+745	+1025	+1150
Emitted as longwave radiation by the surface	−380	−465	−570	−615
Leaving for removal as sensible heat	−60	−280	−455	−535
Conduction–convection upward from interface	−25	−190	−330	−420
Conduction downward from interface	−35	−90	−125	−115
Fraction of total conduction	0.60	0.33	0.26	0.22

[a] Data given in watts per square meter.
[b] Latitude 40°05′N. Data from Aizenshtat (1960). Latent-heat conversion equals zero. Thermal admittance of sandy soil is 510 J m^{-2} deg^{-1} sec$^{-1/2}$ (comparable with admittance in the Peruvian desert (Stearns, 1969), 570 J m^{-2} deg^{-1} sec$^{-1/2}$ at −5 mm depth).

greater thermal admittance (930 J m^{-2} deg^{-1} sec$^{-1/2}$) (Vehrencamp, 1953; Brooks and Rhoades, 1954) gave a peak heat flux into the clay-silt substrate of 210 W m^{-2}, almost double that into the sandy substrate at Shafrikan. The total daytime heat intake was greater, 4.2 MJ m^{-2}, and supported a larger return flow at night.*

In both desert ecosystems the downward heat flux prevailed for less than half of the daily cycle, i.e., 10–11 hr. The upward flux lasts longer, and it is slower because the small nocturnal radiation deficit does not produce a steep gradient of soil temperature.

Moist Ecosystems

Substrate heat fluxes are large in desert ecosystems because much of the heat-storing soil body lies exposed to incoming radiation and so

---

* These heat intakes are about the same size as those of solar-energy systems during the winter in Wisconsin per square meter of collector area (Duffie and Beckman, 1976, Fig. 6). The movement of heat into storage in active solar installations, however, is speeded by pumps in water systems and fans in air–rock systems, evaluated by a volumetric heat transfer coefficient that is analogous to conductivity per unit frontal area of a rock bed (Persons et al., 1980). In a sense, passive solar systems with massive walls or floors for heat storage represent an acceleration of the diurnal substrate heat flux.

becomes the active surface. In most ecosystems much of the soil is covered by litter and by a foliage canopy that is not in close thermal contact with the soil. The increase in soil temperature gradient is small because the soil–air interface is shaded, and since its surface temperature rises less through the day, less heat enters even a soil body that has a large coefficient of thermal admittance, as most agricultural soils do.

Furthermore, the latent-heat conversion now enters the surface energy budget; remembering its large energy demands as described in Chapter XV, it is clear that little sensible heat is left to be partitioned between the upward (Chapter XVI) and downward fluxes. The return flow of heat from the soil during the night, however, is still a factor in interpreting canopy temperature patterns (Byrne *et al.*, 1979) since it bulks relatively large in the nocturnal budget.

The combined effect of the shade and latent-heat conversion of a plant cover is illustrated in a grassland ecosystem in Denmark, the soil of which took in 1.8 MJ m^{-2} on a sunny day in summer (Kristensen, 1959). This was less than in desert sites, in spite of the denser, more conductive soil (thermal admittance of 1900 J m^{-2} deg^{-1} sec$^{-1/2}$ under the grass layer). Absorption of radiant energy through a layer several centimeters deep and conversion of much of it into latent heat, as well as shading of the soil itself, kept the soil-surface temperature low; a steep temperature gradient into the soil never developed.

Temperature Range

The diurnal range in surface temperature indexes the quantity of energy available as sensible heat relative to the sum of the admittance to soil and air, and so is affected by soil admittance. Brooks's Riverside example of nocturnal cooling cited earlier in this chapter illustrates this effect. Remote sensing of surface temperature at the times of its diurnal extremes defines the range; sensing of surface albedo and measurement or estimate of incoming solar radiation approximate the radiative forcing function, or radiant-energy intake, when allowance is made for variation in downward longwave radiation. At a dry surface, in weather when atmospheric thermal admittance can be considered to be small, the analyst now has data sufficient to determine substrate thermal admittance. This parameter can be used to delineate near-surface geology or moisture by such methods as those developed by scientists at Newcastle University in Australia (Pratt and Ellyett, 1979), or to analyze urban ecosystems (Carlson and Boland, 1978). In moist ecosystems spatial variation in latent-heat flux still presents a problem (Pratt *et al.*, 1980).

APERIODIC VARIATIONS

Soil-heat flux at Hamburg in June 1958 displayed an average daytime intake of 0.9 MJ m^{-2} and an average nocturnal outflow of 0.5 MJ m^{-2}. The intake exceeds the outflow because in June the soil is still warming on the average, but in seven 24-hr periods outflow exceeded the intake because of cold advection.

Large heat fluxes occur if rain accompanies the change in air temperature (Landsberg and Blanc, 1958) since conductivity is raised by convective transport of heat in the percolating water. Periods of warm advection following cold weather also alter the soil-heat flux pattern. Such day-to-day variations in weather, especially in spring before trees leaf out and shade the soil, produce substantial variations in soil-heat flux in the Hubbard Brook forest (Federer, 1973) and deviations of as much as 5°C in soil temperature at -0.025 m depth.

HEAT STORAGE IN THE ANNUAL CYCLE

Fluxes on Days in Spring

Storage of sensible heat in the soil does not occur in one continuous 6-month wave, but rather is the accumulation of hundreds of daytime intakes and nocturnal withdrawals of heat. In June 1958, for example, heat moved into the soil at Hamburg during 11.5 hr at a mean flux density of 21 W m^{-2}, and out of it during 12.5 hr at a mean flux density of 11 W m^{-2}. Daytime intakes averaged 0.9 MJ m^{-2} and nocturnal outflows 0.5 MJ m^{-2}, and total flux in and out, 1.4 MJ m^{-2}, was several times greater than the net flux (Frankenberger, 1962). However, because more heat entered the soil by day than emerged by night, the net amounts of energy remaining produced the general spring warming. The same process had been operating in April and May and would continue through July and August, accumulating 50 MJ in each square meter column of the soil.

Nocturnal heat flows to the grass from beneath increased only two to three times, from 0.4 MJ m^{-2} during long winter nights to 1.0 MJ m^{-2} during spring nights, a steadiness that probably reflects steadiness in the net loss of energy at night by exchange of longwave radiation, noted in Chapter X. In contrast, the daytime intakes of energy into the substrate grew from negligible amounts in the dark winter days to 1.5 MJ m^{-2} in sun-warmed April, May, and June. Their summer–winter

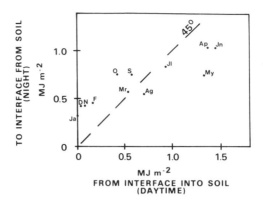

Fig. XVII.2.   Mean nocturnal pulse of heat to grass from soil in months from September 1953 to August 1954 (MJ m^{-2}), plotted against mean daytime pulse from grass into soil (MJ m^{-2}). The daytime pulse varies from zero in winter to 1.5 MJ m^{-2} in early summer while the nocturnal return flow varies from 0.4 MJ m^{-2} to 1.1 MJ m^{-2}. [Data from Frankenberger (1955, p. 6 and Appendix Table 2).]

increase is much larger than the change in the more steady nocturnal outflows (Fig. 2).

The nocturnal movement of energy to a grass surface in interior North America (Argonne National Laboratory) (Carson, 1963) also remains steady in all months outside midwinter at values between 1.0 and 1.5 MJ m^{-2} each night. In contrast, the daytime intakes of energy into the soil vary with radiation absorbed at the surface and reached 2.5 MJ in early summer. These quantities are double the nocturnal return flow, so that each 24-hr cycle ends with 1.0–1.5 MJ of sensible heat remaining in the soil column as the day's contribution to the heating phase of the annual cycle.

Threshold Soil Temperatures.   Melting the snow cover and thawing frozen soil are steps in the arrival of midlatitude spring that may claim 50–100 MJ m^{-2}. Shallow snow means deep freezing of the soil, and heat that is not needed to melt the snow after a dry winter may be needed to thaw the soil. Both represent a carryover of winter cold into spring. However, a warm spring can wipe out these memories of a severe winter so completely that biological energy conversion in summer is entirely unaffected; three cold or snowy winters in Wisconsin were each followed by record crop yields in 1977, 1978, and 1979.

Wet soils in spring are poor in oxygen and may be too cold for cell division and other activities of root growth; much of the radiant energy they absorb is converted to latent heat and does not warm them much. Well-drained soils take in more sensible heat than poorly drained soils, and their smaller thermal capacities mean a greater temperature response for each joule admitted and quicker arrival at the optimum thermal environment for seed germination. Clay soils are usually a degree or two colder than sandy soils in spring, and peat soils, with small heat intake and high evaporation, average three or four degrees colder than mineral soils.

Cultigens of middle-latitude origin, like wheat and barley, require soil temperatures just above freezing for germination, while those domesticated in the low latitudes, i.e., corn and sorghum, require 8–12°C, and their optimal temperatures are 8–10°K higher than wheat (Marshall and Holmes, 1979, p. 281). Where such different crop ecosystems are grown in the same region, small grain is planted weeks before corn. The delay depends on how fast the soil warms in April and May and represents a waste of photosynthetically active solar radiation of special concern to climatologists in such countries of food urgency as China.

Summer

The downward progression of the wave of sensible heat, as shown by the first harmonic in a Fourier analysis of soil temperatures at Argonne, is slow; the mid-July peak at 0.01 m is delayed to 5 August at -0.5 m and 10 October at -3.0 m. The depth in the soil at which the range between summer and winter temperatures become negligible (defined here as 0.01 of the range in surface temperature) depends on the square root of thermal diffusivity K (or to thermal admittance divided by ρc), and is equal in meters to $4400\sqrt{K}$ (Geiger, 1961). Measured depths agree with those calculated from this Fourier relation (Chang, 1957) if K is taken as 0.5×10^{-6} m^2 sec^{-1}, a value between those for dry and wet soils (see Table I).

Summer soil temperatures limit the occurrence of fossorial mammals, those that spend most of their lives in closed burrows. Large fossorial mammals (>80 g mass) exhibit low basal rates of metabolism as compared with other mammals, which also reduce their demand on the limited oxygen supply; small fossorial mammals exhibit a complicated trade-off between mass, basal metabolic rate, and thermal conduction to prevent overheating. Mammals living in open burrows, especially if they forage outside them when burrows are hottest, i.e., during the night, do not display these trade-offs (McNab, 1979).

The Declining Phase of the Yearly Cycle

As the days shorten in late summer, the daytime intakes of sensible heat into the substrate eventually become less than the more constant nocturnal outflows, which cushion the decline in surface temperature through assessments laid on successively deeper layers of the soil.* Underground houses take the same kind of advantage of warmth in the deeper soils (LaNier, 1976), often combining burial of one part of the house with exposure of another part for passive solar heating. Energy savings for space heating may be as much as 0.6 in parts of the Midwest (Landsberg, 1977). Many Canadian cities have followed the example of Montreal and have built underground shopping malls (Mercer, 1979).

As the wave of cooling descends into the soil, biological organisms adjust to it. Japanese beetles spend the summer near the −0.1 m layer of the soil in a temperature between 20 and 28°C and overwinter as grubs that burrow more deeply as the soil cools. When the soil temperature falls to about 10 C they stop eating and become inactive; when the soil begins to warm in spring, they resume activity and move upward "about 5 days behind the corresponding temperature change, resume feeding to complete their third larval stage and prepare to emerge as adult beetles" (Bourke, 1961). Soils that are inhospitable to this insect lie in northwest Europe, where they are too cold in summer (below 20°C), and in the Soviet Union, too cold in winter (below −2°C). Several species of desert rodents similarly occupy burrows at different depths in different seasons (Kenagy, 1973).

Grain crops that are sown in fall to winter over have to be chilled, but their tillering nodes, about −0.03 m deep, cannot survive extremely low temperatures. Temperature at this depth is also important for perennial grasses (Strashnaia, 1980) and is used to forecast survival in spring. If the minimum −0.03 m temperature averages warmer than −12°C, grain has few overwintering problems, but if it averages colder than −16°C, it is necessary to plant cold-resistant strains of wheat and take measures to keep the soil as warm as possible (Kulik and Sinel'shchikov, 1966, p. 288), often by managing depth and conductivity (i.e. density) of the insulating snow blanket.

The insulating power of snow cover is familiar in the shallow freezing of soil in winters or regions when snow is deep; a snow-covered plot gave up 48 MJ m^{-2} of heat and froze to only 0.11 m, while a bare plot gave up 137 MJ m^{-2} and froze to 1.5 m (Pavlov, 1965). In some Wisconsin winters the pattern of snow depth has been such that northern counties

---

* This transport of energy is increased by use of heat pumps as a mode of house heating, which extract heat from a large volume of substrate. In some installations the summer heating of the soil also is increased by a pumped circulation.

had almost no frozen soil at a time when southern counties were frozen deeply.

The removal of heat from the soil body in winter is accelerated in the special case of groundwater to melt snow and to warm surface soil in the Black Forest (Fezer, 1971); molecular conduction of heat through the soil is augmented by convection. Heat pumps in the Midwest utilize groundwater similarly (Connelly, 1979). Heat pipes containing ammonia multiplied the upward transfer of soil heat to a bridge deck in Wyoming to a rate of 173 W m^{-2} (Cundy *et al.*, 1979) without indication of any "permanent depression" of temperature in the weathered granite at 15 m depth.

Annual Heat Turnover

The half-cycle intake of energy at Argonne, found by multiplying the thermal capacity of each soil layer by the warming that it experienced, is 170 MJ m^{-2} (Carson and Moses, 1963) and claims about a tenth of the radiation surplus from April through August.

Two-thirds of the total intake of heat in a grassland ecosystem in Denmark occurs in May, June, and July when the 24-hr average is 12 W m^{-2}, a fraction of the net surplus of all-wave radiation that decreases from 0.11 in May to 0.07 in August (Kristensen, 1959). Most of the energy is admitted to the soil early in summer, in a regime that leads the regime of soil temperature by more than a month, as heat-flow theory indicates it should.

The annual intake of sensible heat in the whole soil column averages 125 MJ m^{-2} (Aslyng and Jensen, 1966), but in a summer when drought-injured grass covered the soil less completely than usual (1955), it was about 10% greater than the mean. Although heat turnover is only 0.02 of annual radiant-energy intake, it is important in the ecosystem energy budget because it transfers energy from the season that is rich in the radiant form—which cannot be stored—to the dark season and helps reduce the effects of the annual fluctuation in radiant-energy income. However, it has greatest value to ecosystems in winter if it is coupled with a large downward flux of longwave radiation, as it is in Denmark. It is less biologically significant in the Midwest (although useful in earth-sheltered dwellings).

SPATIAL PATTERNS OF SOIL-HEAT FLUX

Small-Scale Spatial Patterns

The size of the diurnal soil-heat flux varies spatially as a function of differences in radiant-energy inputs at the soil surface, soil moisture,

atmospheric ventilation, and thermal admittance. Rockiness and porosity affect soil thermal admittance, as described earlier in this chapter,
and the relative values of thermal admittance in substrate and air determine how much of the available sensible energy at the soil surface
goes downward. The extreme spatial variability of soil moisture produces like variability in thermal admittance and soil-heat flux.

Daytime heat intakes into different substrates of southern Finland
(60°N latitude) are the most reliable components in the first energy budgets ever cast (Homén, 1897). The diurnal heat wave penetrated below
the deepest measurements (-0.7 m) in granite and died out in sand at
about -0.5 m. In peat soil, with very low conductivity in spite of its
high water content, the diurnal heat pulse did not penetrate below
-0.2 m. The daily turnovers of heat in these substrates, averaged over
a 3-day period, are shown in Table III. So great a depth of high-density
granite participated in this oscillation that it took in more heat by day
and gave out more by night than the other substrates. Daily turnover
in the sandy soil was a little larger than in the desert at Shafrikan mentioned earlier in this chapter (3.1 MJ m^{-2}), and a little less than in the
silt of the dry lake (4.2).

The turnover of heat in the peat was small and confined to the upper
layers because evaporation left only a third as much heat available for
partitioning as was partitioned at the granite surface between soil and
air. Although a moderately high thermal admittance in the peat assured
the intake of a reasonable fraction—about 0.3—of the heat available at
the surface, this amount was small, 1.6 MJ m^{-2}. Nocturnal cooling exhausted it and brought the soil surface to 6°C as compared with 15°C
at the granite; its grass came close to freezing.*

Microtopography, such as represented by differences in soil drainage
or windthrow mounds in forest, produces admittance and soil-heat flux
contrasts. Mounds in Hubbard Brook Experimental Forest average 1–2°C
colder in winter than a standard soil-temperature site and correspondingly warmer in summer (Federer, 1973). Farmers and ecologists are
aware of these microscale differences in soil temperature, but few studies
of differences in soil-heat fluxes have yet been made. However, the daily
regimes of surface temperature can be interpreted in terms of thermal
admittance of the substrate, as mentioned earlier. Mesoscale temperature differences in an agricultural region are due, for example, to soil-
moisture contrasts (Nixon and Hales, 1975), which influence admittance.

---

* The hazard of summer night frosts was the reason Homén began to study daily intake
and nightly outgo of heat.

TABLE III

Diurnal Heat Turnover in Three Substrates in Southern Finland in Summer[a]

Substrate	Thermal admittance $(J\ m^{-2}\ deg^{-1}\ sec^{-1/2})$	Heat turnover $(MJ\ m^{-2})$
granite	2900	6.8
sandy heath	1510	3.7
peat moor	1550	1.6

[a] Homén (1897, pp. 47, 48, 65, 83).

[b] For comparison, grassland soil near Copenhagen with $1900\ J\ m^{-2}\ deg^{-1}\ sec^{-1/2}$ admittance took in 1.8 MJ m^{-2} (Kristensen, 1959).

Managing Soil Temperature by Means of Heat-Flux Modification

Crops on muck soils in the Midwest are occasionally frozen in August or even near the summer solstice (as in 1965), following an invasion of an unseasonably cold, dry polar air stream that radiates little heat to the underlying ecosystems. When during a calm night little sensible heat can be extracted from the air, the smallness of soil-heat storage is plain. Mixing in sand increases thermal admittance and increases the turnover of heat. When an 8 cm-deep layer of sand was mixed into peat soil in Finland [Pessi (1956) quoted in Geiger (1961, pp. 158–159)], thermal admittance increased from $1010\ J\ m^{-2}\ deg^{-1}\ sec^{-1/2}$ to 1390. The formula given earlier for nocturnal fall in temperature during an 8-hr night and assuming a net radiation deficit of 40 W m^{-2} yields a temperature drop of $-7.8°C$ on untreated peat and $-5.7°C$ on treated, an improvement of 2.1°C; the minimum air temperature observed in oats on these soils showed an improvement of 1.9°C.

Flooding cranberry marshes represents primarily energy transport to the threatened sites, but also increases thermal admittance. Sand is also used in these marshes.

Mulching by Vegetation Residues   In spring and early summer, when the soil body is taking in heat from the surface, insulation is undesirable and mulched soil is a degree or more colder than unmulched. The same mulching practice that produced favorable results in South Carolina produced unfavorable results in Iowa and Minnesota (van Wijk and Derksen, 1963), because in South Carolina soil temperature at -0.1 m remained above the optimum for maize and mulching had no unfavorable effect. Trial-and-error experimentation had to give way to physical and biological reasoning.

Cultivated Soil. Tilling the upper layer of soil usually reduces thermal admittance and increases the range in temperature at its surface and in its upper layers, with overheating by day and frost hazard at night. When soil temperature is rising in spring, cultivation speeds the heating of the upper layers; the opposite effects in deeper layers are unimportant to the still shallow-rooted young plants.

Thickening the Top Layer of Soil or Ecosystem. Increasing soil depth, even without changing thermal admittance, reduces the heat flux reaching a given layer. The mallee fowl, by mounding dirt over a hole in which incubating eggs are buried, maintains an amazing constancy of temperature at 33°C in the eggs month after month (Frith, 1956, 1957, 1962) in a climate (interior New South Wales) of large daily and substantial seasonal fluctuation (Marshall and Holmes, 1979, p. 280). The bird also opens the mound in hot weather to allow excess soil heat to escape or spreads out sand for the sun to warm in cool weather, practices that Frith measured by heat-flux meters. Energy demands on the mother to produce large eggs and on the father to toil unceasingly in managing the soil–heat flux leave them no time to be parents; after the long incubation the chicks are on their own after they hatch and dig their way up to the surface.

Dense vegetation separating the active outer surface of an ecosystem from the soil surface itself also restricts soil-heat flux by reducing the temperature gradient in the soil. The daytime downward flux is reduced from 0.2 of the net surplus of all-wave radiation in bare soil to as little as 0.02 under dense vegetation (Thom, 1975), especially where the active surface is as distant from the soil as it is in forest stands. Soil-heat fluxes measured at three sites at Zagorsk, near Moscow, in early summer showed the morning intake of energy to be 4.5 MJ m^{-2} into bare soil, 2.8 into soil under meadow grass, and 1.0 into forest soil. Peak rates of heat flux in midmorning were 175, 105, and 45 W m^{-2}, respectively (Pavlov, 1965; Rauner, 1960).

These spatial differences are less marked in the annual cycle. Springtime and summer intake into the forest soil is three-quarters or more that into the meadow soil, which suggests that in a long cycle vegetation cover insulates less than does snow cover, which can sustain temperature differences of 10 or even 20° between its top and bottom over a long period. Forest and snow cover are often managed jointly to reduce the depth of soil freezing and alleviate its effects on infiltration of water in spring [e.g., Weitzman and Bay (1963)].

Larger-Scale Contrasts

Substrate storage of heat is a clear expression of seasonality, which is a large-scale variable in the energy climate of a region. For example, a small summertime intake into peat in the Amur region allows deep wintertime freezing of the peat and underlying clay, which blocks percolation and creates large areas of water-logged land and bogs (Kholoden, 1978). Modelers of atmospheric circulations find it profitable to include thermal capacity (Bhumralkar, 1975), seasonal heat storage in land and ocean (Taylor, 1976), or computations of soil-heat flux (Deardorff, 1978) in order to specify atmospheric heating.

Sensible heat stored in the substrates of terrestrial ecosystems provides both the thermal environment of their below-ground members and organisms, and also carries energy over into night or the cold season. Its magnitude in the time-limited cycles of day and year depends on the product of the thermal capacity of the substrate and thermal conductivity, or thermal admittance, a property that explains spatial differences in the role of substrate-heat fluxes in ecosystem energy budgets. Variations in exposure and in other fluxes in the budget also cause the substrate-heat flux to vary, and daily and yearly cycles in absorbed radiant energy cause it regularly to reverse direction.

By removing heat from the surface in daytime and in summer and returning it at night and in winter, this flux reduces fluctuations in surface temperature caused by variations in other components of the ecosystem energy budget and couples the surface and substrate parts of an ecosystem. This coupling is to be understood in the context of the complete ecosystem energy budget, which our discussions are' approaching. One way to examine the complete budget is in a case of close substrate–surface coupling and large substrate heat flux, both due to a substrate of large thermal admittance; we will look at aquatic ecosystems in Chapter XVIII.

REFERENCES

Aizenshtat, B. A. (1960). "The Heat Balance and Microclimate of Certain Landscapes in in a Sandy Desert" (transl. by G. S. Mitchell). U.S. Weather Bur., Washington, D.C.

Aslyng, H. C., and Jensen, S. E. (1966). Radiation and energy balances at Copenhagen 1955–1964. R. Vet. Agric. Coll. (Copenhagen), Yearb. 1965, pp. 22–40.

Bagdonas, A., Georg, J. C., and Gerber, J. F. (1978). Techniques of frost prediction and methods of frost and cold protection. WMO Tech. Note No. 157.

Bakalov, S. A., Deriugin, B. A., and Sychev, K. A. (1961). "Radiation and the Heat Balance of the Earth's Surface in the Arctic." U.S. Weather Bur., Washington, D.C. [Transl. by I. A. Donehoo from Radiatsionny i Teplovoi Balans Poverkhnosti Sushi v Arktike. Tr. Glav. Geofiz. Observ. **92**, 102–126 (1959).]

Bhumralkar, C. M. (1975). Numerical experiments on the computation of ground surface temperature in an atmospheric general circulation model. *J. Appl. Meteorol.* **14**, 1246–1258.

Bourke, P. A. (1961). Climatic aspects of the possible establishment of the Japanese beetle in Europe. *WMO Tech. Note* No. 41.

Bridley, S. F., Taylor, R. J., and Webber, R. T. J. (1965). The effects of irrigation and rolling on nocturnal air temperatures in vineyards. *Agric. Meteorol.* **2**, 373–383.

Brooks, F. A. (1959). "An Introduction to Physical Microclimatology." Associated Student Store, Davis, California.

Brooks, F. A., and Rhoades, D. G. (1954). Daytime partition of irradiation and the evaporation chilling of the ground. *Trans. Am. Geophys. Union* **35**, 145–152.

Brooks, F. A., Kelly, C. F., Rhoades, D. G., and Schultz, H. B. (1952). Heat transfers in citrus orchards using wind machines for frost protection. *Agric. Eng.* **33**, 74–78, 143–147, 154.

Businger, J. A., and Buettner, K. J. K. (1961). Thermal contact coefficient. *J. Meteorol.* **18**, 422.

Byrne, G. F., and Davis, J. R. (1980). Thermal inertia, thermal admittance, and the effect of layers. *Rem. Sens. Environ.* **9**, 295–300.

Byrne, G. F., Begg, J. E., Fleming, P. M., and Dunin, F. X. (1979). Remotely sensed land cover temperature and soil water status—a brief review. *Rem. Sens. Environ.* **8**, 291–305.

Carlson, T. N., and Boland, F. E. (1978). Analysis of urban–rural canopy using a surface heat flux/temperature model. *J. Appl. Meteorol.* **17**, 998–1013.

Carson, J. E. (1963). Analysis of soil and air temperature by Fourier techniques. *J. Geophys. Res.* **68**, 2217–2232.

Carson, J. E., and Moses, H. (1963). The annual and diurnal heat-exchange cycles in upper layers of soil. *J. Appl. Meteorol.* **2**, 397–406.

Chang, J.-H. (1957). Global distribution of the annual range in soil temperature. *Trans. Am. Geophys. Union* **38**, 718–723.

Chase, S. C., Miner, E. D., Morrison, D., Münch, G., Neugebauer, G., and Schroeder, M., (1974). Preliminary infrared radiometry of the night side of Mercury from Mariner 10. *Science* **185**, 142–145.

Chiu, H. H., Rodgers, L. W., Saleem, Z. A., Ahlwalia, R. K. (1979). Mechanical energy storage systems: Compressed air and underground pumped hydro. *J. Energy* 3(3), 131–139.

Connelly, J. P. (1979). Residential heating and cooling with ground water. Univ. Wisconsin, Geol. Nat. Hist. Surv.

Crofts, F. C. (1969). Nitrogen fertilizers for balancing pasture production with animal needs. *In* "Intensive Utilization of Pastures" (B. J. F. James, ed.), pp. 76–89. Angus & Robertson, Sydney, Australia.

Cundy, V. A., Nydahl, J. E., and Pell, K. M. (1979). Geothermal heating of bridge decks. *In* "Snow Removal and Ice Control Research," Nat. Res. Counc., Transp. Res. Bd., *Spec. Rep.* **185**, pp. 169–175, Nat. Acad. Science, Washington, D.C.

Deardorff, J. W. (1978). Efficient prediction of ground surface temperature and moisture, with inclusion of a layer of vegetation. *J. Geophys. Res.* **83**, 1889–1903.

Duffie, J. A., and Beckman, W. A. (1976). Solar heating and cooling. *Science* **191**, 143–149.

Federer, C. A. (1973). Annual cycles of soil and water temperatures at Hubbard Brook. *U.S. For. Serv., Res. Note* **NE-167**.

Fezer, F. (1971). Kuppenlandschaft mit Wässerwiesen im Biotitgranit des Nordschwarzwaldes. *Erde* **102**, 1–5.

Frankenberger, E. (1955). Über vertikale Temperatur-, Feuchte- und Windgradienten der Atmosphäre, den Vertikalaustausch und den Wärmehaushalt an Wiesboden bei Quickborn/Holstein 1953/1954. *Ber. Dtsch. Wetterdienstes,* **3,** No. 20.

Frankenberger, E. (1962). "Contributions to the International Geophysical Year, 1957–58. 1. Measurement Results and Computations of the Heat Balance of the Earth's Surface" (transl. by A. F. Spano). U.S. Weather Bur., Washington, D.C.

Frith, H. J. (1956). Temperature regulation in the nesting mounds of the mallee-fowl, *Leipoa ocellata* Gould. *Austral. CSIRO, Wildl. Res.* **1,** 79–95.

Frith, H. J. (1957). Experiments on the control of temperature in the mound of the mallee-fowl, Leipoa ocellata Gould (Megapotiidae). *Austral. CSIRO, Wildl. Res.* **2,** 101–110.

Frith, H. J. (1962). "The Mallee-Fowl. The Bird that Builds an Incubator." Angus & Robertson, Sydney, Australia.

Geiger, R. (1961). "Das Klima der bodennahen Luftschicht," 4th ed. Vieweg, Braunschweig.

Gilliland, M. W. (1975). Energy analysis and public policy. *Science* **189,** 1051–1056.

Haltiner, G. J., and Martin, F. L. (1957). "Dynamical and Physical Meteorology." McGraw-Hill, New York.

Heilman, P. E. (1966). Change in distribution and availability of nitrogen with forest succession on north slopes in interior Alaska. *Ecology* **47,** 825–831.

Hershfield, D. M. (1979). Freeze–thaw cycles, potholes, and the winter of 1977–78. *J. Appl. Meteorol.* **18,** 1003–1007.

Hicks, B. B., Hyson, P., and Moore, C. J. (1975). A study of eddy fluxes over a forest. *J. Appl. Meteorol.* **14,** 58–66.

Homén, T. (1897). Der tägliche Wärmeumsatz im Boden und die Wärmestrahlung zwischen Himmel und Erde. *Acta Soc. Sci. Fenn.* **23,** No. 3.

Huntley, D. (1978). On the detection of shallow aquifers using thermal infrared imagery. *Water Resour. Res.* **14,** 1075–1083.

Kalhammer, F. R. (1979). Energy-storage systems. *Sci. Am.* **241**(6), 56–65.

Kenagy, G. J. (1973). Daily and seasonal patterns of activity and energetics in a heteromyid rodent community. *Ecology* **54,** 1201–1219.

Kholoden, Ye. E. (1978). Role of the thermal factor in the formation of bogs in the Amur region. *Sov. Hydrol.* **17**(1), 6–10. Transl. from *Tr. Gosud. Gidrolog. Inst.* **236,** 96–105, 1977. Issued 1980.

Kimball, B. A., and Jackson, R. D. (1975). Soil heat flux determination: A null-alignment method. *Agric. Meteorol.* **15,** 1–9.

Kristensen, K. J. (1959). Temperature and heat balance of soil. *Oikos* **10,** 103–120.

Kulik, M. S., and Sinel'shchikov, V. V. (1966). "Lektsii po Sel'skokhoziaistvennoi Meteorologii." Gidrometeorol. Izd., Leningrad.

Landsberg, H. E. (1977). Climate and shelter. *EDS (Environ. Data Serv., Natl. Ocean Atmos. Adm.)* May, 7–11.

Landsberg, H. E., and Blanc, M. L. (1958). Interaction of soil and weather. *Soil Sci. Soc. Am., Proc.* **22,** 491–495.

LaNier, R. (1976). Earth covered buildings and environmental impact. *In* "Alternatives in Energy Conservation: The Use of Earth Covered Buildings," pp. 269–278. Natl. Sci. Found., pp. 269–278. U.S. Gov. Print. Off., Washington, D.C.

Lettau, H. H. (1952). Synthetische Klimatologie. *Ber. Dtsch. Wetterdienstes (US-Zone)* **38,** 127–136.

Lettau, H. H. (1954). A study of the mass, momentum, and energy budget of the atmosphere. *Arch. Meteorol., Geophys. Bioklimatol., Ser. A* **7,** 133–157.

Marshall, T. J., and Holmes, J. W. (1979). "Soil Physics". Cambridge Univ. Press, London and New York.

McNab, B. K. (1979). The influence of body size on the energetics and distribution of fossorial and burrowing mammals. *Ecology* **60**, 1010–1021.

Meinel, A. B., and Meinel, M. P. (1976). "Applied Solar Energy: An Introduction." Addison-Wesley, Reading, Massachusetts.

Mercer, J. (1979). On continentalism, distinctiveness, and comparative urban geography: Canadian and American cities. *Canad. Geogr.* **23**, 119–139.

Molz, F. J., Parr, A. D., Andersen, P. F., and Lucido, V. D. (1979). Thermal energy storage in a confined aquifer: Experimental results. *Water Resour. Res.* **15**(6), 1509–1514.

Nikiforoff, C. C. (1959). Reappraisal of the soil. *Science* **129**, 186–196.

Nixon, P. R., and Hales, T. A. (1975). Observing cold-night temperature of agricultural landscapes with an airplane-mounted radiation thermometer. *J. Appl. Meteorol.* **14**, 498–505.

Outcalt, S. I. (1971). Field observations of soil temperature and water tension feedback effects on needle ice nights. *Arch. Meteorol., Geophys. Bioklimatol., Ser. A* **20**, 43–53.

Pavlov, A. V. (1965). Teplovoi balans nekotorykh vidov deiatel'noi poverkhnosti v Podmoskov'e. *In* "Teplovoi i Radiatsionnyi Balans Estestvennoi Rastitel'nosti i Sel'skokhoziaistvennykh Polei" (Iu. L. Rauner, ed.), pp. 106–116. Izd. Nauka, Moscow.

Persons, R. W., Duffie, J. A., and Mitchell, J. W. (1980). Comparison of measured and predicted rock bed storage performance. *Solar Energy* **24**, 199–201.

Petterssen, S. (1969). "Introduction to Meteorology," 3rd ed. McGraw-Hill, New York.

Pratt, D. A., and Ellyett, C. D. (1979). The thermal inertia approach to mapping of soil moisture and geology. *Rem. Sens. Environ.* **8**, 151–168.

Pratt, D. A., Foster, S. J., Ellyett, C. D. (1980). A calibration procedure for Fourier series thermal inertia models. *Photogram. Engr. Rem. Sens.* **46**(4), 529–538.

Rauner, Iu. L. (1960). "Zakonomernosti Formirovaniia Teplovogo Balansa i Mikroklimata v Zasushlivykh Usloviiakh." Izd. Akad. Nauk SSSR, Moscow.

Reuss, J. O., and Innis, G. S. (1977). A grassland nitrogen flow simulation model. *Ecology* **58**, 379–388.

Rohweder, D. A., and Smith, D. (1978). Winter injury to forages. *Univ. Wis. Coll. Agric., Ext. Publ.* A-2905.

Saltzman, B., and Pollack, J. A. (1977). Sensitivity of the diurnal temperature range to changes in physical parameters. *J. Appl. Meteorol.* **16**, 614–619.

Sclater, J. G., Jaupart, C., and Galson, D. (1980). The heat flow through oceanic and continental crust and the heat loss of the earth. *Rev. Geophys. Space Sci.* **18**, 269–311.

Serova, N. V. (1959). "An Investigation of the Heat Regime in Soil During Winter" (transl. by G. S. Mitchell). U.S. Weather Bur., Washington, D.C. [From: Issledovanie teplovogo rezhima pochvy v zimnee vremia. *Meteorol. Gidrol.* **2**, 24–27 (1958).]

Stearns, C. R. (1969). Application of Lettau's theoretical model of thermal diffusivity to soil profiles of temperature and heat flux. *J. Geophys. Res.* **74**, 532–541.

Strashnaia, A. I. (1980). Vliianie agrometeorologicheskikh uslovii na perezimovku mnogoletnykh bobovykh trav v tsentral'nykh oblastiakh ETS. *Meteorol. Gidrol.* **1**, 88–94.

Sutton, O. G. (1953). "Microclimatology. A Study of Physical Processes in the lowest layers of the Earth's Atmosphere." McGraw-Hill, New York.

Sverdrup, H. U. (1942). "Oceanography for Meteorologists." Prentice-Hall, New York.

Taylor, K. (1976). The influence of subsurface energy storage on seasonal temperature variations. *J. Appl. Meteorol.* **15**, 1129–1138.

Thom, A. S. (1975). Momentum, mass and heat exchange of plant communities. *In* "Vegetation and the Atmosphere" (J. L. Monteith, ed.), Vol. 1, pp. 57–109. Academic Press, New York.

van Straaten, J. F. (1967). "Thermal Performance of Buildings." Am. Elsevier, New York.

van Wijk, W. R., and Derksen, W. J. (1963). Sinusoidal temperature variation in a layered soil. *In* "Physics of Plant Environment" (W. R. van Wijk, ed.), pp. 171–209. North-Holland Publ., Amsterdam.

Vehrencamp, J. E. (1953). Experimental investigation of heat transfer at an air-earth interface. *Trans. Am. Geophys. Union* **34**, 22–30.

Washburn, A. L. (1980). Permafrost features as evidence of climatic change. *Earth-Science Rev.* **15**, 327–402.

Weitzman, S., and Bay, R. R. (1963). Forest soil freezing and the influence of management practices, northern Minnesota. *U.S. For. Serv., Res. Pap.* **LS-2**.

White, D. E. (1965). Geothermal energy. *U.S. Geol. Surv., Circ.* No. 519.

Chapter XVIII

SUBSTRATE ENERGY STORAGE IN AQUATIC
ECOSYSTEMS AND ITS PLACE IN THEIR
ENERGY BUDGETS

In this chapter we will discuss energy relations in ecosystems that possess a large internal capacity to store heat. Ecosystem members immersed in water live in a steady thermal environment: Penetration of solar energy supports a deep zone of photosynthetic energy conversion in a buoyant transparent medium; the mobility of water permits rapid intake of sensible heat from the surface; water itself has a high specific heat.* In seasonal climates particularly, the large release of heat upon freezing and the low density of water at the freezing point protect against winter cold, and the large heat of vaporization buffers summer heat. Temperatures in aquatic ecosystems of middle and high latitudes seldom exceed 30°C and display a limited diurnal range. The role of internal heat storage is so great in these ecosystems that in describing it we are casting virtually the entire energy budget, a task for which discussions of its major components in the preceding chapters have prepared us.

ECOSYSTEM STRUCTURE

Aquatic ecosystems characteristically are deep, and this implies a storage capacity that is large even if stratification develops. Whether its components are emergent plants, submerged macrophytes, or algae and other microphytes, the large depth of a supporting fluid distinguishes aquatic from terrestrial ecosystems.

* Aquatic ecosystems also are distinguished by the constant availability of water for metabolic processes; they benefit from aqueous circulation of nutrients and the buoyancy of water and suffer the drawback of low oxygen concentration around their roots. These factors have energy connotations that are largely internal to the plants of the ecosystem.

Most of the biologically useless solar infrared radiation is absorbed in the top layer of water while photosynthetically active radiation penetrates to a considerable depth, especially the wavelengths near 0.46 μm (Kondratyev, 1969, p. 518). Over the whole solar spectrum about 0.05 penetrates to a depth of 3 m in Lake Mendota, Wisconsin, according to early (1912) measurements by Birge, the same fraction that penetrates a clover sward only a few centimeters (Chang, 1968, p. 39). The distributions of plants within this volume varies a great deal, and different ecosystems are "unique aggregations of similar components" (Richey et al., 1978). Spatial patterns of aquatic communities in many basins, often concentric zones associated with water depth, control species diversity, productivity, and wildlife (Weller, 1978).

Mobility of Liquid Water

The fluidity of water, its low viscosity, and large specific heat make it an ideal heat carrier, and turbulence in a water body efficiently diffuses sensible heat, especially in the upper photosynthesizing zone. Thermal admittance ranges up to 500 kJ m^{-2} deg^{-1} sec$^{-1/2}$ in conditions of strong wind mixing and far surpasses the admittance in soil (1–2 kJ m^{-2} deg^{-1} sec$^{-1/2}$). As a result the daytime intake of heat amounts to 5–10 MJ m^{-2} and is spread to depths from which "its ultimate expenditure [can be] postponed" (Mortimer, 1956).

Mixing this heat deeper is hampered by buoyancy forces between the warm upper layer and the heavier bottom water, but these are not large. Where the wind can be brought to bear, mixing is accomplished by only a "few per cent" of its kinetic energy (Hutchinson, 1957, p. 508). Wind effect increases with area of water on which it acts, i.e., the fetch, and depth to the thermocline is a function of fetch (Arai, 1965).

Strata

Wind action dies out downward from the surface, and in summer the thermocline (see Hutchinson, 1957, Chap. 7) lies at a depth of a few meters, depending on wind mixing. Below it is a zone of darkness, cold, and oxygen lack, but adequate nutrients. It contrasts with the upper layers—"warm and filled with abundant plant and animal life" (Mortimer, 1956, p. 177). The most important grazers "are excluded from the lower water by the accumulation in it of products of the decomposition of the plankton plants and animals," which exhaust its oxygen (Birge, 1897, p. 423). Eliminating summer stratification by bubbling air up through a California reservoir cooled the surface and reduced evaporation substantially (Koberg and Ford, 1965). Correspondingly, stratification in the storage tank of a solar energy system increases its overall efficiency (Sharp and Loehrke, 1979). Temperature contrast in a reservoir is equivalent to a large quantity of gravitational potential energy, perhaps more than can be generated as hydropower (McNichols et al., 1979; Hall, 1980).

Winter stratification of a cold top layer over warmer water can be destroyed by stirring, which raises enough heat to the surface to forestall freezing. An unobtrusive air pump keeps a duck lagoon in Milwaukee open all winter, and by an ingenious energy conversion the kinetic energy of the wind over a lake in central Wisconsin pumps air to the lake bottom, resulting in a stirring that keeps 100 m^2 open for oxygenation to forestall fish kill. Some devices for oxygenating bottom water stir the water enough to keep ice from forming (Dunst et al., 1974, p. 18).

Stratification can result from a density gradient caused by a high solute concentration in bottom water, which in some deep lakes in the Antarctic does not freeze. A similar storage of heat is effected in solar ponds; the deep layers can be heated to as much as 90°C by absorbing solar radiation and generate thermal energy convertible into electrical at a rate approximating 4 W m^{-2} in Israel (Tabor, 1980), certainly comparable with biomass conversion.

COUPLING WITH THE SUN

Penetration of Solar Radiation

Penetration of solar radiation of a given wavelength depends on the density and distribution of plants—emergent stems and leaves, floating and submerged leaves, algae and other plankton. Algae can find the level of optimum light because gas vacuoles give them the ability to sink or rise in the water (Walsby, 1977), and a seasonal change in the profile of photosynthetic production in Lake Tahoe depends on algal adaptations to the prevailing conditions of light flux (Tilzer and Goldman, 1978). Larger plants have been controlled by dyes that reduce light penetration (Dunst et al., 1974, p. 23). In general, 0.01 of solar radiation incident on the surface of Lake Mendota reaches a depth of 5 m, which delimits the layer of water "in which the phytoplankton can grow and multiply" (Mortimer, 1956, p. 171).

Penetration is deepest when the sun is high and little of the incoming beam is scattered upward by water molecules eventually to escape from the lake. Water bodies absorb 0.94–0.95 of the solar flux and are closely coupled with the sun.

Snow that accumulates on winter ice cover reduces the penetration of solar radiation so much that photosynthesis stops, and with it the generation of oxygen needed by fish. The result is winter-kill, which in particular seasons, e g , 1969–1970 in Wisconsin following a 0.3 m December snowfall (Brynildson, 1970), brings massive biological destruction.

As in forest penetration of solar energy into aquatic ecosystems deepens the zone in which biological conversions take place and allows it to expand in an open structure with maximum ventilation by the CO_2-transporting medium, whether air or water. Accompanying the flux of photosynthetically active radiation, however, is a flux of ultraviolet radiation, which may be destructive to reef organisms without protective pigments (Jokiel, 1980).

Biological Conversion of Solar Energy

The large quantity of thermal energy stored in aquatic ecosystems is matched by the large quantity fixed by photosynthesis; wetlands and marshes display high biological productivity (Auclair et al., 1976). Prairie glacial marshes in the upper Midwest, cattail and riverine marshes with large nutrient inflow, are most productive (de la Cruz, 1978), and sedge meadows and northern bogs about a third as much. Total production above and below ground in Theresa Marsh near Milwaukee is reported as 3–4 kg m^{-2} yr^{-1}, to which emergent species contribute the most (Klopatek and Stearns, 1978). The density of energy conversion over a 150-day growing season is 4–5 W m^{-2}. Still higher annual values are found in such low-latitude ecosystems as papyrus (Westlake, 1975).

Wetland productivity at the higher trophic levels of finfish, shellfish, and waterfowl is correspondingly high. Lake Mendota produces as much energy in the form of fish as the adjacent alfalfa and corn fields, even with fossil-energy support, produce as milk or meat. A census of perch caught by ice-fishermen came to 1,500,000 in one winter (Frey, 1963, p. 68), about 4 tons a day. The energy equivalent of this catch, 6–8 mW m^{-2}, is consistent with other studies of mean annual yield of nearby Lake Waubesa [from data of Dunst et al. (1974, p. 15)].

Trout ponds in the northern United States support 100 kg of fish per hectare without fertilization (Marriage et al., 1971), an approximate yield of 2 mW m^{-2}. In farm ponds, as in recreational lakes, the biological productivity of macrophytes is not always welcome because weed removal has a dollar cost of \$100 ha^{-1} (Dunst et al., 1974, p. 21) that amounts to an energy cost as high as 100 mW m^{-2}, considerably more than the cost to harvest corn. Biological harvesting of aquatic ecosystems is still far from the developed stage of grass-and-cow association of a typical terrestrial ecosystem.

A basin of aquatic ecosystems, Horicon Marsh, where native Americans built the largest group of mounds in the state, is downstream from Theresa Marsh. Until ceded by the Winnebago Nation in 1832, this mosaic of marsh, open water, and slightly higher lands produced fish, wild rice, game, root crops, and maize (Hanson, 1977, p. 6). Since then it has been a lake, then unproductive farms (Gard and Mueller, 1972), and after the 1930s again a marsh with an enormous population of muskrats, ducks, and Canada geese—about 200,000 of the latter, sought by thousands of hunters and hundreds of thousands of bird watchers.

Aquaculture without fossil-energy supplementation has long been a part of traditional agriculture, especially for protein supply; ponds in China and Bangladesh are well-known. Small aquatic ecosystems in Mesoamerica intermingled with raised beds, drained fields, or chinampas as terrestrial ecosystems, were integral parts of an intensively farmed land-and-water landscape. Wilken (1970) details the watercress and other plants and aquatic animals serving as food, and the plants used for fodder and rush mats, from aquatic ecosystems that form a "web of swamp-like environments over much of the basin floor" in Tlaxcala. Such ecosystems also were mingled with raised fields in the Mayan lowlands (Turner, 1974; Matheny, 1976, 1978) and may have been vital in Mayan culture (Willey, 1978).

Such aquacultural ecosystems as catfish farms are supplemented by fertilizers and fish foods in lands of mechanized agriculture and reach high levels of production, especially of protein (Lovell, 1979). Other aquatic ecosystems in these mechanized lands inadvertently receive excess nutrients and sediment from polluting farmland, ecosystems under substandard management practices (Karr and Schlosser, 1978). These nutrients might increase biological productivity, though not necessarily in marketable species. An " 'explosion' of aquatic weeds" (Holm et al., 1969) is a threat to freshwater bodies and a "symptom of our failure to manage our resources." Eutrophicated aquatic ecosystems might yet have long-term value for stable energy conversion; the high biomass productivity of such aquatic plants as *Scirpus* and duckweed can, with careful management (Stearns, 1978), clean water and salvage its energy-conversion potential (Woodwell, 1977; Hillman and Culley, 1978).

The cattail and wild-rice marshes and trout and catfish ponds of North America, the rice fields of Asia, and the taro (Fig. 1) and fish fields of Polynesia (Kikuchi, 1976) are ecosystems that overcome the handicap of low oxygen in the root zone, enjoy thermal stability, and efficiently convert solar energy.

Fig. XVIII.1. A variety of aquatic ecosystems form this mosaic of riparian and variously flooded taro fields on the floor of Hanalei Valley on Kauai.

Nonbiological Transformation of Solar Energy

Exposed vegetation reflects much of the solar infrared radiation and reduces the total solar energy absorbed by the ecosystem. In contrast, late-planted rice in southern Japan that has formed only a sparse cover by midsummer allows soil and water to be heated by the sun to temperatures (exceeding 35°C) that impair root function (Inoue *et al.*, 1965). Streams in clear-cut forest in the eastern United States overheat by 5–8 degrees in summer (Corbett *et al.*, 1978; Patric, 1980), even in New England, and injure fish metabolism, hatching, development, and patterns of migration; riparian strips of forest can alleviate these problems.

The dominant solar coupling is shown in the midday energy budget of a small stream in Virginia (Table I), in which water was warmed 3.4°C in traversing an unshaded reach 0.33 km long. Of the intake of absorbed radiation, 945 W m^{-2}, almost half went to warm the water. Displacing the locus of absorption away from the outer surface of the ecosystem forestalled the overheating of a thin outer layer, from which energy would easily be lost. Rather, the water surface remained slightly colder than the air and extracted a little heat from it. An energy budget of a

TABLE I

Midday Energy Fluxes at Stream Surface[a]

Absorbed solar radiation	+585
Absorbed longwave radiation	+360
Radiant energy intake	+945
Emitted longwave radiation	−445
Net surplus of all-wave radiation	+500
Exchanges with the air	
sensible	+15
latent	−55
Net exchange	−40
Exchanges with substrate	
stream bed	−17
warming of water	−443
Net exchange	−460

[a] Data from Pluhowski (1972). Data given in watts per square meter.

stream traversing a powerline clearing in West Virginia displays similar flux densities, except that more heat was taken into the bed [Lee (1978, pp. 200–210) from data of C. G. Day]. Such transient heating can have biological impact (Day and Carvell, 1978). If high temperatures can be accepted, as in an aqueduct, the emitted longwave radiation and latent-heat conversion in a dry environment remove as much energy as is absorbed from solar and downward longwave radiation [data from Jobson (1980)].

COUPLING WITH THE ATMOSPHERE

The midday energy budgets presented show solar coupling at its maximum. Atmospheric coupling is important in daily and yearly average conditions.

Energy Inputs

One atmosphere–ecosystem coupling that has already been noted is the effect of kinetic energy of the wind in mixing water, especially important in fall and winter when it tends to warm the surface. Incoming longwave radiation, another coupling, plays about the same role in aquatic ecosystems as in terrestrial since both have high absorptivity.

Aquatic ecosystems extract sensible heat from the atmosphere more easily than terrestrial ones because their daytime surface temperature is generally lower than that of the air in summer and almost always so in spring. A transfer coefficient for this downward flux of sensible heat is approximately 8 W m^{-2} deg^{-1} (Inoue et al., 1965; Lee, 1978, p. 207), which for typical temperature differences indicates 20–40 W m^{-2} as a likely flux level in summer.

Aquatic ecosystems also extract vapor from moist air. A downward gradient of vapor pressure is most likely to occur over melt-water streams (Inoue et al., 1965) or lakes in spring after surrounding terrestrial ecosystems become vapor sources.

Emergent vegetation tends to reduce turbulent heat inputs because its foliage warms to a point that produces upward gradients of temperature and vapor concentration, but its roughness acts in the opposite sense. Ecosystem area is also a factor; a small lake in a dry Australian landscape extracted more heat from the air than a large reed swamp (Linacre et al., 1970) where a long fetch allowed the air stream to come into equilibrium with the swamp.

Energy Outputs

The outward flux of longwave radiation is a function of ecosystem surface temperature, which depends on the degree of mixing and turbidity. An algal bloom, for example, absorbed solar energy and heated a thin surface layer in a shallow pond in Arizona to a temperature above 35°C, and surface vapor pressure reached 60 mbar (Idso and Foster, 1974). This effect occurs only by day (Uchijima, 1976b), but it is clear that errors in determining surface temperature affect the accuracy of data on the energy fluxes (Malevskii-Malevich, 1965).

The outward fluxes of sensible and latent heat are distinguished from those from terrestrial ecosystems by their greater intensity, steadiness in the face of the diurnal forcing regime of the sun, and lagging the seasonal regime of the sun. The size of these fluxes suggests that, from an energetic standpoint, aquatic ecosystems outperform terrestrial ones just as forest ecosystems outperform low ecosystems, and for about the same reasons: greater absorption of solar energy and better access to water.

Evapotranspiration

With respect to vaporization of water aquatic ecosystems have been the object of a great deal of myth-making, especially outside the parts of the world where they are familiar food producers. Reed swamps in California, potholes on the Great Plains, and lake vegetation in the South are suspected to waste water, but these myths are not based on energy-budget analyses but on transfer of greenhouse pot experiments in which transpiration is supported by inputs of advective energy larger than ever found in plant communities. One such transpiration rate is given (Benton et al., 1978), for example, as 14 mm day^{-1}, which is equivalent to 410 W m^{-2} when solar radiation was 240 W m^{-2} and little stored heat was released. Heat advection that would supplement absorbed radiation enough to attain an evaporation rate of 14 mm day^{-1}, never observed in natural ecosystems, could be realized only in single plants or tiny stands, as Idso (1979) comments.

Advectively supported evapotranspiration decreases as ecosystem area increases, and in the large Australian swamp that Linacre et al. (1970) studied, midday evapotranspiration measured accurately by the eddy correlation method, at 160 W m^{-2}, was only two-thirds that from a small lake, which gained more heat by advection than the swamp. Evapotranspiration from hydrophytes in potholes in North Dakota, 570 mm over the summer, was 120 mm less than evaporation from open potholes (Eisenlohr, 1966), partly as a result of the shading of the water, and a similar finding was reached for a cedar swamp in Ontario (Munro, 1979), which generated a large sensible-heat flux in morning hours.

Linacre (1976) feels that in typical conditions swamp evapotranspiration is roughly equal to that from open lakes; the smaller absorption of solar energy by emergent plants is balanced by their greater extraction of sensible heat from the air. Floating plants exhibit only the first effect; they evaporate less than open water (Cooley and Idso, 1980).

Evapotranspiration from rice, which usually claims 0.8 of the net surplus of all-wave radiation (Uchijima, 1976a), increases to 1.0 or more in advection, the strength of which can be estimated by the radiative index of dryness [see Budyko (1974, pp. 324, 426)] at the surrounding terrestrial ecosystems. Evapotranspiration measured in a peat bog in northern Minnesota (Bay, 1966), where heat advection was minor, approximated potential evapotranspiration as calculated by the Thornthwaite method. Evapotranspiration from aquatic ecosystems on summer days is undoubtedly larger than from terrestrial ecosystems, which are often restricted by moisture stress, but the physics of their energy budget gives no reason whatever to ascribe rates of transpiration two to three times open-water evaporation.

ADVECTION OF UPSTREAM HEAT OR COLD IN ECOSYSTEM ENERGY BUDGETS

Stream Water

In stream ecosystems the continued inflow of cold water at night lowers the water temperature that was raised during the day by solar heating, as in the Virginia stream (Table I). In headwater streams in Japan (Nishizawa, 1967) extraction of heat from the air contributes approximately 50 W m^{-2} in the 24-hr average and the net surplus of all-wave radiation 100 W m^{-2}. Along with a small contribution of heat from the ground, these sources of energy equal the advection of cold water from upstream. Organic matter that falls or washes into streams from terrestrial ecosystems dominates their energy conversions, in contrast to grazing in still-water ecosystems. Flowing-water ecosystems in forested landscapes have four times as much gross primary production as still-water ecosystems (Brown *et al.*, 1978).

Wetlands

Where meltwater direct from high mountains flows into rice fields, as it does in Italy, Japan, and California, crop damage occurs (Raney, 1963; Raney and Mihara, 1967), unless the water is warmed somehow to 23°C or warmer, because water temperatures lower than 19°C prevent rice from forming grain (Inoue *et al.*, 1965). Ideally, warming practices are based on energy-budget analysis. A typical equation for water temperature T_w that includes the net surplus of all-wave radiation R_n, is

$$T_w = T_a + \frac{(R_n/h) - 2D}{1 + 2S}$$

in which T_a represents air temperature, h is the transfer coefficient for sensible heat ($=$ sensible-heat flux/($T_w - T_a$), D the vapor-pressure deficit, and S the slope of the curve relating saturation vapor pressure to temperature at $T = T_a$ (Inoue *et al.*, 1965). Warming basins built without energy analysis are inefficient, waste land, and may even be "irrational" as heat exchangers. Suppression of evaporation by spreading of monolayers on the water is practiced by rice farmers (Inoue *et al.*, 1965) and raises water temperature by amounts that indicate a substantial shift in the partitioning of energy at the water surface.

Reservoirs

Reservoirs differ from most natural lakes in respect to advection of cold or hot water (Arai, 1967), and density currents of the inflowing water seek different depths at different seasons. The heat content of warm inflow water in one reservoir in early summer was equivalent to 350 W m^{-2} of reservoir surface.

Low-level outlets in some dams tap a layer of cold water formed by winter inflows and have been found responsible for damage to downstream rice fields; the closing of Shasta Dam in 1946 brought about a several-degree drop in the temperature of the Sacramento River in the rice-field area (Raney and Mihara, 1967).

A different kind of advection into aquatic ecosystems is the injection of hot water from power plants (Gibbons and Sharitz, 1974). Rates of heat addition are approximately 200 W m^{-2} in lakes and 450 in cooling ponds in the U.S. (Jirka and Harleman, 1979) and 250 W m^{-2} in lakes in Poland (Gadkowski and Tichenor, 1980). This added heat has biological effects, some not yet well understood, and alters the energy budget although the effects may be mitigated and be less irreversible than injection by cooling towers into the atmosphere (Reynolds, 1980). Evaporation equations for a stream receiving such heat (temperature 33–41°C) have to include effects of free convection (DeWalle, 1976). An energy budget model of the Chattahoochee River found that the injected heat was offset by the release of cold bottom water from an upstream reservoir (Faye et al., 1979). However, this fortuitous availability of winter cold cannot be counted on if many synfuel plants are built (Harte and El-Gasseir, 1978).

VARIATIONS IN THE ENERGY BUDGET

The thermal inertia of aquatic ecosystems is illustrated in the time variations of their energy budgets. We shall consider the regular variations of day and year and the aperiodic fluctuations.

The Diurnal Energy Cycle

The capacity of a water body to store daytime peak flows of incoming energy results from its ability to spread the conversion of photosynthetically active radiation through a deep layer and to mix downward the heat generated near the surface by absorption of solar infrared radiation or extraction from warm air. Aquatic ecosystems, taking most of the morning solar impulse into their substrates, experience only a small rise in surface temperature and the surface-to-atmosphere energy fluxes. Emission of longwave radiation increases about 10 W m^{-2} for a typical rise of 2°C; the sensible-heat and latent-heat fluxes increase little over their nocturnal rates.

The aquatic system at night receives only atmospheric radiation and, to meet the demands to emit longwave radiation, draws on its own thermal resources to the extent of 2–3 MJ m^{-2}, but once vertical circulation develops in the water (Anderson, 1968), these resources are large, being the heat (10 MJ m^{-2} or more) that entered storage during the midday hours of the preceding day. Even a slight cooling can call forth the 2–3 MJ m^{-2} needed to add to the incoming longwave radiation to meet the emission demand.

Ample heat remains available in the ecosystem through the night to sustain a relatively high surface temperature, and heat is transferred to the now cool inflowing land air. Much of the energy that went into storage in the water during the day "is later released as evaporation during the night" say Stewart and Rouse (1976) in reporting on the energy balance of a lake on the coastal lowland of Hudson Bay, even though storage was limited by its shallowness (mean depth 0.6 m). The warmth and moistness of lake air is familiar to people rowing or sailing on a small lake in the evening, and lake air in gradient flow can ward off frost in Florida orange groves for several hundred meters downwind (Bill et al., 1978).

Irregular Variations

The thermal stability of an aquatic ecosystem gives it the ability to accelerate its outputs of energy if conditions of the overlying air change from day to day in temperature and vapor concentration. Variations in wind speed affect the turbulent heat fluxes and by stirring the water increase the availability of stored heat. Variations in solar radiation from day to day, in contrast, are not as important in the energy budget of aquatic as in that of terrestrial ecosystems.

When variations in radiation are taken into consideration by expressing nonradiative fluxes as fractions of the net surplus of all-wave radiation, the daily sums of the turbulent fluxes from a shallow pond averaged 0.80 of the net radiation surplus [from data presented in Stewart and Rouse (1976)], but large interdiurnal variations produced a standard deviation of 0.97. Transfers to the atmosphere were as small as 0.19 of the net radiation surplus on one day, and as large as 1.23 on another—a cold day in which the turbulent fluxes drew heavily upon substrate heat storage.

Rice fields are flooded more deeply when cold days occur in summer (Inoue et al., 1965) because water not only brings its own warmth but also the capacity to carry daytime radiant heat over into the critical night hours. The threat of cold damage to reproductive cells is met by 0.1–0.15

m of water. Cranberry bogs are flooded or sprinkled (2–3 mm hr^{-1}) when summer frost threatens these low-lying areas of small thermal capacity, and can release sensible heat of fusion sufficient to protect down to $-7°C$ (Dana and Klingbeil, 1966, p. 18).*

The Annual Energy Cycle

Aquatic ecosystems convert seasonal changes in solar energy into flows of heat in and out of storage, which delay the transformation into evaporation and sensible-heat flow. The range in internal temperature is small, and winter has a different meaning to an aquatic ecosystem in the middle latitudes than to a terrestrial one, which has developed more elaborate ways of getting through the low-energy part of the year.

Most of the heating occurs in spring and early summer and at its peak is characterized by an almost exclusive transformation of solar energy into sensible heat of water, which crowds the turbulent fluxes between water and air nearly out of the picture. This transformation is seen in May (Table II) in Lake Mendota in southern Wisconsin when both turbulent fluxes total only 20 W m^{-2}, less than one-tenth of the absorbed solar energy. It is interesting that while the water is avidly taking energy into storage at this season the biological energy stored in rhizomes of cattail ecosystems in Horicon Marsh is being drawn down toward its late-June low point of the year (Linde et al., 1976).

The annual regime is epitomized in the turn-around between May and November (Table II). In May the budget expresses the warming of water in the sun; in November the sun has almost vanished from the scene, which is now dominated by the transfer of heat (155 W m^{-2}) from the water body into the cold, dry, receptive air.

Several months separate the initial transformation of solar energy from the subsequent transfer from ecosystem to atmosphere. In terrestrial ecosystems, in contrast, heat is transferred into the air almost immediately after solar energy is absorbed by the leaves. The heat-storing capacity of an aquatic ecosystem inserts a lag of several months between the solar forcing function and the atmospheric response, and in the process the ecosystem remains warm well into autumn: 8° at the surface in November and 12° at depth. Large tanks serve to store heat over

---

* Flooding was formerly done in Wisconsin by gravity flow of water from reservoirs that were sometimes far away; now it is done more quickly by high-capacity pumps that move water—and hence heat—from ground water. Thermal energy gained by pumping 1 m^3 of water that is 5°C warmer than the ecosystem is about 20 MJ; the mechanical energy to lift the water 5 m is 0.05 MJ and the fossil-energy operating cost perhaps 0.4 MJ (Cf. Chapter XIII; energy cost of equipment not included). The energy trade-off is advantageous.

TABLE II

Energy Fluxes at Lake Mendota, Wisconsin[a,b]

	May	November
Radiation		
absorbed shortwave	225	60
absorbed longwave	310	290
radiant-energy intake	535	350
Profile		
Temperature:		
Air	14°	2°
Water		
surface	12°	8°
at 10 m depth	10°	12°
Vapor Pressure:		
air	11 mbar	5 mbar
surface	14 mbar	11 mbar
Dependent Fluxes		
emitted longwave radiation	−375	−350
exchange of surface with deeper layers	−140	+155
sensible heat	+30	−65
latent heat	−50	−90
sum of turbulent fluxes	−20	−155

[a] Latitude 43°N, altitude 260 m, area 40 km², mean depth 12 m.

[b] Data show mean conditions. In a stream of Canadian air in November, however, the fluxes of latent and sensible heat increase to at least 400–500 W m^{-2}. This figure is consonant with the fact that a flux of 350 W m^{-2} over a longer fetch across Lake Huron (Dewey, 1975) is necessary to generate convective snow showers, which are frequent winter phenomena on Lake Mendota. Data given in watts per square meter. Data from Birge *et al.* (1928), Dutton and Bryson (1962), Scott (1964), and U. S. Weather Bureau records.

similarly long portions of the annual cycle in some solar energy installations.

Thermal storage in the annual regime would be expected to be 19 times as large ($\sqrt{365} = 19$) as that in the diurnal cycle, but actual heat storage in Lake Mendota, including its bed, approximately 1050 MJ m^{-2},* is two orders of magnitude greater than that in the diurnal cycle. The reason lies in the equinoctial physical overturning of the water, a

* This value depends on lake depth and wind exposure, as well as seasonality in energy inputs. In the Sea of Galilee it is 1420 MJ m^{-2}, and in much-studied Lake Geneva 1550 (Hutchinson, 1957, pp. 496–500).

phenomenon absent in the diurnal cycle. The autumn circulation makes the warmth below the thermocline accessible for transfer to the atmosphere in the shortening, darkening days of October and November.

Overturning postpones the date at which the newly formed top layer will be cooled to the freezing point. After ice cover forms over the ecosystem, heat transfer into the atmosphere diminishes, and the bottom water remains at 4°C. Release of 150–200 MJ m^{-2} of energy into the lake budget by ice formation reduces further cooling.* As a result, the annual range of temperature in the deep water, from $+4$ to $+21$, is about half the temperature range that nearby terrestrial ecosystems undergo. Aquatic ecosystems rely on this small temperature range, and water-level manipulation that exposes them to winter freezing and drying (Dunst et al., 1974, p. 22; Cooke, 1980) is an effective management practice. Winter ice blankets, created by flooding cranberry bogs, prevent drying of the vines and reduce temperature fluctuations (Dana and Klingbeil, 1966, p. 17).

Changes from Year to Year

Year-to-year changes in the surroundings of an aquatic ecosystem, as well as in solar radiation or rainfall, influence its energy budget. In drought years wetlands that are intermingled with uplands are subject to dry, hot air streams from the terrestrial ecosystems, which accelerate wetland evapotranspiration. For example, a 50-ha wetland in southern Wisconsin, fed by cold through-flowing ground water, which evaporated 800–850 mm in the summers of 1974 and 1975 (Novitzki, 1978, p. 15), evaporated 1000 mm in 1976, a drought summer when adjacent upland ecosystems were short of water by about 200 mm.

On a longer time scale, if decomposition in aquatic ecosystems slows down and if hydrodynamic energy cannot carry away organic material, a matter of topographic situation (Hulme, 1980) peat begins to accumulate. The rate of through-flow and frequency of overbank flooding express the "hydrodynamic energy gradient" (Gosselink and Turner, 1978), which by controlling sediment and nutrient flows, determines species composition, primary productivity, and the accumulation of organic matter. The organic matter in turn alters the movement of water through the ecosystem. In low-energy ecosystems the net effect is to accumulate

---

* Ice harvesting from Wisconsin lakes, still extant in a few places, was a large industry prior to mechanical refrigeration based on fossil energy. By a kind of reverse energy transfer, it facilitated such industries as brewing and meat packing and shipping (Lawrence, 1965). At one time the Milwaukee River, in cold weather, yielded up to 250,000 tons of ice in several cuttings per winter.

a large amount of chemical energy, perhaps two orders greater than is fixed in an annual cycle, in long-term storage.

Spatial Contrasts

Different water depths in a basin create a suite of aquatic communities that ranges from open-water submerged or floating types through emergent forms with soil as a nutrient source (Klopatek and Stearns, 1978), to wetmeadow sedges. The intermingling of ecosystems gives a spatial structure of great diversity of species and habitats of birds and mammals (Weller, 1978). Fire, which creates diversity among terrestrial ecosystems, does so also among aquatic ones. When organic soils in south Florida dry and burn, the "fire-created depressions" (Tiedemann et al., 1979, p. 18) later fill with water that harbors aquatic species and become recolonization centers when wide flooding occurs.

Lakes also differ from one another, as a study of carbon flow indicates (Richey et al., 1978), each one showing unique characteristics. These differences usually are reflected in solar-energy conversions. Water movement also differentiates ecosystems, and streams are particularly a distinctive and heterogenous class (Blum, 1972).

Aquatic and terrestrial ecosystems differ in radiation absorptivity and energy storage, as we have seen. Where they occupy only a small fraction of a region, their energy situation is at odds with the regional atmosphere, which is largely formed by interaction with terrestrial ecosystems. The aquatic ecosystems cool more in winter and heat more in summer than they otherwise would, by mesoscale atmospheric advection of sensible heat.

Air moving over a small pond or a narrow belt of riparian vegetation will incur little modification and support a large downward flux of sensible heat and rapid evapotranspiration. Air moving over larger wetlands is transformed and tends toward equilibrium with the underlying surface, as was described in Chapter XV on latent-heat conversion. This transformation was found to be regular and exponential over a rice ecosystem of Australia (Lang et al., 1974). Spatial scale influences the energy budgets of aquatic ecosystems.

On a larger scale coastal marsh ecosystems display clear zoning that is caused in part by seaward increase in salinity and tidal effects on water level. The resulting differences in soils and nutrients in Louisiana affect biological energy conversion (Hopkinson et al., 1978), which averages 3 W m^{-2} annually in *Spartina patens*, a high figure due in part to river-transported nutrients and the long growing season. Their productivity is reduced if water circulation is disturbed, e.g., by road construction (Hoffnagle, 1980).

Aquatic ecosystems receive most of their energy as terrestrial ecosystems do, by absorbing solar and longwave radiation, but they differ in being able to store a large fraction of this energy. Considering the intermittent nature of solar radiation, their quick intake and large heat capacity forestall excess heating and carry energy into the following period of darkness or winter. The magnitude of this storage is shown by the warmth of a pond or marsh at night and, more dramatically, by the extravagant outpouring of heat and moisture from a pond or lake in the fall. The initial delivery of solar energy is separated by as much as a third of the annual or diurnal cycle from the eventual transfer of energy into the atmosphere, and in this long period the stored energy provides a buffered environment for photosynthetic energy conversion.

REFERENCES

Anderson, D. V. (1968). Nocturnal heat loss of a lake and seasonal variation in its vertical thermal structure. *Int. Assoc. Hydrol. Sci., Bull.* **13**(3), 33–40.

Arai, T. (1965). Some relations between the thermal property of lake and its scale or fetch size. *Tokyo J. Climatol.* **2**, No. 2. [Abstr. of paper in *Geogr. Rev. Jpn.* **37**, 131–137 (1964).]

Arai, T. (1967). Heat budget studies on the water temperature of the artificial reservoir (Part 1). *Shigen Kagaku Kenkyusho Iho* No. 68, 31–54.

Auclair, A. N. D., Bouchard, A., and Pajaczkowski, J. (1976). Plant standing crop and productivity relations in a *Scirpus-Equisetum* wetland. *Ecology* **57**, 941–952.

Bay, R. R. (1966). Evaluation of an evapotranspirometer for peat bogs. *Water Resour. Res.* **2**, 437–442.

Benton, A. R., Jr., James, W. P., and Rouse, J. W., Jr. (1978). Evapotranspiration from water hyacinth (*Eichhornia crassipes* (*Mart.*) *Solms*) in Texan reservoirs. *Water Res. Bull.* **14**, 919–930.

Bill, R. G., Jr., Sutherland, R. A., Bartholic, J. F., and Chen, E. (1978). Observations of the convective plume of a lake under cold-air advective conditions. *Boundary-Layer Meteorol.* **14**, 543–556.

Birge, E. A. (1897). Plankton studies on Lake Mendota: II, The crustacea of the plankton from July, 1894, to December, 1896. *Trans. Wisc. Acad. Sci., Arts Lett.* **11**, 274–448.

Birge, E. A., Juday, C., and March, H. W. (1928). The temperature of the bottom deposits of Lake Mendota; a chapter in the heat exchanges in the lake. *Trans. Wis. Acad. Sci., Arts Lett.* **23**, 166–213.

Blum, J. L. (1972). Plant ecology in flowing water. In "River Ecology and Man" (R. T. Oglesby, C. A. Carlson, and J. A. McCann, eds.), pp. 53–65. Academic Press, New York.

Brown, S., Brinson, M. M., and Lugo, A. E. (1978). Structure and function of riparian wetlands. In "Strategies for Protection and Management of Floodplain Wetlands and Other Riparian Ecosystems." *U.S. For. Serv., Gen. Tech. Rep.* **WO-12**, 17–31.

Brynildson, C. (1970). Winterkill: Cause and solutions. *Wis. Conserv. Bull.* **35**(6), 12–13.

Budyko, M. I. (1974). "Climate and Life" (transl. ed. by D. H. Miller). Academic Press, New York.

Chang, J.-H. (1968). "Climate and Agriculture." Aldine, Chicago, Illinois.

Cooke, G. D. (1980). Lake level drawdown as a macrophyte control technique. *Water Resour. Bull.* **16**, 317–322.

Cooley, K. R., and Idso, S. B. (1980). Effects of lily pads on evaporation. *Water Resour. Res.* **16**, 605–606.

Corbett, E. S., Lynch, J. A., and Sopper, W. E. (1978). Timber harvesting practices and water quality in the eastern United States. *J. For.* **76**, 484–488.

Dana, M. N., and Klingbeil, G. C. (1966). Cranberry Growing in Wisconsin. *Univ. Wis. Coll. Agric., Circ.* No. 654.

Day, C. G., and Carvell, K. L. (1978). Effects of power line corridor clearance and maintenance on stream habitat. *In* "Strategies for Protection and Management of Floodplain Wetlands and Other Riparian Ecosystems." *U.S. For. Serv., Gen. Tech. Rep.* **WO-12**, 383–385.

de la Cruz, A. A. (1978). Primary production processes: Summary and recommendations. *In* "Freshwater Wetlands" (R. E. Good, D. F. Whigham, and R. L. Simpson, eds.), pp. 79–86. Academic Press, New York.

DeWalle, D. R. (1976). Effect of atmospheric stability on water temperature predictions for a thermally loaded stream. *Water Resour. Res.* **12**, 239–244.

Dewey, K. F. (1975). The prediction of Lake Huron lake-effect snowfall systems. *J. Appl. Meteorol.* **14**, 3–7.

Dunst, R. C., *et al.* (1974). Survey of lake rehabilitation techniques and experiences. *Wis. Dep. Nat. Resour., Tech. Bull.* No. 75.

Dutton, J. A., and Bryson, R. A. (1962). Heat flux in Lake Mendota. *Limnol. Oceanogr.* **7**, 80–97.

Eisenlohr, W. S., Jr. (1966). Water loss from a natural pond through transpiration by hydrophytes. *Water Resour. Res.* **2**, 443–453.

Faye, R. E., Jobson, H. E., and Land, L. F. (1979). Impact of flow regulation and powerplant effluents on the flow and temperature regimes of the Chattahoochee River—Atlanta to Whitesburg, Georgia. *U.S. Geol. Surv., Prof. Pap.* **1108**.

Frey, D. G. (1963). "Limnology in North America." Univ. of Wisconsin Press, Madison.

Gadkowski, M., and Tichenor, B. A. (1980). Performance of power plant cooling lakes in Poland. *ASCE, J. Energy Division*, **106** (EY-1), 1–8.

Gard, R. E., and Mueller, E. G. (1972). Wild Goose Marsh: Horicon Stopover. Wisconsin House, Ltd., Madison, Wis.

Gibbons, J. W., and Sharitz, R. R., eds. (1974). "Thermal Ecology," AEC Symposium Series, AEC, Washington, D.C.

Gosselink, J. G., and Turner, R. E. (1978). The role of hydrology in freshwater wetland ecosystems. *In* "Freshwater Wetlands" (R. E. Good, D. F. Whigham, and R. L. Simpson, eds.), pp. 63–78. Academic Press, New York.

Hall, E. H. (1980). Thermocline temperature differences and realizable energy. *Science* **208**, 1292.

Hanson, I. A. (1977). Horicon Marsh, a history of change. *Lore* (*Milwaukee, Wis.*) **27**(2), 2–42.

Harte, J., and El-Gasseir, M. (1978). Energy and water. *Science* **199**, 623–634.

Hillman, W. S., and Culley, D. D., Jr. (1978). The uses of duckweed. *Am. Sci.* **66**, 442–451.

Hoffnagle, J. R. (1980). Estimates of vascular plant primary-production in a West Coast saltmarsh-estuary ecosystem. *Northwest Sci.* **54**, 68–79.

Holm, L. G., Weldon, L. W., and Blackburn, R. D. (1969). Aquatic weeds. *Science* **166**, 699–709.

Hopkinson, C. S., Gosselink, J. G., and Parrondo, R. T. (1978). Aboveground production of seven marsh plant species in coastal Louisiana. *Ecology* **59**, 760–769.

Hulme, P. D. (1980). The classification of Scottish peatlands. *Scot. Geogr. Mag.* **96**, 46–50.

Hutchinson, G. E. (1957). "A Treatise on Limnology. Vol. 1: Geography, Physics and Chemistry." Wiley, New York.

Idso, S. B. (1979). Discussion. "Evapotranspiration from water hyacinth (*Eichhornia crassipes* (Mart.) Solms) in Texas reservoirs," by A. R. Benton, Jr., W. P. James, and J. W. Rouse, Jr. *Water Resour. Bull.* **15**, 1466–1467.

Idso, S. B., and Foster, J. M. (1974). Light and temperature relations in a small desert pond as influenced by phytoplanktonic density variations. *Water Resour. Res.* **10**, 129–132.

Inoue, E., Mihara, Y., and Tsuboi, Y. (1965). Agrometeorological studies on rice growth in Japan. *Agric. Meteorol.* **2**, 85–107.

Jirka, G. H., and Harleman, D. R. F. (1979). Cooling impoundments: Classification and analysis. ASCE, *J. Energy Div.* **105**, (EY2), 291–309.

Jobson, H. E. (1980). Thermal modeling of flow in the San Diego aqueduct, California, and its relation to evaporation. *U.S. Geol. Surv., Prof. Pap.* **1122**.

Jokiel, P. L. (1980). Solar ultraviolet radiation and coral reef epifauna. *Science* **207**, 1069–1071.

Karr, J. R., and Schlosser, I. J. (1978). Water resources and the land-water interface. *Science* **201**, 229–234.

Kikuchi, W. K. (1976). Prehistoric Hawaiian fishponds. *Science* **193**, 295–299.

Klopatek, J. M., and Stearns, F. W. (1978). Primary productivity of emergent macrophytes in a Wisconsin freshwater marsh ecosystem. *Am. Midl. Nat.* **100**, 320–332.

Koberg, G. E., and Ford, M. E., Jr. (1965). Elimination of thermal stratification in reservoirs and the resulting benefits. *U.S. Geol. Surv., Water-Supply Pap.* **1890-M**.

Kondratyev, K. Ya. (1969). "Radiation in the Atmosphere." Academic Press, New York.

Lang, A. R. G., Evans, G. N., and Ho, P. Y. (1974). The influence of local advection on evapotranspiration from irrigated rice in a semi-arid region. *Agric. Meteorol.* **13**, 5–13.

Lawrence, L. E. (1965). The Wisconsin ice trade. *Wis. Mag. Hist.* **48**, 257–267.

Lee, R. (1978). "Forest Microclimatology." Columbia Univ. Press, New York.

Linacre, E. (1976). Swamps. *In* "Vegetation and the Atmosphere" (J. L. Monteith, ed.), Vol. 2, pp. 329–347. Academic Press, New York.

Linacre, E., Hicks, B. B., Sainty, G. R., and Grauze, C. (1970). The evaporation from a swamp. *Agric. Meteorol.* **7**, 375–386.

Linde, A. F., Janisch, T., and Smith, D. (1976). Cattail—the significance of its growth, phenology and carbohydrate storage to its control and management. *Wis. Dep. Nat. Resour., Tech. Bull.* No. 94.

Lovell, R. T. (1979). Fish culture in the United States. *Science* **206**, 1368–1372.

McNichols, J. L., Ginell, W. S., and Cory, J. S. (1979). Thermoclines: A solar thermal energy resource for enhanced hydroelectric power production. *Science* **203**, 167–168.

Malevskii-Malevich, S. P. (1965). O sistematicheskoi oshibke pri raschete sostavliaiushchikh teplovogo balansa vodoemov. *Tr. Gl. Geofiz. Obs.* **167**, 140–143.

Marriage, L. D., Borell, A. E., and Scheffer, P. M. (1971). Trout ponds for recreation. *U.S. Dep. Agric., Farmers' Bull.* No. 2249.

Matheny, R. T. (1976). Maya lowland hydraulic systems. *Science* **193**, 639–646.

Matheny, R. T. (1978). Northern Maya water-control systems. *In* "Pre-Hispanic Maya Agriculture" (P. D. Harrison and B. L. Turner, II, eds.), pp. 185–210. Univ. of New Mexico Press, Albuquerque.

Mortimer, C. H. (1956). An explorer of lakes. *In* "E. A. Birge. A Memoir" (G. C. Sellery, ed.), pp. 163–211. Univ. of Wisconsin Press, Madison.

Munro, D. S. (1979). Daytime energy exchange and evaporation from a wooded swamp. *Water Resour. Res.* **15**, 1259–1265.

Nishizawa, T. (1967). River water temperature in Japan. *Shigen Kagaku Kenkyusho Iho* No. 68, 55–61.

Novitzki, R. P. (1978). Hydrology of the Nevin Wetland near Madison, Wisconsin. *U.S. Geol. Surv., Water Res. Invest.* **78–48**.

Patric, J. H. (1980). Effects of wood products harvest on forest soil and water relations. *J. Environ. Qual.* **9**(1), 73–80.

Pluhowski, E. J. (1972). Clear cutting and its effect on the water temperature of a small stream in northern Virginia. *U.S. Geol. Surv., Prof. Pap.* **800-C**, 257–262.

Raney, F. (1963). Rice water temperature. *Calif. Agric.* **7**(9), 6–7.

Raney, F. C., and Mihara, Y. (1967). Water and soil temperature. In "Irrigation of Agricultural Lands" (R. M. Hagan, H. R. Haise, and T. W. Edminster, eds.), pp. 1024–1036. Am. Soc. Agron., Madison, Wisconsin.

Reynolds, J. Z. (1980). Power plant cooling systems: Policy alternatives. *Science* **207**, 367–372.

Richey, J. E., *et al.* (1978). Carbon flow in four lake ecosystems: a structural approach. *Science* **202**, 1183–1186.

Scott, J. T. (1964). A comparison of the heat balance of lakes in winter. *Univ. Wis. Meteorol. Dep., Tech. Rep.* No. 13.

Sharp, M. K., and Loehrke, R. I. (1979). Stratified thermal storage in residential solar energy applications. *J. Energy* **3**(2), 106–113.

Stearns, F. (1978). Management potential: Summary and recommendations. In "Freshwater Wetlands" (R. E. Good, D. F. Whigham, and R. L. Simpson, eds.), pp. 357–363. Academic Press, New York.

Stewart, R. B., and Rouse, W. R. (1976). A simple method for determining the evaporation from shallow lakes and ponds. *Water Resour. Res.* **12**, 623–628.

Tabor, H. (1980). Non-convecting solar ponds. *Phil. Trans. R. Soc. Lond. A* **295**, 423–433.

Tiedemann, A. R., Conrad, C. E., Dieterich, J. H., Hornbeck, J. W., Megahan, W. F., Viereck, L. A., and Wade, D. D. (1979). Effects of fire on water. *U.S. For. Serv., Gen. Tech. Rep.* **WO-10**.

Tilzer, M. M., and Goldman, C. R. (1978). Importance of mixing, thermal stratification and light adaptation for phytoplankton productivity in Lake Tahoe (California–Nevada). *Ecology* **59**, 810–821.

Turner, B. L., II (1974). Prehistoric intensive agriculture in the Mayan lowlands. *Science* **185**, 118–124.

Uchijima, Z. (1976a). Maize and rice. In "Vegetation and the Atmosphere" (J. L. Monteith, ed.), Vol. 2, pp. 33–64. Academic Press, New York.

Uchijima, Z. (1976b). Microclimate of the rice crop. In "Climate and Rice," pp. 115–140. Int. Rice Res. Inst., Los Baños, Philippines.

Walsby, A. E. (1977). The gas vacuoles of blue-green algae. *Sci. Am.* **237**(2), 90–97.

Weller, M. W. (1978). Management of freshwater marshes for wildlife. In "Freshwater Wetlands" (R. E. Good, D. F. Whigham, and R. L. Simpson, eds.), pp. 267–284. Academic Press, New York.

Westlake, D. P. (1975). Primary production of freshwater macrophytes. In "Photosynthesis and Productivity in Different Environments" (J. F. Cooper, ed.), pp. 189–206. Cambridge Univ. Press, London and New York.

Wilken, G. C. (1970). The ecology of gathering in a Mexican farming region. *Econ. Bot.* **24**, 286–295.

Willey, G. R. (1978). Pre-Hispanic Maya agriculture: A contemporary summation. In "Pre-Hispanic Maya Agriculture" (P. D. Harrison and B. L. Turner, II, eds.), pp. 325–335. Univ. of New Mexico Press, Albuquerque.

Woodwell, G. M. (1977). Recycling sewage through plant communities. *Am. Sci.* **65**, 556–562.

Chapter XIX

POTENTIAL AND KINETIC ENERGY IN ECOSYSTEMS

Many gate functions in ecosystems are performed by conversions of potential or kinetic energy in the ecosystems themselves or in their environments, and at relatively small energy cost these control the larger fluxes described in the preceding chapters. For example, wind kinetic energy stirs a lake to release hundreds of watts per square meter from the deep water; the potential energy of groundwater supports circulations that control biological functions and energy storage.

POTENTIAL ENERGY IN ECOSYSTEMS

The upward and downward growth of ecosystems into a thick zone, necessary for intake of CO_2, water, light, and nutrients and the discharge of vapor and oxygen, endows upper layers more than lower ones with potential energy. Upward movement of water from roots to leaves, which adds potential energy to several kilograms of water per square meter daily, requires an energy input; downward movement of water is powered largely by potential energy of the intercepted rain.

Intercepted Rain and Snow

Rain loading is seldom important in forest ecosystems, but can cause lodging in small grains or soybeans, which reduces production and makes harvesting difficult. It is often combined with wind effects (Grace, 1977, p. 120).

Intercepted snow is more serious; wind packing or thaw-and-freeze weather causes masses reaching 20 kg or more per square meter of ground area to load evergreen canopy. The silvicultural effects of snow-break are worst when a rare snowstorm occurs where tree species have not developed the structural strength of those in snowy regions, which represents a substantial energy cost to the trees.

Freezing rain, which adheres more tenaciously than snow, can accumulate to destructive weights and flattens or tears the tops out of forest stands, but seldom comes when crop ecosystems would be affected. Ice storms were only a minor problem in southern Ontario (except in woodlots) before 1920, but now present a threat to thousands of kilometers of electric lines (Hewitt and Burton, 1971, p. 53). The March 1976 ice storm in southern Wisconsin (Fig. 1) stopped rural traffic and shut off electrical energy to dairy farms, which virtually ceased to operate and suffered losses in their herds.

Intercepted snow or rain has to be further energized before its potential energy can move it out of the canopy. Clumps of dry snow can be blown apart by the wind (kinetic energy); wet masses can lose adhesion to supporting branches by partial melting (thermal or radiative energy) or may entirely melt away at greater energy cost; in rare cases intercepted snow evaporates at a very large cost in thermal energy (Miller, 1967).

Fig. XIX.1.   Ice-storm destruction 13 months earlier is still evident in trees near Mineral Point in southwestern Wisconsin.

Detained Snow and Rain

Potential energy removes meltwater from snow cover, so that it does not accumulate and choke off the melting process. Snow cover creeps downslope in a quiet, slow, and almost irresistible mass movement. An obstacle bears the weight of a large upslope sector in a bridging action that can drag down any obstruction or tear branches out of trees. Snow creep has geomorphic significance, creating stresses in the soil that are as great as those of glaciers (Thorn, 1978).

Liquid water in the detention layer applies its potential energy to breaking up soil aggregates and carrying them out of an ecosystem, but does perhaps more work through kinetic energy. Subsurface runoff, unseen but common in forest ecosystems, carries off soil solutes and nutrients. In some soils an accumulation of water during prolonged rain causes a small slab of soil to slide out, disintegrate, and flow as a debris avalanche (Kesseli, 1943). At rainfalls exceeding 5 mm hr^{-1}, following regolith saturation by 250 mm or more, such shallow soil slips were common in the 1969 storm in southern California (Campbell, 1975). Mass movement to the point of slope failure occurs in clear-cut areas of Oregon (Gresswell et al., 1979, p. 12).

Air Processes

Vapor from transpiration is removed by the help of buoyancy imparted to the air by sensible heat and the low density of vapor. Thermal convection becomes dominant at a short distance above the surface during periods of greatest ecosystem activity, i.e., on summer days, and not only removes vapor (and oxygen, another residue) but also brings CO_2 down into the ecosystem.

Quiet and unrelenting, potential energy gives a common direction to ecosystem functions[*] and aids in removal of waste products and excess water. When leaves drop and trees eventually fall to earth, gravity assembles them in the decomposition zone of the ecosystem, where their nutrients remain within reach.

POTENTIAL ENERGY IN ECOSYSTEM ENVIRONMENTS

Energy generated in the earth drives convection that moves the tectonic plates across the surface and raises mountain ranges.[†] Dissection

---

[*] Aided by the coincidental action of phototropism that in general operates at 180° from the gravitational force.

[†] Upthrust of the edge of the Pacific plate in the New Zealand Alps adds potential energy (above sea level) at a rate that can be calculated as roughly equivalent to 5 mW m^{-2}. It is eroded at the same rate, as shown by data on river load and offshore deposits (Adams, 1980).

of these highlands creates potential energy differences that support continuous movement in the water and air environing the ecosystems that mantle these slopes. Energy fluxes in water and air are far greater than those that raise up the rock itself.

Potential Energy of Water in Highland Ecosystems

By virtue of the altitudes, 1–3 km, at which most condensation of atmospheric vapor occurs, condensed water in the atmosphere possesses a great deal of potential energy, and after a short fall to mountain ecosystems still possesses much of this energy. Winter snowfall to a headwaters basin in the Sierra Nevada, 2.1 km above sea level, more than a ton per square meter, has potential energy (P.E.) equal to:

$$P.E. = 1000 \text{ kg m}^{-2} \times 2100 \text{ m} \times 9.8 \text{ m sec}^{-2} = 20.6 \text{ MJ m}^{-2}.$$

This quantity is not much greater than the daily pulse of latent heat from an active ecosystem in summer, but is in a form far more amenable to being converted into useful work: the power of falling water. The meltwater from the 10-km^2 basin of the Central Sierra Snow Laboratory, which runs off in 2 months, releases about 40,000 kW (flux density 4000 mW m^{-2}) before it reaches the Central Valley. Much of this energy, once used in hydraulic mining of gold, is now captured in turbines and enters the power grid to operate urban industries and if the water had been stored in Lake Van Norden just downstream from the laboratory basin, would in late summer run the electric pumps of farmers irrigating with ground water.

Avalanche Snow. Snow accumulates on an avalanche source area, and such areas in Colorado at slope angles at 30–45° possess enormous potential energy. That on 1000 m^2 of land 200 m above a valley is 2000 MJ, and avalanches can be classified by their potential energy, expressed as the logarithm of ton-m^2-sec^{-2} or log kilojoules (Martinelli, 1966). The quantity is less on steeper slopes, which retain less snow (Perla and Martinelli, 1976, p. 71), and is opposed where forest ecosystems anchor the snow.

Timberline forest in the Alps was widely cut in the Middle Ages to provide energy for making cheese from the milk of cows on summer pasture and so to reduce the transport cost of the biological production of these remote pastures. Avalanches increased from these high ecosystems, but people stayed out of their tracks until after World War II, when skiers moving into the high mountains increased the population at risk. As a result, extensive energy-budget research on timberline ecosystems and radiative, soil, micrometeorological, snow cover, and wind aspects of their environments has been carried out in western

Austria (Aulitzky, 1961) and Switzerland (Turner, 1971; Turner et al., 1975) as a basis for engineering* and silvicultural works necessary to reestablish forest ecosystems on these difficult sites. Similar efforts to hold snow and soil on high slopes are made in the Soviet Union by protecting and planting forests on sites where snow avalanches and mudflows start (Gerasimov and Zvonkova, 1974).

Overland Runoff. Overland runoff is a less spectacular potential-energy conversion than an avalanche, but more widespread. Under mechanized agriculture it may be producing the greatest volume of sheet erosion ever experienced in the United States, an indication that the potential energy of water on moderate slopes is not being well managed (U.S. Council on Environmental Quality, 1978).

Overland runoff converts potential into kinetic energy more visibly than mass movement and in some ecosystems is more important. In Australia, slope wash was found to be several times as prevalent as soil creep (Williams, 1973).

Channel Runoff. Overland runoff in most ecosystems collects itself into rills, in which increased depth reduces the friction energy cost per unit mass. This economy of scale frees potential energy to carve channels and move soil particles. A stabilized stream may follow a path that will "minimize its time of potential energy expenditure" and the theoretical result is a sequence of pools and riffles (Yang, 1971; Davy and Davies, 1979, 1980). The total power actually developed on a 0.34 km reach of a small river in Italy, 15 kW, was the smallest of several modeled alternatives (Cherkauer, 1973). Nine times as much power was expended per unit length in the riffles as in the pools; the mean is 45 W m^{-1}. This energy moves dissolved and particulate organic matter from terrestrial into aquatic ecosystems, where it is a major support.

Hydropower has supplied energy for human activities for many years, beginning with mill streams of small flow and small heads, and going on, with improved technology, to large rivers like the Merrimack by 1830. Recent developments in automation make it again economic to operate relatively small generating plants, often in the 3–5 kW size class (Clegg, 1975).

Upland and mountain ecosystems yield large quantities of this valuable energy. The Alps of New Zealand, Australia, and Japan and the mountains of Taiwan have national value for power, and special concern is felt for the welfare of ecosystems that produce a yield so valuable.

---

* These works are expensive to erect in the remote starting zones, as much as $50 m^{-2} of slope stabilized (Ives et al., 1976), and have a correspondingly high energy cost.

Their biological production is important less for itself than because it takes healthy ecosystems to generate clean water.

Soil and Soil Water

Potential energy is released in avalanches, mudslides, and landslides, and enters formulations for such transport (Ahnert, 1977) as a term for sine of the slope, with various exponents. About a quarter of the transport measured in a study in Scotland is caused by freeze-and-thaw cycles (51 per year) and by diurnal thermal expansion and contraction, both energy related. Another cause, the filling of burrows and wormholes from their upslope sides, results from the earlier expenditure of biological energy. Creep here was 25 times as great as transport by surface runoff, but indicates a flux density of energy conversion of only 1 nW m^{-2} (Kirkby, 1967).

Dry-weather slides in the mountains of southern California occur on slopes steeper than the theoretical angle of repose (0.65) (Anderson *et al.*, 1959). On many slopes soil is held only by the chaparral ecosystems, and fire in them may release soil so fast that fire-control vehicles cannot move in until bulldozers have cleared roads of debris. These slopes also suffer shallow landslides in long rainstorms, grass ecosystems being more vulnerable (0.17 of area lost in the storms of 1969) than chaparral (0.06 of area lost) (Rice and Foggin, 1971).

The movement of subsurface water differentiates the upper and lower parts of many hillsides. The movement of deeper groundwater affects many riparian ecosystems and controls the accumulation of peat in wetlands (Gosselink and Turner, 1978).

The Local Air

Differences in the density of air in dissected terrain create differences in potential energy, one result being the familiar process of cold-air drainage, which is expressed with a slope term (Yoshino, 1975, pp. 410–412). When air in a radiating slope ecosystem cools more than air at the same altitude above the valley, since air alone is not an efficient radiator, the difference in temperature, hence density, provides potential energy. There are two consequences: the slope ecosystem enjoys a supply of warm air sinking from aloft, and the receiving valley ecosystems lie in a cold pool that deepens during the night. As their fuel moisture rises, their daytime fire potential is lowered (Furman, 1978). Mean summer temperatures that define Douglas-fir site quality in Oregon reflect cold-air drainage (Minore, 1972), and spores carried by air drainage down a relict fluvial system near Lake Superior carry a disease fatal to red pine (Dorworth, 1978). Frost pockets may eliminate tree growth entirely. Removing cold air ponded in Alpine valleys requires the energy of strong foehn winds (Urfer-Henneberger, 1979, p. 410).

Potential Energy in the Landscape

Potential-energy differences are created in the environment of an ecosystem as water dissects the landscape and steepens its slopes. Ecosystems live with the steady pressure that potential energy exerts on their soil and plants and with the slow movements of soil* and cold air powered by it.

Contemporary hillside ecosystems hold material against potential energy and allow prolonged weathering, but prior to the spread of terrestrial ecosystems weathering products were rapidly transported into valleys and piedmonts (Schumm, 1977, p. 84), and the landscape had an entirely different look.

Tides

Tides confer potential energy on a large mass of water and have important effects on estuarine and coastal marsh ecosystems. These enormous volumes of water twice a day excavate channels larger than those found among terrestrial ecosystems (Geyl, 1968) and operate so as to bring in organic debris as a substantial energy subsidy to a marsh ecosystem.

KINETIC ENERGY OF WATER IN ECOSYSTEMS

Impact of Raindrops and Hailstones

Particles of precipitation soon reach terminal speeds of about 1 m $\sec^{-1}$ for snowflakes and 6 m $\sec^{-1}$ for raindrops, depending on size, in their fall to lowland ecosystems. Their impact is a function of kinetic energy:

$$K.E. = \tfrac{1}{2}mv^2,$$

and at a moderate rate of rainfall, 10 mm hr^{-1} (10 kg m^{-2} hr^{-1}), is

$$K.E. = \tfrac{1}{2}(10 \times 6^2) = 180 \text{ J m}^{-2} \text{ hr}^{-1} = 50 \text{ mW m}^{-2}.$$

* "Repeated slow landslides" in southwestern Japan have over the years produced gentle slopes that provide the main sites for agricultural ecosystems in this rugged area (Nakano, 1974).

This is not a high rate of energy conversion, but it is in a form capable of doing mechanical work. The impact of large raindrops breaks up soil aggregates unprotected by vegetation in an effect quite separate from that of overland runoff (Young and Wiersma, 1973); soil loss is reduced 0.08 when raindrop energy is reduced 0.10. Furthermore, the downslope component of splash transports soil particles as a function of sine of slope angle (Ahnert, 1977).

Hailstones exhibit extreme concentration of mass with consequent amplifying of kinetic energy* in a relatively few stones falling at terminal speeds exceeding 10 m sec^{-1} (Morgan and Towery, 1976; Matson and Huggins, 1980). The speed, double that of a large raindrop, quadruples the kinetic energy, and mass is several orders of magnitude greater; vertical kinetic energy as estimated by hail pads ranges from as little as 4 mW m^{-2} to as much as 5 W m^{-2} (Waldvogel et al., 1978). Few ecosystem canopies can stand up to these energy impacts, and stems as well as leaves are damaged. The fact that hail stripes are usually narrow (15 30 m) (Towery and Morgan, 1977) is no comfort to the individual ecosystem.

Water in Gullies and Streams

The fact that gully flow is ephemeral hardly detracts from the energy it develops at high speeds, which may happen often enough to keep the slopes above undercut, crumbling, and difficult for an ecosystem to colonize. Where the kinetic energy of streams erodes the toe of the slope, it increases the available potential energy of the soil and water of ecosystems above, which increases their instabilitity.

Stream kinetic energy supports transport of a large sediment load, which itself can produce damage. It also creates new ecosystems: aquatic in the channel itself and those on the "flood plain which is formed by the kinetic energy of the stream during periods of high flow" (U.S. Water Resources Council, 1978, p. 239). These ecosystems differ in many ways from upland ecosystems. They have access to water more of the time, but are liable to swamping or flooding. They often benefit from a more complete mix of nutrients because the stream integrates solute outflows from diverse geological formations and ecosystems in its drainage basin. These effects result from low potential energy relative to surrounding upland ecosystems, and confer definite hydrologic, biotic and cultural values on them (U.S. Water Resources Council, 1979, Chap. 5).

* Hail pads measure kinetic energy by the size of dents in a plastic foam pad.

Kinetic energy of the stream is manifested not only in the rush of water* but also in turbulence, viscous dissipation, erosion of banks and bed, shifting bed material, and transportation of suspended material and bedload, in a set of energy transfers not yet well understood (Davy and Davies, 1979, 1980). Two similar floods in Wales, for example, performed work quite differently: one tore up slopes, the other scoured out channels (Newson, 1980).

Avalanching Snow and Soil

Rapid conversion of potential energy generates high speeds in the lower track and runout zone of an avalanche, cited as 10–30 m sec^{-1} for wet snow and greater than 30 m sec^{-1} in dry or air-borne powder (Perla and Martinelli, 1976, p. 86). Dust avalanches of snow particles entrained into the air become "a self-energizing flow" independent of lower flow at densities nearly twice air density (Mellor, 1978). Dry avalanches in western Colorado that reach 60 m up the valley side trim and bend trees as is shown in reaction wood in their trunks (Ives et al., 1976). Trees in the upper parts of a track may survive, but much depends on species; spruce is preferred to pine for planting in the Alps because of dominance of its terminal bud (Schönenberger, 1978).

Kinetic energy in the runout zone is converted into frictional work, flow work, particle resistance, and partly back into potential energy in a debris cone. Application of energetics analysis (Mellor, 1978) allows the runout distance, i.e., the area of destruction, to be calculated (Leaf and Martinelli, 1977, p. 13) and assesses the relative effects of slab thickness and external and internal friction (Lang et al., 1979, p. 27).

Saturated debris avalanches in southern California produce damage by high-velocity impact and inundation (Campbell, 1975). Mudflow deposits in central California, dated by "successive root systems of redwood trees that developed higher on the tree after the previous root system was buried by mud" (Smith, 1978, p. 17) at Big Sur, were found to occur at a yearly frequency of 0.007. The Van Duzen basin in northern California experienced a 0.002-frequency storm in 1964 that produced a large mass movement (Kelsey, 1980) equivalent to 3 mW m^{-2} change in potential energy. From a 35-year budget a mean loss of 0.2 mW m^{-2} potential energy can be calculated, reflecting recent uplift and abuse of the forest ecosystem.

Debris flows occur in the Appalachians in ordinary wet weather

---

* Although Blum (1972) points out that many aquatic plant species do not resist this movement at all.

(Briggs *et al.*, 1975). In the 700-mm rain (in 8 hrs!) that hurricane Camille dumped on central Virginia (Miller, 1977, pp. 50, 76, 426) slopes of 0.5–0.6 produced debris avalanches, which, averaged over small drainage basins, removed 4 cm of soil in great scars, the equal of "several thousand years of normal degradation" (Williams and Guy, 1973, p. 78). Most of the 150 persons lost were killed by the impact of the debris flows or in the surges produced when they entered the racing streams.

ATMOSPHERIC KINETIC ENERGY IN ECOSYSTEMS

Although energetics is the base of many branches of the atmospheric sciences, it would take us far afield to discuss here all atmospheric energy transformations other than to say that except as generated by local differences in pressure, most of the kinetic energy in the local air comes from a cascade by which energy of large-scale atmospheric motion systems is handed over to smaller systems, then to turbulent eddies of diminishing scale. Such a source accounts for episodic bursts of kinetic energy to which ecosystems must be accommodated.

Even the common downslope thermal winds develop high speeds when they combine with synoptic-scale systems. Strong mistrals in the lower Rhone valley produce a flow, against which cypress, poplar, or other windbreaks as close as 20 m protect intensive horticultural ecosystems established after about 1830 (Gade, 1978). The bora, a fall wind from a cold air pool on the plateau above the Dalmatian coast, is classified as occurring in cyclonic or anticyclonic flow (Yoshino, 1975, p. 363), and its speed correlates with wind-shaping of trees (Yoshino *et al.*, 1976).

Expressions of Energy Conversions

Energy *content* in a unit cross section of the air stream varies over three orders of magnitude (Table I). Energy *conversion* in a unit cross section, such as takes place in a windmill, requires another length dimension (meters) to express wind travel through the section and another time dimension (per second) to express rate. Wind energy conversion is therefore

$$(\tfrac{1}{2}\rho u^2)(\text{m sec}^{-1}) = \tfrac{1}{2}\rho u^2(u) = \tfrac{1}{2}\rho u^3 \ (\text{kg sec}^{-3} = \text{W m}^{-2}).$$

Cubing wind speed amplifies its sporadic nature. Energy conversion in prevailing conditions in the western United States (Table I) is 40 W m^{-2}, but in the cyclonic storm 16 times as much (2.5^3).

TABLE I

Illustrative Values of Kinetic Energy for Unit Wind Travel

	Speed u (m sec^{-1})	$u^2/2$ (m^2 sec^{-2})	Unit-travel kinetic energy content[a] $\rho u^2/2$ (kg m^2 sec^{-2})(m^{-2})(m^{-1})
Mean speed in sheltered mountain valley[b]	0.8	0.3	0.4
Mean speed in western U.S.[c]	4	8	10
Typical cyclonic storm	10	50	65
Storm gust	20	200	260
Upper limit for wind power[d]	27	365	485
Threshold for structural damage	30	450	585
Columbus Day storm, 1962	40	800	1040

[a] K.E. per unit area cross section and unit travel. Density ρ taken as 1.3 kg m^{-3}.
[b] Central Sierra Snow Laboratory during interstorm periods of winter and spring.
[c] At NOAA stations, mostly in basins.
[d] Baker *et al.* (1978). The lower limit is 4.5 m sec^{-1}.

Wind Stress

The rates of energy conversion just given occur in horizontal flow through a unit-area cross section of the air stream (1 m^2), and in order to see the effect on a unit area of the earth's surface, we must consider another parameter, wind stress τ or the force, per unit area of ecosystem surface, equal to kinetic energy times a drag coefficient and including skin friction and form drag (Grace, 1977, p. 14). It can be transformed (Thom, 1975) to the downward flux (per unit time and area) of the momentum of the air, a property analogous to its temperature or vapor concentration.

This vertical transport can be expressed in flux-gradient form, as for sensible heat (Chapter XVI), with wind shear $\partial u/\partial z$ taking the place of temperature gradient:

$$\tau = \rho K_M(\partial u/\partial z), \text{ kg m}^{-1}\text{ sec}^{-2}$$
$$= \text{kg m}^{-3}\text{ m}^2\text{ sec}^{-1}\text{ m sec}^{-1}\text{ m}^{-1}.$$

The coefficient of eddy diffusion for momentum K_M is a "ventilation characteristic" of an ecosystem (Lemon, 1967) that controls the gas exchange with the atmosphere.

Energy Dissipation. Wind stress on an ecosystem can be converted into rate of energy dissipation per unit area to give an expression dimensionally like those for the other energy fluxes. In the absence of energy generation by horizontal pressure and temperature patterns unit-area stress τ times speed is equal to the unit-area rate of change of kinetic energy with time, i.e., the rate of energy conversion per square meter of the underlying ecosystem surface (Palmén and Newton, 1969, p. 527). The yearly and global average, which runs about 1 W m^{-2},* is a minor component of the energy budget of any ecosystem and largest in the middle latitudes (Lettau, 1954): 1.4 W m^{-2} at 30° and 2.2 at 60° lat.† Its significance lies in its control function over other energy fluxes.

Effects of Conversion of Kinetic Energy

Both expressions of energy conversion, per unit of air stream and per unit area of ecosystem surface, are useful in examining ecosystem energy budgets. Individual trees (or windmills), shelterbelts, and isolated forest stands are subject to impact on their sides, producing "aeromechanical behavior" (Holbo et al., 1980), and alleviated by deformation in tree or ecosystem shape (Fig. 2) (Weischet, 1955, p. 105). These deformations can be mapped as an index to wind exposure (Yoshino, 1964; Grace, 1977, p. 155) and power potential (Baker et al., 1978). Wind-borne salt also affects coastal ecosystems (Grace, 1977, p. 182; Reitz, 1978).

Rates of energy conversion expressed per unit surface area, i.e., a function of wind stress, are appropriate for low ecosystems, bare soil, and large forest stands, and are related to surface roughness of the ecosystem through the K_M term in the stress equation. Roughened systems absorb large quantities of kinetic energy by swaying and tossing of branches, converting atmospheric energy to sound and heat, sometimes at a higher frequency than that of variations in the shear force (Holbo et al., 1980).

Sound. Ecosystems generate acoustic energy in high winds‡; they also absorb and diffuse it as they do kinetic energy. Measurements of the attenuation of highway noise by different shelterbelts (Cook and Haverbeke, 1971) found that ecosystem dimensions were more impor-

---

* Dissipation of energy by viscous processes in the boundary layer but above the surface has about the same size (Kung, 1966).

† Extremes occur at nonbiological surfaces: 0.05 W m^{-2} under stable conditions at Antarctica (Lettau and Dabberdt, 1970) and 23 W m^{-2} at the sea surface in a hurricane (Riehl and Malkus, 1961).

‡ This is also true in ice storms. Persons caught in the ice storm of March 1976 in southern Wisconsin were astonished by an almost continuous roar from the breaking branches and trunks—in windless weather!

Fig. XIX.2. Wind-flagged trees near the timberline in Boulder Canyon, Colorado, are a visible evidence of the effect of atmospheric kinetic energy.

tant than leaf size, shape, or branch deployment. Shelterbelt density and width index the numbers of absorbing and diffusing elements, and height indexes the surface area presented to the advancing sound waves.

Mesoscale Transport.   Local transport of water by atmospheric kinetic energy has ecological significance as a form of redistribution among adjacent ecosystems. Snow eroded from the snow cover and supported by the turbulent energy of the wind is carried until evaporated or deposited where the wind is slowed by slope angle or trees.

Episodes of High Kinetic Energy

Much of the total impact of atmospheric kinetic energy on ecosystems is concentrated in brief episodes, in prolonged cyclonic storms with heavy rain that weakens soil support of trees, and in shorter convective storms.

Coastal Storms.   Large storms off the ocean (Blumenstock, 1959, Chap. 4), whether midlatitude or low-latitude in origin, bring destruction to terrestrial ecosystems. The New England hurricane of 1938 blew down thousands of square kilometers of forest and created a hillocky microrelief (mounds of dirt carried up by the roots, depressions where the roots had been). This relief was then recognized elsewhere in the region, confirming the occurrence of hurricanes in past centuries, e.g., in 1635 and 1815 (Brooks, 1940). Destruction in coastal cities in Japan correlates with total storm energy over a range up to 61,000 in the number of houses destroyed in individual storms up to kinetic energy of 3×10^{12} MJ (Takahashi, 1955). Storms can also be classified by a rate of energy conversion ($\sim 10^6$ MW), but damage is proportional to a higher power of wind speed, the fifth power. Frequency analysis of hurricane winds indicates that the mean annual cost of repairing damage ranges up to 0.33% in Australian ports; it is 0.24% at Townsville, for instance (Lehane, 1980).

Rice ecosystems suffer from kinetic energy that causes excess fluttering of leaves that disturbs water ascent in the plants, decreases CO_2 assimilation, and allocation of photosynthate to grain (Inoue et al., 1965). The plants can be protected by deep flooding of the paddies, by nets that reduce the waving motion (honami) of the leaves, or by genetic improvement of mechanical strength.

The Columbus Day (1962) storm, which began in a dying hurricane in the western Pacific Ocean, arrived at the northern California coast at an angle such that its winds paralleled the north and south grain of the coastal mountains. Squall activity that coupled upper and surface winds brought high kinetic energy down into the local air: "No longer was there any shallow layer of cold air shielding the earth's surface" (Lynott and Cramer, 1966). Gusts in coastal valleys in California and Oregon reached kinetic energy of 20 to 25 kW m^{-2} of cross-wind area (Decker et al., 1962). Windfall in experimental cuttings of old-growth redwood in northern California was 0.11 in a shelterwood cutting in which vigorous trees had been left, 0.38 in a selection cutting leaving groups of trees, and 0.30 on the north, or downwind, side of clearcuttings (Boe, 1965). Forest ecosystems were sheltered neither by topography nor by the planetary boundary layer, which is usually "the protective zone for the earth's surface" (Vorontsov, 1959).

Storm kinetic energy exerted on the sea surface is converted to wave energy that is expended on coastal landforms and their ecosystems at mean rates of 6 kW per meter of coastline (Inman and Brush, 1973) and at episodic rates that approach the 2–3 MW m^{-1} of a fire front. In contrast, on low-energy coasts, like the Gulf Coast of Florida, net littoral energy westward is about 4×10^{-6} μW m^{-1} (Tanner, 1971). Tidal and subtidal ecosystems are adapted to energy expenditures, and coral, one of the most complex of these ecosystems, is affected by wave-induced turbidity that reduces the input of solar energy, as well as by direct impact. Coral growth tends to be porous in sites of moderate wave energy and "slow but compact" in sites of high energy (Adey, 1978). Low wave energy in the Caribbean permits upward growth three times as fast as in the stormy Pacific.

Coastal dunes form and reform under the flows of kinetic energy in swells from distant storms, waves of nearby storms, and the sea breeze. Combined with the influence of colonizing ecosystems, these flows of kinetic energy build a sequence of beach ridges (Alexander, 1969). Six beach types studied in New South Wales express contrasts in wave energy (Anonymous, 1979). Grasses root deeply and extend their stems to keep leaves afloat in the shifting sand; shrub and tree ecosystems like *Pinus contorta* (shore pine) live in dune-sheltered hollows. Inland dunes on the move near the Columbia River have carried juniper ecosystems 26 km in 12,000 years as individual trees grow up and are overridden by the next dune (Long *et al.*, 1979).

Continental Frontal Systems. Friction of terrestrial ecosystems dissipates kinetic energy of the air continuously, and strong winds are not as commonplace as on coasts. However, instability is more likely, and gusts following a cold front in spring pick up, from bare soil, particles already detached by impact of the wind itself and bouncing along the ground and carry them aloft in suspension. The mass of soil removable by wind erosion is proportional to $\tau^{2.5}$ (Chepil, 1958, p. 13), and typical loss of 40 kg m^{-2} from sandy soils in southwestern Kansas reduces wheat yield by 13 g m^{-2} (Anonymous 1977) or 30 mW m^{-2} energy equivalent.

Many vigorous continental storms—the blizzards of winter—occur when natural ecosystems are lying low or dormant, protected by snow cover or snow they trap during the storm (Tabler, 1975). Human ecosystems, in contrast, expose themselves and try to keep their circulatory apparatus going throughout the long periods of low or zero visibility [16 hr in the Upper Great Plains blizzard of January 1975, see Babin, (1975)], strong wind energy, and drifted roads.*

Downslope winds to the lee of the Rockies and other ranges transverse to upper-air flow penetrate at times to the surface of the earth. Boulder, Colorado, for example, experiences them once a year on the average, with speeds exceeding 22 m sec^{-1} and gusts 33 m sec^{-1} (Brinkmann,

---

* In this storm Canadians heeded the warning "Do not move—stay where you are," and only two persons died (Babin, 1975), whereas on the U.S. side 35 died (Graff and Strub, 1975).

1974). Speeds are still higher in the defiles of valleys leading out of the mountains, and in the lee of the New Zealand Alps railroad bridges are protected by windbreaks (Fig. 3). Model experiments explain the high wind speeds experienced in the Yugoslavian bora (Yoshino, 1975, p. 405).

Thunderstorms. Thunderstorm downdrafts spread at the earth's surface into gust fronts with speeds up to 20–30 m sec^{-1}. Urban systems of mobile homes are particularly vulnerable, because gusts as small as 23 m sec^{-1} can roll and smash them (American Meteorological Society, 1975); several thousand are lost this way every year. Gust energy amplifies the destructive power of hailstones (Towery *et al.*, 1976), and the sheltering power of trees at the edge of a field appears in the spatial pattern of damage.

Downbursts were identified by Fujita (Fujita and Caracena, 1977) in the storm of 4 July 1977 that leveled forests across northern Wisconsin; hemlock ecosystems that had survived 3 centuries of wind and storm went down under speeds that reached 70 m sec^{-1}.

Fig. XIX.3.   High bridges across valleys on the east side of the New Zealand Alps are fenced to protect trains against strong winds.

Tornadoes destroy annual ecosystems in swaths or circular tracks and defoliate forest stands. Fortunately, they have narrow paths, and the expectancy that a given ecosystem in southern Ontario will be struck is only 0.0005 compared to 0.02 per year for hurricane winds (Hewitt and Burton, 1971, pp. 61, 71).

Lightning. Lightning strokes dissipate energy at a rate of approximately 10^4 J m^{-1} (Hill, 1977, 1979), depending on the degree of branching and multiple strokes. Taking this figure, a 2-km* stroke from cloud to ground would dissipate as heat, radiation, and work on the surrounding air (producing thunder as acoustic energy), 20 MJ of energy, all in a period measured in milliseconds—an enormous pulse of power.

Lightning fires in forest ecosystems—0.97 of them burning areas smaller than 4 ha (Taylor, 1971)—create small openings that fill in with younger trees or different species and may become centers "for the spread of insects, disease, windthrow, and future fires" (Taylor, 1971).

Rules to avoid being struck by lightning (Mogil *et al.*, 1977) apply also to ecosystems: Do not stand out above the surroundings! Trees are more vulnerable than low ecosystems. The elevated network that distributes electrical energy over the landscape is often struck, and design or operations failures have far-reaching consequences, for example, the 1977 blackout of New York City (Boffey, 1978).

Converting Atmospheric Kinetic Energy for Human Ecosystems

The wind-driven ship as an energy-converting device was at the forefront of technology in the prefossil energy period, and much of our early knowledge about atmospheric motion systems world-wide came from studying its source of energy. Wind technology that successfully lowered the water level in crop soils in the Netherlands was exported to lowlands elsewhere. Windmills that pumped ground water into stock-watering tanks on the Great Plains made it possible to use domesticated animals to harvest the biological productivity of grassland ecosystems (Webb, 1931). Plains sites are favorable for energy conversion (Justus *et al.*, 1976), and of the 6×10^6 windmills sold in the U.S. after 1850, a large fraction went onto the Great Plains (Metz, 1977). Coastal sites are also favorable (Fig. 4), up to about 10 km inland (Benesch and Jurksch, 1978, p. 14). Urban sites are unfavorable (U.S. House of Representatives, 1980, p. 22).

Kinetic energy extracted from moving air is proportional to u^3 (the u^2

---

* Longer strokes have been measured, e.g., one that extended 7.6 km vertically and then 6.5 km horizontally (Uman *et al.*, 1978).

Fig. XIX.4. A 19th-century windmill from western Estonia, now in the Outdoor Agricultural Museum near Tallinn. The mill, which turns to face the wind, was again in use during World War II.

factor of kinetic energy multiplied by u to express the volume of air going through the fan). Conversion in good sites is 200 W (Gustavson, 1979) per square meter of swept area,* comparable with the unit-area input of solar energy, but in a form that is more effective "in extracting work from the natural medium" (Metz, 1977) and "can be readily converted to either electrical energy or heat energy" (Diesendorf, 1979), both being of high quality in Second-law terms.

In remote areas (outside the electric-energy network; having high oil

---

* Swept area of the Great Plains windmills was 3–6 m², of the large turbine that operated in Vermont in the 1940s about 1900 m², and of a later Danish mill at Gedser 450 m².

costs) even low wind speeds are useful for pumping water for irrigation of small fields. Filling the soil reservoir serves as storage; water 40–100 mm deep in rice ecosystems can "cushion the effect of nonavailability of wind energy for 3 to 4 days at a stretch" (Trewari, 1978). Water tanks* served the same cushioning purpose for the Great Plains windmills.

Analysis of energy requirements of a small community in the semiarid low latitudes showed that conversion of wind energy at a mean annual rate of 7 kW would do such jobs as lighting, radio, threshing, and water supply better than energy from crop residues or solar collection (Golding and Thacker, 1956).

Units of sizes up to 25–30 kW present few problems in high winds and are not as restrictive in siting (Elderkin *et al.*, 1977) as large units (Liu and Yocke, 1980). Large units also produce wakes that may affect downwind ecosystems. Where ecosystem roughness is small ($z_0 = 0.01$ m), elevating the mill from 10 m above the surface to 15 m yields an 0.8 increase in its output, but in rougher country ($z_0 = 0.1$ m) yields only 0.3 increase in output (Laikhtman, 1964, p. 170).

Converted energy usually does not need to be stored over long periods: "storage that can replace the average power output for about 10 hours makes a wind energy system as dependable as one large nuclear plant" (Sørensen, 1976), and hydropower can provide such storage (Baker *et al.*, 1978). Wind energy in Denmark has a winter maximum and a late-summer minimum, complements the seasonal regime in solar energy, and meets the seasonal demand curve (Sørensen, 1975).

VARIATIONS OF POTENTIAL AND KINETIC ENERGY

Conversions of potential and kinetic energy in ecosystems tend to be sporadic, even catastrophic, in occurrence, and instead of mean rates, incidence is more likely to be reported (such as the 0.04 frequency flood). Many of these conversions are site-specific, and ecosystems are more or less accommodated to them.

Variations in Time

The accumulation of potential energy in the source areas of avalanches and landslides can be monitored, and the prediction of periods when avalanches will run is standard procedure in Switzerland and Colorado (Judson, 1976). Individual avalanches are treated in frequency terms if

---

* If elevated to 6–8 m, as was common in California farmsteads, they also store potential energy.

the history is long enough, for example, 0.05 probability in Ophir, Colorado (Ives *et al.*, 1976), from which hazard areas can be mapped (Lang *et al.*, 1979, p. 1).

Streams have always experienced floods that damage natural ecosystems, but the increased perception of flash floods results from urban invasion of the floodplains of streams* so apparently insignificant that their flow regime has never been gauged and may be changing because of urbanization.

Availability of latent energy in a deeply moistened air mass and the likelihood of such a triggering energy-release mechanism as thermal convection or squall-line convergence can be monitored as precursors to kinetic-energy impacts. The frequency of tornadoes, damaging hail, and windstorms can be approximated by climatological analysis.

Ecosystem energy budgets have evolved in environments of specific expectancies of ice storms, wind storms, soil movement, and other energy-conversion episodes. Those that occur every year or so, such as meltwater flooding, moderate thunderstorm gust fronts, and avalanches that run every winter exert virtually constant pressure on an ecosystem, to which it can accommodate if the energy cost of such accommodation lies within its capabilities.

Damaging episodic energy releases often occur seasonally, especially where water is the lubricating or energy-transporting medium. Large-scale atmospheric generation of kinetic energy is seasonal (Palmén and Newton, 1969; Kung, 1966), especially in the northern hemisphere, and synoptic-scale storms are weak and less frequent in summer. However, convective latent energy releases are strong in this season.

Spatial Variations

Within Ecosystems.   Ecosystem foliage is deployed for maximum uptake of CO_2 and photons, yet must be secured against excessive loading by freezing rain or snow or wind impact. At the same time the vegetaion component protects the soil component from the direct impact of atmospheric kinetic energy, whether raindrops or wind. Overgrazing sparse grass and cutting willow shrubs in Iceland exposed a dark surface to heating and generates thermal convection that carries off the normally erosion-resistant soil (Ashwell, 1966). Crop residues on the Great Plains absorb enough kinetic energy to protect soil from wind erosion (Chepil, 1965). Trees afford substantial protection in savanna ecosystems in western New South Wales (Stannard, 1958). These remnants of open wood-

---

* Some 15,000 urban systems are at risk in the United States (American Meteorological Society, 1978).

land cast little shade and do not reduce the growth of grass, but do absorb kinetic energy.

Contrasts among Ecosystems.   A slope ecosystem is subject to snow and soil creep, and the spatial pattern of potential energy can be read from a contour map. Kinetic energy varies with the pattern of the wind field, amplified by squaring or cubing, in an invisible and complex (Liu and Yocke, 1980) pattern of exposure and shelter that determines impact on and ventilation of ecosystems, and can be very important in hurricanes (Lehane, 1980).

Mosaic-Scale Linkages among Ecosystems.   Flows of snowflakes, vapor, fog droplets, pollen, nutrients, and many other substances, as well as sensible heat and sometimes fire that link the contrasting ecosystems in a mosaic landscape are powered by potential and kinetic energy, acting in an integrating role. These linkages, to be described in more detail in Chapter XXI, have important control functions in ecosystem energy budgets.

The kinetic energy extracted by forest stands is not immediately re-placed from higher levels of an air stream, and downwind ecosystems are sheltered. If one stands to the lee of a grove of trees, one hears a gust of several hundred meters diameter strike the grove and, after a few seconds, feels only a weak impulse. Even thin shelterbelts absorb and filter kinetic and turbulent energy, reducing diffusivity K_H as well as speed (Rosenberg, 1975).

Windfall in forest ecosystems depends on topographic situation, and while lee slopes occasionally experience significant eddy flows, those in the Rockies experience half as much damage as ecosystems on slopes that face storm winds. Cut edges on slopes lose 20 trees km^{-1}, those in saddles lose 150 km^{-1} (Alexander, 1964). Cutting strategy is to start at the bottom of a lee (east) slope, cut strips in successive decades work-ing into the wind and parallel to the slope contour if possible, up to the ridge, and then down the windward slope "without exposing a leeward boundary to the wind" (Alexander, 1964). The strategy in the redwoods is to create a wind-firm edge and work toward the south, the quarter from which damaging winds come (Boe, 1965). In these strategies of ecosystem harvesting the terrain-bound pattern of kinetic energy be-comes a part of the management pattern. Similarly, the terrain-bound pattern of fair-weather up-slope and up-valley winds enters into the daily regime of an ecosystem. It shapes the trees in the Rhone Valley (Yoshino, 1964, 1975, p. 445), carries fire, powers downward heat fluxes to melting snow, and controls fluxes of sensible and latent heat from foliage. The "vectorial elements," beam radiation and wind, dominate the pattern of high-altitude ecosystems (Turner, 1968) and are important to ecosystems everywhere. Conversions of potential and kinetic energy of the atmospheric water and soil media in and around ecosystems can be destructive in brief episodes. Their everyday activity in powering circulations of heat and matter into and out of ecosystems is just as important for ecosystem energetics and illustrates ways that small energy conversions control large fluxes in the energy budget.

REFERENCES

Adey, W. H. (1978). Coral reef morphogenesis: a multidimensional model. *Science* **202**, 831–837.

Adams, J. (1980). Contemporary uplift and erosion of the Southern Alps, New Zealand: Summary. *Geol. Soc. Am. Bull. Pt. I,* **91**, 2–4.

Ahnert, F. (1977). Some comments on the quantitative formulation of geomorphological processes in a theoretical model. *Earth Surf. Processes* **2**, 191–201.

Alexander, R. R. (1964). Minimizing windfall around clear cuttings in spruce–fir forest. *For. Sci.* **10**, 130–142.

Alexander, C. S. (1969). Beach ridges in northeastern Tanzania. *Geogr. Rev.* **59**, 104–122.

American Meteorological Society (1975). Policy statement of the American Meteorological Society on mobile homes and severe windstorms. *Bull. Am. Meteorol. Soc.* **56**, 466–467.

American Meteorological Society (1978). Flash floods—a national problem. *Bull. Am. Meteorol. Soc.* **59**, 985–986.

Anderson, H. W., Coleman, G. B., and Zinke, P. J. (1959). Summer slides and winter scour . . . dry–wet erosion in Southern California mountains. *U.S. For. Serv., Pac. Southwest For. Res. Expt. Stn., Tech. Pap.* No. 36.

Anonymous (1977). Wind erosion equals lost food. *U.S. Dep. Agric., Agric. Situation* December, p. 4.

Anonymous (1979). The problem of disappearing beaches. *Ecos (Melbourne)* **19**, 26–31.

Ashwell, I. Y. (1966). Glacial control of wind and of soil erosion in Iceland. *Ann. Assoc. Am. Geogr.* **56**, 529–540.

Aulitzky, H., ed. (1961). Ökologische Untersuchungen in der subalpinen Stufe zum Zwecke der Hochlagenaufforstung. *Mitt. Forstl. Bundes-Versuchsanst. Mariabrunn* **59**, No. 1.

Babin, G. (1975). Blizzard of 1975 in western Canada. *Weatherwise* **28**, 71–75.

Baker, R. W., Hewson, E. W., Butler, N. G., and Warchol, E. J. (1978). Wind power potential in the Pacific Northwest. *J. Appl. Meteorol.* **17**, 1814–1826.

Benesch, W., and Jurksch, G. (1978). Die Windverhältnisse in der Bundesrepublik Deutschland im Hinblick auf die Nutzung der Windkraft. Teil I: Binnenland. *Germany. Wetterdienst, Ber.* **147**.

Blum, J. L. (1972). Plant ecology in flowing water. *In* "River Ecology and Man" (R. T. Oglesby, C. A. Carlson, and J. A. McCann, eds.), pp. 53–65. Academic Press, New York.

Blumenstock, D. I. (1959). "The Ocean of Air." Rutgers Univ. Press, New Brunswick, New Jersey.

Boe, K. N. (1965). Windfall after experimental cuttings in old-growth redwood. *Proc. Soc. Am. For.* 59–63.

Boffey, P. M. (1978). Investigators agree N.Y. blackout of 1977 could have been avoided. *Science* **201**, 994–998.

Briggs, R. P., Pomeroy, J. S., and Davies, W. E. (1975). Landsliding in Allegheny County, Pennsylvania. *U.S. Geol. Surv., Circ.* No. 728.

Brinkmann, W. A. R. (1974). Strong downslope winds at Boulder, Colorado, *Mon. Weather Rev.* **102**, 592–602.

Brooks, C. F. (1940). Hurricanes into New England. Meteorology of the storm of September 21, 1938. *Smithson. Rep., 1939* Publ. No. 3563, 241–251.

Campbell, R. H. (1975). Soil slips, debris flows, and rainstorms in the Santa Monica mountains and vicinity, Southern California. *U.S. Geol. Surv., Prof. Pap.* No. 851.

Chepil, W. S. (1958). Soil conditions that influence wind erosion. *U.S. Dep. Agric., Tech. Bull.* No. 1185.

Chepil, W. S. (1965). Transport of soil and snow by wind. *Meteorol. Monog.* **6**(28), 123–132.

Cherkauer, D. S. (1973). Minimization of power expenditure in a riffle-pool alluvial channel. *Water Resour. Res.* **9**, 1613–1628.

Clegg, P. (1975). "New Low-Cost Sources of Energy for the Home." Garden Way, Charlotte, Vermont.

Cook, D. J., and van Haverbeke, D. F. (1971). Trees and shrubs for noise abatement. *Nebr. Agric. Exp. Stn., Res. Bull.* No. 246.

Davy, B. W., and Davies, T. R. H. (1979, 1980). Entropy concepts in fluvial geomorphology: A reevaluation. *Water Resour. Res.* **15**, 103–106, **16**, 251.

Decker, F. W., Cramer, O. P., and Harper, B. P. (1962). The Columbus Day "big blow" in Oregon. *Weatherwise* **15**, 238–245 and cover photo.

Diesendorf, M. (1979). Recent Scandinavian R and D in wind electric power—implications for Australia. *Search* **10**(5), 165–173.

Dorworth, C. E. (1978). Presence of a late Pleistocene drainage system manifested through depredation of red pine by *Gremmeniella abietina*. *Ecology* **59**, 645–648.

Elderkin, C. E., Ramsdell, J. V., and Tennyson, G. P. (1977). Meeting review/Wind characteristics workshop, 2–4 June 1976, Boston, Mass. *Bull. Am. Meteorol. Soc.* **58**, 45–51.

Fujita, T. T., and Caracena, F. (1977). An analysis of three weather-related aircraft accidents. *Bull. Am. Meteorol. Soc.* **58**, 1164–1181.

Furman, R. W. (1978). Wildfire zones on a mountain ridge. *Ann. Assoc. Am. Geogr.* **68**, 89–94.

Gade, D. W. (1978). Windbreaks in the lower Rhone valley. *Geogr. Rev.* **68**, 127–144.

Gerasimov, I. P., and Zvonkova, T. V. (1974). Natural hazards in the territory of the USSR: Study, control, and warning. *In* "Natural Hazards" (G. F. White, ed.), pp. 243–251. Oxford Univ. Press, London and New York.

Geyl, W. F. (1968). Tidal stream action and sea level change as one cause of valley meanders and underfit streams. *Austral. Geogr. Stud.* **6**, 24–42.

Golding, E. W., and Thacker, M. S. (1956). The utilization of wind, solar radiation and other local energy sources for the development of a community in an arid or semi-arid area. *In* Wind and Solar Energy. *Arid Zone Res.* **7**, 119–126.

Gosselink, J. G., and Turner, R. E. (1978). The role of hydrology in freshwater wetland ecosystems. *In* "Freshwater Wetlands" (R. E. Good, D. F. Whigham, and R. L. Simpson, eds.), pp. 63–78. Academic Press, New York.

Grace, J. (1977). "Plant Response to Wind," Experimental Botany, No. 13. Academic Press, New York.

Graff, J. V., and Strub, J. H. (1975). The great Upper Plains blizzard of January 1975. *Weatherwise* **28**, 66–69, 83, 103.

Gresswell, S., Heller, D., Swanston, D. N. (1979). Mass movement response to forest management in the central Oregon Coast Ranges. *U.S. For. Serv. Resour. Bull.* **PNW-84**, 19–20.

Gustavson, M. R. (1979). Limits to wind power utilization. *Science* 204, 13–17.

Hewitt, K., and Burton, I. (1971). The hazardousness of a place. *Univ. Toronto, Geogr. Res. Publ.* 6.

Hill, R. D. (1977). Energy dissipation in lightning. *J. Geophys. Res.* 82, 4967–4968.

Hill, R. D. (1979). A survey of lightning energy estimates. *Rev. Geophys. Space Phys.* 17, 155–164.

Holbo, H. R., Corbett, T. C., and Horton, P. J. (1980). Aeromechanical behavior of selected Douglas-fir. *For. Res. Lab. Oregon State Univ., Pap.* No. 1259, Corvallis, Ore.

Inman, D. L., and Brush, B. M. (1973). The coastal challenge. *Science* 181, 20–32.

Inoue, E., Mihara, Y., and Tsuboi, Y. (1965). Agrometeorological studies on rice growth in Japan. *Agric. Meteorol.* 2, 85–107.

Ives, J. D., Mears, A. I., Carrara, P. E., and Bovis, M. J. (1976). Natural hazards in mountain Colorado. *Ann. Assoc. Am. Geogr.* 66(1), 129–144.

Judson, A. (1976). Colorado's avalanche warning program. *Weatherwise* 29, 268–277.

Justus, C. G., Hargraves, W. R., and Yalcin, A. (1976). Nationwide assessment of potential output from wind-powered generators. *J. Appl. Meteorol.* 15, 673–678.

Kelsey, H. M. (1980). A sediment budget and an analysis of geomorphic process in the Van Duzen River basin, north coastal California 1941–1975: Summary. *Geol. Soc. Am. Bull., Pt. I,* 91, 190–195.

Kesseli, J. E. (1943). Disintegrating soil slips of the Coast Ranges of central California. *J. Geol.* 51, 342–352.

Kirkby, M. J. (1967). Measurement and theory of soil creep. *J. Geol.* 75, 359–378.

Kung, E. C. (1966). Large-scale balance of kinetic energy in the atmosphere. *Mon. Weather Rev.* 94, 627–640.

Laikhtman, D. L. (1964). "Physics of the Boundary Layer of the Atmosphere." Isr. Program Sci. Transl., Jerusalem. (Transl. by I. Shectman from "Fizika Pogranichnogo Sloia Atmosfery." Gidrometeord. Izd., Leningrad, 1961.)

Lang, T. E., Dawson, K. L., and Martinelli, M., Jr. (1979). Numerical simulation of snow avalanche flow. *U.S. For. Serv., Res. Pap.* RM-205.

Leaf, C. F., and Martinelli, M., Jr. (1977). Avalanche dynamics: Applications for land use planning. *U.S. For. Serv., Res. Pap.* RM-183.

Lehane, R. (1980). Cyclones—assessing the risk. *ECOS (Melbourne)* 23, 3–7.

Lemon, E. (1967). Aerodynamic studies of CO_2 exchange between the atmosphere and the plant. *In* "Harvesting the Sun" (A. San Pietro, F. A. Greer, and T. J. Army, eds.), pp. 263–290. Academic Press, New York.

Lettau, H. H. (1954). A study of the mass, momentum, and energy budget of the atmosphere. *Arch. Meteorol., Geophys. Bioklimatol., Ser. A* 7, 133–157.

Lettau, H. H., and Dabberdt, W. F. (1970). Variangular wind spirals. *Boundary-Layer Meteorol.* 1, 64–79.

Liu, M.-K., and Yocke, M. A. (1980). Siting of wind turbine generators in complex terrain. *J. Energy* 4(1), 10–16.

Long, J. N., Schreiner, E. G., and Mannwal, N. J. (1979). The role of actively moving sand dunes in the maintenance of an azonal, juniper-dominated community. *Northwest Sci.* 53, 170–179.

Lynott, R. E., and Cramer, O. P. (1966). Detailed analysis of the 1962 Columbus Day windstorm in Oregon and Washington. *Mon. Weather Rev.* 94, 105–117.

Martinelli, M., Jr. (1966). Avalanche technology and research—recent accomplishments and future prospects. *Weatherwise* 19, 233–239, 270–271.

Matson, R. J., and Huggins, A. W. (1980). The direct measurement of the sizes, shapes and kinematics of falling hailstones. *J. Atmos. Sci.* 37, 1107–125.

Mellor, M. (1978). Dynamics of snow avalanches. *In* "Rockslides and Avalanches. I: Natural

Phenomena" (B. Voight, ed.), Dev. Geotech. Eng. Vol. 14A, pp. 753–792. Am. Elsevier, New York.

Metz, W. D. (1977). Wind energy: large and small systems competing. Science 197, 971–973.

Miller, D. H. (1967). Sources of energy for thermodynamically-caused transport of intercepted snow from forest crowns. In "Forest Hydrology" (W. E. Sopper and H. W. Lull, eds.) pp. 201–211. Pergamon, Oxford.

Miller, D. H. (1977). "Water at the Surface of the Earth." Academic Press, New York.

Minore, D. (1972). A classification of forest environments in the South Umpqua Basin. U.S. For. Serv., Res. Pap. PNW-129.

Mogil, H. M., Rush, M., and Kutka, M. (1977). Lightning—a preliminary assessment. Weatherwise 30, 192–200.

Morgan, G. M., Jr., and Towery, N. G. (1976). On the role of strong winds in damage to crops by hail and its estimation with a simple instrument. J. Appl. Meteorol. 15, 891–898.

Nakano, T. (1974). Natural hazards: report from Japan. In "Natural Hazards" (G. F. White, ed.), pp. 231–243. Oxford Univ. Press, London and New York.

Newson, M. (1980). The geomorphological effectiveness of floods—A contribution stimulated by two recent events in Mid-Wales. Earth Surf. Processes 5, 1–16.

Palmén, E., and Newton, C. W. (1969). "Atmospheric Circulation Systems." Academic Press, New York.

Perla, R. I., and Martinelli, M., Jr. (1976). Avalanche handbook. U.S. Dep. Agric., Agric. Handb. No. 489.

Rice, R. M., and Foggin, G. T., III (1971). Effect of high intensity storms on soil slippage on mountainous watersheds in southern California. Water Resour. Res. 7, 1485–1496.

Reitz, G. (1978). Windschur oder Salzschur? Erdkunde 32, 1–10.

Riehl, H., and Malkus, J. (1961). Some aspects of hurricane Daisy, 1958. Tellus 13, 181–213.

Rosenberg, N. J. (1975). Windbreak and shelter effects. In "Progress in Biometeorology," Division C (L. P. Smith, ed.), pp. 108–134. Swets & Zeitlinger, Amsterdam.

Schönenberger, W. (1978). Ökologie der natürlichen Verjüngung von Fichte und Bergföhre in Lawinenzügen der nördlichen Voralpen. Mitt. Schweiz. Anst. Forstl. Versuchswes. 54, 217–362.

Schumm, S. A. (1977). "The Fluvial System." Wiley (Interscience), New York.

Smith, G. I., ed. (1978). Climate variation and its effects on our land and water. Part B. Current research by the Geological Survey. U.S. Geol. Surv. Circ. 776-B.

Sørensen, B. (1975). Energy and resources. Science 189, 255–260.

Sørensen, B. (1976). Dependability of wind energy generators with short-term energy storage. Science 194, 935–937.

Stannard, M. E. (1958). Erosion survey of the south-west Cobar peneplain. Part III—Erosion. J. Soil Conserv. Serv. N.S.W. 14, 137–156.

Tabler, R. D. (1975). Estimating the transport and evaporation of blowing snow. In Snow Management on Great Plains. Proc. Great Plains Agric. Counc. Publ. No. 73, 85–104.

Takahashi, K. (1955). On the relationship between typhoon energy and damage amount. Proc. UNESCO Symp. Typhoons, Tokyo, 1954 pp. 23–30.

Tanner, W. F. (1971). Net kinetic energy in littoral transport. Science 172, 1231–1232.

Taylor, A. R. (1971). Agent of change in forest ecosystems. J. For. 68, 477–480 and cover illustration.

Thom, A. S. (1975). Momentum, mass and heat exchange of plant communities. In "Vegetation and the Atmosphere" (J. L. Monteith, ed.), Vol. 1, pp. 57–109. Academic Press, New York.

Thorn, C. E. (1978). The geomorphic role of snow. Ann. Assoc. Am. Geogr. 68, 414–425.

Towery, N. G., and Morgan, G. M., Jr. (1977). Hailstripes. Bull. Am. Meteorol. Soc. 58, 588–591.

Towery, N. G., Morgan, G. M., Jr., and Changnon, S. A., Jr. (1976). Examples of the wind factor in crop-hail damage. *J. Appl. Meteorol.* **15**, 1116–1120.

Trewari, S. K. (1978). Economics of wind energy use for irrigation in India. *Science* **202**, 481–486.

Turner, H. (1960). Reliefbedingte Mikroklimate und ihr Einfluss auf die Vegetationsver teilung im Hochgebirge. *Meteorologie* Nos. 10/11, 43.

Turner, H. (1971). Mikroklimatographie und ihre Anwendung in der Ökologie der subalpinen Stufe. *Ann. Meteorol.* **5**, 275–281.

Turner, H., Rochat, P., and Streule, A. (1975). Thermische Charakteristik von Hauptstandorten im Bereich der obern Waldgrenze (Stillberg, Dischmatal bei Davos). *Mitt. Schweiz. Anst. Forstl. Versuchswes.* **51**, 95–119.

Uman, M. A., *et al.* (1978). An unusual flash at Kennedy Space Center. *Science* **201**, 9–16, and cover illustration.

U.S. Council on Environmental Quality (1978). "Environmental Quality", Ann. Rep., 9th.

U.S. House of Representatives (1980). Compact cities: Energy saving strategies for the Eighties. Rep. Subcomm. on the City, H. S. Reuss, Chmn. Comm. on Banking, Finance and Urban Affairs, 96th Congress, 2d Session. U.S. Govt. Printing Office, Washington, D.C.

U.S. Water Resources Council (1978). "The Nation's Water Resources. The Second National Water Assessment. Part III. Functional Water Uses." Washington, D.C.

U.S. Water Resources Council. (1979). "A Unified National Plan for Flood Plain Management." Water Resour. Counc. Washington, D.C.

Urfer-Henneberger, C. (1979). Temperaturverteilung im Dischmatal bei Davos mit Berücksichtigung typischer Witterungslagen. *Mitt. Schweiz. Anst. Forstl. Versuchswes.* **55**(4), 299–412.

Vorontsov, V. A. (1959). "Certain Problems in Aerological Investigations of the Boundary Layer of the Atmosphere." U.S. Weather Bur., Washington, D.C. (Transl. by N. A. Stephanova and G. S. Mitchell of Nekotorye voprosy aerologicheskikh issledovanii pogranichnogo sloia atmosfery. *In* "Sovremennye Problemy Meteorologii Prizemnogo Sloia Vozdukha" (M. I. Budyko, ed.). Gidrometeorol. Izd., Leningrad, 1958.)

Waldvogel, A., Schmid, W., and Federer, B. (1978). The kinetic energy of hailfalls. Part I: Hailstone spectra. *J. Appl. Meteorol.* **17**, 515–520.

Webb, W. P. (1931). "The Great Plains." Ginn, Boston, Massachusetts.

Weischet, W. (1955). Die Geländeklimate der Niederrheinischen Bucht und ihrer Rahmenlandschaften. Eine geographische Analyse subregionaler Klimadifferenzierungen. *Muench. Geogr. Hefte* No. 8.

Williams, G. O., and Guy, H. P. (1973). Erosional and depositional aspects of hurricane Camille in Virginia, 1969. *U.S. Geol. Surv., Prof. Pap.* No. 804.

Williams, M. A. J. (1973). The efficacy of creep and slopewash in tropical and temperate Australia. *Austral. Geogr. Stud.* **11**, 62–78.

Yang, C. T. (1971). Formation of riffles and pools. *Water Resour. Res.* **7**, 1567–1574.

Yoshino, M. M. (1964). Some local characteristics of the winds as revealed by wind-shaped trees in the Rhone Valley in Switzerland. *Erdkunde* **18**, 28–39.

Yoshino, M. M. (1975). "Climate in a Small Area. An Introduction to Local Meteorology." Univ. of Tokyo Press, Tokyo.

Yoshino, M. M., *et al.* (1976). Bora regions as revealed by wind-shaped trees on the Adriatic coast. *In* "Local Wind Bora" (M. M. Yoshino, ed.), pp. 59–71. Univ. of Tokyo Press, Tokyo.

Young, R. A., and Wiersma, J. L. (1973). The role of rainfall impact in soil detachment and transport. *Water Resour. Res.* **9**, 1629–1636.

Chapter XX

ENERGY BUDGETS AT DIFFERENT DEPTHS IN ECOSYSTEMS

Stratification is characteristic of ecosystems because solar and long-wave radiation and atmospheric ventilation all tend to differentiate upper from lower layers. As a result, energy budgets vary with depth.

When driving past Wisconsin maize ecosystems in August, one sees a brown mantle over the fields—the dry tassels of a layer 0.2 m thick that is so dense as to be opaque in slant vision. A view from a lower angle shows other layers: the dark green leaves in a zone about a meter thick; below that a largely leafless layer of numerous stalks like the trunk space of a young forest; and at the bottom the soil, which one knows to be full of those essential members of the ecosystem, the roots. Indeed, the root–shoot antithesis is so basic an element of ecosystem structure as to require analysis in an assessment of possible impacts of environmental change (Erickson, 1979, p. 168). The downward fluxes of radiant energy and atmospheric ventilation that reach these four strata result in distinctive energy budgets.

An analog is the stratification of sites of water storage (Miller, 1977, Chap. I), which are connected by fluxes that move water predominantly in a vertical direction. Integration of these storages and fluxes depicts the pattern of water in an ecosystem from delivery to eventual runoff, percolation, or evapotranspiration, and a similar integration of energy storages and fluxes at different depths depicts conversions of energy. The energy budgets in each layer combine the radiation fluxes of Chapters III–VII, the carbon conversions of Chapters XI and XII, and the nonradiative fluxes of Chapters XV–XVII that are in part controlled by such gate functions as that of kinetic energy (Chapter XIX).

396

THE ECOSYSTEM AS ENVIRONMENT OF ITS MEMBERS

To a large degree ecosystems become stratified because their environment is one-sided: The forcing fluxes of radiant-energy input and most of that of CO_2 come from above and are depleted as they progress downward. The energy-converting apparatus changes accordingly; in particular, leaf-area density (square meters of leaf area per cubic meter of space) varies with depth, as was shown in Chapter II. Geiger (1927) long ago pointed out that species with horizontal leaves develop different interior microclimates than those with vertical leaves, and Dirmhirn (1964, p. 232) showed further canopy differences. In some species leaves are distributed uniformly over a thick layer, and in others the canopy is thin,* and energy conversions are concentrated in a profile that changes with time; for example, the maximum leaf area density in a maize field varied as shown in Table I. The level of maximum leaf deployment remained about the same distance (0.3–0.4 m) below the top of the ecosystem as the canopy thickened, but its density increased substantially, which reduced penetration of radiation and CO_2. Instead of leaf area, canopy mass may be used, e.g., in pines, to characterize the vertical dimension of an ecosystem and explain the depth of maximum temperature (Gary, 1974; Bergen, 1974).

Ecosystem strata are not always continuous. Many Australian pastoral ecosystems are comprised of shrubs and grass, in which the palatable species "may depend for their existence on micro-site conditions produced or maintained by the more perennial, often woody, and perhaps

TABLE I[a]

Date (1963)	Maximum leaf area density (m^2 m^{-3})	Maximum height (m)	Crop height (m)
12 July	3.2	0.20	0.50
25 July	4.5	0.37	0.82
3 August	4.6	0.75	1.15
13 August	8.0	1.15	1.60
20 August	11.5	1.40	1.80

[a] Data from Ross and Nil'son (1966, Fig. 1).

* An extreme case is a desert winter annual in which "the single layer of prostrate leaves" forms the canopy and yet fixes solar energy at a very high rate (Ehleringer et al., 1979). The plant builds its "canopy structure with less construction costs" yet greater efficiency than shrubs with a much larger leaf area index.

unpalatable species" (Costin and Mosley, 1969). The shrubs are especially important during strong wind episodes that would remove organic
matter and nutrients from the ecosystem. Their sheltering effect depends on the relation of height to spacing; equal erosion protection is
provided by 0.6 m saltbush (*Atriplex vesicaria*) 2.0 m apart or 1.2 m blue-
bush (*Kochia sedifolia*) 3.5 m apart (Marshall, 1970). Tall, thin shrubs are
the most effective roughness elements. In high latitudes spruce trees
spaced rather widely above a lower stratum of lichens intercept most
of the low-angle direct beam (Fuller and Rouse, 1979).

Forest ecosystems typically have an understory, which may follow a
different productivity regime than the upper story and have less biomass
than the trees with their large stems. The above-ground biomass of an
aspen forest in northern Wisconsin was 10^4 g m^{-2} in the trees and 118
g m^{-2} in ground vegetation (including 37 g m^{-2} in forbs, 52 in ferns,
20 in rushes, and tiny amounts in grass and seedlings of woody species)
(Zavitkovski, 1976). An inverse relation between the layers is suggested
by the fact that understory biomass is large where solar radiation reaches
it ($r \approx 0.6$); in other stands of Wisconsin, understory productivity also
varies with canopy openness (Crow, 1978). The same situation is shown
when riparian tree canopy over trout streams was removed (Hunt, 1979,
p. 13) and the net surplus of midday all-wave radiation increased from
100 to 415 W m^{-2}; most of the increase was solar radiation, and it stimulated a substantial increase in the macrophytes of the aquatic ecosystem, without excessively warming the cold groundwater outflow in the
streams, as occurred in the Appalachian streams described in Chapter
XVIII. Strata in forest are not necessarily trees, shrubs, and grass, or
trees of different ages; in some stands certain species are in the second
or third layer down even though the same age as the dominants that
are shading them (Oliver, 1980). Their chance comes later.

RADIANT-ENERGY ABSORPTION AT DIFFERENT LEVELS

The absorption of solar and longwave radiation by leaves and skeletal
material at different depths in an ecosystem depends on the penetration
of radiation and the density and absorptivity of surfaces at each depth.
The profile often optimizes productivity summed over all depths, as in
a model of a reconstructed rainforest (Wiseman, 1978).

Solar Energy Input and Absorption

Some sunshine reaches the ground in even the densest spruce forest,
but at a flux density two orders less than in the incoming beam. The

fraction penetrating to the ground is important for photosynthetic activity in ground flora and for melting snow cover (Miller, 1959), but sparsity of data on canopy architecture beyond the "canopy closure" traditional in silviculture forced use of such parameters as stand age, species, stem density (sum of diameters of all tree stems in a hectare of stand), and the like, that have a sketchy physical rationale but are being replaced by more pertinent measurements (Ross, 1975a, p. 13, 1975b).

The downward attenuation of solar radiation in an ecosystem canopy more or less conforms to Beer's law:

$$I_z/I_0 = e^{-vz},$$

where I is intensity at the top I_0 and at any given depth I_z and v is an extinction coefficient combining absorption and scattering and is approximately 0.8 m^{-1} in corn (Allen et al., 1964), 0.4 in rice (Chang, 1968, p. 39), 0.1 in pines in general (Miller, 1959), and 0.15 in a 30-m red pine stand near Yale University (Reifsnyder, 1967). Leaf area density gives a more precise result (McCaughey and Davies, 1974) and modeling leaf shapes and orientation (geometric models) or their stratified distribution is still better (Lemeur and Blad, 1974); Anderson (1975) points out that different coefficients should be used for direct and diffuse radiation.

The direct beam makes the familiar sun flecks on the forest floor, which tend to be nearly circular with shadowy edges or penumbrae due to refraction, and are difficult to measure (Saeki, 1975). Flecks under lodgepole pine in the Sierra Nevada brought 44 W m^{-2} in the daily means, four times the contribution of radiation scattered by the needles (Muller, 1971, p. 9). As flecks move across the forest floor, they bring, for a few minutes, a burst of energy to ground plants or snow. The physiological response to this energy or leaf heating has, however, "not been investigated" (Bazzaz, 1979). At the human scale "the play of light and shadow across the campgrounds largely determines their attractiveness" (Reifsnyder and Lull, 1965, p. 94).

The spikes of sun-fleck energy are imposed on a base flow of two components: (1) solar radiation scattered in the atmosphere or clouds and delivered in diffuse form to the top surface of the ecosystem, and (2) that scattered out of the solar beam by reflection from leaves and in deciduous forest, by transmission through them. (Differences between deciduous and conifer stands in Wisconsin were significant (Federer and Tanner, 1966) for this reason.) This base flow is usually low in flux density (18 W m^{-2} in Sierra pines (Muller, 1971, p. 9), but is continuous in the lower part of any ecosystem canopy and highly important in photosynthesis. Its flux density in rice is about 7 W m^{-2} for each unit leaf-area depth (Uchijima, 1976). A geometric analysis of row-crop ecosystems adds the direct beam reaching the soil between rows to that transmitted through the rows themselves, a function of leaf density and orientation, plant size and spacing, and row size and azimuth. The model calculates these fluxes as sun height and azimuth change through the day (Mann *et al.*, 1980).

Scattering and reflection are selective as to wavelength, as is shown by an extinction coefficient of ~ 1 m^{-1} for visible radiation and 0.6 for solar infrared in one crop ecosystem (Lemon, 1963). (1) The atmosphere scatters in inverse proportion to the fourth power of the wavelength by Rayleigh's law, and some aerosols are also selective. (2) Leaves reflect strongly in the green and solar infrared wavelengths. Scattering and transmission are commonly modeled (e.g., Torssell and McPherson, 1977).

Diffuse radiation entering an ecosystem is strong in violet and blue wavelengths, but as these are strongly absorbed by leaves, its spectral composition changes with depth. It is least attenuated in solar infrared, as is apparent in infrared photographs taken in the trunk space. The absorption coefficient for solar infrared is 0.18 and for photosynthetically active radiation 0.55, at a depth in maize equivalent to 1 m^2 leaf area; it is 0.43 and 0.92, respectively, at a depth (Tooming, 1966) equivalent to 5 m^2 leaf area below the top. The infrared has no photosynthetic value, but warms the soil, melts snow cover, and dries litter and soil. Multiple reflection tends to make the shortwave radiation field less directional than it was on entry.

Change in leaf area through a growing season affects penetration and is visibly and geometrically expressed in increased shading of the soil surface (Lopukhin, 1963). Beam radiation at the floor of a forest in Tennessee dropped abruptly when leaves first came out, and diffuse radiation responded somewhat more slowly (Hutchison and Matt, 1977). Less marked changes were measured at the midcanopy level. The profile shape is determined by measuring at different depths, often by traversing sensors that average out the spatial sun and shade pattern, as in maize (Kyle *et al.*, 1977) and forest (Hutchison, 1979b).

Longwave Radiation

The downward flux of longwave radiation comes from an area source and readily penetrates ecosystem gaps. Strongly absorbed by leaves, it is also augmented by radiation the leaves emit in about the same spectral range and also in the 8–12 μm window. The downward component of the leaf flux augments the downward component of radiation from the hemisphere, and while the two fluxes are indistinguishable, they can be calculated at successive depths (Ross, 1975a, p. 307). Most of the flux emerging from the bottom of a dense canopy comes from the canopy itself, 0.80–0.85 in a Canadian pine stand (Mukammal, 1971). Autumn daytime flux densities measured by Reifsnyder (1967) in 30-m pines in Connecticut show 290 W m^{-2} entering the stand and 370 W m^{-2} at the floor; augmentation at night was smaller, averaging 50 W m^{-2}.

Radiation emitted by the ecosystem floor has an upward component about the same by day as the downward flux and 5–10 W m^{-2} greater by night (Reifsnyder, 1967, p. 158). This upward flux is partly absorbed and partly augmented at successively higher levels of the canopy. Each level emits at its own temperature, and since leaf temperature in midday increases with nearness to the top, the upward component of the long-wave flux increases until it bursts from the outer surface at a flux density integrating the surface temperature of all the leaves that look at the sky. This upward augmentation in a linden stand in the daytime was about 50 W m^{-2} (Rauner, 1976).

Longwave radiation exchanges tend to equalize the temperatures of different layers in an ecosystem, countering the differentiating effect of solar energy. Its very large flux density, 300–400 W m^{-2} either upward or downward, smooths out spatial differences in leaf temperature and renders more uniform the generation of sensible and latent heat.

All-Wave Radiation

The changes in different spectral regimes of radiation entering a forest ecosystem are quite marked (Table II). Little photosynthetically active radiation emerges from the bottom of the canopy, and longwave radiation makes up 0.94 of the energy flux received at the forest floor. Leaves are discriminating and effective absorbers of radiant energy.

TABLE II

Daytime Radiation Fluxes in Pine Forest near Yale University[a]

Downward flux	Flux density (W m^{-2})	Fractional distribution		
		Visible	Solar infrared	Longwave
At top	800	0.30	0.34	0.36
At forest floor	390	0.01	0.05	0.94

[a] Data from Reifsnyder (1967, p. 158). Visible portion estimated. Mean of 10 hr on 1 October 1964.

VENTILATION

Turbulent Exchanges

Turbulence also tends to smooth out differences in temperature, but its vigor decreases with depth and leaf area. Local temperature differences, e.g., between sun and shade leaves, generate thermal convection, which is present even on fairly calm days and supplies CO_2 and removes water vapor. Thermally induced flows, too slow to become turbulent, also occur, for example, threads of cold air descending from radiatively chilled leaves at night.

Mechanical turbulence is generated from atmospheric kinetic energy* as a function of ecosystem roughness and penetrates the canopy, in a sense decreasing its effective depth [as it also does in the pelage of deer (Moen, 1973, p. 261)], and in windy sites ecosystems often grow close to the ground (Grace, 1977, p. 171). The mechanics of penetration by wind are not well understood (Hutchison, 1979b, p. 15), partly because of sparsity of data on canopy porosity, on elasticity of leaf petioles and small branches, which affects their ability to absorb turbulent energy (Fig. 1), and on skin friction and form drag.

The extinction coefficient for turbulence in maize is greater than in rice because maize leaves tend to have a more horizontal habit (Inoue *et al.*, 1975). Penetrations of turbulence and solar radiation into ecosystems are analogous (Budyko, 1974, p. 408) because both are absorbed by leaves. Coefficients indexing the downward transport of momentum and CO_2 are proportional to those of upward diffusion of vapor and sensible heat and are related to the exponential decay of wind speed with depth (Inoue and Uchijima, 1979, p. 46).

VERTICAL PROFILES

Temperature

The daytime profile of temperature through an ecosystem usually has a maximum at a shallow depth. The top itself is too well ventilated to be overheated, but the layer just below it receives strong sunlight and is somewhat sheltered from the wind, and from it sensible heat moves upward into the free air and downward into the ecosystem. This layer is located at a depth equivalent to the accumulation of 1 m^2 m^{-2} leaf-area index in deciduous forest and at a depth of 3–4 m^2 m^{-2} leaf-area

---

* Turbulent flows also arise underneath forest canopy, but the relations among above-canopy, in-canopy, and below-canopy winds and turbulence are still unclear.

Fig. XX.1.   A forest ecosystem that has been established west and north of a farmstead on the prairie of southwestern Minnesota shows a deep zone of energy conversion.

index in pine which has drier leaves and more nontranspiring skeletal matter (Rauner, 1972, p. 103), hence a deeper layer generating sensible heat. This zone of warmth is fostered in Arctic and alpine plants by a low growth habit, especially in windy locations (Grace, 1977, pp. 172–174), as well as in winter annuals in the desert (Ehleringer *et al.*, 1979).

The trunk space below the canopy, with few absorbing surfaces, is generally the coldest layer. The soil surface, absorbing solar energy (insofar as it penetrates to the ground) and longwave radiation, is slightly warmer than the air above but not enough to generate much thermal convection. The trunk space displays a stable density stratification with large, slow eddies (Zolotokrylin and Konyayev, 1978).

The vertical profile of temperature in the plant–air space is, like that in the surface boundary layer, "the integrated result of turbulent mixing and radiative transfer" (Munn, 1966, p. 89), and radiative means can be used in determining it (van Meurs, 1976). Fluxes of heat in biomass and the soil are small in comparison with the transformations of energy in the upper layers of warmth and light, a fact pointed out by Geiger (1927, Fig. 53) after many days of observing the top of a forest stand from a tower. By recognizing the true active levels of an ecosystem he brought a new approach to the study of vegetation in its environment.

Water Vapor

The daytime source of atmospheric vapor lies in the evaporating soil and transpiring foliage of ecosystems. The upper canopy has a large solar input, perhaps too large for efficient photosynthetic conversion, and generates excess sensible heat, which, carried downward, helps support transpiration associated with photosynthesis in deeper layers. The photosynthesizing zone is deepened.* The soil under a dense canopy generates a base flow of vapor that, ascending through the ecosystem, joins the usually larger flows generated in the canopy.

CO$_2$ Concentration

The shape of the profile shows the direction in which CO_2 is being transported and identifies the sources and sinks. It indicates, for example, that only at midday is the flux downward throughout a rice canopy (Uchijima, 1976). The rest of the time both atmosphere and soil are sources. Decomposition in the soil generates CO_2 that diffuses upward into the canopy. A deep ecosystem traps a considerable quantity during the night and holds it for daytime uptake by the leaves (Hellmers, 1964).

The sinks of CO_2 are the layers of active daytime photosynthesis, and their relative strengths, i.e., the rates of photosynthetic energy conversion, are shown by flux divergence in the turbulent transport of CO_2, which is a function of profile steepness and turbulence. Because these layers are also sites of maximum extraction of photons from the flow of solar energy, the two approaches can be joined to determine where biological productivity occurs under different external conditions. Intercropping gives a complex ecosystem high capacity to convert radiant energy since a fast-growing crop "temporarily uses the empty space between larger slow-growing plants" and the ecosystem as a whole makes maximum "use of the available solar energy" (Innis, 1980). Moisture conditions affect the closure and productivity of different layers in the Santa Catalina Mountains in Arizona; tree crowns are the level of greatest production, but in sites of greater dryness shrub and grass strata take over (Lieth, 1975).

In many environments tree stands with their thick, absorbent canopy

---

* A reversed order occurs in an ecosystem of dry grass and scattered 8-m trees that have access to water and are transpiring. In this situation sensible heat generated by the grass moves into the tree foliage (Garratt, 1978).

and access to deep moisture and nutrient storage outperform thinner ecosystems. This ability, combined with soil-conserving characteristics that stimulated the geographer J. Russell Smith (1929, 1950) to promote tree crops, suggest the importance of what is now called agroforestry for various national purposes: food production in China (Anonymous, 1980), biomass for industrial energy in Sweden, halting desertification in the low latitudes (Anonymous, 1979). Because these deep ecosystems have been harder to study than grass or grainfields, research is needed on both radiation exchanges and ventilation (Hutchison, 1979a; Reifsnyder, 1979) as well as CO_2 assimilation. Both radiation and ventilation affect cambium temperature, which in low-energy sites becomes "a directly limiting factor through its control of respiration and assimilation processes" (Fritts, 1976, p. 227).

SUBSTRATE ENERGY

Beneath the leaf canopy both radiative and convective fluxes are small, and in the soil body conduction takes over to link above-ground energy transactions with the heat-storage capacity of the substrate. This zone is, except in aquatic ecosystems, rather quiescent in comparison with the canopy zone. Substrate-heat fluxes in well-developed ecosystems are small, the daytime heat intake being of the order of 0.5 MJ m^{-2}. The substrate is more a storehouse of energy (and of water and nutrients) than an arena of intensive and fluctuating energy conversions. Decomposition on and in the soil goes on at about the same overall rate as net photosynthesis, but more steadily. It is often depth-dependent (Chapter XII), and especially so in wetlands; in a coastal marsh little decomposition occurred below 0.2 m (Hackney and de la Cruz, 1980). Faunal activity may concentrate decomposition at certain sites like mallee fowl mounds (Chapter XVII) or ant hills (Coenen-Stass et al., 1980) to which organic matter is brought. In the ant hill studied, animal metabolism was more intensive* at 2.6 W per kilogram of mass of ants, than the organic decomposition at 0.25 W per kilogram, but the large quantity of decomposing matter contributed most of the excess heat generation (20–25 W m^{-2}).

STRATIFICATION OF ENERGY BUDGETS

Studies of the vertical differentiation of energy budgets in grassland and crop ecosystems indicate that primary productivity can only be

---

* Disturbed ants at the surface cause a sharp rise in remotely measured surface temperature.

determined from the actual fluxes of light and CO_2 at photosynthesizing surfaces in each microenvironment, otherwise the shifting factors that govern photosynthesis cannot be satisfactorily modeled.

The interplay of radiation and ventilation in an ecosystem can be analyzed by casting the energy budget at each level. For example, in an Australian pine ecosystem the unit-volume conversion of all-wave radiation was 170 W m^{-3} in a layer 1.5 m below the top, and ventilation removed this as sensible heat (source strength -80 W m^{-3}) and latent heat (-90 W m^{-3}). In a deeper layer (-2.5 m) radiation brought 115 W m^{-3}, a third less than in the higher layer, and most of it (70 W m^{-3}) was removed by latent-heat conversion (Denmead, 1964). The significant factor is the quantity of biomass above each level.

Budgets at Different Levels in an Ecosystem

Joint analysis of the profiles and vertical fluxes of CO_2 and photosynthetically active radiation down through an ecosystem shows flux divergence and hence the numbers of CO_2 molecules and photons extracted in each layer. Photosynthetic energy conversion being associated via stomatal resistance with the conversion into latent heat, the source strength of latent heat in each layer can now be determined, and it determines the shape of the profiles of vapor flux and vapor concentration.

The source strength for vaporization has to be supported by inputs of other forms of energy, and casting the energy balance of each layer shows which inputs are important in a specific case. These might be the absorption of solar radiation not photosynthetically converted, absorption of longwave radiation, or turbulent transport of sensible heat.

These inputs into a given layer originate in other layers, and the energy budgets of these layers must also balance. Vertical transport among layers keeps all these budgets in balance and takes place largely by exchange of longwave radiation and turbulent transport of sensible heat, which are associated with the vertical profiles of leaf and air temperature. The flux of sensible heat is carried by the same turbulent eddies that carry CO_2, heat, and vapor upward or downward to or from any individual layer, and it integrates the whole system. Transports between layers are expressed in flux-gradient relationships, and the energy budget of each layer balances its incoming and outgoing fluxes.

The number of layers to be individually analyzed depends on the detail desired because they need not be differentiated by visible structure, which indeed might not really occur in some forests (Oldeman, 1978), but can simply be identified geometrically by depth. The analysis of a thin but dense tundra ecosystem, for example, employed 7 levels

(Miller *et al.*, 1976) and of a pine stand 9; 19 levels added little precision (Waggoner *et al.*, 1969).

Ecosystem Models

The layers of an ecosystem are interlocked by the interchange of energy from layer to layer. Balancing each layer's energy budget demands certain vertical flux densities, e.g., of sensible heat or vapor, which may or may not initially be specified at the soil surface (Waggoner, 1975). A model of a stratified canopy permits a "layer-by-layer calculation of the energy balance and thus of the vertical distribution of evaporation and sensible heat exchange" (Waggoner and Reifsnyder, 1968). Set up for red clover, the model was confirmed in barley, and then in pine forest (Waggoner *et al.*, 1969), and modified to continuous form by letting the levels "become infinitely thin" (Furnival *et al.*, 1975). Energy budgets at several levels in rice were used to determine values of diffusivity, which were consistent with other profile measurements (Inoue and Uchijima, 1979, pp. 23, 47).

Such models show how ecosystem profiles of temperature, wind, CO_2 concentration, and so on, change either when ecosystem characteristics (like leaf area) change or when forcing functions (like flux density of solar radiation) change. This approach to ecosystem operations is useful in forest and in many crop ecosystems, including sunflowers (Saugier, 1976), maize, and rice, as well as tundra and grassland (Ripley and Redmann, 1976). As interest grows in the efficiency of crops in fixing solar energy and of forests in growing biomass for fuel, these transfer studies will increase.

Layer models correspond in a general way to models of the water storages and fluxes in an ecosystem (Miller, 1977), but are more easily evaluated because the fluxes can be determined more accurately. Water fluxes in an ecosystem in the forms of snow and liquid water are episodic, difficult to measure, and powered by diverse forms of energy, so that energy analysis principally serves to suggest that a certain flux, e.g., evaporation of intercepted snow (Miller, 1967), is unlikely. Moreover, heat storage in each layer of an ecosystem is indicated by leaf and air temperature, while water storages are hard to measure. For these reasons stratified budgets are most used in energetic and productivity studies of ecosystems.

Variation

The increase in radiant-energy intake each morning activates energy conversions in upper layers of an ecosystem, and as the sun rises higher,

TABLE III

*Upward and Downward Fluxes of Sensible
and Latent Heat at Several Depths in a 5.5-m
Pine Forest near Canberra, Australia, 20 May
1963[a]*

Depth (m)	Sensible heat $(W\ m^{-2})$	Latent heat $(W\ m^{-2})$
0	-70	-280
-1.5	-60	-180
-3.2	$+35$	-45

[a] Data from Denmead (1964). Minus sign
indicates upward flux; plus sign indicates
downward flux.

deeper layers are activated. System ventilation increases as the local
wind speed picks up, turbulent interchanges of energy between layers
increase, and intake of CO_2 from the atmosphere outweighs that from
the soil.

Midday fluxes of sensible and latent heat in a pine plantation illustrate
the interlocking of different levels by upward or downward fluxes of
energy (Table III). Vapor ascends from the soil through the whole ecosystem in a turbulent flux that grows, particularly in the upper layers,
as shown by the source strengths noted earlier. Sensible heat at 3.2 m
depth, on the other hand, is moving downward. The sunlit upper canopy is supplying heat downward as well as into the free air; the downward flux provides much of the thermal support for vaporization in the
lower part of the canopy. This interchange, which exemplifies a common
source–sink distribution (Thom, 1975, p. 77), diffuses heat from the
strongly absorbing upper canopy, increases the net evapotranspiration
from the ecosystem, and diminishes its overall output of sensible heat.
Later in the day warm air that came over the pine stand caused the
sensible-heat flux to be directed downward throughout the depth of the
stand, and extraction of sensible heat from the atmosphere supported
0.4 of the latent heat generation. (This advective process is discussed
further in Chapter XXI.)

The responses of energy conversions and fluxes to an increase in absorbed radiant energy are most obvious in a thin system, such as a
desert, but in most ecosystems absorption and response take place
through an appreciable depth, so that deviations in the rates of energy
transformation are smoothed out. As the vertical profiles of energy flux
densities change, they shape the profiles of surface and air temperature

and concentrations of water vapor and CO_2. The top of the forest remains differentiated from the lower canopy, but the functions that are localized in different layers are integrated into a whole system, as also happens in soil bodies and other objects in physical geography (Neumeister, 1979).

All ecosystems have a vertical dimension, which reaches as much as 100 m and represents an energy cost to the system, but improves absorption of solar radiation, extraction of CO_2 from the air, and access to moisture deep in the soil. Stratification of an ecosystem offers a means of analyzing where light is absorbed in fixing CO_2, how this fixing allows leaves to leak vapor, how energy is provided to each layer for vaporization, and how vapor and excess sensible heat are removed from the layers that comprise the ecosystem.

The radiative and turbulent energy fluxes change with depth, and their size in each layer creates a distinctive microclimate. The sunlit, windy top of the forest is a different world, in energetic terms, than the dark trunk space or the soil body. Life processes and animal habitats are different at each level, and the localization of biological phenomena enhances the overall functioning of the system. The fluxes within the ecosystem bring about an integration that the fluxes that enter and emerge from the top surface epitomize as they couple the whole ecosystem with its environment.

REFERENCES

Allen, L. H., Yocum, C. S., and Lemon, E. R. (1964). Photosynthesis under field conditions. VII. Radiant energy exchanges within a corn crop canopy and implications in water use efficiency. *Agron. J.* **56**, 253–259.

Anderson, M. C. (1975). Solar radiation and carbon dioxide in plant communities—conclusions. *In* "Photosynthesis and Productivity in Different Environments" (J. P. Cooper, ed.), pp. 345–354. Cambridge Univ. Press, London and New York.

Anonymous (1979). Food from forests. *Mazingira (Oxford)* **7**, 43–46.

Anonymous (1980). Restructuring the agricultural economy. *Beijing Rev.* **23**(25), 4–5.

Bazzaz, F. A. (1979). The physiological ecology of plant succession. *Annual Rev. Ecol. Systematics* **10**, 351–371.

Bergen, J. D. (1974). Vertical air temperature profiles in a pine stand: Spatial variation and scaling problems. *For. Sci.* **20**, 64–73.

Budyko, M. I. (1974). "Climate and Life" (transl. ed. by D. H. Miller). Academic Press, New York.

Chang, J.-H. (1968). "Climate and Agriculture." Aldine, Chicago, Illinois.

Coenen-Stass, D., Schaarschmidt, B., and Lamprecht, I. (1980). Temperature distribution and calorimetric determination of heat production in the nest of the wood ant, *Formica polyetena* (Hymenoptera, Formicidae). *Ecology* **61**, 238–244.

Costin, A. B., and Mosley, J. G. (1969). Conservation and recreation in arid Australia. *In*

"Arid Lands of Australia" (R. O. Slatyer and R. A. Perry, eds.), pp. 158–168, Austral. Nat. Univ. Press, Canberra.

Crow, T. R. (1978). Biomass and production in three contiguous forests in northern Wisconsin. *Ecology* **59**, 265–273.

Denmead, O. T. (1964). Evaporation sources and apparent diffusivities in a forest canopy. *J. Appl. Meteorol.* **3**, 383–389.

Dirmhirn, I. (1964). "Das Strahlungsfeld im Lebensraum." Akad. Verlagsges., Frankfurt a.M.

Ehleringer, J., Mooney, H. A., and Berry, J. A. (1979). Photosynthesis and microclimate of *Camissonia claviformis*, a desert winter annual. *Ecology* **60**, 280–286.

Erickson, P. A. (1979). "Environmental Impact Assessment/Principles and Applications." Academic Press, New York.

Federer, C. A., and Tanner, C. B. (1966). Spectral distribution of light in the forest. *Ecology* **47**, 555–560.

Fritts, H. C. (1976). "Tree Rings and Climate." Academic Press, New York.

Fuller, S. P., and Rouse, W. R. (1979). Spectral reflectance changes accompanying a postfire recovery sequence in a subarctic spruce lichen woodland. *Rem. Sens. Envir.* **8**, 11–23.

Furnival, G. M., Waggoner, P. E., and Reifsnyder, W. E. (1975). Computing the energy budget of a leaf canopy with matrix algebra and numerical integration. *Agric. Meteorol.* **14**, 405–416.

Garratt, J. R. (1978). Transfer characteristics for a heterogeneous surface of large dynamic roughness. *Q. J. R. Meteorol. Soc.* **104**, 491–502.

Gary, H. L. (1974). Canopy weight distribution affects windspeed and temperature in a lodgepole pine forest. *For. Sci.* **20**, 369–371.

Geiger, R. (1927). "Das Klima der bodennahen Luftschicht." Vieweg, Braunschweig.

Grace, J. (1977). "Plant Response to Wind." Academic Press, New York.

Hackney, C. T., and de la Cruz, A. (1980). In situ decomposition of roots and rhizomes of two tidal marsh plants. *Ecology* **61**, 226–231.

Hellmers, H. (1964). An evaluation of the photosynthetic efficiency of forests. *Q. Rev. Biol.* **39**, 249–257.

Hunt, R. L. (1979). Removal of woody streambank vegetation to improve trout habitat. *Wisconsin Dept. Nat. Resour., Tech. Bull.* No. 115.

Hutchison, B. A. (1979a). Forest meteorology, research needs for an energy and resource limited future, 28–30 August 1978, Ottawa, Ontario, Canada. *Bull. Am. Meteorol. Soc.* **60**, 331.

Hutchison, B. A., ed. (1979b). Forest meteorology—Research needs for an energy and resource limited future, proceedings of a workshop August 28–30, 1978, Ottawa. *U.S. Dep. Energy, Atmos. Turb. Diffus. Lab., Contrib.* **79/7**.

Hutchison, B. A., and Matt, D. R. (1977). The distribution of solar radiation within a deciduous forest. *Ecol. Monogr.* **47**, 185–207.

Innis, D. Q. (1980). The future of traditional agriculture. *Focus* (New York) **30**(3), 1–8.

Inoue, K., and Uchijima, Z. (1979). Experimental study of microstructure of wind turbulence in rice and maize canopies, *Jpn. Natl. Inst. Agric. Sci., Bull. A* **26**, 1–88.

Inoue, K., Uchijima, Z., Horie, T., and Iwakiri, S. (1975). Studies of energy and gas exchange within crop canopies. (10) Structure of turbulence in rice crop. *J. Agric. Meteorol.* **31**, 71–82.

Kyle, W. J., Davies, J. A., and Nunez, M. (1977). Global radiation within corn. *Boundary-Layer Meteorol.* **12**, 25–35.

Lemeur, R., and Blad, B. L. (1974). A critical review of light models for estimating the shortwave radiation regime of plant canopies. *Agric. Meteorol.* **14**, 255–286.

Lemon, E. (1963). Energy and water balance of plant communities. *In* "Environmental

Control of Plant Growth̄"" (L. T. Evans, ed.), pp. 55–78. Academic Press, New York.

Lieth, H. (1975). Some prospects beyond production measurement. In "Primary Productivity of the Biosphere," (H. Lieth and R. H. Whittaker, eds.), pp. 285–304. Springer-Verlag, Berlin and New York.

Lopukhin, E. A. (1963). Osnovoi sposob rascheta vsekh sostavliaiushchikh radiatsionnogo balansa khlopkovogo polia. Tr. Sredneasiat. Nauchno-Issled. Gidrometeorol. Inst. 11(26), 114–116.

McCaughey, J. H., and Davies, J. A. (1974). Diurnal variation in net radiation depletion within a corn crop. Boundary-Layer Meteorol. 5, 505–511.

Mann, J. E., Curry, G. L., DeMichele, D. W., and Baker, D. N. (1980). Light penetration in a row-crop with random plant spacing. Agron. J. 72, 131–142.

Marshall, J. K. (1970). Assessing the protective role of shrub-dominated rangeland vegetation against soil erosion by wind. Proc. 11th Internat. Grassland Congr., pp. 19–23.

Miller, D. H. (1959). Transmission of insolation through pine forest as it affects the melting of snow. Mitt. Schweiz. Anst. Forstl. Versuchswes. (Festschr. Hans Burger) 35 No. 1, 57–79.

Miller, D. H. (1967). Sources of energy for thermodynamically-caused transport of intercepted snow from forest crowns. In "Forest Hydrology" (W. E. Sopper and H. W. Lull, eds.), pp. 201–211. Pergamon, Oxford.

Miller, D. H. (1977). "Water at the Surface of the Earth." Academic Press, New York.

Miller, P. C., Stoner, W. A., and Tieszen, L. L. (1976). A model of stand photosynthesis for the wet meadow tundra at Barrow, Alaska. Ecology 57, 411–430.

Moen, A. N. (1973). "Wildlife Ecology: An Analytical Approach." Freeman, San Francisco, California.

Mukammal, E. I. (1971). Some aspects of radiant energy in a pine forest. Arch. Meteorol., Geophys. Bioklimatol., Ser. B 19, 29–52.

Muller, R. A. (1971). Transmission components of solar radiation in pine stands in relation to climatic and stand variables. U.S. For. Serv., Res. Pap. PSW-71.

Munn, R. E. (1966). "Descriptive Micrometeorology." Academic Press, New York.

Neumeister, H. (1979). Das "Schichtkonzept" und einfache Algorithmen zur Vertikalverknüpfung von "Schichten" in der physischen Geographie. Petermanns Geog. Mitt. 123, 19–23.

Oldeman, R. A. A. (1978). Architecture and energy exchange of dicotyledonous trees in the forest. In "Tropical Trees as Living Systems" (P. B. Tomlinson and M. H. Zimmerman, eds.), pp. 535–560. Cambridge Univ. Press, London and New York.

Oliver, C. D. (1980). Even-aged development of mixed-species stands. J. For. 78, 201–203.

Rauner, Iu. L. (1972). "Teplovoi Balans Rastitel'nogo Pokrova." Gidrometeoizd., Leningrad.

Rauner, Iu. L. (1976). Deciduous forest. In "Vegetation and the Atmosphere" (J. L. Monteith, ed.), Vol. 2, pp. 241–262. Academic Press, New York.

Reifsnyder, W. E. (1967). Forest meteorology: the forest energy balance. Int. Rev. For. Res. 2, 127–179.

Reifsnyder, W. E. (1979). WMO Symposium on Forest Meteorology. In "Symposium on Forest Meteorology," (W. E. Reifsnyder, ed.) pp. xii–xiii, WMO No. 527, World Meteorological Organization, Geneva.

Reifsnyder, W. E., and Lull, H. W. (1965). Radiant energy in relation to forests. U.S. Dep. Agric., Tech. Bull. No. 1344.

Ripley, E. A., and Redmann, R. E. (1976). Grassland. In "Vegetation and the Atmosphere" (J. L. Monteith, ed.), Vol. 2, pp. 349–398. Academic Press, New York.

Ross, Iu. K. (1975a). "Radiatsionnyi Rezhim i Arkhitektonika Rastitel'nogo Pokrova." Gidrometeorol Izd., Leningrad.

Ross, J. (1975b). Radiative transfer in plant communities. In "Vegetation and the Atmosphere" (J. L. Monteith, ed.), Vol. 1, pp. 13–55. Academic Press, New York.

Ross, Iu., and Nil'son, T. (1966). Vertikal'noe raspredelenia biomassy v posevakh. *In* "Fotosinteziruiushchie Sistemy Vysokoi Produktivnosti" (A. A. Nichiporovich, ed.), pp. 96–108. Nauka, Moscow.

Saeki, T. (1975). Distribution of radiant energy and CO_2 in terrestrial communities. *In* "Photosynthesis and Productivity in Different Environments" (J. P. Cooper, ed.), pp. 297–322. Cambridge Univ. Press, London and New York.

Saugier, B. (1976). Sunflower. *In* "Vegetation and the Atmosphere" (J. L. Monteith, ed.), Vol. 2, pp. 87–119. Academic Press, New York.

Smith, J. R. (1929). "Tree Crops: A Permanent Agriculture." Harcourt, New York.

Smith, J. R. (1950). "Tree Crops: A Permanent Agriculture." Devin-Adair, New York.

Thom, A. S. (1975). Momentum, mass and heat exchange of plant communities. *In* "Vegetation and the Atmosphere" (J. L. Monteith, ed.), Vol. 1, pp. 57–109. Academic Press, New York.

Tooming, Kh. (1966). Priblizhennyi metod opredeleniia oslableniia i ostrazheniia FAR i blizhnei infrakrasnoi radiatsii v poseve kukurzy po izmereniiam integralnoi radiatsii. *In* "Fotosinteziruiushchie Sistemy Vysokoi Produkhtivnosti," (A. A. Nichiporovich, ed.), pp. 126–141. Nauka, Moscow.

Torssell, B. W. R., and McPherson, H. G. (1977). An improved model for simulating the penetration, propagation and absorption of radiation within plant canopies. *Austral. J. Ecol.* **2**, 245–256.

Uchijima, Z. (1976). Microclimate of the rice crop. *In* "Climate and Rice." pp. 115–140. Int. Rice Res. Inst., Los Baños, Philippines.

van Meurs, B. (1976). A method of estimating leaf temperatures in a plant community. *Arch. Meteorol., Geophys. Bioklimatol., Ser. B* **24**, 77–83.

Waggoner, P. E. (1975). Micrometeorological models. *In* "Vegetation and the Atmosphere" (J. L. Monteith, ed.), Vol. 1, pp. 205–228. Academic Press, New York.

Waggoner, P. E., and Reifsnyder, W. E. (1968). Simulation of the temperature, humidity and evaporation profiles in a leaf canopy. *J. Appl. Meteorol.* **7**, 400–409.

Waggoner, P. E., Furnival, G. M., and Reifsnyder, W. E. (1969). Simulation of the microclimate in a forest. *For. Sci.* **15**, 37–45.

Wiseman, F. M. (1978). Agricultural and historical ecology of the Maya lowlands. *In* "Pre-Hispanic Maya Agriculture" (P. D. Harrison and B. L. Turner, II, eds.), pp. 63–115. Univ. of New Mexico Press, Albuquerque.

Zavitkovski, J. (1976). Ground vegetation biomass, production, and efficiency of energy utilization in some northern Wisconsin forest ecosystems. *Ecology* **57**, 694–706.

Zolotokrylin, A. N., and Konyayev, K. V. (1978). Vertical movements in a stratified surface air layer below a coniferous forest canopy. *Izv. Acad. Sci. USSR, Atmos. Oceanic Phy.* **14**, 691–695.

Chapter XXI

ECOSYSTEM CONTRASTS

To understand how a whole ecosystem works, it is necessary to probe the differentiation of its strata and also compare it with another ecosystem. We did the former in Chapter XX and now will examine the horizontal contrasts of ecosystems that make up a mosaic landscape.

The vertical profiles of energy fluxes culminate in a set of exchanges at the outer surface of an ecosystem that epitomize its relation with its environment. These are expressed in an outer-surface budget that encompasses the radiative fluxes of the early chapters (Chapters III–VII) of this book and the nonradiative conversions and fluxes of the middle chapters (Chapters XI, XII, XIV–XVII). It summarizes the temperature-independent absorption of solar and incoming longwave radiation and the temperature-dependent emitted radiation, substrate-heat, and turbulent energy fluxes.

A unique set of relations might seem to lock the ecosystem's energy budget in to the energetic characteristics of its environment, yet in an area where sun and atmosphere seemingly deliver the same quantities of energy to underlying ecosystems, different ecosystems display different energy budgets. Under the same sun and sky a woodlot develops an energy budget that is distinctly different from that of the cornfield next to it. How much and in what ways does the energy budget of the cornfield differ from that of the woodlot? Why do they differ?

THE ORIGINS OF CONTRAST

The natural factors that differentiate the landscape are geology, terrain, and soils. Geologic formations vary, indicating the environments in which delta sands accumulated or coral reefs grew, and subsequent structural changes introduced more spatial contrast in what is now the outer surface of the earth. Dissection by excess water produces contrast of finer texture. Soil-forming processes continue these geological and geomorphic differentiations (Schmidt, 1980) and augment them where contrasts in moisture affect soil development. Vegetation contrasts also bend the course of soil development.

Vegetation classes in New England mountains are influenced by factors "associated with glacial history and soil materials" (Leak, 1978, p. 1); in New Hampshire glacial drift of 3 mineralogical types was deposited in 14 geomorphic classes that are distinct enough to define ecologic types. In Idaho, one suite of ecosystems is found in large valleys, another suite on the interfluves (Daubenmire, 1980).

The removal of excess water from ecosystems produces contrast in availability of soil moisture, and its collection into a self-created network of drainage channels dissects the surface into ridges and valleys. Low areas tend to have saturated soils and become wetlands or lakes, and a mixture of contrasting uplands and wetlands takes form.

Most lightning fires, unless carried by strong kinetic energy, burn relatively small areas and create (Arno, 1980) more diversity, as do insect population explosions and blowdowns. Gaps in low-latitude forest vary in size and constitute an important phase in the cycle of growth (Whitmore, 1978).

Superimposed on these natural forces are the results of human activity, for instance, cutting lodgepole pine in small, irregular patches to increase water yield (Alexander and Edminster, 1980). Much economic activity is associated with the area that fits a family's labor resources and subsistence needs; ecosystem diversity increased in central Europe with agriculture 6000 years ago (Schlüter, 1980). Other activity seems to be an atavistic re-creation of ancient mosaics in which the human species evolved; we plant trees on the prairies and open up clearings in the forest. In the Middle Ages the power of human activity to change the physical environment was understood, and official policy kept forests and cultivated lands in balance (Glacken, 1967, p. 334).

Where several crops are grown in the same management unit, it tends to be divided into 10- to 50-ha fields, more or less in accord with existing differences in soil and moisture. The entire landscape for 100 km becomes a quilt or mosaic of ecosystems of mixed cultural and natural origin that tend to be of about the same spatial scale (Miller, 1978) but are characterized by different properties, or "biophysical descriptors" (Davis 1980) and so display different energy budgets.

While the energy budget of each ecosystem in a mosaic landscape is in balance at all times, this balance is attained in different ecosystems at a different level of energy intake, a different surface temperature, and a different mix and strength of the energy-removing fluxes. Spatial contrasts have been described in chapters on the separate energy fluxes; now we examine contrasts in the whole budget.

CONTRASTS IN TEMPERATURE-INDEPENDENT FLUXES

Radiant-energy intake, or the absorption of solar beam, solar diffuse, and incoming longwave radiation, being the algebraic sum of the individual fluxes, partakes of their patterns of spatial variation.

Delivery of Radiant Energy

The Solar Beam. This is the only strongly directional flux in the ecosystem energy budget and induces marked contrasts in a mixture of slope facets at varying orientation and steepness. East or west orientation changes the timing of energy input through the day; alpine east slopes bear ecosystems that get their energy in a burst early in the day. All relevant factors can be mapped by computer (Friedel, 1979), a method applied in the alpine areas of Austria being reforested for avalanche control. Ecosystem attitude proved more important in explaining hourly variation of surface temperature at five sites in Alaska than did substrate condition (Brazel and Outcalt, 1973a) confirmed in later studies calculating incoming solar and longwave radiation (Dozier and Outcalt, 1979).

The greatest solar-beam contrast is found between north-facing and south-facing facets. Daily sums in the growing season in middle latitudes range from an excess of a tenth or so on sun-facing slopes above the flux density on flat land to a deficiency of two to three tenths on the most shaded slopes. Contrasts are larger at low sun heights and at high altitudes.

Diffuse Solar Radiation. Contrast in dissected terrain is reduced by area-source scattered solar radiation, especially when clouds or large particles are the diffusers. Under a clear sky, however, much of the diffuse flux is directional.

Incoming Longwave Radiation. This energy flux has an area source in the whole hemisphere, but in particular, the sky near the horizon. Except in special sites incoming longwave radiation plays a minor differentiation role. This role, where it occurs, is usually due to other ecosystems; a crop near a shelterbelt that remains warm and radiating at night is also warmed.

Absorption of Radiant Energy

Coefficients of ecosystem absorptivity that apply to these radiation fluxes vary among ecosystems. Absorptivity of longwave radiation is so near unity that only the largest contrasts, as between an exposed quartz sand surface and vegetation, are likely to exceed 20–30 W m^{-2}.

Albedo contrasts have greater significance, but among terrestrial ecosystems in summer seldom exceed 0.2, equivalent to 100–200 W m^{-2} in midday energy flux densities. The largest contrasts indicate the presence or absence of chlorophyll, which reflects much of the solar infrared, and also occur between deep and thin ecosystems, e.g., dark coniferous tree stands and grassland. Forest stands themselves differ substantially; spruce and birch in the Leningrad area absorb 81–83 W m^{-2} of photosynthetically active radiation, and pine only 73 (Alekseev, 1975, p. 92). Aquatic ecosystems and snow cover lie, respectively, above and below the 0.7–0.9 absorptivity range characteristic of most terrestrial ecosystems.

Radiant-Energy Intake.   These contrasts in absorptivity are multiplied by contrasts in incidence of the three incoming radiation fluxes [described in Kondratyev (1977)]. One can visualize, for example, a late-spring snowbank on the north side of a ridge, just across the ridge crest from a south-facing chaparral stand; the difference in absorbed radiant energy at midday in May would be approximately 500 W m^{-2}. The chaparral ecosystem has much more radiant energy to work with and to dispose of than the other system. Nearly as great a contrast might occur between an aquatic and adjacent dryland ecosystems.

CONTRASTS IN SURFACE TEMPERATURE

Contrasts in surface temperature of ecosystems depend on differential intakes of radiant energy and heat-removing conversions and fluxes, especially those that operate at relatively low rises of surface temperature. Water conversions, for instance, sequester large quantities of energy with little or no rise in surface temperature. The radiant-energy intake of an irrigated cotton field is larger than that of a desert ecosystem, but the desert surface heats up simply because its heat-removing processes cannot do their job unless surface temperature is high.

Surface temperature T_0 is the hinge in energy-budget analysis, as described in Chapter VIII, hence a fulcrum in models to be discussed in Chapter XXIII. Such models (e.g., Halstead *et al.*, 1957; Myrup, 1969) use an iterative technique in the energy budget, expressed as

$$R_a = L\uparrow + 2.5 \times 10^6 E + H + G \quad \text{W m}^{-2},$$

in which R_a is absorbed all-wave radiation, $L \uparrow$ upward longwave (σT_0^4), E is evapotranspiration in kg m^{-2} sec^{-1}, H is sensible-heat flow to atmosphere, and G is sensible-heat flow to soil. All terms on the right-hand side are functions of T_0, which is approximated successively until the residual becomes smaller than 0.7 W m^{-2} (Brazel and Outcalt, 1973a).

Surface temperature is a long neglected parameter in ecology and climatology. Remote sensing technology promises to overcome the lack of observations, however, and as surface temperature comes to be employed as an indicator of radiant-energy loading and heat removal processes, energy-budget analysis can contribute to the physical rationale of these sciences.

CONTRASTS IN EMITTED RADIATION AND ENERGY CONVERSION

Radiation

Emission of longwave radiation as a unique function of surface temperature displays the same pattern of spatial contrast, which can be measured in a thermal scanning overflight. One can distinguish poorly insulated houses, water bodies, and even highways warmed by heat from auto tires. Scanning is a powerful tool for studying the scale and grain of ecosystems in a landscape mosaic and can supply a great volume of information about thousands of ecosystems and the contrasts among them. A typical range among ecosystems in this flux is 50–100 W m^{-2}.

Latent Energy Conversions

Two ecosystems having the same intake of radiant energy may yet differ in how they use it. A snow-covered meadow may absorb about as much solar and longwave radiation by day as a gravelly sun-facing slope bordering it, but emit half of it and convert all the rest to melting, with little or no rise in surface temperature, whereas the slope emits a great deal of radiation and heats its rocks and local air. Water conversion in ecosystems depends on availability and hence displays mesoscale spatial differentiation.

Melting. Melting snow is the archetypical heat sink, unable to raise its surface temperature above 273°K and in melting weather extracting heat from the atmosphere as well as from absorption of radiation. Its disbursal of energy by longwave radiation is constant.

Melting rate is the variable energy conversion and follows radiant-energy inputs closely, in spite of the apparently weak coupling of snow and sun that high albedo often indicates. An east-northeast slope in Ungava caught the sun and began rapid melting about an hour before a southwest slope, but the latter caught up before noon and thereafter melted faster (Price and Dunne, 1976). Patchy snow interspersed with bare soil or pavement creates large contrasts in energy conversion, and surface temperature differs 20 degrees or more in a few centimeters. Snow may melt on one slope when its meltwater is freezing as it runs into the shade.

Evapotranspiration. Vaporizing water is nearly as effective as melting ice at preempting energy that remains available after emission of radiation appropriate to the surface temperature. It also has the capacity, due in part to the high heat of vaporization of the water molecule, to ingest a large amount of energy and take a large place in the surface energy budget; observed values at tundra sites in Alaska agree with values calculated from a budget model solved by simulation of surface temperature (Brazel and Outcalt, 1973b).

Differences in moisture availability, due perhaps to differences in soil texture and storage capacity, affect this energy conversion. It is usually large from well-watered ecosystems (several hundred watts per square meter during daytime hours), and its reduction in a drying crop disturbs the suite of other heat-removing processes. Deserts seem hot, not because size of the energy budget is unusually large, for it is not, but because the absence of evapotranspiration throws a large quantity of energy into thermal forms.

Fields unirrigated because of water shortage or infertile soil, which lie scattered through an irrigated district, experience these same modified energy budgets. Conversely, riparian or irrigated ecosystems lying amid dry ones maintain typical low-temperature, high-transpiration budgets. Deserts display moisture contrasts, hence energy contrasts, at four scales, at least: 10^3, 10^2, 10, and 1 m critical dimension (Graetz and Cowan, 1979).

More subtle contrasts in ecosystem energy budgets also express differences in moisture availability caused by position on a hillside, local differences in rain delivery or the catching of extra drifting snow the winter before, and differences in soil depth or rooting depth or soil capacity to hold moisture. Sun-facing slopes often hold less soil moisture than those facing away, but not always, as was found under drought-resisting chaparral near San Diego (Ng and Miller, 1980). Ecosystem vigor is also a factor; sparse saltcedar generated 60 W m^{-2} less latent heat than a denser part of the same community (Gay and Fritschen, 1979).

These differences constantly vary with time. For example, a pine plantation in the Crimea evaporated 1.0–1.2 mm day^{-1} less than a forest opening in early spring and in a moist summer, but 0.8 mm more at the end of a dry summer (Ved', 1978); the energy equivalents of these differences are 20–30 W m^{2}.

Conversions of Carbon

Photosynthesis. This process converts relatively small quantities of energy, but has budget implications through its association with transpiration. Spatial differentiation is usually due less to differences in CO_2 concentration in the atmosphere than in light or in leaf temperature and moisture.

Contrasts in biomass production may be substantial. Forest ecosystems tend to outproduce crop systems and store energy in highly visible form. A landscape of particular diversity has been created along the margins of lakes of the Valley of Mexico and nearby basins (Wilken, 1969), in which mingled raised fields and small water bodies offered a diversity of product significant to human development as early as 6000 BC (Niederberger, 1979).

Decomposition and Fire. Spatial differences in decomposition of biomass are inconspicuous except where ground water potential energy is small and anaerobic conditions permit biomass to remain immobilized for very long periods. Peat bogs display distinctive energy budgets in comparison with mineral soil or the preexisting aquatic ecosystem.

Sites vulnerable to fire display a spatial pattern related to fuel volume or fuel moisture. Fuel volume clearly differs among ecosystems, and the moisture content of fine fuels is related to aspect and cold-air drainage (Furman, 1978).

Burns develop special energy budgets that may persist for years; where lichen mulches the surface in lowlands near Hudson Bay, its effect is removed by fire (Rouse and Kershaw, 1971). Albedo contrasts become large on burning—a drop from 0.19 to 0.05 (Rouse, 1976). The large radiant-energy intake in the burns produces high surface and soil temperatures in summer, hence a larger emission of energy as longwave radiation and less for turbulent heat fluxes. Shallow snow cover in later years allows late-summer dryness that reduces evaporation, raises soil temperature, and increases the sensible-heat flux, with observable effects on downwind ecosystems. Eroding of the charcoal lightens the surface and changes spectral reflectance (Fuller and Rouse, 1979).

Intense conversion of biomass of the current year occurs in animal feedlots, on which great tonnages of grains converge. Wood-heated cities before 1850 generated large quantities of sensible heat from poorly insulated houses.

Coal, and later gas and oil, accepted as a substitute for wood, intensified industrial conversions, and urban ecosystems grew more numerous and larger. They came to process energy more intensively as buildings grew larger and more mechanized, as industrial processes grew more energy-intensive on cheap fuel and as the internal urban circulation became mechanically powered after 1890. The contrast between their altered energy budgets and those of ecosystems processing current flows of solar energy sharpened. It is particularly large in winter when natural systems slow down but cities do not.

CONTRASTS IN ECOSYSTEM COUPLINGS

Coupling of Surface with Substrate

Spatial contrasts among ecosystems often derive from their substrate capacity to store captive heat, buffer the day-and-night variation in radiant-energy intake, and thereby affect surface temperature. The large thermal admittance and heat-storage capacity of aquatic ecosystems is apparent in their steady surface temperature.

Contrasts in substrate heat storage of flux are particularly plain when wetland ecosystems in glaciated terrain—ponds and marshes of differing depths, wind exposure, and nutrient status—are intermingled with upland ecosystems. The differences in heat storage are an order of magnitude. The contrast reverses, however, when wetland basins fill with organic matter that becomes peat. Even wet peat has small thermal conductivity because it blocks convection, and dry peat has very small thermal capacity and admittance. These ecosystems are now set apart from adjacent uplands by their small energy budgets, wide fluctuations in surface temperature, and liability to summer frost.

Atmospheric Ventilation of Ecosystems

The atmosphere bathes the above-ground layers of an ecosystem, supplies sensible heat on occasion, and carries away heat and vapor differently in different exposures under the control function of its kinetic energy (Chapter XIX).

Shelter. At the mesoscale persistent differences in wind speed and turbulence and in kinetic-energy conversion are caused by differences in topographic shelter, which varies with wind direction. However, in terrain-bound motion systems like the lake or up-valley breeze, direction itself becomes a steady function of topography.

Contrasts in shelter may be subtle; sheep have an unerring skill at finding the best-sheltered parts of an apparently uniform paddock for their night camping, and even individual trees have an effect (Anon., 1979). Other contrasts are more obvious. Farmsteads in the northern hemisphere are located on the southeast side of a ridge; small wind-energy installations can be sited by persons with little training in meteorology (Elderkin et al., 1977).

Human action has for centuries introduced shelter into the landscape on exposed plateaus, broad valleys, or interior and coastal plains. Examples are the steppes, the northern Great Plains, the *bocage* or *Heckenlandschaften* of the coastal plains near the North Sea, the Canterbury Plain of New Zealand, and the Rhone Valley. These linear ecosystems shelter crop ecosystems and farmsteads, reduce wind erosion and crop damage [to nearly zero where shelterbelt ecosystems occupy 5% of a mosaic (Zakharov, 1965, p. 132)]. They stabilize snow cover, reduce dry-weather transpiration, and reduce heat loss from animals and buildings in cold winds, each objective implying a rearrangement of the energy budget of an organism or ecosystem.

The turbulent flux of sensible heat to sheltered ecosystems is reduced, as well as the removal of water vapor, because the advective term in the combination equation for evapotranspiration is decreased (see Chapter XV); the reduction in kinetic energy depends on belt permeability (Konstantinov and Struzer, 1965, p. 5). Radiative components of the energy budget usually become stronger, and surface temperatures are higher by day and lower by night (Guyot and Seguin, 1978) with increased likelihood of dew or frost. Whether or not higher midday temperatures are beneficial to an ecosystem depends on the level of regional temperature in relation to the temperature of maximum net photosynthesis in the major species. An overall production benefit (Aslyng, 1958) is found in a generally cool region like western Denmark, where grassland ecosystems are sheltered from strong marine air flow. Grain yield in the Soviet Union is increased by 0.30–0.35 behind a 5-m shelterbelt and 0.50 behind a 10-m one (Konstantinov and Struzer, 1965, p. 70).

Roughness. Wide shelterbelts along northern coasts of Japan extract fog droplets by the mechanical turbulence they create in the air flowing over their rough top surface (Hori, 1953). Clearing of a 100-m layer of foggy air passing over a few hundred meters of forest canopy means that inland rice ecosystems have a more favorable radiation budget.

Rough ecosystems extract more from the air than smooth ones: fog droplets, carbon dioxide, raindrops and even more, snowflakes, as well as sensible heat during warm-air advection. This superior coupling with the resources of the atmosphere distinguishes their energy budgets from those of low, smooth ecosystems. Eddy diffusivity over pines was an order of magnitude greater than over wheat near Canberra (Denmead, 1969); the forest had lower resistance to removal of vapor and heat, was cooler, photosynthesized at a higher rate, and transpired more. Pines in South Australia transpire so much more than the former grassland as to cause concern for groundwater recharge. The direction of wind flow over a lineated ecosystem like a row crop or vineyard determines roughness length and affects the partition between sensible and latent heat (Hicks, 1973).

Ventilation and radiation as major antagonists in climate were identified by Voeikov (1883) in contrasting convex and concave topography and account for differences of rough and smooth ecosystems. Even if radiant-energy intake is the same in a rough and a smooth ecosystem, the energy-removal fluxes are different. A rough, well-ventilated ecosystem can move heat easily into the air, its surface temperature does not need to rise so much and it emits less longwave radiation.

Ecosystems closely coupled with the atmosphere are good sources or receptors for airborne substances and thus are linked with other ecosystems, near or far. Rough ecosystems launch pollen easily and receive pollen and other substances by impaction as well as gravitational settling. They catch reentrained snow and also ozone and other pollutants, with diverse energy-budget implications. Rough ecosystems generate and receive sensible heat easily, and this transfer reduces temperature contrasts.

CONTRASTS IN THE OUTPUTS FROM ECOSYSTEMS

Contrasts among the energy transactions of ecosystems are found in absorbed radiation, internal energy conversions and storage, and outputs, three phases that form one whole. A contrast in radiation absorbed or internal conversion of energy causes a contrast in one or more of the energy outputs, for example, in the atmospheric couplings just described. The mix of outputs is a signature to the sequence of energy intakes and conversions.

Contrasts in Individual Energy Outputs

Biological productivity is small in comparison with, for example, the hundreds of watts per square meter in longwave radiation, but it is a very high quality product. Contrasts in it are contrasts in life itself, and we attach special value to barren or fertile ecosystems, to vegetation above and below the irrigation ditch, to eroded versus well-managed land. Aquatic ecosystems have risen in national perception as their true production value has become recognized.

Contrasts in the associated flux of latent heat are also significant and suggest the possibility of moisture stress that might cripple biomass production. Two moor ecosystems in northwestern Germany that had almost identical mean values of the net surplus of all-wave radiation (145 W m^{-2}) apportioned it differently: the low moor put 100 W m^{-2} into latent-heat flux, the high moor only 85 W m^{-2} (Miess, 1968, p. 176).

Contrasts in the output from aquatic ecosystems show up as islands of coolness in the midday heat of surrounding terrestrial ecosystems and pour out heat and moisture at night long after the terrestrial systems have settled down to their small nocturnal energy budgets. Contrasts in longwave radiation emitted by different ecosystems derive from surface temperature differences, as do those in sensible-heat outflows, which are more effective in destroying stable stratification of the local air. These stability contrasts can be followed as thermal plumes from a lake in cold conditions (Bill et al., 1978) or as cold plumes, when cold air over a lake in summer is entrained in the local air (Businger and Frisch, 1972; Holmes, 1970).

Unequal heating of the local air powers horizontal flows that form thermal circulations of many kinds. Cold-air drainage ponded in frost hollows and fog-filled basins (Puigdefábregas Tomás, 1970) has profound effects on ecosystem functioning and survival. Local winds shape trees and shrub ecosystems (Yoshino, 1978).

Contrasts in the Level of Outputs

The level of outputs differs in ecosystems that differ in radiant-energy absorption and, in low-energy conditions especially, those that differ in access to heat in the substrate or in the free air (ecosystems on ridges). Total energy outputs may differ by a hundred or so watts per square meter in midday conditions, as for instance between forest and dry grassland. Contrasts are smaller at night and in winter.

Linkages tend to develop among neighboring ecosystems at times when contrasts are sharp and when a large energy output is available to power them. They are an integrating factor of considerable interest in the landscape (Miller, 1978) that warrants fuller treatment than is relevant here, but which appear to give landscapes a functional quality that some students (Mori, 1977) ascribe only to ecosystems.

Local* linkages, especially of sensible heat but also of CO_2 and other properties carried in the air stream, make difficulties in such one-dimensional models as those for CO_2 transfer (Inoue, 1974) and may require three-dimensional modeling (Inoue, 1977). Similarly, "the horizontal movement of energy, called advection, has been especially disconcerting when measurements are made above the canopy for calculating evaporation" (Waggoner, 1975). Problems of advection in the energy budget were pinpointed by Hofmann (1960), and Munn (1966, p. 107) devotes a chapter of his book to the transitional zones where air flow is "readjusting itself to a new set of boundary conditions," that is, to an ecosystem surface with a different roughness (Bradley, 1968) or energy budget, which in the process is itself modified. These edges are particularly vulnerable to impact by human activity (Erickson, 1979, p. 174).

Contrasts in the Mix of Outputs

The mix of outputs varies among adjacent ecosystems. For example, a dry forest opening has a simple mix of outputs—longwave radiation and sensible heat—in contrast with added latent heat in the forest budget. If the season is spring, the opening is putting out only longwave radiation and meltwater. Cold ecosystems emit longwave radiation to the exclusion of all other outgoing fluxes.

Contrasts in mix occur where soil–moisture differences shift the fractions of available energy allotted to sensible versus latent heat, a partition established as the useful Bowen (1926) ratio. They also occur where differences in substrate admittance shift the fractions of sensible heat allotted to substrate versus atmosphere, another partitioning to be discussed in Chapter XXIII. The mosaic of contrasting ecosystem outputs over the earth's surface mirrors the intricate spatial pattern of soil and vegetation that makes the surface so interesting.

TIME VARIATIONS IN CONTRAST

All the conditions that bring about contrasts among ecosystems change with time, and the pattern of contrast among ecosystems in a mosaic landscape changes accordingly. The scales of variation are those associated with the forcing functions, particularly the varying height of

---

* Sensible-heat advection also occurs on a regional scale, particularly on days of strong wind (Brakke *et al.*, 1978), and presents a different set of problems.

the sun through the day, the varying midday sun height and day length over the year, and the aperiodic variations when atmospheric systems pass over.

Night is a period of muted contrasts among ecosystems when differences in access to heat in the atmosphere or substrate show up. The intake of absorbed radiation comes to exceed emitted radiation following sunrise and then begins to support heat flows into the soil and air, but crossover times occur at different hours in different ecosystems. Darker ones and those with vertical surfaces respond first to the rising sun; for example, vultures in India take to the air an hour or two earlier from towns than from open country (Cone, 1962) because thermals are generated earlier.

Spatial contrasts in the annual cycle are largest in summer when the high rate of incoming solar energy emphasizes absorptivity contrasts and rapid evapotranspiration brings about local soil-moisture shortages that create contrast among ecosystems in latent and sensible heat outputs.

Spatial contrasts occur in the length of the growing season. Early-starting annuals have an advantage over those waiting for warmer soil, and deep-rooted forest stands operate through dry spells that halt photosynthesis in adjacent crop fields. Forest ecosystems experience summer differently than crops or prairie; Rauner (1972, p. 195) found this to be a reason for 27 MJ m^{-2} annual biomass production in deciduous stands, 22 in steppe, and 19 from short-lived barley fields in central Russia. Mean absorption of solar energy corresponded: 33 W m^{-2} by forest, 24 by steppe, and 14 by barley.

In the transitional seasons contrasts in ecosystem energy budgets due to substrate heat storage become large. Deep mixing of heat in spring leaves the surface of aquatic ecosystems cold, and the recovery of stored heat in late fall gives them an unseasonable warmth with resulting acceleration of the turbulent heat fluxes.

Winter is the period of least contrast among terrestrial ecosystems. Incoming fluxes are small, and the fact that surface and air are equally cold reduces the effect of differences in ecosystem ventilation, except where incoming radiation is very small and cold-air drainage at a maximum.

Passing clouds and cloud systems in the atmosphere tend to reduce contrast because they suppress the differentiating influence of the solar beam and enhance longwave radiation, which is absorbed equally by most ecosystems. Energy turnover in adjacent ecosystems tends to come to the same level. Storms are not times to look for large contrasts in surface temperature of ecosystems or in their energy budgets.

THE VARIEGATED MANTLE OF THE EARTH

Under more or less uniform inputs of radiation and precipitation, the principal factors of climate, it is of the nature of ecosystems to display contrast in the way they take in radiant energy, have access to different capacities of heat and moisture storage, convert energy as they process carbon and water, and are coupled with the atmosphere. It is not surprising that these contrasts are portrayed in spatial contrasts in surface temperature, which is the intermediary of all the energy fluxes, and the energy conversions in ecosystems and the outputs they contribute to the rest of the earth.

REFERENCES

Alekseev, V. A. (1975). "Svetovoi Rezhim Lesa." Izd. Nauka, Leningrad.
Alexander, R. R., and Edminster, C. B. (1980). Lodgepole pine management in the central Rocky Mountains. *J. For.* **78**, 196–201.
Anon. (1979). Requiem for the rural gum tree? *Ecos (Melbourne)* **19**, 10–15.
Arno, S. F. (1980). Forest fire history in the northern Rockies. *J. For.* **78**, 460–465.
Aslyng, H. C. (1958). Shelter and its effect on climate and water balance. *Oikos* **9**, 282–310.
Bill, R. G., Jr., Sutherland, R. A., Bartholic, J. F., and Chen, E. (1978). Observations of the convective plume of a lake under cold-air advective conditions. *Boundary-Layer Meteorol.* **14**, 543–556.
Bowen, I. S. (1926). The ratio of heat losses by conduction and by evaporation from any water surface. *Phys. Rev.* **27**, 779–787.
Bradley, E. F. (1968). A micrometeorological study of velocity profiles and surface drag in the region modified by a change in surface roughness. *Q. J. R. Meteorol. Soc.* **94**, 361–379.
Brakke, T. W., Verma, S. B., and Rosenberg, N. J. (1978). Local and regional components of sensible heat advection. *J. Appl. Meteorol.* **17**, 955–963.
Brazel, A., and Outcalt, S. I. (1973a). The observation and simulation of diurnal surface thermal contrast in an Alaskan alpine pass. *Arch. Meteorol., Geophys. Bioklimatol., Ser. B* **21**, 157–174.
Brazel, A. J. and Outcalt, S. I. (1973b) The observation and simulation of diurnal evaporation contrast in an Alaskan alpine pass. *J. Appl. Meteorol.* **12**, 1134–1143.
Businger, J. A., and Frisch, A. S. (1972). Cold plumes. *J. Geophys. Res.* **77**, 3270–3271.
Cone, C. D., Jr. (1962). Thermal soaring of birds. *Am. Sci.* **50**, 180–209.
Daubenmire, R. (1980). Mountain topography and vegetation patterns. *Northwest Sci.* **54**, 146–152.
Davis, L. S. (1980). Strategy for building a location-specific, multi-purpose information system for wildland management. *J. For.* **78**, 402–406, 408.
Denmead, O. T. (1969). Comparative micrometeorology of a wheat field and a forest of *Pinus radiata. Agric. Meteorol.* **6**, 357–371.
Dozier, J., and Outcalt, S. I. (1979). An approach toward energy balance simulation over rugged terrain. *Geogr. Analysis* **11**, 65–85.
Elderkin, C. E., Ramsdell, J. V., and Tennyson, G. P. (1977). Meeting review/Wind characteristics workshop, 2–4 June 1976, Boston, Mass. *Bull. Am. Meteorol. Soc.* **58**, 45–51.
Erickson, P. A. (1979). "Environmental Impact Assessment." Academic Press, New York.

Friedel, H. (1979). Kleinklima-Kartographie (Die Kartierung und specielle Berechnung potentieller Einstrahlungssummen). *Wetter u. Leben* **31**, 169–188.

Fuller, S. P., and Rouse, W. R. (1979). Spectral reflectance changes accompanying a post-fire recovery sequence in a subarctic spruce lichen woodland. *Rem. Sens. Environ.* **8**, 11–23

Furman, R. W. (1978). Wildfire zones on a mountain ridge. *Ann. Assoc. Am. Geogr.* **68**, 89–94.

Gay, L. W., and Fritschen, L. J. (1979). An energy budget analysis of water use by saltcedar. *Water Resour. Res.* **15**, 1589–1592.

Glacken, C. J. (1967). "Traces on the Rhodian Shore. Nature and Culture in Western Thought from Ancient Times to the End of the Eighteenth Century." Univ. of California Press, Berkeley.

Graetz, R. D., and Cowan, I. (1979). Microclimate and evaporation. *In* "Arid-land Eco-systems: Structure, Functioning and Management" (D. W. Goodall, R. A. Perry, and K. M. W. Howes, eds.), Vol. 1, pp. 409–433. Cambridge Univ. Press, London and New York.

Guyot, G., and Seguin, B. (1978). Influence du bocage sur le climat d'une petit région: Résultats des mesures effectuées en Bretagne. *Agric. Meteorol.* **19**, 411–430.

Halstead, M. H., Richman, R. L., Convey, W., and Merryman, J. D. (1957). A preliminary report on the design of a computer for micrometeorology. *J. Meteorol.* **14**, 308–325.

Hicks, B. B. (1973). Eddy fluxes over a vineyard. *Agric. Meteorol.* **12**, 204–215.

Hofmann, G. (1960). Wärmehaushalt und Advection. *Arch. Meteorol., Geophys. Bioklimatol., A* **11**, 474–502.

Holmes, R. M. (1970). Meso-scale effects of agriculture and a large prairie lake on the atmospheric boundary layer. *Agron. J.* **62**, 546–549.

Hori, T., ed. (1953). "Studies on Fogs in Relation to Fog-Preventing Forest." Tanne, Sapporo. [Summarized in Miller, D. H. Coastal fogs and clouds. *Geogr. Rev.* **47**, 591–594 (1957).]

Inoue, K. (1974). Numerical experiments of effects of advection on CO_2 environment and photosynthesis of crop fields. *Jpn. Nat. Inst. Agric. Sci., Bull. A* **21**, 1–25.

Inoue, K. (1977). Numerical experiments about three-dimensional transfer of CO_2 over a finite model rice field in relation to canopy photosynthesis. *Jpn. Natl. Inst. Agric. Sci., Bull., A* **24**, 19–44.

Kondratyev, K. Y. (1977). Radiation regime of inclined surfaces. *WMO Tech. Note* No. 152 (WMO-467).

Konstantinov, A. R., and Struzer, A. R. (1965). "Lesnye Polosy i Urozhai." Gidrometeorol. Izd., Leningrad.

Leak, W. B. (1978). Relationship of species and site index to habitat in the White Mountains of New Hampshire. *U.S. For. Serv., Res. Pap.* **NE-397**.

Miess, M. (1968). Vergleichende Darstellung von meteorologischen Messergebnisse und Wärmehaushaltsuntersuchungen an drei unterscheidlichen Standorten in Norddeutschland. *Tech. Univ. Hannover, Inst. Meteorol. Klimatol., Ber.* **2**.

Miller, D. H. (1978). The factor of scale: Ecosystem, landscape mosaic, and region. *In* "Sourcebook on the Environment" (K. A. Hammond, G. Macinko, and W. B. Fairchild, eds.), pp. 63–88. Univ. of Chicago Press, Chicago, Illinois.

Mori, A. (1977). Classification et cartographie du paysage sur base écologique avec application à l'Italie. *Geoforum* **8**, 327–340.

Munn, R. E. (1966). "Descriptive Micrometeorology." Academic Press, New York.

Myrup, L. (1969). A numerical model of the urban heat island. *J. Appl. Meteorol.* **8**, 908–918.

Ng, E., and Miller, P. C. (1980). Soil moisture relations in the Southern California chaparral. *Ecology* **61**, 98–107.

Niederberger, C. (1979). Early sedentary economy in the Basin of Mexico. *Science* **203**, 131–142.

Price, A. G., and Dunne, T. (1976). Energy balance computations of snowmelt in a subarctic area. *Water Resour. Res.* **12**, 686–694.

Puigdefábregas Tomás, J. (1970). Características de la inversión térmica en el extremo oriental de la depresión interior altoaragonesa. *Pirineos* **96**, 21–50.

Rauner, Iu. L. (1972). "Teplovoi Balans Rastitel'nogo Pokrova," Gidrometeoizd., Leningrad.

Rouse, W. R. (1976). Microclimate changes accompanying burning in subarctic lichen woodland. *Arct. Alp. Res.* **8**, 357–376.

Rouse, W. R., and Kershaw, K. A. (1971). The effects of burning on the heat and water regimes of lichen-dominated subarctic surfaces. *Arct. Alp. Res.* **3**, 291–304.

Schlüter, H. (1980). Biotische Diversität und ihr Regenerationsvermögen in der Landschaft. *Petermanns Geog. Mitt.* **124**, 19–22.

Schmidt, R. (1980). Die Heterogenität der Bodendecke und ihr Einfluss auf Bodenfruchtbarkeit, Melioration und Landeskultur. *Petermanns Geog. Mitt.* **124**, 11–18.

Ved', I. P. (1978). Sezonnye osobennosti radiatsionnogo, teplovogo i vodnogo rezhimov meliorativnykh nasazhdenii sosny krimskoi. *Izv. Akad. Nauk SSSR, Ser. Geogr.* No. 2, 79–84.

Voeikov, A. (1883). Über die Grösse der täglichen Wärmeschwankung in ihrer Abhängigkeit von den Localverhältnisse. *Z. Oesterr. Ges. Meteorol.* **18**, 211–220, 241–248.

Waggoner, P. E. (1975). Micrometeorological models. *In* "Vegetation and the Atmosphere" (J. L. Monteith, ed.), Vol. 1, pp. 205–228. Academic Press, New York.

Whitmore, T. C. (1978). Gaps in forest canopy. *In* "Tropical Trees as Living Systems" (P. B. Tomlinson and M. H. Zimmerman, eds.), pp. 639–655. Cambridge Univ. Press, London and New York.

Wilken, G. C. (1969). Drained field agriculture: An intensive farming system in Tlaxcala, Mexico. *Geogr. Rev.* **59**, 215–241.

Yoshino, M. M. (1978). Local winds in the tropics and subtropics. *Tsukuba Univ., Inst. Geosci., Annu. Rep.* **4**, 57–63.

Zakharov, P. S. (1965). "Pyl'nye Buri." Gidrometeorol. Izd., Leningrad.

Chapter XXII

ENERGY CONVERSIONS AT NODES

Ecosystems survive and flourish over the centuries in places subject to transient episodes of energy conversion in hurricanes and tornadoes, landslides, lightning, and wildfires, as noted in Chapters XII and XIX, not to mention infrequent seismic events, volcanic ashfalls, and human disturbances of many kinds. Some ecosystems maintain themselves in shifting sand, beaches under wave attack, avalanche tracks, creeping hillsides, and swift-flowing rivers in which episodic energy conversions recur in the same place. These events produce many of the contrasts among ecosystems described in Chapter XXI. Finally, some ecosystems exist in sites in which intensive energy conversions are both site-specific and continuous. One example would be hot-spring or geothermal area communities; more numerous examples are systems in which energy conversion is supported by the transport of biomass from a wide area, like feedlots, or by the collection of biomass from a long expanse of time—the concentration of fossil fuels in industries and cities. In these nodes the energy budget is augmented beyond the levels normally supported by solar energy by virtue of the artificial transport of biomass energy of contemporaneous origin and fossil energy that was accumulated over a million years. We now examine such augmented energy budgets and the contrast they form with more natural ecosystems.

POINTS OF ENERGY CONCENTRATION

Concentration of Biomass

Grain and hay that represent solar energy converted in outlying fields and brought in to a barnyard to be converted into animal products and metabolic energy exemplify a common form of concentration of energy.

Industrialization of this transport has created the feed lot, where grain and hay from thousands of square kilometers of crop land are carried to beef cattle brought together from even larger areas of range land, where they have built bone structure to underpin the weight they will gain on feed. The ratios of concentration at one large lot in Colorado (Mather, 1972) are 30,000 to 1 for the feed and 100,000 to 1 for the animals. In addition to solar-energy content, the inflows of feed also carry the fossil-energy content of fertilizer, fuel, and irrigation, and the animals carry additional fossil-energy content from their herding and trucking to the feed lot; the lot itself receives fossil energy to operate its feed mills and materials-handling equipment—equal perhaps to two-thirds of the energy of the feed itself (Pimentel et al., 1975).

Metabolic energy conversion on the Colorado lot (area 3.2 km^2) is about 100 W m^{-2}, which substantially augments the natural inputs in the energy budget. Outputs of the energy budget are correspondingly augmented: more evaporation (water is supplied at double the tonnage of feed), sensible heat and ammonia in the air, animal tissue in amounts equivalent to about 5 W m^{-2}, and several times as much energy in animal wastes, which might be converted into biogas or, reaching streams, support algal pollution.

Sawmills and other forest industries, as well as power plants using mill or logging residues (Burlington, Vermont; Mt. Gambier, South Australia) (Fig. 1), also exemplify biomass concentration. Again, fossil energy accompanies these flows of biomass: the fuel consumed in growing the trees, in logging and chipping, and in transporting the chips to the power plant. This last cost is a major obstacle to wider use of forest residues.

Concentration of Fossil Fuels

The high transport costs of wood, charcoal, cane, or other biomass crops are one reason for the temporary eclipse of these energy sources during the past 100 years. Coal, oil, and natural gas were cheaper to harvest and transport because their deposits represent a compression of time as well as space. Intensities of energy conversion are greatest where they are converted into other forms of high-quality, marketable energy, a costly process carried on in refineries, synthetic fuel plants, and electricity-generating stations. For example, refineries consume 0.15 of their input (Cook, 1977), and a power plant of typical 10^6 kW electrical output takes in 3 × 10^6 kW equivalent of oil or coal. Since it occupies only a few square kilometers, an energy flux density of thousands of watts per square meter is dumped into the environment.

Fig. XXII.1. Electricity-generating plant at Mt. Gambier, South Australia, burns residues from logging and milling of plantation pines that are more or less contemporaneous biological energy.

The flux density of the dump energy is relatively smaller if it goes through a lake or cooling pond (Chapter XVIII), more if through cooling towers (Fig. 2). For example, the 2000 MW Liddell plant of the Hunter Valley in Australia uses a cooling pond 10 km^2 in area, on which the average surface energy budget is augmented by $(2 \times 2 \times 10^9 \text{ W})/(10 \times 10^6 \text{ m}^2)$ or 400 W m^{-2}.* In contrast, a new plant to be built nearby will dispose of its waste heat through cooling towers only 1000 m^2 or so in area. The enormous injections of heat, both sensible and latent, into the lower atmosphere by cooling towers have destabilizing effects that are visible as long plumes rising to 2–3 km (Hanna, 1977) and produce snow (Kramer *et al.*, 1976).

Fossil fuels were exploited in converting materials from one form to another, as, for instance, when biomass fuels, i.e., charcoal, ran low in

---

* Much of this loss tends to be concentrated near the hot-water inflow. In an Illinois cooling pond eddy-correlation flux measurements in the first of five sectors found 590 W m^{-2} latent-heat flux and 260 W m^{-2} sensible heat for a sum of 850 W m^{-2} (Hicks *et al.*, 1977). Another 460 W m^{-2} was emitted as longwave radiation.

Fig. XXII.2.   An electricity-generating station near open-pit mines of brown coal produces power for Melbourne and its rail mass transit. The cooling tower speeds energy long ago stored in organic matter out of the boilers and turbines and into the present-day atmosphere.

16th century England and accelerated the mining, in spite of its recognized dirtiness, of coal for salt boiling, brickmaking, copper smelting, brewing, and many other products (Harris, 1974; Nef, 1977). Later it came to be used for mechanical power in the Newcomen coal-fired pumps to dewater coal mines (1712), which coexisted with windmills and waterwheels until Watt added the condenser in 1780 (Hawthorne, 1980). The coal–iron combination was then applied to railway transport, which expanded and intensified the nodes of fossil-energy conversion.

Most of the nationwide mean flux density (15 mW m^{-2}) in the U.S. coal–iron–steel complex is concentrated in nodes like that at the southern end of Lake Michigan. This node collects fossil energy from distant areas in amounts three or four times as great as the total quantity of energy used by the skilled industries of the state of Wisconsin (Shaver *et al.*, 1974, p. 19) and generates a flux density of 500 W m^{-2} over its approximately 50 km^2 area (Miller, 1971). Approximately the same flux density is found at a steel mill in Hamilton, Ontario (Oke and Hannell, 1970). The switch from natural ores to pellets has made a change (Miller,

1970): the making of pellets near the mines requires an additional 3.5 MJ added there per kilogram of iron content, but using them as blast-furnace charge reduces energy conversions at the mill by 5.0 MJ per kilogram of molten iron (Kakela, 1978) or about 50 W m^{-2}; some of the energy conversion and concomittant pollution is transferred north to Lake Superior.

Other industries are also nodes of augmented energy conversion. Any metal-using industry involves large energy conversions that are added to those embodied in the machinery it operates. Energy-flow studies that have been made for the American food industry as a whole (Steinhart and Steinhart, 1974) and in detail for its beverage component (Hannon, 1972) show fossil-energy conversions far exceeding the energy content of the food or drink being processed, packaged, stored, and transported. Costs in the beverage industry (beer or soft drinks in bottles or cans) are smaller for returnable than for one-use containers since both glass and aluminum containers embody a large energy content. Much of the total conversion of energy in these industries occurs in factories, storage warehouses (especially if refrigerated), and supermarkets, energy nodes that are mostly located in cities.

CHARACTERISTICS OF THE URBAN INTERFACE

Cities are artifacts that concentrate three artificial flows of energy into a node of intricate texture made up of a multitude of ecosystems. (1) The first inflow is that of energy-rich materials (at an areal density of 8200 MJ m^{-2} of single-family housing in the U.S., for example (Hannon et al., 1978), and 19,000 MJ m^{-2} of office space). The corresponding flux densities, divided by a 40-year life span, are, respectively 7 and 16 W m^{-2} per unit floor area; growth and amortization of the physical structure of Hong Kong is 6 W m^{-2} over the whole urban area (Newcombe et al., 1978). These materials change the natural absorption, storage, and transformation of energy by the earth. (2) A second energy flow is that of biomass, and a simple calculation indicates its flux density: 125 W per capita plus about the same energy cost for cooking food shows that a population density of 5000 km^{-2} means energy conversion at a rate of 1.2 W m^{-2} of urban area. (3) Finally, there is the larger import of fossil energy that powers modern cities and makes them habitable. These three processes form the urban interface.

Texture

The first thing we notice when we analyze energy in a city is that ecosystems are smaller than those of wildlands or croplands. Yards, roofs, and gardens create a fine texture that is difficult to study not only because of the numbers of individual energy-converting systems, but also because of their interactions with one another. Houses shade yards, trees change the turbulence structure of the air moving over lower surfaces, and surfaces at different angles exchange radiation in a geometrically determined but complicated manner.

City ecosystems are smaller than most rural ecosystems because this scale fits the need for human mobility, indicated by dedication of 0.2 of the whole area to streets. Access to human-scale habitats and workplaces accounts for the intricate texture of a city, which forms one of its visual attractions.

Urban ecosystems, though numerous, belong to only a few types: paved surfaces, low structures, tall structures, low vegetation, and trees. These differ in absorptivity of radiant energy, moisture status, rates of fossil-energy conversion, and coupling with the atmosphere and the substrate. These characteristics are necessary for modeling urban energy budgets, but unfortunately have seldom been measured.

Depth and Stratification.   Ecosystem thickness or depth and stratification have been discussed in Chapters II and XX and apply with particular force to urban systems. Urban effluvia turn the lower part of the overlying air into a special atmosphere, and cities disturb their substrates more deeply than natural ecosystems.

The interface, a layer between the ground and the level of roofs and tree canopies (Kratzer, 1956, p. 67; Oke, 1976; Carlson and Boland, 1978), is sheltered and to a degree separated from the free air above, as shown by diffusion of lead released at street level (Assaf and Biscaye, 1972). It is partitioned into separate volumes: urban canyons (Nunez and Oke, 1976, 1977); courts among buildings, particularly in densely built-up sections and in 19th-century cities like Wien (Mahringer, 1963); tree overhangs (Heisler, 1974; Morgan *et al.*, 1977)*; and low ecosystems sheltered by trees or structures. This layer corresponds to the plant–air space in vegetation and in some cases even to the sheltered trunk space in a forest ecosystem.

The layer of air just above roof level is under the influence of heat and mass exchanges at the canopy top. Tall buildings and trees that protrude into it, as well as such heat and pollutant sources as factory stacks, also influence this boundary layer (Kratzer, 1956, p. 62; Gutman and Torrance, 1975).

---

* The death of thousands of elms along streets of eastern North American cities that were once truly urban forests will change energy conversions substantially, mostly for the worse.

The Urban Atmosphere

The warm, dirty urban atmosphere, "the dust dome" (Geiger, 1942, p. 356) visible from great distances and of varying depth, is full of debris of many kinds, sizes, and compositions, which, moreover, differ from city to city as a product of different industrial processes, and even from place to place over a city.

Aerosols variously scatter incoming solar photons, depending on their refraction index and size. Bridgman (1979a) measured fewer small particles over Milwaukee than are predicted in the usual statistical distributions of particle size and found a day-to-day variation in the vertical distribution above an industrial area that contributes particles in the size class from 0.11 to 0.35 μm. A study in St. Louis showed that small particles have quite different scattering and absorbing properties than large ones do (Bergstrom and Peterson, 1977), and that the effect differs on clear and hazy days, as well as with solar altitude. Their vertical distribution is also important (Idso, 1974). Gases in the Denver "brown cloud" absorb more strongly than natural aerosols (DeLuisi et al., 1977); high CO_2 concentrations over Tokyo (Arai et al., 1975) and cities in New Jersey (McRae and Graedel, 1979) affect radiation.

Because chemical makeup, size, and vertical concentration of aerosols and gases in the atmosphere are not generally observed, we have little information on emissivity, and the apparent emissivities deduced from longwave radiation measurements at the underlying surface, as, for example, at Montreal (Oke and Fuggle, 1972) compared with Hamilton (Rouse et al., 1973) or Windsor (Brazel and Osborne, 1976) tend to differ. Assumptions for modeling radiation have not been "related to a particular urban situation in any way" (Ackerman, 1977); "most often the aerosol properties are not measured and in order to predict the solar flux one is forced into unrealistic assumptions" (Bergstrom and Peterson, 1977).

The Urban Interface

Absorption of Radiant Energy. Absorption of solar energy at the urban interface may be greater or less than in typical rural ecosystems. Cities have more snow-free land in winter, (10–20 days) or a third less in Wien (Böhm, 1979a), and airplane measurements over Madison, Wisconsin, (Kung et al., 1964) found absorptivities of 0.6 compared with rural ones of 0.5 or less. Far northern settlements may enjoy twice as much all-wave radiation surplus as rural areas (Gorbacheva, 1974).

The many vertical surfaces of a city are especially important at low sun angles, and morning heating begins early (Fig. 3); however, different

Fig. XXII.3.   East-facing ends of Milwaukee houses lit by the morning sun at 1030, 23 April 1976, as seen from 150 m. This orientation is efficient from a standpoint of solar geometry.

wall materials differ in absorptivity as much as from 0.24 to 0.80 in Wien (Mahringer, 1961). Comparing indoor and outdoor temperatures suggests that passive solar heating in cold weather amounts to 5–10 W m^{-2}. The overall absorptivity of Adelaide, 0.75–0.80, is about the same as that of dry grass and ripe grain in this dry-summer climate and less than that of pine stands, vineyards, and irrigated pasture (Coppin et al., 1978). The absorptivity of St. Louis, in a humid-summer climate, is about the same as that of adjacent cropland but less than that of forest or aquatic ecosystems (White et al., 1978).

Spatial contrasts are due to concentrations of trees or metal roofs in different parts of the city; the eastern suburbs of Adelaide, with shrubs and trees, have absorptivities of about 0.8, whereas commercial and industrial centers run about 0.7 (Coppin et al., 1978). Absorption by only the upper parts of tall office buildings in Dallas prevents warming of the lower part of the below-roof layer (Ludwig, 1970). The degree to which blackening Los Angeles would increase absorption in order to inject more sensible heat into the atmosphere was briefly explored by Neiburger (1957) with negative results. The increase in absorption by either active or passive solar energy utilization does not seem likely to become large (U.S. House of Representatives, 1980, pp. 21, 23).

Absorptivity for incoming longwave radiation has seldom been measured. Walls and pavements that face one another, as in an urban canyon (Nunez and Oke, 1977), create a cavity-rich interface that might compensate for the low absorptivity of metal roofs.

Evapotranspiration Opportunity. Streets and roofs are usually kept as dry as possible. The detention film on streets, a millimeter or less, vanishes in a short time in summer, and walls dry off even faster. The role of trees is not well understood (Heisler, 1974) nor is the effect of diversity in such small agroecosystems (Ehler, 1978). Urban lawns are well watered and display rural-type energy budgets unless near large, dry surfaces (Suckling, 1980).

Energy Conversion. Conversion of imported biomass and fossil energy differs with ecosystem function. Those providing vegetation for amenity convert such energy at rates not much different from rural crop land; those providing microenvironments for human beings convert biomass in proportion to population density and fossil energy in accordance with the extent to which the desired microenvironment departs from the external environment. Most workplaces convert great quantities of fossil energy, depending on the kind of materials and processing, and the circulation network converts energy as a function of traffic density.

Coupling of Interface and Atmosphere

Turbulent outflows of energy from the urban interface reflect coupling of the interface and atmosphere, which depends on the exposure and roughness of individual ecosystems and also, because ecosystems are small and interacting, on the effects of upstream ecosystems on wind speed, stability, and turbulence.

Cities have large coefficients of roughness (Gutman and Torrance, 1975), as would be expected from the prevalence of trees and bluff obstacles and the local occurrence of taller buildings. Overall roughness in Sacramento and Baltimore is of the order of half a meter, indicating good atmospheric coupling of roof-level surfaces; low-density residential areas in Sacramento were calculated by Lettau's (1969) method to have roughness lengths of 0.3 m in winter, 1.0 m when trees are in leaf (Morgan et al., 1977), and 0.7 m in Baltimore (Nicholas and Lewis, 1980, p. 20). The central business district of Sacramento has a roughness length of 3.2 m, that of Baltimore 4.0 m overall and up to 8.0 m in its core.

The layer below roof level may have poor atmospheric coupling, like the below-canopy space in forest. Deep courts in densely built-up areas are only weakly coupled with the free air in most weather conditions, with a smaller range through the day and between days than the air at roof level (Mahringer, 1963). Low areas of cities are poorly ventilated and in anticyclonic weather develop extreme microenvironments, which may be further worsened by polluted fog.

Coupling with the Substrate

The substrates of urban green areas receive, store, and return heat and water much as do those of rural ecosystems. In contrast, the substrates of the extensive (up to 0.4 of the total area) paved lands store little water but much heat since values of thermal admittance approach those of rock ledges.

The space under a roof is a special kind of substrate that is seldom met in nature, distinguished by zero capacity to take in and store water, a small capacity to store heat,* and usually containing heat-generating processes. The coupling of building interiors with their interfaces depends on insulation; poor insulation in summer makes overheated houses, from which heat has to be removed by attic fans or air conditioners. Good insulation uncouples the interface and substrate like deep snow and helps maintain favorable microenvironments for living organisms.

The thermal admittance of the urban layer below roof level depends on specific heat, density, and thermal conductivity of structural members, especially walls, a matter of building practice in different cities and countries.† Brick and concrete walls combine large heat capacity with ready access to surface energy. European structures, in general, are more massive and store more heat (Banham, 1969, pp. 22, 101), a style found only in older commercial and office buildings in American cities and in climates with a large diurnal range in temperature. Heat-storage walls are being built into newer buildings as an aid in passive solar energy utilization, and other devices (water tanks, pebble beds, fans, or pumps) in active solar houses also change thermal admittance (Chapter XVII), as does the practice of construction underground.

A map of a small Japanese city (Kawamura, 1965; Yoshino, 1975, p. 91) displays a range from 870 to 990 J m^{-2} K^{-1} sec$^{-1/2}$ between residential

---

* Empty unventilated compartments trap solar heat and reach temperatures briefly as high as 88°C (Daniels, 1973, p. 228).

† Concrete and aggregate in the built-up environment of Hong Kong have a mass of 17 tonnes per capita (Newcombe et al., 1978), nearly 1 tonne m^{-2} of urban land or a heat capacity of 0.8 MJ K^{-1} m^{-2}.

and commercial areas, values that seem reasonable for the construction materials. In an urban canyon the mean admittance is 1730 (Nunez and Oke, 1976).

Overall admittance of a low-density city, Los Angeles, is estimated at 910 J m^{-2} K^{-1} sec$^{-1/2}$ (Carlson and Boland, 1978) and of a nearby small town at 790 (Ackerman, 1977). These values reflect the density of building materials per unit area and the amount of exposed soil.

The green area of a city indexes access to soil moisture, and wetness fractions of 0.4–0.6 have been estimated in low–density residential districts, which suggests the desire of urban people to surround themselves wherever possible with microenvironments of grass, shrubs, and trees. Trees overhang roofs and streets to the extent of 0.10–0.15 of the area of some parts of Sacramento (for the West an uncharacteristically forested city) (Morgan et al., 1977). Whole-city wetness fractions (0.46 in Sacramento, for instance, fed by meltwater from the Sierra Nevada) are larger than often is assumed in modeling studies (Chandler, 1976, p. 21) and are most important in summer.

The Urban Landscape

The close-grained urban interface, lying between its self-inflicted atmosphere and its compacted substrate, presents diverse spatial patterns of absorptivity, roughness, thermal admittance, and wetness, to name the properties found to exert the most direct influence on energy transformations. Their effect varies with time; wetness, for example, is significant during the hours of transpiration and hardly at all at night and varies with the episodic occurrence of rains. Absorptivity is important only by day, but thermal admittance affects the small energy budget at night more than it does the larger daytime budget. Fossil energy conversions in industry and circulation vary with weekdays and weekends; those in space heating and cooling tend to vary counter-seasonally.

The tiny, interacting ecosystems in the city landscape can be grouped for analysis most easily with digitized scanning of the upward fluxes of shortwave and longwave radiation (Pease et al., 1976, 1980). This procedure, or delineation of ecological types (Brady et al., 1979), gives land-use types having overall values of absorptivity and other characteristics that index the fractions of paved and green areas, roofs, and mass of construction materials, which tend to be reasonably uniform over fairly large areas in consequence of land-use concepts in city planning as applied over several decades.

Population density is another surrogate for detailed data on absorptivity, admittance, and so on because in a way it indexes structural mass and green and wet areas, as well as heat from combustion. In the long run, however, only direct measurements of these characteristics of the urban landscape will adequately serve energy analysis and modeling.

MODIFIED RADIANT-ENERGY INTAKE

Solar Radiation

The effect of the urban atmosphere on solar energy reaching the urban interface has been known for more than a century, and many measurements of its depleting influence, especially in the heating season, have been made (Munn, 1966, p. 196). Turbidity is much greater at urban stations (Flowers et al., 1969); the Los Angeles atmosphere, for example, reduces ultraviolet income by at least 0.15–0.20 and total shortwave radiation by 0.06–0.08 (Peterson et al., 1978). Depletion in the 0.40–0.467 μm range is 0.4 more in Milwaukee than at a rural site (Bridgman, 1978); in the 0.55–0.625 μm range it is 0.2 more, again demonstrating the selective action of aerosols against shorter wavelengths.

Scattering produces a shift from directional to diffused solar energy.* Urban green areas lose in the selective depletion of short wavelengths, but might gain from the shift to a diffused source (Lister and Lemon, 1976). At Newcastle, 0.7 of the energy scattered out of the beam is recovered as diffuse radiation (Bridgman, 1979b).

The altered solar receipts, when combined with the difference in albedo between cities and rural ecosystems, as modified, e.g., for Toronto (Yamashita, 1979), may result in more or less equal quantities of absorbed solar radiation in city and country, depending on the particular city (Chandler, 1976, p. 10; Oke, 1978, p. 247). Within-city variations over Baltimore showed a range of 350 W m^{-2} at midday in summer (Pease et al., 1976) due to albedo differences, as was true also in St. Louis (White et al., 1978). A seasonal variation is caused by the high solar infrared reflectance of tree canopy when leafed out; the tree effect on the energy budget of houses is most favorable also in summer (DeWalle, 1979).

---

* Beam radiation at Moscow University is 15–50 W m^{-2} less than that in the center of the city when the university station is downwind, but 35–55 W m^{-2} more when it is upwind of the center (Dmitriev and Bessonov, 1969, p. 148). A similar effect of wind direction is found in St. Louis (Peterson and Stoffel 1980). Diffuse radiation at the university station averages 5–10 W m^{-2} over the summer greater than that outside the city (Dmitriev and Bessonov, 1969, p. 158).

Longwave Radiation

Absorption of solar radiation in urban haze contributes, along with other factors, to atmospheric warming, which helps augment the downward flux of longwave radiation by as much as 30–40 W m^{-2} (Oke and Fuggle, 1972; Rouse et al., 1973) and about half as much on clear winter nights in Tokyo (Aida and Yaji, 1979).

The increased flux is strongly absorbed by the cavity-filled urban interface, and while its biological effects are less familiar than those of the depleted solar input, they do extend the frost-free season in fall and ameliorate winter cold. Conversely, this incoming flux aggravates the stress of summer heat waves, not only by its size, approaching 400 W m^{-2}, but also by its day and night continuity. In a winter night in Christchurch, New Zealand, a person in the business district receives 25 W m^{-2} more radiant energy than in the residential area and 40 W m^{-2} more than in the rural surroundings (Tuller, 1980).

Radiant-Energy Intake

Absorption of solar and longwave radiation added 5 and 20 W m^{-2}, respectively, to total radiant-energy intake on clear days in Cincinnati (Bach and Patterson, 1969) an increase of about .02 over a rural site. A similar difference can be calculated from summer airborne observations in daytime over St. Louis reported by White et al (1978). An extreme midday difference of 170 W m^{-2} is reported at Columbia, Maryland (a new city), the urban site here being a dark parking lot (Landsberg and Maisel, 1972; Landsberg, 1979).

The general level of radiant-energy intake at midday in summer by the Baltimore urban surface was 1050 W m^{-2}, and peaks of 1190 W m^{-2} were measured in the central business district and a steel-mill area (Pease et al., 1976). Variation was principally due to variation in albedo; values in parks were not much lower than the overall average, and those of water bodies exceeded 1200 W m^{-2}.

FOSSIL-ENERGY AUGMENTATION

It is hard to believe when walking on a cold winter street in Milwaukee that the city energy budget gets much increase by energy from fossil sources. When thousands of air conditioners are dumping heat on a sweltering day in summer, however, and trucks and factories radiate energy, it is easier to visualize fossil fuel in the urban energy budget and to see it always working in the same direction, always adding 0.1 or more to the radiation energy intake.

The flux density of fossil fuel conversion varies through the day, week, and year and differs from one ecosystem to another because it derives from that ecosystem's particular activity. Following city planning practice, we speak of activities associated with land use—industrial, commercial, and residential—and with the circulations that bind these land uses into the functioning city.

Land Uses

Cities provide workplaces and shelter, and both require fossil energy.

Industrial Land Uses. Some industries were noted earlier as nodes that convert large quantities of energy, and most are located in cities. The working of metals is done against friction; electrical energy in a drill room is converted into enough heat to keep the room hot in the coldest weather. A coal-and-iron city like Chorzow in upper Silesia gives off 62 W m^{-2} from its whole urban area (Kraujalis, 1975). Although lighter material requires less brute force and a cotton mill generates less heat than a machine shop, even a textile-dominated city like Hong Kong that is congested adds as much as 15 W m^{-2} (Newcombe, 1975) to its energy budget from industrial land uses.

Commercial Centers. Offices and shops concentrate large numbers of people and need energy to support the required microenvironments—heating, cooling, lighting, ventilating—and to power business machines. So much metabolic, lighting, and equipment energy is released in schools and office buildings that if well designed they may need spend no additional energy for space heating. Such energy-conscious design has not been prevalent in the recent past; while office buildings in New York City that were built before World War II convert, for all functions, 22–26 W m^{-2} of floor area, those built in the glass-wall period convert 42 W m^{-2}. Fortunately, those under design are back to 24 W m^{-2} (New York Times, 1978). A single-story neighborhood library in Milwaukee that modified its ventilation system now saves 0.56 of its former conversion of 64 W m^{-2}. A tall building is a node of high flux density at 85 W m^{-2}, double that of general office buildings (Spielvogel, 1980).

Large shopping centers convert up to 150 W m^{-2}. The overall U.S. average conversion of 42.8 GJ commercial energy annually per capita (Schipper and Lichtenberg, 1976) works out to 130 W m^{-2} (at 10 m^2 floor area per capita), including heat wasted in electricity generation. The floor-area flux density is 90 W m^{-2} in Sweden, where less energy is lost as escaping heat and unnecessary lighting. Parking lots at shopping centers add the indirect costs of energy-rich vehicles depreciating on them at a rate equal to several watts per square meter.

Shelter   Cities bring human beings into convenient contact for the exchange of services, goods, and work, and these people need individual microenvironments, in which fossil energy is converted for heating, cooling, lighting, humidifying or dehumidifying, and household operations that chiefly revolve around food and clothing. Where these microenvironments are stacked up to heights of 15–20 floors, as in Hong Kong (Liang, 1973, p. 21), site-area flux densities are large. Poorly designed houses in cold climates leak warmed air in winter and require enormous inputs of fuel to make them habitable. People have frozen in their own houses, and the public intent not to permit this has focused attention on the large quantities of fossil energy wasted in American cities. The Wisconsin "lifeline" level, 16×10^3 MJ per month, works out to 6 kW per household or 10 W m^{-2} from a 600-m^2 lot.

One of the most power-hungry of the environment-forming mechanisms is air conditioning, which removes solar heat load, excess longwave radiation on hot nights, and usually excess moisture or latent heat. Badly designed houses are heat traps in summer and need this mode of heat removal. The expenditure of 5 W m^{-2} of site area on a hot August day in Milwaukee also occasions the dumping into Lake Michigan of heat equivalent to 10 W m^{-2} of house site area. Energy conversion in a more densely populated city, Hong Kong, over the whole summer is 3 W m^{-2} electric energy (Newcombe, 1976) and in certain neighborhoods up to 310 W per capita and 10–15 W m^{-2}, plus waste heat at the power plant. By displacing and generating sensible heat they add to the thermal stress of an already difficult environment (Newcombe, 1979).

Fossil energy is more efficiently converted to heat houses in winter, but the demand, particularly in poorly built structures, is also greater. A typical Milwaukee house of the 1950s could go from a demand of 110 W m^{-2} floor area down to 32 simply by adding 10 cm of insulation and storm windows (Landsberg, 1954), but many did not do this much. The FHA design value for insulation reducing wall transmission to 0.7 W m^{-2} for each °K difference between indoor and outdoor air temperature is double the heat flow out of Swedish houses (Schipper and Lichtenberg, 1976). In addition, heat is lost through the roof and in warming and moistening cold air that penetrates into the house.

Both conduction and infiltration must be considered, as for analysis of heat loss through clothing (Campbell et al., 1980). Moistening (latent-heat flux) is not insignificant; raising the moisture content of polar air from 1–2 grams of vapor per kilogram of air to 8–10 g kg^{-1} requires 8–10 MJ for each change of air in a house, about as much as needed to warm the penetrating air from 0 to 18°C. Infiltration of cold dry air is powered by pressure differences on a building that reflect atmospheric kinetic energy and can be determined for various building groupings by wind tunnel studies (Lee, *et al.*, 1980).

Caulking windows and doors and turning the blank side of the house toward the north in North America are obvious ways to reduce the energy toll, although Landsberg (1975) has pointed out the possible increased indoor air pollution. The effect of construction is seen in a comparison of annual space-heating costs in Minneapolis, 4.4 W m^{-2} mean over the whole city (Karkheck *et al.*, 1977), with those in Stockholm, 2.9 W m^{-2}, much of the latter being efficiently supplied by district heating. Per capita energy conversion for space heating ranges from the United States at 36 GJ year^{-1}* to much colder Sweden at 25 GJ (Schipper and Lichtenberg, 1976) and Japan at 15 GJ year^{-1} per household (Doernberg, 1978) or perhaps 5 GJ year^{-1} per capita, which can be converted to flux density if population density is known. Changes in fuel consumption in a Milwaukee house (Fig. 4) show what is possible.

Household Operations.   Houses are microenvironments but also workplaces where energy is converted in storing and cooking food and caring for clothing. Operational costs in the U.S., 19 GJ year^{-1} per capita (Schipper and Lichtenberg, 1976), work out in a 4-person family on a 600-m^2 lot to 4 W m^{-2}. As much energy is expended on heating water as space in mild climates, and hot-water solar installations have a strong economic appeal; better design of nonsolar heaters can reduce the cost by 0.32 (Hirst and Carney, 1978).

Green Areas.   Many city people set off their immediate microenvironments by vegetation on which they lavish energy in fertilizer, irrigating, and mowing. Fossil energy inputs into one suburban lawn averaged 0.24 W m^{-2} over a year, about half of its primary production of biomass (Falk, 1976). A Milwaukee oak woods produced 0.34 W m^{-2} at a much lower fossil cost (Herte *et al.*, 1971).

Urban Circulations

Moving materials, wastes, water, food, people, and even energy to and from urban ecosystems takes energy, which is expended in the 0.2–0.3 of city area dedicated to circulation systems.

* 1 GJ year^{-1} = 32 W.

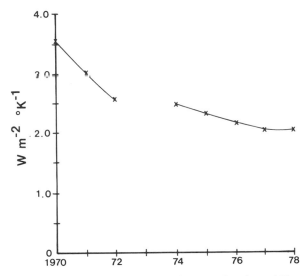

Fig. XXII.4. Decreasing energy conversion for space heating a Milwaukee house (W m⁻² floor area and degree depression of air temperature below 18°C). The initial use of fuel, 60 GJ per capita, was cut in half by a series of small structural changes and two remodelings—"all that was easy to do" [Data from Hayes (1977), and personal communication (1979).]

Water. One of the great changes in urban health in the late 19th century was the provision of clean water and the water-borne removal of domestic and industrial wastes. Pumping water to the consumer is efficient at energy costs less than 0.25 W m⁻² of urban land in Hong Kong and much less in most other cities. Pipes and their installation below the frost level represent an energy cost amortized over many years.

Removal of wastewater is mostly accomplished by gravity flow and with treatment involves energy costs less than 0.1 W m⁻² of urban area. Removal and treatment of storm water, polluted by salt, lead, grease, and other wastes of the automobile circulation system, probably has similar costs. Even the enormous pumps that lift rainwater out of New Orleans (Lewis, 1976) can handle a 150-mm rain at a cost of 25 MW of electric energy, or less than 1 W m⁻² of protected area during a short period.

Recycling of nutrients in wastewater, as in the drying of sludge into a lawn fertilizer (Milorganite), requires fuel, but land application distant from the city involves trucking costs. Methane is generated at many treatment plants in quantities that meet operating needs.

Industrial Materials.   Manufacturing and commercial activity of a city requires moving raw materials and finished and waste products, as well as commodities going into or out of storage. These costs in Hong Kong are 0.3 W m^{-2} (Newcombe, 1975), increased by congestion but lowered by the light nature of the industry.

Collecting urban solid wastes requires larger quantities of fossil energy as landfills become more distant [although landfills also generate methane; a 5-m-deep fill would yield 0.5 W m^{-2} over 5 years (James, 1977)]. Many cities are experimenting with techniques to obtain industrial heat from the combustible fraction of solid wastes, which may be as great as 1 W m^{-2} of city area. Recycling metals, glass, and perhaps eventually paper will reduce the nationwide energy costs of materials.

Energy.   Electrical energy is lost in transit to the amount of 10% or more, and the same is true in distribution of natural or manufactured gas, as leakage and pumping costs. District heating involves circulation of hot water through the city; in Moscow 4000 km of pipes serve this energy-transfer function (USSR, 1978) and must experience some energy losses.

When food and its packaging are transported from store to household by automobile, the energy cost is as great as the energy content of the food itself. Transportation of food to stores and home in Hong Kong requires about 0.5 W m^{-2}, and the store-to-home link alone is nearly 1 W m^{-2} in Sydney (Millington, 1974).

People.   Cities were originally pedestrian societies, and their inhabitants circulated at an areal cost smaller than 1 W m^{-2} metabolic energy. Large cities later supplemented human with animal metabolic energy that was supplied by imported biomass. The hay lands of upstate New York were impoverished to operate New York City's circulation system, and hay was transported from the Hunter Valley floodplain to Sydney by steamboat or train for many years.

Rural circulation systems ran on fossil energy long before urban systems, which only here and there copied steam railroads; instead, two different technologies evolved. One was the transfer in 1873 of cable trams from the mines of California to San Francisco's hills. This mode developed to its greatest extent in the 1880s (Keating, 1970) on the long straight streets of such flat cities as Chicago or Melbourne, where 80 km of line carried 50 million passengers by 1891 (Cannon, 1975, p. 60). The other technology, more efficient on winding streets, was the electric tram, which after 1890–1900 revolutionized cities all over the world by making urban circulation a mass process that extended areal limits and promoted spatial specialization in land uses (McKay, 1976). All this was achieved at a modest energy cost: 1100 J m^{-1} per passenger in American and Swedish cities and only 200 J m^{-1} in Hong Kong (Newcombe, 1975), about the energy cost of walking (160 J m^{-1} for a 70 kg person).

Subsequent shifts to buses and automobiles increased the energy costs of urban circulation to 500 J m^{-1} and 2400 J m^{-1}, respectively, in Hong Kong (Newcombe et al., 1978) and much more in the United States, plus an additional investment in energy content of the vehicles (equivalent of 3.5 years fuel cost; see Newcombe et al., 1978). The later shift from streets to freeways in Los Angeles increased energy flux density from 1 W m^{-2} of city area in 1940 to 4 W m^{-2} in 1967 (Miller, 1971*). Peak rates in a similar city, Melbourne, reach 14 W m^{-2} in rush hours (Sharpe, 1978).† One cause of these high flux densities is the large tare weights of the vehicles in comparison with payload. Americans travel 0.7 farther than Swedes and do so at a unit cost 0.7 greater (Schipper, 1978). We need not dwell upon the energy demands of automobile circulation since their effect on foreign trade has brought them to public notice.

Operating the Circulation System. The 0.2–0.3 of a city devoted to movement of goods and people and storage of vehicles incurs operations and maintenance costs that can be large at times. Street lighting requires large quantities of energy and, in fact, can be used in satellite imagery to estimate total energy use of a city (Welch, 1980).

Although rainwater is removed largely by gravity, storm sewers and pollutant removal entail energy costs. Snow requires mechanical energy to push it to the curbs or bodily haul it to the edge of town in densely built-up cities like Montreal, where energy conversion during the snow-fighting campaign exceeds 1000 W m^{-2} [calculated from data of O'Connor (1975)]. If the snow is disposed of at oil-fired melters, energy costs are very large. Mechanical energy to move the snow comes from thousands of men and women shoveling their walks and drives and from snowblowers and plows burning gasoline and distillate. The quantities involved are not known with precision, but are obviously large; 200,000 men and women each working at the rate of 300–400 W amounts to 70 MW in Milwaukee, and this figure is exceeded by the fuel consumption of snowblowers and jeeps that plow private parking lots and drives.

Salt on streets involves a small direct, but a large indirect, energy cost because it shortens the life of an automobile, calling for earlier replacement, hence earlier conversion of energy in steel and other materials.

---

* These figures are derived from detailed records of fossil energy conversion kept by the Air Pollution Control District Los Angeles County (1969).

† Flux densities are several times greater when referred to street area alone, and perhaps an order of magnitude greater on freeways.

This indirect cost may perhaps average 1 W m^{-2} over a city. Energy invested in initial construction of streets is amortized at as much as 1 W m^{-2} [from data in Scott, (1975)].

Interaction of Circulation and Land Use

The contemporary city concentrates fossil energy for its industrial and commercial activities, for support of microenvironments, and for the circulation systems that link these land uses. Interaction of these components of the urban landscape changed after the 1890s, when the circulation system became powered by electrical energy, as documented for the Boston area by Vance (1952) and Warner (1962). The energized circulation allowed expansion of residential areas and differentiation and intensification of commercial and industrial activities, all of which increased total energy conversions. For example, sites outside Reading, England, need an increase of 0.25–0.30 in fuel (Parry, 1957) in addition to increased transportation costs.

This revolution in urban structure therefore did little to diminish the overall unit-area densities of energy conversion. Flux densities in London in the 1870s were about 20 W m^{-2} of fossil energy (Eaton, 1877), almost entirely for land-use activities since the circulation system was largely pedestrian or horse-powered. Berlin and Wien at the time of World War I displayed about the same flux density (Schmidt, 1917). The effect of urban expansion is illustrated by comparing clustered versus sprawl development (Scott, 1975). Clustered development requires less energy investment per dwelling unit and per unit area in infrastructure (1.3 versus 1.9 W m^{-2} of total area, including roads and open space) because there is less unproductive travel. Operation and maintenance of homes required 3.3 W m^{-2} in the clustered site versus 6.0 W m^{-2} in the sprawl condition. The overall difference (4.6 versus 7.9 W m^{-2}) is 3.3 W m^{-2}. Energy costs of sprawling urban edges appear in extensions of water and wastewater circulation systems, as well as in the operating costs noted (U.S. House of Representatives, 1980, pp. 13, 39).

This comparison suggests why the decreased density of activities and population when western cities expanded in land area did not produce a corresponding decrease in energy flux density, which, rather, on a whole-city basis integrating land uses and circulation,* has remained at

* The overall means include green and derelict lands in the city, where energy budgets receive little augmentation from fossil energy, as well as industries, shopping centers, and office districts that convert energy in great quantities; for example, the City of Sydney, 50 W m^{-2}, or the central parts of Victoria and Kowloon in Hong Kong, 70 W m^{-2} (Kalma and Newcombe, 1976). These areas display a larger diurnal range, the peak being about three times as large as the minimum, than do other parts of the city.

about 20 W m^{-2}. The situation is analogous to the concentration of airborne pollutants; as a city extends itself and lowers the unit-area source strength of pollutant emission, the atmospheric concentration does not decrease because more automobiles are on the roads and the air stream is exposed longer (McCormick, 1968). The vulnerability of cities is also apparent in the uncertainties of fuel-oil supply and the blackouts of New York City and Paris. Freezing rain interrupts the supply of electrical energy almost every winter; summer rains in Chicago cause 0.6 of all electrical outage time (Bertness, 1980).

TEMPERATURE-DEPENDENT ENERGY FLUXES

Fossil-energy conversion and altered radiant-energy intake change the quantity of energy available in an urban ecosystem, usually increasing it and thereby raising surface temperature and expediting the removal of energy. Moreover, the modified urban surface changes the partition of outgoing energy.

Surface Temperature

Fossil-energy augmentation and the radiant-energy intake are unevenly distributed. Streets unshaded by overhanging trees absorb much solar radiation, but large thermal admittance restrains the rise in their daytime temperature and keeps them warm at night. The effect of substrate was shown in a daily range of 31°K in street surface temperature but of 39°K in that of a 2-cm thick sheet of asphalt insulated from the ground (air-temperature range was 13°K) (Mahringer, 1961). The maximum street temperature, 51°C, exceeded air temperature by 22°K. A surface temperature of streets and sidewalks near zero in winter affects their use; occupational-injury and hospital admission records show that falls on ice often cause permanent injury. Two downtown streets in Wien are icy on 0.07 of the days in an average winter (Fischer, 1977) and 0.27 in a severe one; they are snow-covered on 0.12 of the days.

Roofs, being better insulated than streets from their substrates and receiving fossil energy from below in the heating season, probably are as warm as streets on the average. Poorly insulated roofs in winter show up by their relative warmth in a thermal scan while well-insulated roofs retain a mantle of snow that further isolates the building interior.

House walls of brick or masonry display surface temperatures that change with the incidence of direct sun during the day (van Straaten, 1967, p. 83). A west-facing brick wall took in 230 W m^{-2} at 1800 hr, and gave off 180 W m^{-2} at 2100; 32 W m^{-2} peak flow emerged at the inner surface at 2200, keeping it warmer than the interior air. Walls, like streets, store large quantities of daytime heat, which keeps both interior and exterior surfaces warm during the following night. "Megatherm, or massive, exposed buildings deviated the most from the temperature of the air, displaying at midday in early October a range of 41C, from 3C to 44C depending on direction" (Freiburg, 1974, p. 39). Weischet found that "mesotherm" structures in an open layout with yards and street trees varied in surface temperature at midday from 8 to 29°C, a range about half that of massive freestanding structures. "Microtherm," or closely built-up sections of the city, some dating from medieval days (i.e., pedestrian or human in scale), are controlled by diffuse rather than beam solar radiation, and at midday their walls varied from 5 to 25°C. From noon on, all these walls were colder than the air (Freiburg, 1974). Orientation obviously affects wall surface temperature and excesses over air temperature of walls in Wien showed a consistent relation to solar radiation received (Mahringer, 1961).

Small parks, narrow boulevards, and parks with few trees in Moscow displayed less cooling than large parks (Dmitriev and Bessonov, 1969, p. 253). Parks appear cool at midday on thermal scans, as also do older residential areas with many trees, e.g., in Ann Arbor (Outcalt, 1972b) or the new city of Columbia, Maryland (Landsberg, 1979), where midday surface temperature rise is related to built-up fraction. The green-area fraction in a city has a major effect on its overall temperature (Rauner and Chernavskaia, 1972), particularly in summer (Outcalt, 1972a). The association is not one-sided; urban green areas are warmer than rural ecosystems in fall and enjoy a longer growing season (Kratzer, 1956, p. 55).

The diurnal march of surface temperature has the same shape at urban as at rural sites, but is more peaked—from 10 to 37°C at the urban site on a summer day and only from 8 to 24°C at the rural site (Dabberdt and Davis, 1978).

Longwave Radiation

As in all ecosystems, a rise in surface temperature increases the removal of energy by longwave radiation by the fourth-power law. Where surfaces are dry, the primacy of this mode of energy flow becomes marked; it reached midday peaks of 550 W m^{-2} from the central business district and a steel-mill area in Baltimore (Pease et al., 1976), 100 W m^{-2} larger than the flux from a large park. High rates of longwave radiation from hot streets—perhaps 100 W m^{-2} more than from grass—depress the net surplus of both longwave and all-wave radiation; on Manezhnaya Square in Moscow the whole-day all-wave surplus in summer averaged 113 W m^{-2}, 15 less than at a grass plot at Moscow University (Dmitriev and Bessonov, 1969, pp. 174, 253). In winter longwave flux, although smaller, is yet a major drain on the warmth of a city.

Many urban surfaces face each other more or less directly, and internal transfer of energy among themselves by longwave radiation forms a heat trap that adds to the summer burden on human beings. The upward escape of radiation from these cavities is readily measured at roof level (Fuggle and Oke, 1976), and remote sensing can provide hitherto unobtainable detail on surface temperature patterns.

Substrate Heat Flux

This periodically reversing heat flux is large in many urban ecosystems, where soil is compacted or covered by concrete and walls afford a large, accessible storage capacity. The heat removed from the active surface in midday is captive and returns at night slowing the rate of cooling to 1°C hr^{-1}, half or less of that in rural ecosystems (Oke, 1978, p. 256), and worsening the thermal environment. Nocturnal temperatures in a small city in Japan showed a good relation to the spatial pattern of thermal admittance (Kawamura, 1965). The return flux of heat in residential districts of Sacramento, even at the end of the night, was 5–10 W m^{-2} (Morgan et al., 1977), and perhaps as much as 50 W m^{-2} to the paved surfaces of industrial and shopping centers.

The transfer of substrate heat from summer to winter is hard to identify in the total energy budget, except when it is artificially accelerated by forced circulation, as in heat-pump installations used to reduce overall fossil-energy costs (Hirst and Hannon, 1979). Sporadic use of such convective substrate-heat transfer for melting snow in mild-winter climates can produce flux densities of 100–200 W m^{-2} (Winters, 1970), or even more on bridge ramps, as described in Chapter XVIII.

Turbulent Energy Fluxes

Both latent-heat and sensible-heat fluxes are influenced by ventilation, which tends to be reduced by low wind speeds in urban landscapes but increased by surface roughness and represents one of the most difficult aspects of modeling the energy budget. In a specific ecosystem it reflects shelter or exposure relative to neighboring systems and is also affected by strong winds at the surface induced by tall buildings (Oke, 1978, p. 234). The competition of the two turbulent fluxes, in cities as elsewhere, depends on surface wetness.

Latent-Heat Flux. The latent-heat flux is generated by streets and walls only until the millimeter or less of water retained after a rain is evaporated; the annual total depends on how often they are wetted. The small latent-heat flux from industrial areas in St. Louis in dry weather showed up as drier volumes in the mixing layer (Sisterson and Dirks, 1978).

Some urban ecosystems generate vapor from fuel combustion (e.g., ice fog in Fairbanks), from some industrial processes, from air-conditioning systems in summer, and from leaks from humidified buildings in winter. Most gardens are irrigated and transpire as much as rural ecosystems, if not more, since they benefit by heat advected from dry ecosystems (Oke, 1979) depending on neighborhood composition and watering practices (Suckling, 1980). Trees in older residential areas in Sacramento produce a mean latent-heat flux nearly double that from newer areas (Morgan et al., 1977).

The net effect of the typical urban mixture of dry and moist ecosystems is not the same everywhere. Cities like Dallas in formerly dry sites seem to develop a more humid local atmosphere, and in other cities the net result is not clear, as in Columbia, Maryland (Landsberg, 1979), and may vary between day and night (Kopec, 1973).

Sensible-Heat Flux.   No such uncertainty attends the movement of sensible heat from the urban surface. Even in humid climates, where it competes with the flux of latent heat, it is larger than in rural landscapes, and the result, the urban heat island, has been observed and discussed for 150 years. As early as 1927, Schmidt was making traverses with 5000 temperature measurements across Wien (Eckel, 1957). However, the sensible-heat flux is not always well understood (Oke, 1978) and is hard to measure in view of "the complexity of the physics of active surfaces within the city" (Chandler, 1976, p. 10). Controversy has existed over such possible sources as excess absorption of radiant energy or the role of fossil energy, as well as the competitive roles of substrate heat flux and evapotranspiration, and even the effects of roughness and air pollution, factors that are not necessarily additive (Welch et al., 1978).

The fact that the heat island of Uppsala, for instance, is strongest on clear nights suggests the importance of incoming longwave radiation from urban air (Sundborg, 1951); its smaller size in windy weather suggests the importance of turbulent transport of sensible heat upward and out of the city. However, the other factors noted are also significant on occasion, and it is necessary to view the sensible-heat flux in the context of the whole energy budget.

Much fossil-energy conversion takes place in dry ecosystems and tends to favor the sensible-heat flux, especially in winter and when radiant-energy intake is small. Country air moving into Sydney over traffic and industrial heat sources in the morning rush hour developed an internal boundary layer that thickened to 200–300 m and warmed by 4–5°C, calculated and observed values being in agreement (Kalma, 1974). Such a boundary-heating model was also utilized in Calgary (Nkemdirim, 1976) in defining heating above roof level, and identifying wind and stability effects (Nkemdirim and Truch, 1978), especially the role of upwind lapse rate (Nkemdirim, 1980a). Cold-air drainage from mountain valleys into piedmont cities like Denver, Freiburg (Nübler, 1979), and Calgary (Nkemdirim, 1980b) affects the urban energy budget strongly.

The large carryover of "captive" substrate heat (Carlson and Boland, 1978) maintains surface temperature in the central city higher than air temperature through most, if not all, the night, a situation that is uncommon in rural landscapes except at lakes or rock ledges. The continued sensible-heat flux accentuates the nocturnal heat island and changes the thermal structure aloft so that inversions are often found above the city rather than at the surface.

Rural air pushes into the city from all sides (Pooler, 1963) by day as well as night (Shreffler, 1979), raising the dust dome and even producing cumulus clouds (Geiger, 1942, p. 356). But rural air brings insufficient relief in hot weather; continued sensible-heat flux into the air layer below roof level stresses the human heat-control apparatus (Clarke, 1972), and the heat island becomes a death island (Buechley et al., 1972). Excess deaths in July 1966 in New York City (1300 such deaths) and elsewhere displayed a clear relation to population density, i.e., to built-up areas with large thermal storage.

Absorptivity of the urban surface enhances the daytime sensible-heat flux, which reaches rates of 200–300 W m^{-2} from roofs and urban canyons (Yap and Oke, 1974; Nunez and Oke, 1977). These heat emitters show up in satellite imagery of New England (Matson and Legeckis, 1980) much more clearly than residential parts of these cities.

Measurements of temperature in the air below roof level, e.g., by automobile traverses, do not define the magnitude of the sensible-heat flux as well as would measurements of flux and temperature at roof level, i.e., the base of the true boundary layer (Oke, 1976; Pease et al., 1976; Carlson and Boland, 1978). The heating of this boundary layer is sufficient to destabilize conditionally unstable air (Harnack and Landsberg, 1975), and deep mixing by day weakens the urban–rural contrast, e.g., in St. Louis (Vukovich et al., 1979), to less than half its nocturnal intensity, producing the well-documented and remotely sensed (Matson et al., 1978) urban effect on daily minima that is greater than on daily maxima

(Kratzer, 1956, p. 55). This effect also shows up in recent increases of the heat island in Wien (Böhm, 1979b). A similar explanation applies to the more intense heat island of Budapest in winter, as compared with summer (Probald, 1972), and to the occurrence of a daytime cold island in Birmingham (Unwin, 1980). Mixing depth was also important over Montreal (Oke and East, 1971), where helicopter soundings identified both combustion and solar-absorption components in the flux of sensible heat determining the temperature field aloft (East, 1971).

RESPONSES OF URBAN ECOSYSTEMS

The preceding discussion of the sensible-heat flux from urban land-scapes and the heat island it creates touched on several explanations for its size and variations with time, many of them only speculative until set in the framework of the urban energy budget (Yoshino, 1975, p. 93). Modeling efforts to reconstruct the energy budget (Chandler, 1976, p. 36) are becoming more realistic in their assumptions about interface characteristics.

As in other ecosystem energy budgets, surface temperatures mediate change. In this respect urban landscapes differ not at all from rural, though responses may be obscured by interactions among their many ecosystems—cool parkland air entering an urban canyon, for instance. They do differ in fossil-energy augmentation; the response to a polar invasion is to start thousands of furnaces laboring to maintain steady indoor microenvironments in spite of diminished intake of longwave radiation and increased sensible-heat flow into the cold air.

Some models deal with the mixing depth in rural air moving over a city, like Kalma's (1974) for Sydney, Neiburger's (1957) for Los Angeles, and a general one of Henderson-Sellers (1980). Others calculate surface temperature from the interface energy budget at as many as 600 grid points [as in Baltimore, see Pease et al., (1976); Pease et al. (1980)], some-times without inclusion of fossil energy, sometimes with it. Of the in-terface parameters used, thermal admittance and wetness appear to be the most important (Carlson and Boland, 1978), especially in summer nights and days, respectively; fossil energy and absorptivity, especially of low-angle beam radiation (Pease et al., 1976) may be most important in winter.*

* A note on kinetic energy is needed to complete the energy budget. Neiburger (1957) estimated the energy cost of moving polluted air out of Los Angeles at more than 4 W m^{-2}, which seems of the right order for ordinary wind regimes. Kinetic energy increases in streams whose regime is altered by urbanization; Cherkauer (1978) found an increase

The role of fossil energy is large in the small budgets of winter nights, when it augments incoming longwave radiation, the only sizable input, by 0.1 –0.2. It is even larger when compared to winter solar radiation, although this conventional comparison is unrealistic in high-latitude winter because solar radiation is a manyfold smaller contributor to the budget than longwave. In comparison with the turbulent heat fluxes fossil energy is minor by day, when these fluxes total hundreds of watts per square meter, but at night may become of equal size and may be responsible, along with the substrate-heat flux, in maintaining an upward direction of the flux of sensible heat in a significant deviation from rural conditions. Poleward of the 40th parallel in winter when mean rural turbulent fluxes are directed downward to meet a deficit in all-wave radiation, their contribution may be no larger than fossil-energy conversions in urban landscapes. A few models of the urban energy budget take account of fossil energy; one of these, for Hong Kong, is driven by the diurnal marches of fossil and solar energy in several land-use areas (Kalma et al., 1978). Near-surface temperatures varied through a July day from 26 to 29°C in residential areas and from 26 to 37°C over streets, which thus should not be taken as "representative of the remainder of the urban area," as they often are in heat-island traverses.

Fossil-energy conversions in the energy budget are intrusions into the present day of a photosynthesis–decomposition cycle left unfinished in a past epoch, pulses of energy from far outside the contemporary budget. This intrusive character is evident in their time distributions, which are not the regular solar regimes of summer and winter, day and night, nor yet the episodic fluctuations in radiant-energy intake or turbulent fluxes that occur when synoptic systems pass, but rather, the work day and work week in economic activities and urban circulation and the counterseasonal energy conversions to heat our buildings.

This augmentation and the altered urban interface and atmosphere produce a spatially complex set of rhythms in the rates at which incoming solar, longwave, and fossil energy are converted into other forms. Incoming energy fluxes are altered by the urban atmosphere even before they arrive at the interface and are absorbed at surfaces that often differ from natural ones. Conversions are modified by dryness, shelter, thermal admittance, and other properties that occur in a unique mix in the

---

from 1 MJ to 62 MJ. This value is small when referred to the whole drainage basin (approximately 0.4 μW m^{-2}), but when concentrated on the channels, it can bring about substantial geomorphic changes, as have occurred in many cities (Graf, 1975; Dunne and Leopold, 1978, pp. 693–701). Kinetic energy of mass movements is generated when alterations of the urban interface aggravate landslides, e.g., in Seattle (Dunne and Leopold, 1978, pp. 18–22).

city and control the fractions of available energy moving upward or downward from the interface or moving in sensible or latent form. In consequence, the outflows of energy from urban systems differ in form as well as quantity from the outflows from rural ecosystems. More heat goes underground to return in an unnaturally warmed night. More energy is radiated in the long wavelengths. More sensible heat is given off, creating a heat island and affecting atmospheric quality and downwind processes, which demonstrate the magnitude of the alterations in the urban budget. Local effects shape our environment: the long duration of the sensible-heat flux, the increased field of longwave radiation, the heat traps of dwellings and streets, the noise of air-conditioners that blots out the noise from the freeways, the treasuring of green areas for the services they provide for human beings (Stearns and Montag, 1974, p. 11). The city is a fascinating artifact; even if it were inhabited by ants rather than human beings, it would still be a worthy object of study (Miller, 1965) by reason of its complicated structure and functions. Its energy budget reveals some of its workings and shows our vulnerability to the costly and unreliable flow of fossil energy that operates it and makes it habitable.

REFERENCES

Ackerman, T. P. (1977). A model of the effect of aerosols on urban climates with particular applications to the Los Angeles Basin. *J. Atmos. Sci.* **34**, 531–547.

Aida, M., and Yaji, M. (1979). Observations of atmospheric downward radiation in the Tokyo area. *Boundary-Layer Meteorol.* **16**, 453–465.

Air Pollution Control District Los Angeles County (1969). "Profile of Air Pollution Control in Los Angeles County" (E. E. Lemke, G. Thomas, and W. E. Zwincher, eds.). Air Poll. Cont. Dist., Los Angeles, California.

Arai, T., Takayama, S., Takamura, H., Sekine, K., Tateishi, Y., Kobayashi, T., and Shyoda, M. (1975). Space and time variations of atmospheric CO_2 in Tokyo. *Geogr. Rev. Jpn.* **48**, 412–417.

Assaf, G., and Biscaye, P. E. (1972). Lead-212 in the urban boundary layer of New York City. *Science* **175**, 890–894.

Bach, W., and Patterson, W. (1969). Heat budget studies in greater Cincinnati. *Proc. Assoc. Am. Geogr.* **1**, 7–11.

Banham, R. (1969). "The Architecture of the Well-Tempered Environment." Univ. of Chicago Press, Chicago, Illinois.

Bergstrom, R. W., and Peterson, J. T. (1977). Comparison of predicted and observed solar radiation in an urban area. *J. Appl. Meteorol.* **16**, 1107–1116.

Bertness, J. (1980). Rain-related impacts on selected transportation activities and utility services in the Chicago area. *J. Appl. Meteorol.* **19**, 545–556.

Böhm, R. (1979a). Meteorologie und Stadtplanung in Wien—ein Überblick (Teil I: Temperatur, Niederschlag). *Wetter Leben* **31**, 1–11.

Böhm, R. (1979b). Stadentwicklung und Trend der Wärmeinselintensität. *Arch. Meteorol., Geophys. Bioklimatol., Ser. B* **27**, 31–46.

Brady, R. F., Tobias, T., Eagles, P. F. J., Ohrner, R., Micak, J., Veale, B., and Dorney, R. S. (1979). A typology for the urban ecosystem and its relationship to larger biogeographical landscape units. *Urban Ecol.* **4**, 11–28.

Brazel, A. J., and Osborne, R. (1976). Observations of atmospheric thermal radiation at Windsor, Ontario, Canada. *Arch. Meteorol., Geophys. Bioklimatol., Ser. B* **24**, 189–200.

Bridgman, H. A. (1978). Direct visible spectra and aerosol optical depths at urban and rural locations during the summer of 1975 at Milwaukee. *Sol. Energy* **21**, 139–148.

Bridgman, H. A. (1979a). Measured and theoretical particle size distribution over industrial and rural locations at Milwaukee; April 1976. *Atmos. Environ.* **13**, 629–638.

Bridgman, H. A. (1979b). Urban-rural shortwave radiation budget comparisons at Newcastle. *In* "Newcastle Studies in Geography" (J. C. R. Camm and R. J. Loughran, eds.), pp. 15–21, Univ. Newcastle Dept. Geog., Newcastle, New South Wales, Australia.

Buechley, R. W., van Bruggen, J., and Truppi, L. E. (1972). Heat island = death island? *Environ. Res.* **5**, 85–92.

Campbell, G. S., McArthur, A. J., and Monteith, J. L. (1980). Windspeed dependence of heat and mass transfer through coats and clothing. *Boundary-Layer Meteorol.* **18**, 485–493.

Cannon, M. (1975). "Australia in the Victorian Age. 3: Life in the Cities." Nelson, Melbourne and Sydney.

Carlson, T. N., and Boland, F. E. (1978). Analysis of urban-rural canopy using a surface heat flux/temperature model. *J. Appl. Meteorol.* **17**, 998–1013.

Chandler, T. J. (1976). Urban climatology and its relevance to urban design. *WMO Tech Note* No. 149 (WMO No. 438).

Cherkauer, D. S. (1978). The effect of urbanization on kinetic energy distributions in small watersheds. *J. Geol.* **86**, 505–515.

Clarke, J. F. (1972). Some effects of the urban structure on heat mortality. *Environ. Res.* **5**, 93–104.

Cook, E. (1977). Energy: The ultimate resource? *Assoc. Am. Geogr. Resour. Pap.* **77-4**. Washington, D.C.

Coppin, P. A., Forgan, B. W., Penny, C. L., and Schwerdtfeger, P. (1978). Zonal characteristics of urban albedos. *Urban Ecol.* **3**, 365–369.

Dabberdt, W. F., and Davis, P. A. (1978). Determination of energetic characteristics of urban–rural surfaces in the greater St. Louis area. *Boundary-Layer Meteorol.* **14**, 105–121.

Daniels, G. E., ed. (1973). Terrestrial environment (climatic) criteria guidelines for use in aerospace vehicle development, 1973 revision. *U.S. Nat. Aeronaut. Space Agency, NASA Tech. Memo.* TM X-64757.

DeLuisi, J. J., Bonelli, J. E., and Shelden, C. E. (1977). Spectral absorption of solar radiation by the Denver brown (pollution) cloud. *Atmos. Environ.* **11**, 829–836.

DeWalle, D. R. (1979). Microclimate modification using urban forests for residential energy conservation in space heating and cooling. *In* Reifsnyder, W. E. (dir.), *Symp. Forest Meteorol., Proc.* pp. 84–86, World Meteorological Organization, Geneva.

Dmitriev, A. A., and Bessonov, N. P., eds. (1969). "Klimat Moskvy (Osobennosti Klimata Bol'shogo Goroda)." Gidrometeorol. Izd., Leningrad.

Doernberg, A. (1978). Energy use in Japan and the United States. *In* "International Comparisons of Energy Consumption" (J. Dunkerly, ed.), Res. Pap. R-10, pp. 56–81. Resources for the Future, Washington, D.C.

Dunne, T., and Leopold, L. B. (1978). "Water in Environmental Planning." Freeman, San Francisco, California.

East, C. (1971). Chaleur urbaine à Montréal. *Atmosphere* **9**, 112–122.

Eaton, H. S. (1877). Presidential address. *Q. J. R. Meteorol. Soc.* **3**, 309–317.

Eckel, O. (1957). Die Temperaturverteilung in Wien und Umgebung. *In* "Klima und Bioklima von Wien," Part II (F. Steinhauser, O. Eckel, and F. Sauberer, eds.), pp. 108–134. Magistrat der Stadt, Vienna.

Ehler, L. E. (1978). Some aspects of urban agriculture. In "Perspectives in Urban Entomology" (G. W. Frankie and C. S. Koehler, eds.), pp. 349–367. Academic Press, New York.

Falk, J. H. (1976). Energetics of a suburban lawn ecosystem. Ecology 57, 141–150.

Fischer, P. L. (1977). Der Winter 1976/77 in der Stadtmitte. Wetter Leben 29, 257–259.

Flowers, E. C., McCormick, R. A., and Kurfis, K. R. (1969). Atmospheric turbidity over the United States, 1961–1966. J. Appl. Meteorol. 8, 955–962.

Freiburg, Interdisziplinäre Arbeitsgruppe (1974). "Untersuchung der klimatischen und lufthygienischen Verhältnisse der Stadt Freiburg i. Br." Stadtplanungsamt Freiburg i. Br.

Fuggle, R. F., and Oke, T. R. (1976). Long-wave radiative flux divergence and nocturnal cooling of the urban atmosphere. I: Above roof-level. Boundary-Layer Meteorol. 10, 113–120.

Geiger, R. (1942). "Das Klima der bodennahen Luftschicht, ein Lehrbuch der Mikroklimatologie," Die Wissenschaft, 2nd ed., Vol. 78. Vieweg, Braunschweig.

Gorbacheva, V. M. (1974). Vliianie gorodskoi zastroiki na meteorologicheskii rezhim territorii v usloviiakh Krainego Severa. In "Klimat i Gorod, Materialy Konferentsii Klimat-Gorod-Chelovek" (V. M. Zhukov, ed.), pp. 146–148. Poligrafist Polygraficheskoe Obedinenie, Moscow.

Graf, W. L. (1975). The impact of suburbanization on fluvial geomorphology. Water Resour. Res. 11, 690–692.

Gutman, D. P., and Torrance, K. E. (1975). Response of the urban boundary layer to heat addition and surface roughness, Boundary-Layer Meteorol. 9, 217–234.

Hanna, S. R. (1977). Predicted climatology of cooling tower plumes from energy centers. J. Appl. Meteorol. 16, 880–887.

Hannon, B. (1972). Bottles, cans, energy. Environment 14(2), 11–21.

Hannon, B., Stein, R. G., Segal, B. Z., and Serber, D. (1978). Energy and labor in the construction sector. Science 202, 837–847.

Harnack, R. P., and Landsberg, H. E. (1975). Selected cases of convective precipitation caused by the metropolitan area of Washington, D.C. J. Appl. Meteorol. 14, 1050–1060.

Harris, J. R. (1974). The rise of coal technology. Sci. Am. 231(2), 92–97.

Hawthorne, Sir W. (1980). Introduction [to discussion on solar energy]. Phil. Trans. R. Soc. Lond. A 295, 345–347.

Hayes, P. G. (1977). Here's how 1 family cut its energy costs. Milwaukee J. Oct. 13, pp. 1, 19.

Heisler, G. M. (1974). Trees and human comfort in urban areas. J. For. 72, 466–469.

Henderson-Sellers, A. (1980). A simple numerical simulation of urban mixing depths. J. Appl. Meteorol. 19, 215–218.

Herte, M., Kobriger, N., and Stearns, F. (1971). Productivity of an urban park. Univ. Wis.-Milwaukee Field St. Bull. 4(2), 14–18.

Hicks, B. B., Wesely, M. L., and Sheih, C. M. (1977). A study of heat transfer processes above a cooling pond. Water Resour. Res. 13, 901–908.

Hirst, E., and Carney, J. (1978). Effects of Federal residential energy conservation programs. Science 199, 845–851.

Hirst, E., and Hannon, B. (1979). Effects of energy conservation in residential and commercial buildings. Science 205, 656–661.

Idso, S. B. (1974). Thermal blanketing: A case for aerosol-induced climatic alteration. Science 186, 50–51.

James, S. C. (1977). The indispensable (sometimes intractable) landfill. Technol. Rev. 79(4), 38–47.

Kakela, P. J. (1978). Iron ore: Energy, labor, and capital changes with technology. *Science* **202,** 1151–1157.

Kalma, J. D. (1974). An advective boundary layer model applied to Sydney, Australia. *Boundary-Layer Meteorol.* **6,** 351–361.

Kalma, J. D., and Newcombe, K. J. (1976). Energy use in two cities: A comparison of Hong Kong and Sydney, Australia. *Environ. Stud.* **9,** 53–64.

Kalma, J. D., Johnson, M., and Newcombe, K. J. (1978). Energy use and the atmospheric environment in Hong Kong: Part II. Waste heat, land use and urban climate. *Urban Ecol.* **3,** 59–83.

Karkheck, J., Powell, J., and Beardsworth, E. (1977). Prospects for district heating in the United States. *Science* **195,** 948–955.

Kawamura, T. (1965). Some considerations on the cause of city temperature at Kumagaya City. *Tokyo J. Climatol.* **2**(2), 38–40.

Keating, J. D. (1970). "Mind the Curve! A History of the Cable Trams." Melbourne Univ. Press, Melbourne.

Kopec, R. J. (1973). Daily spatial and secular variations of atmospheric humidity in a small city. *J. Appl. Meteorol.* **12,** 639–648.

Kramer, M. L., Seymour, D. E., Smith, M. E., Reeves, R. W., and Frankenburg, T. T. (1976). Snowfall observations from natural-draft cooling tower plumes. *Science* **193,** 1239–1241.

Kratzer, P. A. (1956). "Das Stadtklima," 2nd ed. Vieweg, Braunschweig.

Kraujalis, M.-W. (1975). Die räumliche Verteilung der Jahressummen der Wärme aus künstlichen Quellen im Gebiet der Volksrepublik Polen. *Z. Meteorol.* **25,** 312–317.

Kung, E. C., Bryson, R. A., and Lenschow, D. H. (1964). Study of a continental surface albedo on the basis of flight measurements and structure of the earth's surface cover over North America. *Mon. Weather Rev.* **92,** 543–564.

Landsberg, H. E. (1954). Bioclimatology of housing. *Meteorol. Monogr.* **2**(8), 81–98.

Landsberg, H. E. (1975). Housing: Introductory remarks. *Environmental Health Perspectives* **10,** 223–224.

Landsberg, H. E. (1979). Atmospheric changes in a growing community (the Columbia, Maryland, experience). *Urban Ecol.* **4,** 53–81.

Landsberg, H. E., and Maisel, T. N. (1972). Micrometeorological observations in an area of urban growth. *Boundary-Layer Meteorol.* **2,** 20–25.

Lee, B. E., Hussain, M., and Soliman, B. (1980). Predicting natural ventilation forces upon low-rise buildings. *ASHRAE J.* **22**(2), 35–39.

Lettau, H. H. (1969). Note on aerodynamic roughness-parameter estimation on the basis of roughness-element description. *J. Appl. Meteorol.* **8,** 828–832.

Lewis, P. F. (1976). "New Orleans—The Making of an Urban Landscape." Ballinger, Cambridge, Massachusetts.

Liang, C. S. (1973). "Urban Land Use Analysis. A Case Study on Hong Kong." Ernest Publ., Hong Kong.

Lister, R., and Lemon, E. (1976). Interactions of atmospheric carbon dioxide, diffuse light, plant productivity, and climate processes—model predictions. *In* "Atmosphere–Surface Exchange of Particulate and Gaseous Pollutants (1974)." (R. J. Englemann and G. A. Sehmel, coordinators), pp. 112–135. ERDA, Washington, D.C.

Ludwig, F. L. (1970). Urban temperature fields. *In* Urban Climates. *WMO Tech. Note* No. 108, 80–107.

McCormick, R. A. (1968). Air pollution climatology. *In* "Air Pollution" (A. C. Stern, ed.), 2nd ed., pp. 275–320. Academic Press, New York.

McKay, J. P. (1976). "Tramways and Trolleys: The Rise of Urban Mass Transport in Europe." Princeton Univ. Press, Princeton, New Jersey.

McRae, J. E., and Graedel, T. E. (1979). Carbon dioxide in the urban atmosphere: Dependencies and trends. *J. Geophys. Res.* **84,** 5011–5017.

Mahringer, W. (1961). Studie über die Oberflächentemperatur von Gebäuden und Strassendecken in Wien. *Wetter Leben* **13,** 145–155.

Mahringer, W. (1963). Ein Beitrag zum Klima von Höfen im Wiener Stadtbereich. *Wetter Leben* **15,** 137–146.

Mather, E. C. (1972). The American Great Plains. *Ann. Assoc. Am. Geogr.* **62,** 237–257.

Matson, M., and Legeckis, R. V. (1980). Urban heat islands detected by satellite. *Bull. Am. Meteorol. Soc.* **61,** 212 and cover.

Matson, M., McClain, E. P., McGinnis, D. F., Jr., and Pritchard, J. A. (1978). Satellite detection of urban heat islands. *Mon. Weather Rev.* **106,** 1725–1734.

Miller, D. H. (1965). Geography, physical and unified. *Prof. Geogr.* **17**(2), 1–4.

Miller, D. H. (1970). Energy-mass budget and the pelletizing process. *Ann. Assoc. Am. Geogr.* **60,** 406–407.

Miller, D. H. (1971). Kontsepsiia energo- i massoobmena v prirodnoi srede kak metod analiza iavleniia, obuslavlivaemykh deiatel'nost'iu cheloveka. *Izv. Akad. Nauk SSSR, Ser. Geogr.* No. 2, 118–133.

Millington, R. J. (1974). Energy down on the farm. *Rural Res. (CSIRO)* **85,** 4–8.

Morgan, D., Myrup, L., Rogers, D., and Baskett, R. (1977). Microclimates within an urban area. *Ann. Assoc. Am. Geogr.* **67,** 55–65.

Munn, R. E. (1966). "Descriptive Micrometeorology." Academic Press, New York.

Nef, J. U. (1977). An early energy crisis and its consequences. *Sci. Am.* **237**(5), 140–151.

Neiburger, M. (1957). Weather modification and smog. *Science* **126,** 637–645.

Newcombe, K. (1975). Energy use in Hong Kong: Part II, Sector end-use analysis. *Urban Ecol.* **1,** 285–309.

Newcombe, K. (1976). Energy use in Hong Kong: Part III, Spatial and temporal patterns. *Urban Ecol.* **2,** 139–179.

Newcombe, K. (1979). Energy use in Hong Kong: Part IV, Socioeconomic distribution, patterns of personal energy use, and the energy slave syndrome. *Urban Ecol.* **4,** 179–205.

Newcombe, K., Kalma, J. D., and Aston, A. R. (1978). The metabolism of a city: The case of Hong Kong. *Ambio* **7**(1), 3–15.

New York Times (1978). Saving $100 million in energy is termed possible in New York. *N.Y. Times* July 9, p. 41.

Nicholas, F. W., and Lewis, J. E., Jr. (1980). Relationships between aerodynamic roughness and land use and land cover in Baltimore, Maryland. *U.S. Geol. Surv. Prof. Pap.* No. 1099-C.

Nkemdirim, L. C. (1976). Dynamics of an urban temperature field—a case study. *J. Appl. Meteorol.* **15,** 818–828.

Nkemdirim, L. C. (1980a). A test of a lapse rate/wind speed model for estimating heat island magnitude in an urban airshed. *J. Appl. Meteorol.* **19,** 748–756.

Nkemdirim, L. C. (1980b). Cold air drainage and temperature fields in an urban environment: A case study of topographical influence on climate. *Atmos. Environ.* **14,** 375–381.

Nkemdirim, L. C., and Truch, P. (1978). Variability of temperature fields in Calgary, Alberta. *Atmos. Environ.* **12,** 809–822 (see also discussion by H. E. Landsberg on p. 2035).

Nübler, W. (1979). Konfiguration und Genese der Wärmeinsel der Stadt Freiburg. *Freiburger geog. Hefte* 16.

Nunez, M., and Oke, T. R. (1976). Long-wave radiative flux divergence and nocturnal cooling of the urban atmosphere. II: Within an urban canyon. *Boundary-Layer Meteorol.* **10,** 121–135.

Nunez, M., and Oke, T. R. (1977). The energy balance of an urban canyon. *J. Appl. Meteorol.* **16,** 11–19.

O'Connor, D. A. (1975). Snow in Montreal doesn't bring out any white flags. *Wall Street Journal* Feb. 6, pp. 1, 21.

Oke, T. R. (1976). The distinction between canopy and boundary-layer urban heat islands. *Atmosphere* **14**, 268–277.

Oke, T. R. (1978). "Boundary Layer Climates." Methuen, London.

Oke, T. R. (1979). Advectively-assisted evapotranspiration from irrigated urban vegetation. *Boundary-Layer Meteorol.* **17**, 167–173.

Oke, T. R., and East, C. (1971). The urban boundary layer in Montreal. *Boundary-Layer Meteorol.* **1**, 411–437.

Oke, T. R., and Fuggle, R. F. (1972). Comparison of urban/rural counter and net radiation at night. *Boundary-Layer Meteorol.* **2**, 290–308.

Oke, T. R., and Hannell, F. G. (1970). The form of the urban heat island in Hamilton, Ontario. *In* Urban Climates. *WMO Tech. Note* No. 108, 113–126.

Outcalt, S. I. (1972a). A synthetic analysis of seasonal influences in the effects of land use on the urban thermal regime. *Arch. Meteorol., Geophys. Bioklimatol., Ser. B* **20**, 253–260.

Outcalt, S. I. (1972b). A reconnaissance experiment in mapping and modelling the effect of land use on urban thermal regimes. *J. Appl. Meteorol.* **11**, 1369–1373.

Parry, M. (1957). Local climates and house heating. *Adv. Sci.* **13**(52), 326–331.

Pease, R. W., Lewis, J. E., and Outcalt, S. I. (1976). Urban terrain climatology and remote sensing. *Ann. Assoc. Am. Geogr.* **66**, 557–569.

Pease, R. W., Jenner, C. B., and Lewis, J. E., Jr. (1980). Personal communication. (page proofs of *U.S. Geol. Surv. Prof. Pap.* No. 1099A)

Peterson, J. T., and Stoffel, T. L. (1980). Analysis of urban-rural solar radiation data from St. Louis, Missouri. *J. Appl. Meteorol.* **19**, 275–283.

Peterson, J. T., Flowers, E. C., and Rudisill, J. H., III (1978). Atmospheric turbidity across the Los Angeles basin. *J. Appl. Meteorol.* **17**, 428–435.

Pimentel, D., Dritschilo, W., Krummel, J., and Kutzman, J. (1975). Energy and land constraints in food protein production. *Science* **190**, 754–761.

Pooler, F. (1963). Airflow over a city in terrain of moderate relief. *J. Appl. Meteorol.* **2**, 446–456.

Probald, F. (1972). Deviations in the heat balance: the basis of Budapest's urban climate. *In* "International Geography 1972" (W. P. Adams and F. M. Helleiner, eds.), pp. 184–186. Univ. of Toronto Press, Toronto.

Rauner, Iu. L., and Chernavskaia, M. M. (1972). Teplovoi balans goroda i vliianie gorodskogo ozeleniia na temperaturnyi rezhim. *Izv. Akad. Nauk SSSR, Ser. Geogr.* No. 5, 46–53.

Rouse, W. R., Noad, D., and McCutcheon, J. (1973). Radiation, temperature and atmospheric emissivities in a polluted urban atmosphere at Hamilton, Ontario. *J. Appl. Meteorol.* **12**, 798–807.

Schipper, L. (1978). The Swedish-U.S. energy use comparison and beyond: Summary. *In* "International Comparisons of Energy Consumption" (J. Dunkerley, ed.), Res. Pap. R-10, pp. 47–51. Resources for the Future, Washington, D.C.

Schipper, L., and Lichtenberg, A. J. (1976). Efficient energy use and well-being: The Swedish example. *Science* **194**, 1001–1013.

Schmidt, W. (1917). Zum Einfluss grosser Städte auf das Klima. *Naturwissenschaften* **5**, 494–495.

Scott, R. E., ed. (1975). "The Costs of Sprawl: Detailed Cost Analysis. Vol. II. Management and Control of Growth." Urban Land Inst., Washington, D.C.

Sharpe, V. (1978). The effect of urban form on transport energy patterns. *Urban Ecol.* **3**, 125–135.

Shaver, D. B., Pappas, J. L., and Foell, W. K. (1974). A model of industrial energy use in Wisconsin. *Univ. Wis. Inst. Environ. Stud., Rep.* No. 33.

Shreffler, J. H. (1979). Heat island convergence in St. Louis during calm periods. *J. Appl. Meteorol.* **18**, 1512–1520.

Sisterson, D. L., and Dirks, R. A. (1978). Structure of the daytime urban moisture field. *Atmos. Environ.* **12**, 1943–1949.

Spielvogel, L. G. (1980). Building energy performance data. *ASHRAE J.* **22**(1), 46–50.

Stearns, F., and Montag, T. (eds.) (1974). "The Urban Ecosystem: A Holistic Approach." The Institute of Ecology: Dowden, Hutchinson & Ross, Inc., Stroudsburg, Pennsylvania.

Steinhart, J. S., and Steinhart, C. E. (1974). Energy use in the U. S. food system. *Science* **184**, 307–316.

Suckling, P. W. (1980). The energy balance microclimate of a suburban lawn. *J. Appl. Meteorol.* **19**, 606–608.

Sundborg, Å. (1951). Climatological studies in Uppsala with special regard to the temperature conditions in the urban area. *Uppsala Univ. Geogr. Inst., Geographica* No. 22.

Tuller, S. E. (1980). Effects of a moderate sized city on human thermal bioclimate during clear winter nights. *Int. J. Biometeorol.* **24**, 97–106.

USSR (1978). The development of centralized heat supply in the cities of the USSR. *Habitat Int.* **3**, 469–474.

U.S. House of Representatives (1980). Compact cities: Energy saving strategies for the Eighties. Rep. Subcomm. on the city, H. S. Reuss, chmn., Comm. on Banking, Finance and Urban Affairs. 96th Congr., 2d session. U.S. Govt. Printing Office, Washington, D.C.

Unwin, D. J. (1980). The synoptic climatology of Birmingham's urban heat island, 1965–1974. *Weather* **35**(2), 43–50.

Vance, J. E., Jr. (1952). The growth of suburbanism west of Boston: A geographic study of transportation–settlement relationships. Ph.D. Thesis, Clark Univ., Worcester, Massachusetts.

van Straaten, J. F. (1967). "Thermal Performance of Buildings." Am. Elsevier, New York.

Vukovich, F. M., King, W. J., Dunn, J. W., III, and Worth, J. J. B. (1979). Observations and simulations of the diurnal variation of the urban heat island circulation and associated variations of the ozone distribution: A case study. *J. Appl. Meteorol.* **18**, 836–854.

Warner, S. B. (1962). "Streetcar Suburbs: The Process of Growth in Boston, 1870–1900." Harvard Univ. Press, Cambridge, Massachusetts.

Welch, R. (1980). Monitoring urban population and energy utilization patterns from satellite data. *Rem. Sens. Environ.* **9**, 1–9.

Welch, R. M., Paegle, J., and Zdunkowski, W. G. (1978). Two-dimensional numerical simulation of the effects of air pollution upon the urban–rural complex. *Tellus* **30**, 136–150.

White, J. H., Eaton, F. D., and Auer, A. H., Jr. (1978). The net radiation budget of the St. Louis metropolitan area. *J. Appl. Meteorol.* **17**, 593–599.

Winters, F. (1970). Pavement heating. *In* Snow Removal and Ice Control Research. *Highw. Res. Board, Spec. Rep.* No. 115, 129–145.

Yamashita, S. (1979). Shortwave radiation climatonomy of the urban atmosphere in Toronto. *Arch. Meteorol., Geophys. Bioklimatol., Ser. B* **27**, 193–203.

Yap, D., and Oke, T. R. (1974). Sensible heat fluxes over an urban area—Vancouver, B. C. *J. Appl. Meteorol.* **13**, 880–890.

Yoshino, M. M. (1975). "Climate in a Small Area. An Introduction to Local Meteorology." Univ. of Tokyo Press, Tokyo.

Chapter XXIII

INTEGRATING THE ENERGY FLUXES

The energy budget is both the summation of all the energy fluxes and transformations in an ecosystem and, by reason of its always being in balance, a constraint on these fluxes. Energy budgets are useful in analyzing changing thermodynamics in ecosystems that are responding to changes in their environments, and several were presented in earlier chapters (Chapters XX and XXI) on spatial contrasts and vertical structure of ecosystems, in brief budget analyses in chapters primarily devoted to a particular flux like snow-melting (Chapter XIV), and in the chapters on aquatic and urban systems (Chapters XVIII and XXII). At this point we look systematically at ecosystem budgets, their responses to forcing functions or to changes in characteristics, and the interactions among competing fluxes, in particular the major nonradiative ones described in Chapters XV–XVII.

We first examine the effects of (1) changing radiant-energy intake (e.g., in the daily march), (2) changing ecosystem characteristics (e.g., wetness), and (3) changing atmospheric conditions. These changes occur at many time scales, from the momentary shading of an ecosystem by a passing cloud to diurnal, day-to-day, and annual variations, yet the energy budget must at all times be in balance, and the responses, including interactions among the competing fluxes, must occur at the same time scale. The budget faithfully portrays all these adjustments and helps us see farther into the functioning of ecosystems.

MODELS OF ECOSYSTEM ENERGY BUDGETS

The self-balancing property of the energy budget gives us the option of selecting any one or any group of fluxes and equating its flux density to the sum of all the others. For example, incoming solar radiation to a meadow is automatically equal to the algebraic sum of all the other fluxes, radiative and nonradiative. As the sun climbs in the sky and solar input grows, so also, in the same measure, must the sum of the other fluxes. How is this gift of energy partitioned among the outgoing fluxes of reflected sunshine, latent and sensible heat, longwave radiation, and photosynthetic conversion? What are the responses to a forcing function?

Incoming versus Outgoing Fluxes

Incoming energy fluxes are forcing functions external to the system, and outgoing fluxes tend to show the responses (Lettau, 1969). However, groupings of incoming and outgoing fluxes change from time to time; soil-heat flux, for instance, reverses direction. The shifting composition of the groups detracts from the analytical value of this approach.

Radiative versus Nonradiative Fluxes

The problem of shifting group membership is eliminated if fluxes are grouped as radiative or nonradiative. The radiative transports of energy, virtually instantaneous and working at a distance without need for an intervening medium, differ radically from the nonradiative conversions and transports, which are hobbled by time and depend on materials with varying thermal characteristics.

Observationally, this model has the advantage that a single sensor, a net radiometer, can be used to measure the net resultant of all the radiative fluxes. The sum of the nonradiative fluxes is then known and can be partitioned. The disadvantages are uncertainty about net radiometer readings (too often unreplicated) and the fact that one of the radiative fluxes is not external, but is a function of ecosystem temperature.

Radiant-Energy Intake

A narrower grouping of radiative fluxes includes only those that are independent of that most variable characteristic of an ecosystem, its surface temperature.* Temperature mediates the partitioning of the radiant-energy intake to satisfy the emitted flux and to be allocated among the nonradiative fluxes.

Temperature-Dependent Fluxes. Longwave radiation emission is a straightforward exponential function of temperature and of that alone.

---

* In strict logic, the conversion of kinetic energy, which is also independent of surface temperature, should be included; however, it is usually small.

Excepting ecosystem emissivity, a constant, it is independent of any other ecosystem characteristic and of any external resistance. It is always large.

Photosynthetic energy conversion is modeled as a function of cell temperature that differs with species and phenological stage,[*] CO_2 concentration and ventilation, light intensity, ecosystem moisture and nutrient status, and other factors included in crop-yield models such as SPAM (soil–plant–atmosphere model) (Shawcroft et al., 1974) or the Japanese aerodynamic models (Uchijima, 1976). Resistance appears if storage for photosynthate is lacking. Respiration and decomposition are also temperature dependent.

Evapotranspiration is a function of cell-wall temperature, as well as of atmospheric humidity, stomatal state, and canopy ventilation. Atmospheric ventilation that brings carbon dioxide to an ecosystem removes water vapor (with a different resistance in the circuit), and some models make use of this association (Budyko, 1974, pp. 400–421; Lieth and Box, 1972). Evapotranspiration can also be powered by sensible heat extracted from the air and substrate heat extracted from a water body, sources of energy that are valueless for photosynthesis.

Sensible-heat flux is a function of leaf-temperature contrast with air temperature and of canopy diffusion resistance or ventilation, which is enhanced by the flux in an interesting feedback. Energy sharing between latent and sensible heat and between substrate and sensible heat will be discussed later in this chapter.

Substrate-heat flux is a function of surface temperature, compared with temperature deeper in the soil or water, and thermal admittance of the substrate, i.e., an inverse resistance. Depth and time limit the soil capacity to store heat, and modeling down to a meter is usually sufficient. Substrate models are sometimes combined with snow-melting models by adding a temperature constraint (Outcalt et al., 1975). Soil moisture in terrestrial ecosystems and wind stirring in aquatic ones also influence this flux.

Fossil-energy conversions for industrial and circulation purposes are substantially independent of ecosystem temperature; those for space heating and cooling are temperature dependent, as are, in part, those for crop management. Major external factors are economic and demographic rather than environmental.

Model Components. Intake of solar energy depends on atmospheric cloudiness and aerosols, and since these are not well known, attempts

---

[*] Stanhill (1976), for instance, compares energy budgets in three phenological stages of cotton fields in Texas, Israel, and central Asia.

to model the flux are not as reliable as measurements are. The same is true, as noted in Chapter VI, of the downward flux of longwave radiation, which, unfortunately, is measured in few ecosystems. These intakes make up one side of the balance; surface temperature, as described in Chapter VIII, is the mediator and drives the heat-removing fluxes, or "dissipative terms" (Stern et al., 1973, p. 203).

Similarity in the coefficients of diffusion for heat, vapor, momentum, and CO_2 characterizes ecosystem ventilation, but details require much micrometeorological research. Advective conditions, for example, may differ from lapse conditions (Motha et al., 1979). Temperature of the local air, in part dependent on the boundary heat flux, is modeled in boundary-layer studies, which extend upward with variously postulated diffusivities (Kuo, 1968) but tend to be more of regional than ecosystem scale.

The temperature functions and external nonthermodynamic factors differ from one flux to another; the budget equation that includes them all cannot be solved analytically, but only numerically (Estoque, 1963). Simply stated, a trial calculation is made for a selected surface temperature, and the departure from balance of the fluxes in the set of equations is noted; the model tries a new temperature, and so on, until the budget is balanced (Halstead et al., 1957; House et al., 1960; Idso and Baker, 1967; Outcalt, 1972). A short time later, after radiant-energy intake or ventilation has changed, another set of calculations is repeated until balance is achieved. This method also has the advantage of defining the diurnal marches of the energy fluxes and the profiles of temperature, vapor concentration, and wind, like the models used within forest or crop canopies described in Chapter XX [e.g., Denmead (1964), Furnival et al., (1975)].

EFFECTS OF CHANGES IN RADIANT-ENERGY INTAKE

At a steady rate of absorption of solar and longwave radiation the temperature-dependent fluxes compete for the energy made available. Taking a noncompetitive point of view, this process can be looked at as a cooperative sharing of the task of removing the heat load from an ecosystem. Radiation is "the primary external driving mechanism in the physical model" (Estoque, 1963) and changes evoke immediate responses.

Momentary Fluctuations

Brief changes in radiant-energy intake, produced by shading or unshading of a leaf, produce observable changes in the temperature-de-

TABLE I

Partition of Radiant-Energy Intake at Dry Ecosystem in Heating Phase of Diurnal Cycle[a]

	0400	0800	1200	Change from 0400 to 1200
Radiant-energy intake	+300	+745	+1150	+850
Surface temperature (^{0}K)	291	310	332	41
Emitted radiation	−355	−465	−615	−260
Photosynthetic conversion[b]	0	0	0	0
Latent-flux[b]	0	0	0	0
Sensible-heat flux	+20	−190	−420	−440
Soil-heat flux	+35	−90	−115	−150
Sum	−300	−745	−1150	−850

[a] Data from Aizenshtat (1958) and Miller (1972a). Data given in watts per square meter.
[b] Shrubs dead or dormant in July.

pendent fluxes. Knoerr and Gay (1965) measured energy fluxes of a tulip poplar leaf exposed in midair, in conditions where I calculate the radiant-energy loading dropped from 1200 W m^{-2} (both sides) to 800 W m^{-2}; the leaf temperature sank from 12°K above air temperature to 2°K below it, emission of longwave radiation decreased by 200 W m^{-2}, sensible-heat flux changed by 120 W m^{-2} and reversed, and latent-heat flux decreased by 80 W m^{-2}. Ecosystem response is smaller than leaf response, and in a sumac community it was 0.5°K for 100 W m^{-2} change (Lambert, 1970).

The Diurnal March

The Heating Phase. The diurnal march of solar energy is the most familiar change in radiant-energy intake and brings flux density from 300 W m^{-2} or less before dawn (all as downward longwave radiation) to 1000 or more by midday, as at a sparse-shrub ecosystem in Central Asia (Aizenshtat, 1958) (Table I).

The rise in surface temperature did three things: generated an increase by the fourth-power law in radiation emission; changed a small input of sensible heat into a very large flux into the air, destabilizing it; and reversed the soil-heat flux. The first of these quantities is unique, being independent of any external factor. The two reversals can also be expected, but the flux densities attained are governed by deep-soil and free-air temperatures and the thermal admittances of the two media.*

* The rate of change of the substrate-heat flux was 9 mW m^{-2} sec^{-1} and that of the sensible-heat flux 15 mW m^{-2} sec^{-1}.

The early-morning increase is even more sudden on east-facing slopes at high altitude: "The heat load in the alpine tundra advances like an avalanche as the sun bursts upon the mountain slope" (Gates, 1963). On an east-facing slope at 2.2 km altitude in the Swiss Alps the net surplus of all-wave radiation increased to 675 W m^{-2} at 0800 and then declined to 460 W m^{-2} as the sun moved into the south at 1200 [data from Turner et al., (1975)]. In the 0600–0800 period the downward soil-heat flux increased at a rate of +15 mW m^{-2} sec^{-1} and then in late morning diminished at a smaller rate. The sensible-heat flux grew faster at first, at +60 mW m^{-2} sec^{-1}, then, as the solar azimuth angle changed, diminished at a rate of −8 mW m^{-2} sec^{-1}.

The Cooling Phase. Radiant-energy intake decreases slowly after noon, rapidly toward sundown, and slowly during the night, often in an exponential decay curve. Ecosystem responses, mediated through surface temperature, follow this gradual course, which is quite different from their morning regimes, as Lönnqvist (1962) sought to explain.

Sunset is the end of photosynthesis in an active ecosystem, and the latent-heat flux also declines. The stabilizing of the lower layers of the air hampers the downward flux of sensible heat, and it and soil heat bring only modest quantities of energy to the surface to meet the quantity by which incoming radiation falls short of emitted. The decrease in radiant-energy intake of the ecosystem to about 300 W m^{-2} poses an energetic problem that is solved by changes in the temperature-dependent fluxes, which can be determined by submodels involving temperatures of the air and soil away from the interface (Sutherland, 1980). Fighting nocturnal frost in a citrus orchard illustrates ways of augmenting these inputs or adding fossil-energy conversions.

Day-to-Day Fluctuations in Radiant-Energy Intake

Both solar and longwave radiation fluctuate as cloud systems pass over an ecosystem. For example, 1 July 1974 at the Hamburg Observatory was mostly cloudy and 2 July mostly clear; mean radiant-energy intake changed from 461 W m^{-2} (of which 0.17 was absorbed solar energy) to 584 W m^{-2} (0.40 solar). A small part of this augmentation went to increase radiation emission; most of it went into increases in the nonradiative energy fluxes [data from Germany (1974)]. Night-to-night changes in temperature and humidity aloft cause substantial differences in the surface temperature at which the budget comes into balance (Gall and Herman, 1980).

Albedo changes also affect radiant-energy intake. The sudden estab-

TABLE II

Partition of Radiant-Energy Intake at Grass Ecosystem in Heating Phase of Annual Cycle (10-year means)[a]

	December[b]	June[c]	Change[d]
Absorbed solar radiation	+10	+190	+180
Absorbed longwave radiation	+295	+335	+40
Radiant-energy intake	+305	+525	+220
Emitted longwave radiation	−320	−390	−70
Photosynthetic conversion	~0	−3	−3
Latent-heat flux	−5	−62	−57
Sensible-heat flux	+10	−60	−70
Soil-heat flux	+10	−10	−20
Sum	−305	−525	−220

[a] Data from Aslyng and Kristensen (1958), Kristensen (1959), Aslyng (1960), Aslyng and Friis-Nielsen (1960), Aslyng (1966), and Aslyng and Jensen (1966). Data given in watts per square meter. Latitude 56°N.
[b] Mean surface temperature approximately 273°K.
[c] Mean surface temperature 293°K.
[d] Mean surface temperature change 20°K.

lishment of a snow cover marks a threshold in the winter energy budget of an ecosystem, and the more gradual ripening and melting off ("Ausaperung") under the high sun of spring causes a large increase in radiant-energy intake.

Annual Cycle

The annual cycle exhibits a slow change in the structure of the energy budget, illustrated at a grassland near Copenhagen (Table II). The long days and moderately high sun of summer produce a manyfold increase over winter in the intake of solar energy, and the warmer air produces more downward longwave radiation. The grass in June absorbs 220 W m^{-2} more in the 24-hr mean than in December.* Surface temperature is higher, leaf cells generate photosynthate, oxygen, and water vapor, and the biological energy fluxes take 0.27 of the increase in radiant-energy intake. Ecosystem soil, moist and conductive, takes in a moderate flux. Relatively cool marine air accepts a large sensible-heat flux. A third of the 220 W m^{-2} increase in the forcing function goes to increase emitted

* The mean rate of increase is 14 $\mu W\ m^{-2}\ sec^{-1}$, three orders more slowly than the increase in a desert morning.

radiation, and 0.68 is partitioned among the nonradiative fluxes as follows:

to conversions related to biological production   0.27
to reverse the soil-heat flux   0.09
to reverse the heat flux into the air   0.32

These changes in energy fluxes are brought about by a modest 20°K rise in ecosystem surface temperature. The atmosphere is more receptive than the soil and gets three times as much of the increase. Latent-heat flux and photosynthetic conversion claim a quarter of the increase; each month from January on doubles the flux of the previous month, and May and June have long working days of open stomata and active energy transformations. The latent-heat flux can move a large quantity of energy, and a moderate surface temperature suffices. The increase in emitted radiation is only 70 W m^{-2}, compared with 260 W m^{-2} in the desert day (see Table I).

The influence of water on the energy budget is well illustrated here. Soil moisture, by supplying the raw material for the efficient latent-heat flux, favors it over the sensible-heat and emission fluxes, which become large only if surface temperature becomes high. Court (1951) identified the many buffering roles of water* in the energy budget, and its influence on the partitioning process is a major one, which an early determination of the annual march of the energy budget minimized by making measurements in the Gobi Desert (Albrecht, 1941).

Longer-Term Variations

Radiant-energy intake varies from year to year, depending on number of cloud systems, and this variation, which had a range of 80 W m^{-2} in 16 Junes at Hamburg, is reflected in the sensible- and latent-heat fluxes. Over the long term, variations in the solar constant and in atmospheric emission as CO_2 concentration rises will affect the temperature-independent intake of an ecosystem although their principal significance is at regional or global scales. Increases in albedo as a result of overgrazing or drought, or decreases due to irrigation, also alter the radiant-energy intake and hence the temperature-dependent fluxes. An illustration of how albedo and longwave differences can modify the radiant-energy intake is given in Fig. 1: Solar radiation is more intense in Greenland than in Australia, but absorbed shortwave and longwave is so much smaller that the surface temperature is 46°K lower and the

---

* In examining the climatic role of radiation Ångström (1925) first subtracted the latent-heat conversion in order to analyze only the "effective" radiation.

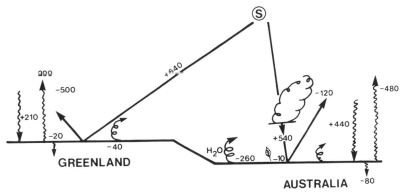

Fig. XXIII.1.   Mean energy flux densities during daytime hours in summer at Eismitte, Greenland (70°N latitude, altitude 3 km) and at Katherine Experiment Station in northern Australia (14°S latitude). Absorbed solar energy at Eismitte 640 − 500 reflected = 140 W m^{-2}, at Katherine 540 − 120 reflected = 420 W m^{-2}; absorbed longwave radiation, 210 and 440 W m^{-2}, respectively; total radiant energy intake, 350 and 860 W m^{-2}, respectively. Outgoes: emitted longwave radiation, −290 W m^{-2} from a −15°C snow surface at Eismitte and −480 W m^{-2} from a 31°C leaf canopy at Katherine; substrate-heat intake, −20 and −80 W m^{-2}, respectively; sensible-heat flux into the air, −40 and −30 W m^{-2}, respectively; no biological energy conversions at Eismitte, but −10 W m^{-2} in photosynthesis and −260 W m^{-2} in evapotranspiration at Katherine. [Eismitte observations made by the Wegener and Victor expeditions analyzed for energy fluxes by Miller (1956), Katherine flux densities from Dyer (1967).]

daytime nonradiative fluxes total only 0.15 of those in the Australian ecosystem.

EFFECTS OF CHANGES IN SURFACE CHARACTERISTICS

Wetness, roughness, substrate thermal admittance, and fossil-energy conversions explain how cities differ from their surroundings, and these characteristics are important in all ecosystems, although incompletely known (Thornthwaite, 1961). They vary with time, and these variations govern the partition of radiant energy absorbed by the ecosystem in such a way that balance is maintained in the total energy budget; whether or not a particular ecosystem is in ecological equilibrium makes no difference (Miller, 1965, p. 178).

Substrate Thermal Admittance

When a snow cover is established on formerly bare ground, thermal admittance drops by nearly an order of magnitude. Substrate-heat flux nearly vanishes, and little heat goes into the air (Priestley, 1959, p. 112); soil warmth is conserved, as Russian soil-energy studies since 1878 have observed.

Admittance changes when an aquatic ecosystem freezes over. Depending on the coldness, dryness, and diffusivity of the air flowing over an unfrozen ecosystem, turbulent heat fluxes may reach 500 W m^{-2} or more, but these losses are cut enormously as soon as ice covers the system, and they drop farther as the ice thickens and supports a snow cover. The energy budget at the snow/air interface runs at a low level because it is able to withdraw but little of the heat stored below.

As an annual ecosystem deploys its leaves, the active surface is moved away from the soil, and a plant–air layer forms that has a small value of thermal admittance. In a desert ecosystem admittance decreases as the exposed soil dries out after rain (Mitchell et al., 1975).

Wetness

Surface moisture has a further effect, for when it is at hand, the latent-heat conversion can outdistance the sensible-heat flux and carry off the lion's share of the available energy—a share that increases as surface temperature approaches 33°C. The effect of changing water access is seen in a comparison of two oak leaves in a greenhouse, one transpiring, the other not (Table III). Without water supply the whole loading of absorbed radiant energy had to be met by emitted radiation and the leaf-to-air flux of sensible heat, and both fluxes require a high surface temperature.

The response of an ecosystem with surface water—a deciduous forest on a clear day following a rain—is shown in Table IV. Since water on leaves evaporates easily (Chapter XV), the latent-heat flux carried off most of the increase in radiant energy. Its flux density grew three times

TABLE III[a]

Ecosystem Characteristic	Transpiring	Nontranspiring
Radiant-energy intake[b]	1660	1660
Leaf temperature	50°C	62°C[c]
Emitted radiation[b]	−1195	−1380
Latent-heat flux[b]	−360	0
Sensible-heat flux[b]	−105	−280

[a] Data from Knoerr and Gay (1965).
[b] Data given in watts per square meter.
[c] Above lethal point.

TABLE IV[a]

Increase in radiant-energy intake from dawn to midday	580
Increase in emitted radiation	50 = 0.09
Reversal in heat flux into soil and forest biomass	50 = 0.09
Increase in latent-heat flux	360 = 0.61
Reversal in sensible-heat flux	120 = 0.21

[a] Data from Rauner (1958). Data given in watts per square meter.

as much as the sensible-heat flux, whereas the two were approximately equal on other days. Almost the same effects of externally wet foliage were found in pines in Australia (Moore, 1976).

The drying of an ecosystem witnesses a gradual shift each day away from the latent-heat flux and toward the sensible-heat flux and emitted radiation. Most of the change in energy partitioning takes place in the ecosystem members above ground because drying soil decreases in admittance. This poststorm shift is illustrated in field studies and models of evapotranspiration and crop energetics discussed in Chapters XI and XV and accompanies a rise in ecosystem temperature. A dry period in summer usually experiences progressively higher surface, plant, and air temperatures with the potential for ecosystem injury.

As a dry spell lengthens into a drought, especially in anticyclonic atmospheric circulation, the increase in sensible heat is often associated with strong solar radiation that heightens thermal stress on both plants and humans. Crops shriveled and people died in Texas in the summer of 1980. It is no coincidence that maximum temperature records were set during the droughts of the 1930s and 1950s, as well as in 1980.

Farmers are well aware of the probabilities of low soil moisture in each part of the growing season. Shortage may come in late spring in some places (Denmark), early summer in others (North China), and later summer in still others (Wisconsin). Whatever the month, the crop energy budget shifts away from the biologically important energy fluxes of photosynthesis and transpiration to the abiotic fluxes of longwave radiation and sensible heat.

Fossil-Energy Conversion

Fossil energy is an additional input into the surface energy budget and forces larger outputs—more emitted radiation, sensible heat, and latent heat. One or another of these fluxes may be favored in particular circumstances, but it appears that in the urban energy budget much of the additional energy takes the form of sensible heat.

Fossil-energy conversion increases in the morning rush hour and industry start-up, and its contributions in different neighborhoods are included in diurnal models of the total energy budgets of Sydney (Kalma, 1974) and Hong Kong (Kalma et al., 1978). Energy conversions for space heating are counterseasonal and reduce the range in the energy budget over the yearly cycle; energy used for air conditioning, on the other hand, peaks in the summer and augments an already excessive energy budget.

Roughness

Ecosystem roughness influences the exchange of sensible heat, water vapor, and CO_2 between it and the air. For example, spines on cactus enter an energy-budget model (Nobel, 1978, 1980), which computes critical tissue temperatures. An ecosystem with a large roughness coefficient, like a savanna, produces large eddy diffusivity in the atmosphere (Garratt, 1978). The small aerodynamic resistance to diffusion of vapor and CO_2 above a rough ecosystem means that stomatal resistance becomes important and that biological control of the ecosystem is strengthened. Forest is a good example in England (Stewart and Thom, 1973), or Denmead's (1969) comparison with wheat in Australia.

Momentary variations in roughness occur when wind deforms vegetation or produces *honami* (waves) in the canopy. Annual variations result from growth in height and canopy closure of low vegetation and from the spring leafing-out of deciduous trees.

Ecosystem Management

These ecosystem characteristics combine with surface temperature to govern the fluxes of energy to or from an ecosystem, and energy models provide guidelines for ecosystem management. Fertilization that increases photosynthetic energy conversion may also increase transpiration, as has occurred in the tablelands of Australia (Dunin and Reyenga, 1978) and in Wisconsin. No-till cultivation intended to improve moisture status also influences the energy budget, not always beneficially (Gersmehl, 1978). Drainage and irrigation affect the partitioning of energy, as does urbanization. Energy-budget models afford means of predicting the probable responses if a proposed land-management practice would alter these surface characteristics because the inputs and conversions of energy would be the object of impacts (Erickson, 1979, p. 115).

EFFECTS OF CHANGES IN ATMOSPHERIC CONDITIONS

The fluid atmosphere is hard to encompass within the limits of an ecosystem, for while we can assign a volume of soil to a given ecosystem, only in quiet conditions in sheltered terrain can we do this for air. The layer of air influenced by and reciprocally influencing an underlying ecosystem is hard to delimit unless the system is very extensive.

For a small ecosystem it is customary to consider atmospheric motion, warmth, vapor concentration, and instability as quantities of the boundary layer. Examples are rice paddy models (Uchijima, 1976), potato fields (Brown, 1976), and forest models (McCaughey, 1978). The gradients of vapor, CO_2 concentration, and temperature along which the fluxes move and postulated values of diffusivity (or admittance) become coefficients in the set of flux-gradient equations (Halstead et al., 1957; House et al., 1960; Outcalt, 1971). The saving grace is that these equations must fit into the overall energy budget, which as "an over-riding physical constraint" on the fluxes of latent and sensible heat (Thom, 1975), is "often invaluable in practice."

On scales larger than most ecosystems most of the daytime boundary layer, the "atmosphere's first mile" (Lettau and Davidson, 1957), is incorporated in models. Surface budgets were used to determine temperatures aloft by Albrecht (1930) and to predict daily maximum temperature by Neiburger (1941), and computers now make it possible to calculate fluxes of heat and vapor at different levels, estimate flux divergences as injections of energy that modify pressure and motion fields, and work back to surface temperature for balancing (Estoque, 1963; Kuo, 1968; Myrup, 1969; Anthes and Warner, 1978). However, surface characteristics are sometimes no more than "arbitrarily prescribed functions of position" (Viskanta and Daniel, 1980).

Surface temperature is the mediator in landscape- and meso-scale models, which, however, usually cannot yield much information at ecosystem scale since grid resolution is of the order of 25 km. Indeed, inputs into some models tend to be sketchy with regard to ecosystem conditions, functioning, depth profiles, and spatial contrasts; detailed outputs can hardly be expected. The nocturnal boundary layer (Oke, 1978, p. 53) is thinner than the daytime layer and less turbulent. However, importance still attaches to modeling the temperature of the underlying surface (Sutherland, 1980).

Temperature of the air above ecosystem canopy, which might come from a large-scale model, provides contrast with ecosystem temperature, and its vertical profile is related to thermal stability or diffusivity K, as shown in Chapters XVI and XX. Thermal admittance (≈ 1300 $\sqrt{K}$) can be compared with substrate admittance.

Water-vapor concentration in the atmosphere is important for its contrast with ecosystem vapor concentration, and the ecosystem is more often a source than a sink, except at night, when a surface temperature model estimates frost deposition (Severini and Olivieri, 1980). Its vertical profile was described in Chapter XV.

Wind direction is especially important if vapor or heat (or cold) is being advected. Wind speed has more general importance since it supports the downward flux of momentum that is associated with other turbulent fluxes; along with atmospheric gradients it determines the coupling between ecosystem and local air. The wind field is included in models that are developed to predict pollutant dispersion, a problem that has occasioned much work at urban scales.

Time Variations

Momentary variations in atmospheric conditions, most evident in wind speed, manifest eddy structure, and eddies that bring down drier air from aloft cause a sudden change in the latent-heat flux from an ecosystem. Diurnal marches of atmospheric temperature, humidity, wind speed, and diffusivity are familiar and act both as external factors and reciprocal conditions in the ecosystem–air duet; some were mentioned in Chapters XI, XIV, XV, and XVI, where associated with the energy fluxes under consideration.

Deviations from the usual regime are significant. For example, an early afternoon increase in vapor pressure deficit above a forest in British Columbia (Spittlehouse and Black, 1979) produces a midafternoon peak in the latent-heat flux several hours after the 1200 peak in the net surplus of all-wave radiation.

Interdiurnal variability in atmospheric conditions indexes the aperiodic component of climate, i.e., the role of atmospheric systems versus solar climate (Landsberg, 1968), and is evident in ecosystem energy budgets. Day-to-day variations in air temperature above a pond near Hudson Bay, for example, produce large fluctuations in the turbulent fluxes (Stewart and Rouse, 1976).

Determinations of heat fluxes at a shrub ecosystem in New Zealand on prefrontal, frontal, and postfrontal days (Greenland, 1973) indicate how day-to-day or synoptic-scale changes in atmospheric conditions

TABLE V

Energy Fluxes at Cass Heat-and-Water Budget Experiment Station of Canterbury University. Type II (Southwest) Fronts[a]

	Substrate-heat flux	Sensible-heat flux	Latent-heat flux
Quantities (Wm^{-2})			
Prefrontal day	-5	-41	-78
Frontal day	$+1$	-35	-43
Postfrontal day	$+4$	-23	-23
As fractions			
Prefrontal day	-0.04	-0.33	-0.63
Frontal day	$+0.01$	-0.45	-0.56
Postfrontal day	$+0.10$	-0.55	-0.55

[a] Data from Greenland (1973).

affect the partitioning of the net surplus of all-wave radiation (Table V). Because the net surplus of all-wave radiation decreased in the course of the weather sequence, there was less energy for the nonradiative fluxes to partition, but in addition, the partitioning itself changed. Soil-heat storage was called upon in frontal and postfrontal days to help meet the increased transfer of sensible heat into the colder airstream moving in from the southwest.

Daytime pulses of sensible heat measured at Quickborn in Germany show large day-to-day changes; for example, in April they form a bimodal distribution with clusters at 2.0 and 5.5 MJ m^{-2} [data from Frankenberger (1960)]. Such thermodynamic data help define local versus external factors in the environment of an ecosystem.

THE STRUCTURE OF ECOSYSTEM ENERGY BUDGETS

The structure of an ecosystem energy budget is delineated by interactions or competition among the temperature-dependent fluxes. The ways these fluxes respond to changes in radiant-energy intake depend on surface and atmospheric conditions, which in different circumstances favor one flux over the others and determine their competitive positions.

Longwave Radiation Emission versus Nonradiative Fluxes

Longwave radiation emission from a cold ecosystem is a major part of the total energy budget: 0.5 at midday but 1.0 at night, as well as in winter. It is the only mode removing energy from winter snow, for instance; all others are toward the snow surface.

Its share diminishes as surface temperature rises unless there is un-usual resistance to the nonradiative fluxes. High resistance to heat flow in sandy soil hampers substrate heat flux, and when leaf resistance to vapor diffusion is large and atmospheric ventilation is weak, the tur-bulent fluxes are weak also. At times of large solar heat loading longwave radiation is left as the main output (Wouters *et al.*, 1980). This compe-tition is plainest when the turbulent and phase-change fluxes drop out (or are constant): at an airless surface like the moon, or in conditions of calm and stable stratification, approached on quiet nights and useful in thermal inertia surveys (Chapter XVII).

Latent Heat of Fusion

The rate of latent-heat conversion at a melting snow surface, inde-pendent of surface temperature (invariant at 273°K), depends solely on heat input, which comes primarily as radiant-energy intake and sec-ondarily in turbulent fluxes dependent on roughness and atmospheric warmth and humidity. So long as the threshold of 315 W m^{-2} required by longwave radiation at 273°K is met, the melting process does not encounter resistances like those that restrict energy conversion in va-porization or photosynthesis. It is a true heat sink.

Substrate-heat flux is negligible in the budgets of new snow with slight thermal conductivity, of old, isothermal spring snow, and of intercepted snow. Although evaporation appears to compete with melting, ther-modynamical analysis shows that it is unimportant in most situations (Miller, 1967), and field measurements confirm this.

Snow cover can be analyzed in conjunction with the substrate heat flux (Outcalt *et al.*, 1975; Anderson, 1976) by a model that computes energy exchanges at the soil/snow interface, within the snow, and at the snow/air interface and that calls for melting when the computed temperature exceeds 273°K. If the local air is included in the model (McKay and Thurtell, 1978), advective melting can be evaluated. It dif-fers from the more predictable radiative melting in timing and hydrologic consequences because it is energized by fluxes in mechanical turbulence driven by the kinetic energy of motion systems outside the solar rhythm.

Partitioning between Substrate- and Sensible-Heat Fluxes

Surfaces without water able to change phase, i.e., very cold snow, deserts, rock, or mulch (Mahrer, 1979), have simple energy budgets distinguished by conceptual partitioning of energy (beyond that radiated away) between the soil and air media, which occurs in accordance with their respective values of thermal admittance (Lettau, 1952). Admittance of the soil is nearly constant in the short run, but that of the air changes

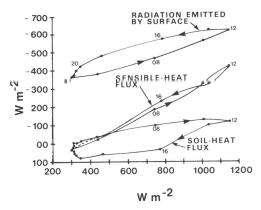

Fig. XXIII.2.   Diurnal marches of three surface-temperature-dependent energy fluxes at a desert ecosystem in central Asia (W m^{-2}) and of radiant-energy intake by the ecosystem (W m^{-2}) (Miller, 1972b). Emission of longwave radiation is the largest of the fluxes that are functions of surface temperature and exhibits the hysteresis noted in Chapter VIII, with afternoon fluxes exceeding morning fluxes at the same intake of radiant energy.

rapidly in the morning after the ground-based inversion is wiped out and turbulent communication reaches more distant layers of air. Growing admittance accelerates the removal of heat from the surface, and the upward flux outstrips the downward. The increase to several thousand J m^{-2} °K^{-1} sec$^{-1/2}$ admittance, which may also further increase with height (Kuo, 1968), is incorporated in deep models, over areas larger than ecosystems, that reach up to 1–2 km and down to several meters into the soil (Chang, 1979).

The three temperature-dependent fluxes at Shafrikan in central Asia (thermal admittance 510 J m^{-2} °K^{-1} sec$^{-1/2}$) are plotted against radiant-energy intake through the day in Fig. 2 (Miller, 1972b), which shows that the sensible-heat flux has the steepest slope, i.e., exhibits the greatest response to a change in radiant-energy intake. Hysteresis in the other curves shows that at the same radiant-energy intake more radiation is emitted by the surface in the afternoon than the morning (because surface temperature is higher in the afternoon). Correspondingly less heat goes into the soil in the afternoon than morning, as described in Chapter XVII. Lönnqvist (1962) examines conditions in which the "non-conductional effects such as evapotranspiration" and longwave radiation sum to zero, when the solar load is removed only by the soil-heat and sensible-heat fluxes, in order to explain the difference between the symmetrical solar regime and the asymmetrical curve of surface temperature.

Bare soil that grows choked with heat by midday may cause heat flow to reverse before sunset. In contrast, the atmospheric heat sink is largest in early afternoon and continues to accept heat for several hours there-

after. The heat taken into the soil during the day has not left the ecosystem and at night is retrieved as the interface cools, when it usually
meets virtually the entire deficit of all-wave radiation (Monteith, 1973,
p. 161). Substrate storage is large in urban systems (Chapter XXII) and
can be enhanced in agricultural ecosystems to stave off nocturnal frost
(Brooks and Rhoades, 1954); in both cases the downward flux of atmospheric heat is small or zero.

Heat stored in the sandy soil at Shafrikan amounts to 0.21 of the total
daytime flux of sensible and soil heat. The fraction is 0.26 at El Mirage
Dry Lake, California (Vehrencamp, 1953; Brooks and Rhoades, 1954),
where a denser silt and clay substrate has larger thermal admittance
($930 \, J \, m^{-2} \, °K^{-1} \, sec^{-1/2}$), and 0.27 in a meadow at Hamburg (Frankenberger, 1962). Pioneer energy-budget measurements by Homén (1897)
show that a granite ledge (admittance $2900 \, J \, m^{-2} \, °K^{-1} \, sec^{-1/2}$) took in
even more heat than the air.

Soil shielded under thick potato leaves takes in only 0.09 of the total
of the soil and sensible-heat fluxes (Frankenberger, 1962), and 0.05–0.10
is typical of most vegetation-covered soils (Monteith, 1973, p. 161),
which cannot take in heat as fast as the air does. A high-altitude plant
in Venezuela minimizes contact with the soil surface by elevating its
rosette, thereby enhancing its coupling with the air and avoiding overheating in hot conditions (Baruch, 1979) and getting above the surface
layer of the coldest air in frost weather (Smith, 1980).

Storage capacity in the annual cycle is limited, and after early summer
the partitioning at the surface favors the atmospheric over the substrate
flux. Soil-heat flux was only 0.06 of the combined soil- and sensible-heat
fluxes in the Iaskan oasis in the Kara-Kum Desert (Kuvshinova, 1965).
Even in a moister climate the atmospheric flux still dominates.

The Bowen Relationship

The way the two turbulent fluxes, those of sensible and latent heat,
partition available energy is significant biologically, meteorologically,
and hydrologically. It was identified by Bowen (1926) in extending to
lake evaporation a small-scale budget of Cummings (1925) that equated
radiative intake to substrate and latent heat, and "a small correction"
that depended on atmospheric conditions, i.e., sensible heat. This partition is simplest, however, when substrate heat is minor, i.e., where
foliage is thermally isolated from the soil by litter, dead or senescent
leaves, or quiet air of the forest trunk space, and except in dense canopy
(Hicks et al., 1975), storage of heat in the ecosystem is small. Stratified
models use the Bowen ratio to calculate source or sink strength of sensible and latent heat in each layer, as described in Chapter XX, and

allocate energy to the sensible- and latent-heat fluxes [Waggoner (1975); Denmead (1964) for pine; Begg (1965) for high-productivity millet; Saugier and Ripley (1978) in grassland with a thick layer of litter].

In a model for a crop and soil ecosystem substrate-heat flux is still small, say 0.1 of the net daytime surplus of all-wave radiation (Dyer, 1967), and easily accounted for. The model sets the net surplus of all radiation fluxes equal to the sum of all nonradiative fluxes.

$$R_{net} = H + E + S,$$

in which H is sensible heat, E is latent heat (equal to 2.5 MJ per kilogram per second mass flux), and S is substrate heat. If S is taken as a small quasi-constant fraction of the sum of the nonradiative fluxes, the energy budget becomes

$$0.9\ R_{net} = H + E$$

Energy partitioning is reduced to the two turbulent transports. In neutral stability mechanical turbulence driven by atmospheric kinetic energy carries H and E equally well, considering different resistances to diffusion.

No difference in resistance occurs at the water surface in an aquatic ecosystem, and the external factors that govern both fluxes are the turbulent transport coefficient and the atmospheric concentrations of vapor and heat, i.e., specific humidity and temperature. Bowen (1926) developed this reasoning for different values of diffusivity (or admittance) as a function of height above the surface and showed that the ratio β of the two fluxes can be expressed as follows:

$$\beta = \frac{H}{E} = \frac{c_p(T_1 - T_2)}{(2.5\ \text{MJ kg}^{-1})(q_1 - q_2)},$$

in which subscripts 1 and 2 represent measurements at two levels in the air, c_p is specific heat of air, T temperature, and q specific humidity. The $c_p/(2.5\ \text{MJ kg}^{-1})$ term provides a dimensional adjustment, and the $(T_1 - T_2)/(q_1 - q_2)$ term expresses the warmth and dryness of the air in a profile above the ecosystem.

After a value of β in a specific situation is measured or otherwise established, either H or E can be expressed in terms of the other:

$$H = E\beta$$

so

$$H + E = E\beta + E = E\ (1 + \beta)$$

and

$$E = (\text{net radiation})(0.9)/(1 + \beta)$$

Net radiation surplus is easily (perhaps deceptively easily) measured and so are air temperature and vapor concentration at two levels in the constant-flux layer as the guide for determining β and partitioning 0.9 R_{net}.

If the Bowen ratio is written in terms of surface temperature T_0 and vapor concentration q_0,

$$\beta = \frac{1.0 \text{ kJ kg}^{-1} \text{ °K}^{-1}}{2500 \text{ kJ kg}^{-1}} \frac{T_0 - T_z}{q_0 - q_z}$$

(where z indicates a level above the canopy), it is evident that raising T_0 from 273°K to 274°K (when T_z is less than 273°K) increases $T_0 - T_z$ by 1°K and $q_0 - q_z$ only by 0.28×10^{-3}, increasing sensible-heat more than latent-heat flux. But if T_0 is initially 303°K and is raised to 304°K, the corresponding change in $q_0 - q_z$ is 1.5×10^{-3}, and latent-heat flux increases more than sensible-heat flux.

More resistance is introduced into the diffusion path of water vapor in a drying ecosystem, and β increases to 1 or even higher, indicating that the sensible-heat flux is winning out in the competition, for example, as forest moisture decreases (McCaughey, 1978). Priestley (1967a) comments that the apportionment between the two fluxes depends "primarily, not on current bulk atmospheric variables, but on the moisture condition of the underlying medium."

The sign of β reverses around sunset, when the fluxes are no longer competing for radiant energy, which is fading; instead, the flux of sensible-heat is directed toward the grass and supports a continuing small outward flux of latent heat. This cooperating rather than competing situation, noted in Chapter XV in particular, corresponds to the familiar wet-bulb thermometer. In oasis and edge phenomena it occurs even in midday and indicates energy advection, which may require a three-dimensional model (Lowry, 1969, p. 141).

β is larger in humid marine air than in continental sites (Jarvis et al., 1976), and the rapid warming of marine air moving inland in summer (Leighly, 1947) demonstrates a large sensible-heat flux. Changes in the turbulent exchange coefficients in large-scale advection on the Great Plains (Verma et al., 1978) are reflected in β.

The Bowen ratio indicates the availability of soil moisture and the canopy resistance to moisture transport and is useful in determining evapotranspiration, for example, in native Australian *Themeda* at Krawarree (Dunin and Reyenga, 1978) (Fig. 3), and has been automated (McIlroy, 1967; Dilley, 1974; Munro, 1980). Alternatively, if H is measured or calculated, E can be found as the residual in the energy budget (Grant, 1975; Stricker and Brutsaert, 1978; Saltzman, 1980).

Fig. XXIII.3.   Bowen-ratio meter above native Themeda grassland at Krawarree in the Southern Tablelands of New South Wales. Shielded wet- and dry-bulb thermometers at two levels are periodically interchanged to avoid systematic error in determining vertical differences in temperature and specific humidity for the Bowen allocation of energy to latent- and sensible-heat fluxes.

Energy-Budget Structure

The three major nonradiative fluxes, if expressed as fractions of the net surplus of all-wave radiation, can be arrayed on a three-phase diagram (Fig. 4) in which a dominant latent-heat flux is plotted near the top corner, dominant substrate heat near the left corner, and dominant atmospheric sensible-heat flux near the right corner. Points for Hamburg (H) show mean noonday fluxes in the growing season months of 1958, which move steadily toward a greater role of latent-heat flux. The New

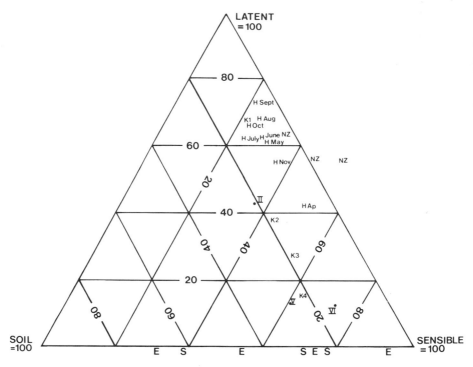

Fig. XXIII.4. Three-phase graph showing relative fractions of energy converted into latent-, sensible-, and soil-heat fluxes at several observation sites. Symbols: S, desert ecosystem at Shafrikan, central Asia, and E, dry bed of El Mirage Lake in California, at daytime hours; NZ, Cass, New Zealand, points to right represent frontal and post-frontal weather; H, noon means of fluxes at Hamburg-Quickborn in the growing-season months indicated; II, V, VI, observation periods in late summer at O'Neill, Nebraska; K, grass of increasing shortness and dryness at Kerang, Australia. See text for details.

Zealand points (NZ) illustrate the shift toward sensible heat after front passage. Points for O'Neill, Nebraska (Roman numbers), near the end of the growing season, move toward a greater fraction of sensible heat. Grassland at Kerang in southern Australia (K) [data from Swinbank and Dyer (1968)] follows a parallel course if followed from depth and live fraction to the dry state. There is no latent-heat flux at the desert sites.

CONCLUSION

This chapter's discussion of the structure of the energy budget brings us to the end of our story. In passing we have noted how scientific and practitioner disciplines have used facets of energy-budget analysis for their own purposes.

Many disciplines use energy-budget analysis in a peripheral way and as a result have been more inclined to make assumptions than measurements and have added little to our scanty store of data. Few of the fluxes have been the subject of systematic observation in networks year after year. Partial analyses and data developed in one discipline often remain unknown in others. We accordingly need investigations centered on the interface and its energy transactions. Whether these might be regarded as foreshadowings of an emerging discipline or simply as an area of shared interdisciplinary effort is not clear, but in any case material benefits have already flowed to research and practitioner disciplines. A glance at these might suggest possibilities for the future of interface energetics.

Benefits of Energetics Analysis

The study of ecosystems has benefitted most from energy-budget analyses, which detected and evaluated the forcing functions ecosystems receive from the sun and atmospheric disturbances and suggested how the systems might respond. The energy budget is also a means of examining internal processes and responses within ecosystems (Waggoner, 1975), especially those that involve CO_2 assimilation, transpiration, and temperature-dependent metabolic functions (Chapter XI). This approach has been foremost in the study of aquatic ecosystems from the very beginning; it has become central in agronomy; it is entering forestry with the rise of forest meteorology and also holds promise for the analysis of urban systems.

The budget describes ecosystem environment and functioning in only one unit of measure (watts per square meter) (Lowry and Chilcote, 1958), but this dimension provides for information transfer between the field and laboratory or phytotron research and between ecosystems studied by different groups of practitioners. Foresters intent on increasing fuel biomass production have adopted many of the techniques acquired in research on food and feed crop ecosystems. Energetic analysis has been particularly successful in these areas because it provides a reliable value of surface temperature, which is also a critical factor in photosynthetic production.

Problems of scale have been overcome in transferring techniques of energetics that originally focussed on the single leaf to plants or animals and to whole ecosystems. The merino sheep and the terrain that it fits so well can equally be analyzed by casting energy budgets (Chapter XII) with due regard for shape factors and surface-volume ratios. Priestley (1967b) notes that while the sheep is a radiator, insects are convectors, "keeping cool purely by virtue of their size." This contribution by a meteorologist to the study of environment-organism relations illustrates a transfer of an energy-budget method. A zoologist, Moen (1973), has similarly analyzed the energy environment of deer in different kinds of native cover. The temperature extremes experienced by apical stems in a cactus ecosystem, which are critical points for the survival of the whole community, has been modeled from the energy budget by a botanist (Nobel, 1978, 1980) and implications drawn for the geographic distribution of the system. Clothing needs of a citizen in the center, residential areas, or rural surroundings of Christchurch, New Zealand, can be determined by energy-budget analysis (Tuller, 1980), though as yet we have no complete analysis for any city as a whole. A truly nested pair of models has been developed, however, in a sphagnum bog in northern Michigan. The outer model determines the regimes of bog surface temperatures, and the inner model uses these values as inputs into an energy budget of a pitcher plant (Kingsolver, 1979), reproducing the habitat it provides for mosquito larvae.

Models of the physical systems of lakes at high latitudes convert external factors into internal processes, which bring about water conditions that shape the chemical and biological models (Fox *et al.* 1979). An overall energy budget of an ecosystem describes the energy environment of special sites like the pitcher plant or pine needles tunneled by needle miners, for which local energy budgets can then be cast. It elucidates interactions among leaves at different depths below the active surface and helps define the allocation of energy, water, and nutrient resources to layers and species within the ecosystem (Chapters XI and XX).

Cities as the most complex of all systems (containing biological as well as artificial ecosystems) are seen to be highly inefficient in the ways they use energy, and spotlight some of the energy crises besetting human society. Making them more efficient as well as more human presents a major opportunity for energy studies (U.S. House of Representatives, 1980), in which separate analyses for individual structures and sites (lawns, urban canyons, parking lots) can be placed within a comprehensive urban budget yet to be undertaken.

Geographic and Geologic Diversity

The focus on a single dimension, flux density, also permits comparative research on ecosystems (Lowry and Chilcote, 1958) that supports comparative studies of water budgets, productive efficiency, nutrient cycling, and other features. Where ecosystems of different ages are juxtaposed in disturbed or fire-prone terrain, energetics can illustrate changes during regrowth or succession (Rouse and Kershaw, 1971).

Ecosystems in dissected topography with unequal exposures to sun, wind, and rain and crop ecosystems in cultivated lowlands differ with respect to their energy budgets, and these differences affect their yields of water and biomass (Chapter XXI). Ecosystem research has been seen (Miller, 1968a) to put "a quantitative footing under areal differentiation in the landscape that is visible but formerly unmeasured," a comment given more point by the recent accomplishments of the International Biological Programme in many countries. These results contribute to the central geographic concern about spatial diversity in the way the earth's surface works. Energy analysis of different surfaces also may uncover subsurface patterns of lithology and structure (Price, 1977), especially if variations in the latent-heat flux of these budgets are considered (Pratt et al., 1980). Geographers and geologists both gain from the techniques of ecosystem energetics.

Interface Energetics and Climate

Ecosystems are small pieces of the surface of our planet and sometimes seem beneath the notice of scientists who look at such planetary-scale phenomena as the circulation systems in the atmosphere. Yet the accumulated knowledge of the planet's energy budget and circulation rests less upon such global-scale direct approaches as those by Meech, Angot, Zenker and others in the 19th century, for example, than it does on the patient work of Voeikov, Homén, Ångström, Geiger, Thornthwaite, Brooks and many others who searched out and measured thermodynamic processes in ecosystems; their small-scale approach has been much more effective than the global-scale (Miller, 1968b). While many of the latter scientists were interested in global patterns, they felt that physical explanations would remain elusive until energy budgets were known at the ecosystem scale. For example, Dzerdzeevskii and Rauner pursued a program of investigating energetics in forest stands in order to understand better the modification of air streams over extensive forest lands and the recycling of water in the U.S.S.R. Budyko similarly led intensive studies of radiative and turbulent fluxes that were required for the study and zonation of climate on the grand scale. Landsberg, Leighly, Lettau and Kung, among others, investigated such ecosystems characteristics as roughness and albedo and ecosystem-scale fluxes as parts of their large-scale vision of climate. For global models of climate, "determination of the surface fluxes is perhaps the most important aspect" (Bhumralkar, 1979) though the problem is made difficult by the coarseness of synoptic model grids (about 100 km) and even of mesoscale grids nested inside them at high computation costs. Current means of estimating surface inputs to the atmosphere in these models cannot be assessed because of lack of field data; one means (the diffusivity approach) is not only "very crude" but also "invalid under unstable or convective conditions" (Bhumralkar, 1979), which as we have seen are truly important times in the energetics life of most ecosystems—the midday hours of the growing season. Further, exemplifying the lack of information transfer among disciplines, the diffusivity approach fails to incorporate knowledge available elsewhere.

Perhaps the impacts of weather and climate on economic functioning of a society less able than it once was to override environmental obstacles by sheer brute-force application of cheap fossil energy (Chapters XIII and XXII) will stimulate studies of the energetics of specific cases of weather impact. It is not unreasonable that civilian design and planning should some day follow the lead of energy-budget analysis in military climatology, in which it was applied, for example, to determine storage life of matériel or operability of equipment in extreme climates.

Resource Applications of Energetics

Energy-budget analysis of the phase changes of water (Chapters XIV and XV) that are so important in ecosystems began in the 1920s and has become standard practice in the era of evapotranspiration after 1948 and in snow hydrology. It rapidly supplanted engineering empiricisms that had little physical basis, i.e., degree-days in melting (McKay and Thurtell, 1978) and an exaggerated wind factor in transpiration. Investigations of evapotranspiration to determine interstorm soil moisture conditions and irrigation requirements and of snow melting to support the design and operation of large dams in the western United States have both been centered at the ecosystem scale. Transfer to operational river basins followed. Hydrology is, perhaps, no more myth-ridden than other disciplines, and some credit for this goes to energetics analysis, which has punctured such folklore as that ascribing unreasonably high evaporation losses to intercepted snow or wetland vegetation.

Hydrologists, foresters, and agronomists like Thornthwaite (1961) who called for geographers to examine the diversified surface of the lands, have applied energy budgets in trying to understand land resources and assess the effects of past or potential changes in land use. Marcus (1979), in a later presidential address to the same society, subsumes many contemporary problems under "energy, environment, and urbanization," which are interdependent—not least through being approachable by way of the study of ecosystem-scale conversions and fluxes of energy. Interventions, whether managed or inadvertent, into environment functions may have to be altered and in some cases re-source development policy reversed, "perhaps repeatedly" (White, 1980), as experience with a development program might dictate. For this purpose environments and resources need to be monitored. This task is most cost-effective when the processes involved are understood, so that it becomes possible to obtain a "broad description from a small set of established points" and observations (White, 1980). Energy-budget analysis can thus make important contributions to the study, monitoring, utilization, and conservation of all our biotic, water, land, and energy resources.

Study of ecosystem energy budgets helps us to analyze the sun-following regimes of day and year and their perturbation by atmospheric storms and advection and to see how thermodynamic interactions at different sites in an ecosystem determine productivity and water use and how surface characteristics affect energy transactions. We can then analyze and compare objectively and quantitatively all the kinds of ecosystems—forests, meadows, wheat fields, wetlands, and cities—that in so great diversity compose the outer active surface of our planet.

REFERENCES

Aizenshtat, B. A. (1958). Teplovoi balans i mikroklimat nekotorykh landshaftov peschanoi pustnyi. In "Sovremennye Problemy Meteorologii Prizemnogo Sloia Vozdukha" (M. I. Budyko, ed.), pp. 67–130. Gidrometeoizd., Leningrad.

Albrecht, F. (1930). Über den Zusammenhaug zwischen täglichen Temperaturgang und Strahlungshaushalt. Gerlands Beitr. Geophys. **25,** 1–35.

Albrecht, F. (1941). Ergebnisse von Dr. Haudes Beobachtungen der Strahlung und des Wärmehaushaltes der Erdoberfläche an den beiden Standlagern bei Ikengüng und am Edsen-Gol 1931/32. Rep. Sci. Exped. NW China Sven Hedin, IX. Meteorology 2, Publ. No. 14.

Anderson, E. A. (1976). A point energy and mass balance model of a snow cover. NOAA (Natl. Oceanic Atmos. Adm.) Tech. Rep. NWS **19.**

Ångström, A. (1925). On radiation and climate. Geogr. Ann. **7,** 122–142.

Anthes, R. A., and Warner, T. T. (1978). Development of hydrodynamic models suitable for air pollution and other mesometeorological studies. Mon. Weather Rev. **106,** 1045–1078.

Aslyng, H. C. (1960). Evaporation and radiation heat balance at the soil surface. Arch. Meteorol., Geophys. Bioklimatol., Ser. B **10,** 359–375.

Aslyng, H. C. (1966). Weather, water balance and plant production at Copenhagen 1955–1964. Denmark: R. Vet. Agric. Coll. (Copenhagen), Yearb. 1965 pp. 1–21.

Aslyng, H. C., and Friis-Nielsen, B. (1960). The radiation balance at Copenhagen. Arch. Meteorol., Geophys. Bioklimatol., Ser. B **10,** 342–358.

Aslyng, H. C., and Jensen, S. E. (1966). Radiation and energy balances at Copenhagen 1955–1964. R. Vet. Agric. Coll. (Copenhagen), Yearb. 1965 pp. 22–40.

Aslyng, H. C., and Kristensen, K. J. (1958). Investigations of the water balance in Danish agriculture. R. Vet. Agric. Coll. (Copenhagen), Yearb. 1958 pp. 64–100.

Baruch, Z. (1979). Elevational differentiation in Espeletia schultzii (Compositae), a giant rosette plant of the Venezuelan Paramos. Ecology 60, 85–98.

Begg, J. E. (1965). High photosynthetic efficiency in a low-latitude environment. Nature (London) 205, 1025–1026.

Bhumralkar, C. M. (1975). Numerical experiments on the computation of ground surface temperature in an atmospheric general circulation model. J. Appl. Meteorol. 14, 1246–1258.

Bhumralkar, C. M. (1979). Atmospheric boundary layer processes and their parameterization in climate models. In "Man's Impact on Climate" (W. Bach, J. Pankrath and W. Kellogg, eds.), pp. 77–97. Elsevier, Amsterdam.

Bowen, I. S. (1926). The ratio of heat losses by conduction and by evaporation from any water surface. Phys. Rev. 27, 779–787.

Brooks, F. A., and Rhoades, D. G. (1954). Daytime partition of irradiation and the evaporation chilling of the ground. Trans. Am. Geophys. Union 35, 145–152.

Brown, K. W. (1976). Sugar beet and potatoes. In "Vegetation and the Atmosphere" (J. L. Monteith, ed.), Vol. 2, pp. 65–86. Academic Press, New York.

Budyko, M. I. (1974). "Climate and Life" (transl. ed. D. H. Miller). Academic Press, New York.

Chang, S. W.-J. (1979). An efficient parameterization of convective and nonconvective planetary boundary layers for use in numerical models. J. Appl. Meteorol. 18, 1205–1215.

Court, A. (1951). Temperature frequencies in the United States. J. Meteorol. 8, 367–380.

Cummings, N. W. (1925). The relative importance of wind, humidity, and solar radiation in determining evaporation from lakes. Phys. Rev. Ser. 2, 25, 721.

Denmead, O. T. (1964). Evaporation sources and apparent diffusivities in a forest canopy. J. Appl. Meteorol. 3, 383–389.

Denmead, O. T. (1969). Comparative micrometeorology of a wheat field and a forest of Pinus radiata. Agric. Meteorol. 6, 357–371.

Dilley, A. C. (1974). An energy partition evaporation recorder. Austral. CSIRO, Div. Atmos. Phys., Tech. Pap. No. 24.

Dunin, F. X., and Reyenga, W. (1978). Evaporation from a Themeda grassland. I. Controls imposed on the process in a sub-humid environment, J. Appl. Ecol. 15, 317–325.

Dyer, A. J. (1961). Measurement of evaporation and heat transfer in the lower atmosphere by an automatic eddy-correlation technique. Q. J. R. Meteorol. Soc. 87, 401–412.

Dyer, A. J. (1967). A combined water and energy balance study at Katherine, Northern Territory. Austral. Meteorol. Mag. 15, 148–155.

Erickson, P. A. (1979). "Environmental Impact Assessment: Principles and Applications." Academic Press, New York.

Estoque, M. A. (1963). A numerical model of the atmospheric boundary layer. J. Geophys. Res. 68, 1103–1113.

Fox, P. M., LaPerriere, J. D., and Carlson, R. F. (1979). Northern lake modeling: A literature review. Water Resour. Res. 15(5), 1065–1072.

Frankenberger, E. (1960). Beiträge zum Berechnungen zum Wärmehaushalt der Erdoberfläche. Ber. Dtsch. Wetterdienstes 10, No. 73.

Frankenberger, E. (1962). "Contributions to the International Geophysical Year, 1957–58. 1. Measurement Results and Computations of the Heat Balance of the Earth's Surface" (transl. by A. F. Spano). U.S. Weather Bur., Washington, D.C.

Furnival, G. M., Waggoner, P. E., and Reifsnyder, W. E. (1975). Computing the energy budget of a leaf canopy with matrix algebra and numerical integration. Agric. Meteorol. 14, 405–416.

Gall, R. L., and Herman, B. M. (1980). The sensitivity of the nocturnal minimum temperature over dry terrain to temperature and humidity changes aloft. *Mon. Weather Rev.* **108**, 286–291.

Garratt, J. R. (1978). Transfer characteristics for a heterogeneous surface of large dynamic roughness. *Q. J. R. Meteorol. Soc.* **104**, 491–502.

Gates, D. M. (1963). The energy environment in which we live. *Am. Sci.* **51**, 327–348.

Germany, Wetterdienst (1974). *Beilage Med.-Meteorol. Ber.* (Hamburg Observatorium, monthly).

Gersmehl, P. J. (1978). No-till farming: the regional applicability of a revolutionary agricultural technology. *Geogr. Rev.* **68**, 66–79.

Grant, D. R. (1975). Comparison of evaporation measurements using different methods. *Q. J. R. Meteorol. Soc.* **101**, 543–550.

Greenland, D. (1973). The surface energy budget and synoptic weather in Chilton Valley, New Zealand Southern Alps. *N.Z. Geogr.* **29**, 1–15.

Halstead, M. H., Richman, R. L., Covey, W., and Merryman, J. D. (1957). A preliminary report on the design of a computer for micrometeorology. *J. Meteorol.* **14**, 308–325.

Hicks, B. B., Hyson, P., and Moore, C. J. (1975). A study of eddy fluxes over a forest. *J. Appl. Meteorol.* **14**, 58–66.

Homén, T. (1897). Der tägliche Wärmeumsatz im Boden und die Wärmestrahlung zwischen Himmel und Erde. *Acta Soc. Sci. Fenn.* **23**, No. 3.

House, G. J., Rider, N. E., and Tugwell, C. P. (1960). A surface energy-balance computer. *Q. J. R. Meteorol. Soc.* **86**, 215–231.

Idso, S. B., and Baker, D. G. (1967). Relative importance of reradiation, convection, and transpiration in heat transfer from plants. *Plant Physiol.* **42**, 631–640.

Jarvis, P. G., James, G. B., and Landsberg, J. J. (1976). Coniferous forest. *In* "Vegetation and the Atmosphere" (J. L. Monteith, ed.), Vol. 2, pp. 171–240. Academic Press, New York.

Kalma, J. D. (1974). An advective boundary layer model applied to Sydney, Australia. *Boundary-Layer Meteorol.* **6**, 351–361.

Kalma, J. D., Johnson, M., and Newcombe, K. J. (1978). Energy use and the atmospheric environment in Hong Kong: Part II. Waste heat, land use and urban climate. *Urban Ecol.* **3**, 59–83.

Kingsolver, J. G. (1979). Thermal and hygric aspects of environmental heterogeneity in the pitcher plant mosquito. *Ecol. Monog.* **49**, 357–376.

Knoerr, K. R., and Gay, L. W. (1965). Tree leaf energy balance. *Ecology* **46**, 17–24.

Kristensen, K. J. (1959). Temperature and heat balance of soil. *Oikos* **10**, 103–120.

Kuo, H. L. (1968). The thermal interaction between the atmosphere and the earth and propagation of diurnal temperature waves. *J. Atmos. Sci.* **25**, 682–706.

Kuvshinova, K. V. (1965). O teplovom balanse peschanoi pustnyi. *In* "Teplovoi i Radiatsionnyi Balans Estestvennoi Rastitel'nosti i Sel'skhoziaistevnnykh Polei" (Iu. L. Rauner, ed.), pp. 136–145. Izd. Nauka, Moscow.

Lambert, J. L. (1970). Thermal response of a plant canopy to drifting cloud shadows. *Ecology* **51**, 143–149.

Landsberg, H. E. (1968). Atmospheric variability and climatic determinism. *In* "Eclectic Climatology. Selected Essays Written in Memory of David I. Blumenstock, 1913–1963" (A. Court, ed.), Yearb. Assoc. Pac. Coast Geogr., Vol. 30, pp. 13–23. Oregon State Univ. Press, Corvallis.

Leighly, J. B. (1947). Profiles of air temperature normal to coast lines. *Ann. Assoc. Am. Geogr.* **37**, 75–85.

Lettau, H. (1952). Synthetische Klimatologie. *Ber. Detsch. Wetterdienstes (US-Zone)* **38**, 127–136.

Lettau, H. (1969). Evapotranspiration climatonomy. I. A new approach to numerical prediction of monthly evapotranspiration, runoff, and soil moisture storage. *Mon. Weather
Rev.* **97,** 691–699.

Lettau, H. H., and Davidson, B. (1957). "Exploring the Atmosphere's First Mile," 2 vols.
Pergamon, Oxford.

Lieth, H., and Box E. (1972). Evapotranspiration and primary productivity: C. W. Thornthwaite memorial model. *Publ. Climatol.* **25**(3), 37–46.

Lönnqvist, O. (1962). On the diurnal variation of surface temperature. *Tellus* **14,** 96–101.

Lowry, W. P. (1969). "Weather and Life. An Introduction to Biometeorology." Academic
Press, New York.

Lowry, W. P., and Chilcote, W. W. (1958). The energy budget approach to the study of
microenvironment. *Northwest Sci.* **32**(2), 49–56.

McCaughey, J. H. (1978). Energy balance and evapotranspiration estimates for a mature
coniferous forest. *Can. J. For. Res.* **8,** 456–462.

McIlroy, I. C. (1967). An energy partition evaporation recorder (EPER). *Austral. J. Instrum.
Control* **23,** 120–122.

McKay, D. C., and Thurtell, C. W. (1978). Measurements of the energy fluxes involved
in the energy budget of snow cover. *J. Appl. Meteorol.* **17,** 339–349.

Mahrer, Y. (1979). Prediction of soil temperatures of a soil mulched with transparent polyethylene. *J. Appl. Meteorol.* **18,** 1263–1267.

Marcus, M. G. (1979). Coming full circle: Physical geography in the Twentieth Century.
Ann. Assoc. Am. Geogr. **69,** 521–532.

Miller, D. H. (1956). The influence of snow cover on the local climate of Greenland. *J.
Meteorol.* **13,** 112–120.

Miller, D. H. (1965). The heat and water budget of the Earth's surface. *Adv. Geophys.* **11,**
175–302.

Miller, D. H. (1967). Sources of energy for thermodynamically-caused transport of intercepted snow from forest crowns. *In* "Forest Hydrology" (W. E. Sopper and H. W. Lull,
eds.), pp. 201–211. Pergamon, Oxford.

Miller, D. H. (1968a). A survey course: The energy and mass budget at the surface of the
earth. *Assoc. Am. Geogr., Comm. Coll. Geog., Publ.* No. 7. Assoc. Am. Geogr., Washington,
D.C.

Miller, D. H. (1968b). Development of the heat budget concept. *In* "Eclectic Climatology.
Selected Essays Written in Memory of David I. Blumenstock, 1913–1963" (A. Court,
ed.), Yearb. Assoc. Pac. Coast Geogr., Vol. 30, pp. 123–144. Oregon State Univ. Press,
Corvallis.

Miller, D. H. (1972a). A new climatic parameter: Radiant-energy intake. *Proc. Int. Geogr.
Congr., 22nd, Montreal.* Univ. of Toronto Press, Toronto.

Miller, D. H. (1972b). On the variations of radiant-energy intake over time, with some
notes on the responses of evapotranspiration and other energy fluxes as functions of
the temperature of the surface of the earth. *Publ. Climatol.* **25**(3), 47–67.

Mitchell, J., Beckman, W., Bailey, R., and Porter, W. (1975). Microclimatic modeling of
the desert. *In* "Heat and Mass Transfer in the Biosphere. Part 1 Transfer Processes in
the Plant Environment" (D. A. deVries and N. H. Afgan, eds.), pp. 275–286. Scripta,
Halsted Press, Wiley, New York.

Moen, A. N. (1973). "Wildlife Ecology: An Analytical Approach." Freeman, San Francisco.

Monteith, J. L. (1973). "Principles of Environmental Physics." Arnold, London.

Moore, C. J. (1976). Energy flux measurements above a pine forest. *Q. J. R. Meteorol. Soc.*
102, 913–917.

Motha, R. P., Verma, S. B., and Rosenberg, S. J. (1979). Turbulence spectra above a veg-
etated surface under conditions of sensible heat advection. *J. Appl. Meteorol* **18,** 317–323.

Munro, D. S. (1980). A portable differential psychrometer system. *J. Appl. Meteorol.* **19**, 206–214.

Myrup, L. (1969). A numerical model of the urban heat island. *J. Appl. Meteorol.* **8**, 908–918.

Neiburger, M. (1941). Insolation and the prediction of maximum temperatures. *Bull. Am. Meteorol. Soc.* **22**, 95–102.

Nobel, P. S. (1978). Surface temperatures of cacti—Influences of environmental and morphological factors. *Ecology* **59**, 986–996.

Nobel, P. S. (1980). Morphology, surface temperatures, and northern limits of columnar cacti in the Sonoran Desert. *Ecology* **61**(1), 1–7.

Oke, T. R. (1978). "Boundary Layer Climates." Methuen, London.

Outcalt, S. I. (1971). A numerical surface climate simulator. *Geogr. Anal.* **3**, 379–393.

Outcalt, S. I. (1972). The development and application of a simple digital surface-climate simulator. *J. Appl. Meteorol.* **11**, 629–636.

Outcalt, S. I., Goodwin, C., Weller, G., and Brown, Jr. (1975). Computer simulation of the snowmelt and soil thermal regime at Barrow, Alaska. *Water Resour. Res.* **11**, 709–715.

Pratt, D. A., Foster, S. J., and Ellyett, C. D. (1980). A calibration procedure for Fourier series thermal inertia models. *Phot. Engr. Remote Sens.* **46**, 529–538.

Price, J. C. (1977). Thermal inertia mapping: A new view of the earth. *J. Geophys. Res.* **82**, 2582–2590.

Priestley, C. H. B. (1959). "Turbulent Transfer in the Lower Atmosphere." Univ. of Chicago Press, Chicago, Illinois.

Priestley, C. H. B. (1967a). Handover in scale of the fluxes of momentum, heat, etc. in the atmospheric boundary layer. *Phys. Fluids* **10**, No. 9, Part 2, S38–S46.

Priestley, C. H. B. (1967b). Microclimates of life. *Sci. J. (London)* **3**(3), 67–73.

Rauner, Iu. L. (1958). Nekotorye rezul'taty teplobalansovykh nabliudenii v listvennom lesu. *Izv. Akad. Nauk SSSR, Ser. Geogr.* No. 5, 79–86.

Rouse, W. R., and Kershaw, K. A. (1971). The effects of burning on the heat and water regimes of lichen-dominated subarctic surfaces. *Arct. Alp. Res.* **3**, 291–304.

Saltzman, B. (1980). Parameterization of the vertical flux of latent heat at the earth's surface for use in statistical-dynamical climate models. *Arch. Met. Geophys. Bioklimatol. Ser. A* **29**, 41–53.

Saugier, B., and Ripley, E. A. (1978). Evaluation of the aerodynamic method of determining fluxes over natural grassland. *Q. J. R. Meteorol. Soc.* **104**, 257–270.

Severini, M., and Olivieri, B. (1980). Experimental evaluation of latent heat flux during night-time radiative hoarfrost. *Boundary-Layer Meteorol.* **19**, 119–124.

Shawcroft, R. W., Lemon, E. R., Allen, L. H., Jr., Stewart, D. W., and Jensen, S. E. (1974). The soil–plant–atmosphere model and some of its predictions. *Agric. Meteorol.* **14**, 287–307.

Smith, A. P. (1980). The paradox of plant height in an Andean giant rosette species. *J. Ecol.* **68**, 63–73.

Spittlehouse, D. L., and Black, T. A. (1979). Determination of forest evapotranspiration using Bowen ratio and eddy correlation measurements. *J. Appl. Meteorol.* **18**, 647–653.

Stanhill, G. (1976). Cotton. *In* "Vegetation and the Atmosphere" (J. L. Monteith, ed.), Vol. 2, pp. 121–150. Academic Press, New York.

Stern, A. C., Wohlers, H. C., Boubel, R. W., and Lowry, W. P. (1973). "Fundamentals of Air Pollution." Academic Press, New York.

Stewart, J. B., and Thom, A. S. (1973). Energy budgets in pine forest. *Q. J. R. Meteorol. Soc.* **99**, 154–170.

Stewart, R. B., and Rouse, W. R. (1976). A simple method for determining the evaporation from shallow lakes and ponds. *Water Resour. Res.* **12**, 623–628.

Stricker, H., and Brutsaert, W. (1978). Actual evapotranspiration over a summer period in the "Hupsel Catchment." *J. Hydrol. (Amsterdam)* **39**, 139–157.

Sutherland, R. A. (1980). A short-range objective nocturnal temperature forecasting model. *J. Appl. Meteorol.* **19,** 247–255.

Swinbank, W. C., and Dyer, A. J. (1968). Micrometeorological expeditions 1962–1964. *Austral. CSIRO, Div. Meteorol. Phys., Tech. Pap.* No. 17.

Thom. A. S. (1975). Momentum, mass and heat exchange of plant communities. *In* "Vegetation and the Atmosphere" (J. L. Monteith, ed.), Vol. 1, pp. 57–109. Academic Press, New York.

Thornthwaite, C. W. (1961). The task ahead [a presidential address]. *Ann. Assoc. Am. Geogr.* **51,** 345–356.

Tuller, S. E. (1980). Effects of a moderate sized city on human thermal bioclimate during clear winter nights. *Int. J. Biometeorol.* **24,** 97–106.

Turner, H., Rochat, P., and Streule, A. (1975). Thermische Charakteristik von Hauptstandorten im Bereich der oberen Waldgrenze (Stillberg, Dischmatal bei Davos). *Mitt. Schweiz. Anst. Forstl. Versuchswes.* **51,** 95–119.

Uchijima, Z. (1976). Maize and rice. *In* "Vegetation and the Atmosphere" (J. L. Monteith, ed.), Vol. 2, pp. 33–64. Academic Press, New York.

U.S. House of Representatives (1980). Compact cities: Energy saving strategies for the Eighties. Rep. Subcomm on the City, H. S. Reuss, Chmn., Comm. on Banking, Finance and Urban Affairs, 96th Congr., 2d session. U.S. Govt. Printing Office, Washington, D.C.

Vehrencamp, J. E. (1953). Experimental investigation of heat transfer at an air–earth interface. *Trans. Am. Geophys. Union* **34,** 22–30.

Verma, S. B., Rosenberg, N. J., and Blad, B. L. (1978). Turbulent exchange coefficients for sensible heat and water vapor under advective conditions. *J. Appl. Meteorol.* **17,** 330–338.

Viskanta, R., and Daniel, R. A. (1980). Radiative effects of elevated pollutant layers on temperature structure and dispersion in an urban atmosphere. *J. Appl. Meteorol.* **19,** 53–70.

Waggoner, P. E. (1975). Micrometeorological models. *In* "Vegetation and the Atmosphere" (J. L. Monteith, ed.), Vol. 1, pp. 205–228. Academic Press, New York.

White, G. F. (1980). Environment. *Science* **209,** 183–190.

Wouters, D. S., Keppens, H., and Impens, I. (1980). Factors determining the longwave radiation exchange over natural surfaces. *Arch. Meteorol. Geophys. Bioklimatol. Ser. B* **28,** 63–71.

INDEX

International Geophysics Series

EDITED BY

J. VAN MIEGHEM
(July 1959–July 1976)

ANTON L. HALES
The University of Texas at Dallas
Richardson, Texas